Electron Spin Relaxation in Liquids

Electron Spin Relaxation in Liquids

Based on lectures given at the NATO Advanced Study Institute held at "Spåtind," Norway, in August 1971

Edited by L. T. Muus
Department of Chemistry
Aarhus University
Aarhus, Denmark
and
P. W. Atkins
Physical Chemistry Laboratory
University of Oxford
Oxford, England

PLENUM PRESS • NEW YORK - LONDON • 1972

Library of Congress Catalog Card Number 72-76022

ISBN 0-306-30588-7

A Division of Plenum Publishing Corporation
227 West 17th Street, New York, N. Y. 10011

United Kingdom edition published by Plenum Press, London
A Division of Plenum Publishing Company, Ltd.
Davis House (4th Floor), 8 Scrubs Lane, Harlesden, London, NW10 6SE, England

Printed in the United States of America

PREFACE

Electron spin relaxation has established itself as an important experimental method for studying the details of molecular motion in liquids, and as a harsh testing ground for theoreticians. The theoretical difficulties are connected with the complexity of the molecular motion, and the theoretical interest lies not only in its important consequences for the interpretation of experiments, but also in the fascination of a system in which a well-defined quantum mechanical component is in interaction with a complex quasi-classical environment. It is because the theories are concerned with such dissimilar but connected systems that the techniques involved are so numerous. Many of the standard manipulations of quantum mechanics must be brought to bear, and at the same time they must be combined with statistical techniques which are often of considerable sophistication. The purpose of this volume is to present a survey of these techniques and their application to spin relaxation problems. No single volume can be exhaustive, but we believe that the contributions to this volume are sufficiently broad to show how those who are concerned with spin relaxation problems think about the subject and circumvent, or expose, its difficulties.

The first few Chapters (I-V) review the basic quantum mechanical and statistical manipulations which are often used. In the following Chapter (VI) the general theory of linear response is presented: this Chapter is central because an electron spin resonance line shape is really a representation of the linear response of a paramagnetic system to an oscillating perturbation. The paramagnetic system is immersed in a rapidly fluctuating bath, and in Chapters VII-IX the separations of the fast fluctuations from the relatively slow spin processes they induce enables one to construct and extend the Redfield equation which is so often at the basis of many calculations and applications.

With the basic theoretical structure in hand we turn to applications. A review of the whole field of applications of the theories is given in Chapter X, and we encourage the reader to consult this Chapter at frequent intervals; a new-comer to the subject might even wish to begin here. Particular points made in this Chapter are developed as examples in Chapter XI (on spin rotation theory) and Chapter XII (on ^6S ions). A recent trend in theories and applications of relaxation theories has been the study of slow and anisotropic motion: such studies are important in spin-labelled systems and in liquid crystals, and the novel features introduced are described in detail in Chapters XIII-XVI. The field of "solid liquids" - amorphous solids - is represented by the slow motion limit of these Chapters and by the Chapter on triplet excitons in ordered systems (XVII). When relaxation is overcome by the stimulated absorption process one encounters saturation, and its theoretical description and application (in the techniques of double resonance) is described in Chapter XVIII which concludes the volume.

The book is based on the lectures given at the NATO Advanced Study Institute which was held at "Spåtind", Norway, in August 1971. A collection of chapters made in this way inevitably suffers from a lack of consistency of notation, but we hope that this will not be a serious impediment to readers of the level we anticipate using this book. The general literature itself is not consistent, and where appropriate we have commented on the different conventions.

A large number of people worked very hard in preparation for the original Advanced Study Institute, and then again in the production of the book. We are especially indebted to the authors without whom this book would not have existed (and in many cases, without whom its subject would not have existed). The principal source of finance of the Advanced Study Institute was the NATO Science Committee, and their generous support is grateful acknowledged. Many participants also benefitted from their national scientific foundations who contributed to their travel expenses. Finally we must thank Mrs. Ruth Buch upon whom rested much of the organisation and preparation of the Advanced Study Institute and this volume.

Aarhus and Oxford, 1971 L.T.M.

P.W.A.

CONTENTS

SUPEROPERATORS, TIME-ORDERING AND DENSITY OPERATORS

L.T. Muus

Aarhus University

I.1. SUPEROPERATORS

A superoperator has the same relation to an ordinary linear operator as the linear operator has to a state vector. The notation "superoperator" was suggested by Crawford[1]. However, a variety of superoperators had been operating for many years before they were classified collectively as "superoperators".

The state vectors j and k of quantum systems form a functional Hilbert space in which a metric is introduced by definition of the scalar product $(j,k)=(k,j)^*$ or in the Dirac notation $\langle j|k\rangle=\langle k|j\rangle^*$. The asterisk indicates complex conjugation.

In a n-dimensional space we may represent a state vector by giving its components $\langle\xi'|j\rangle$ of a base of n linearly independent vectors $|\xi'\rangle$. It is convenient to use an orthonormal set of vectors as a base. A linear operator in this space acts on a state vector to produce another state vector. We define the operator $A^\dagger$ as the adjoint operator to A, i.e. $(j,Ak) = (A^\dagger j,k)$ or $\langle j|A|k\rangle = \langle k|A^\dagger|j\rangle^*$. A hermitian operator is self-adjoint, i.e. $A^\dagger = A$. For a unitary operator we have $A^\dagger A = 1$. The linear operator is represented by its n^2 matrix elements. The hamiltonian has n energy eigenvalues with

n^2 energy differences, including at least n vanishing energy differences.

In analogy with the definition of linear operators over a vector space we may now introduce the following definition for superoperators over an operator space or an operator algebra: A superoperator ч is a quantity which applied to any operator A of the operator space produces another operator B of the space, i.e.

$$чA = B \tag{I.1}$$

Cyrillic script is here used to denote superoperators except the commutator producing superoperators $A^\times$ to be defined later. Linearity, eigenoperators, eigenvalues, the sum and the product of superoperators are defined in close analogy with the definition for operators in vector space, i.e.

$$ч(A+B) = чA + чB$$

$$ч(cA) = c(чA)$$

$$чA = wA$$

$$(ч+ч')A = чA + ч'A$$

$$чч'A = ч(ч'A)$$

In general, superoperators do not commute, i.e.

$$чч'A \neq ч'чA \quad \text{or} \quad [ч,ч'] = чч'-ч'ч \neq 0$$

Any n×n matrix may be represented by a linear combination of n^2 linearly independent n×n matrices, say, the n^2 matrices containing a single non-zero element. The corresponding n^2 operators are said to span the n^2 dimensional operator space or to constitute a complete set. The scalar product of two operators A and B may be defined by the Frobenius trace metric

$$(A,B) = Tr(A^\dagger B) \tag{I.2}$$

Using Eq.(I.2) we may define the adjoint superoperator $ч^{\dagger}$ to the superoperator $ч$ by

$$(A,чB) = (ч^{\dagger}A,B) \quad \text{for all A and B}$$

A superoperator $ч$ is hermitian if $ч = ч^{\dagger}$, unitary if $чч^{\dagger} = ч^{\dagger}ч = 1$.

The complete set of base operators may always be chosen as an orthonormal set U_j defined by

$$Tr(U_j^{\dagger}U_k) = \delta_{jk}, (j,k = 1,2...n^2)$$

and a hermitian operator A may be expanded in terms of this base

$$A = \sum_{j=1}^{n^2} \alpha_j^{*}U_j = \sum_{j=1}^{n^2} \alpha_j U_j^{\dagger}$$

with coefficients given by

$$\alpha_j^{*} = Tr(U_j^{\dagger}A) \text{ or } \alpha_j = Tr(U_j A)$$

Simple superoperators are the two translation superoperators

$$ч^{L}(X) = AX \qquad ч^{R}(X) = XA$$

where L and R stand for left and right translation. Any left translation superoperator commutes with any right translation superoperator.

The derivation superoperator $A^{\times}$ or $ч^{D}$ is the commutator producing superoperator

$$A^{\times}X=ч^{D}X=(ч^{L}-ч^{R})X=AX-XA=[A,X]$$

This superoperator owes its name to the following property which formally is analogous to that for differentiation of a product of functions

$$A^{\times}XY = (A^{\times}X)Y + XA^{\times}Y$$

where the parenthesis indicates that $A^{\times}$ operates on no operator placed on the right-hand side of the parenthesis. The derivation superoperator $A^{\times}$ is hermitian if the corresponding operator A is hermitian and conversely. Proofs for both statements follow from the observations

$$(C,A^{\times}B)=Tr(C^{\dagger}(AB-BA))=Tr(C^{\dagger}AB-AC^{\dagger}B)$$

$$(A^{\times}C,B)=Tr((AC-CA)^{\dagger}B)=Tr(C^{\dagger}A^{\dagger}B-A^{\dagger}C^{\dagger}B)$$

$$\text{i.e. } (C,A^{\times}B)=(A^{\times}C,B) \quad \text{if } A = A^{\dagger} \text{ and conversely.}$$

In the following we shall focus our attention on the Liouville"operator" $H^{\times}$ which is the derivation superoperator associated with a hamiltonian. We want to find the eigenoperator X and eigenvalue w to the Liouville operator

$$H^{\times}X = wX$$

Let $|j>$, ε_j and $|k>$, ε_k be eigenkets and eigenvalues of the hermitian operator H, i.e.

$$H|j>=\varepsilon_j|j>, \quad H|k>=\varepsilon_k|k>$$

Let us introduce the shift operator $S_{jk} = |j><k|$

We now have

$$\begin{aligned} H^{\times}S_{jk}&=H^{\times}|j><k|=H|j><k|-|j><k|H= \\ &\varepsilon_j|j><k|-\varepsilon_k|j><k|= \\ &(\varepsilon_j-\varepsilon_k)|j><k|=(\varepsilon_j-\varepsilon_k)S_{jk} \end{aligned} \tag{I.3}$$

Accordingly, the shift operator $S_{jk}=|j><k|$ is an eigenoperator of $H^{\times}$ with the energy difference $(\varepsilon_j-\varepsilon_k)$ as eigenvalue.

For a n^2 dimensional Liouville operator there are just n^2 linearly independent operators S_{jk}, including the n projection operators S_{kk} with eigenvalues 0, i.e. all eigenoperators of $H^{\times}$ are found. The usefulness of the Liouville operator formalism derives from the fact that the eigenvalues are energy differences. Energy differences -

in contrast to energies - are directly observable in a spectrum.

The Liouville operator may be labeled with four indices related to those of the ordinary operator with which it is associated[2]. We may define

$$\langle jk|A^{\times}|j'k'\rangle = (|j\rangle\langle k|,\ A^{\times}|j'\rangle\langle k'|)$$

Using the definition (I.2) we derive

$$\langle jk|A^{\times}|j'k'\rangle = \mathrm{Tr}(|k\rangle\langle j|(A|j'\rangle\langle k'|-|j'\rangle\langle k'|A))$$

$$\langle jk|A^{\times}|j'k'\rangle = \delta_{kk'}\langle j|A|j'\rangle-\delta_{jj'}\langle k'|A|k\rangle \qquad (I.4)$$

Since

$$((A^{\dagger})^{\times}|j\rangle\langle k|,|j'\rangle\langle k'|) = (|j\rangle\langle k|,\ A^{\times}|j'\rangle\langle k'|)$$

we note that $(A^{\dagger})^{\times}$ is the superoperator adjoint to $A^{\times}$, i.e. $(A^{\dagger})^{\times}= (A^{\times})^{\dagger}$.

From Eq.(I.4) we observe that

$$\langle jk|A^{\times}|j'k'\rangle = -\langle k'j'|A^{\times}|kj\rangle$$

$$\langle jk|A^{\times}|j'k'\rangle = \langle j'k'|(A^{\dagger})^{\times}|jk\rangle^{*}$$

The last equation shows that for hermitian A, i.e. $A^{\dagger}=A$, we have the following relation

$$\langle jk|A^{\times}|j'k'\rangle = \langle j'k'|A^{\times}|jk\rangle^{*}$$

with a formal analogy to a characteristic property of the matrix elements of a hermitian operator.

Useful properties of the Liouville operator are reflected in the following relations

$$\exp(A)B\exp(-A)=\exp(A^{\times})B; \qquad [A^{\times},B^{\times}]=[A,B]^{\times};$$

$$\mathrm{Tr}(AB^{\times}C)=\mathrm{Tr}(CA^{\times}B)=\mathrm{Tr}(BC^{\times}A); \qquad (I.5)$$

$$\exp(A^{\times})B_1^{\times}B_2^{\times}\ldots B_n^{\times}C=(\exp(A^{\times})B_1)^{\times}(\exp(A^{\times})B_2)^{\times}\ldots(\exp(A^{\times})C)$$

The following identity by Kubo[3] is used frequently within linear response theory and may be proved by parameter differentiation[4]

$$A^{\times}\exp(-\beta H) = -\int_0^\beta du\ \exp(-\beta H)\exp(uH^{\times})A^{\times}H$$

where A and H are arbitrary operators and $\beta = (kT)^{-1}$.

Introducing the time-derivative $\dot{A}(t)$ of the operator $A(t)=\exp(iH^{\times}t)A$, where $\hbar = 1$, the identity may be written as

$$A^{\times}\exp(-\beta H) = -i\int_0^\beta du\ \exp(-\beta H)\dot{A}(-iu)$$

Banwell and Primas[5] present several examples taken from high-resolution N.M.R. on the derivation of resonance frequencies directly as eigenvalues of the Liouville operator.

I.2. TIME-ORDERING

Time-ordering is discussed in several texts, e.g. by Roman[6] and Schweber[7]. The differential equations ($\hbar = 1$)

$$i\,\frac{dA'(t)}{dt} = HA'(t) \tag{I.6}$$

$$i\,\frac{dA''(t)}{dt} = A''(t)H \tag{I.7}$$

where H is a time-independent operator and $A'(t)$ and $A''(t)$ are time-dependent operators, have the solutions

$$A'(t) = \exp(-itH)A'(0) \tag{I.8}$$

and

$$A''(t) = A''(0)\exp(-itH) \tag{I.9}$$

Similarly, the equation

$$i\,\frac{dA(t)}{dt} = [H,A(t)] = H^{\times}A(t) \tag{I.10}$$

has the solution

$$A(t)= \exp(-itH)A(0)\exp(itH)=\exp(-itH^{\times})A(0) \tag{I.11}$$

Power series expansions of Eq.(I.11) give the so-called Hausdorff formula

$$A(t) = \sum_{n=0}^{\infty} (-i)^n \frac{t^n}{n!} (H^{\times})^n A(0) = \sum_{n=0}^{\infty} \frac{t^n}{n!} (-i)^n [H,[H,\ldots[H,A(0)]\ldots]]$$

(n commutators)

The term with n commutators multiplied by $(-i)^n$ is the n'th time-derivative of A(t) at t=0. Accordingly, we have

$$A(t) = \sum_{n=0}^{\infty} \frac{t^n}{n!} \left(\frac{d^n A(t)}{dt^n} \right)_{t=0}$$

This relation is just the Taylor expansion for A(t) about t=0.

Time-ordering is necessary when the hamiltonian in the differential equation is time-dependent and $[H(t), H(t')] \neq 0$ for $t \neq t'$

$$i \frac{dA(t)}{dt} = H(t)A(t) \tag{I.12}$$

The formal solution is the Volterra equation

$$A(t) = A(t_o) - i \int_{t_o}^{t} H(t')A(t')dt'$$

which may be expanded in the following way: Setting $A(t')=A(t_o)$, we find in the first approximation

$$A(t) = A(t_o) - i \int_{t_o}^{t} H(t_1)A(t_o)dt_1$$

The second approximation is

$$A(t)=A(t_o)+(-i) \int_{t_o}^{t} H(t_1)A(t_o)dt_1+(-i)^2 \int_{t_o}^{t} dt_1 \int_{t_o}^{t_1} dt_2 H(t_1)H(t_2)A(t_o)$$

Continuing in this manner we obtain the Neumann-Liouville series

$$A(t)= \sum_{n=0}^{\infty} (-i)^n \int_{t_o}^{t} dt_1 \int_{t_o}^{t_1} dt_2 \ldots \int_{t_o}^{t_{n-1}} dt_n H(t_1)H(t_2)\ldots H(t_n)A(t_o)$$

$$= \sum_{n=0}^{\infty} (-i)^n I_n(t)A(t_o)$$

$I_n(t)$ is an integral over the entire time interval between t_o and t with the restriction that $t \geq t_{j-1} \geq t_j \geq t_o$ $(j=1,2\ldots n)$.

Hence, $I_n(t)$ may be rewritten

$$I_n(t) = \int_{t_o}^{t} dt_1 \int_{t_o}^{t} dt_2 \ldots \int_{t_o}^{t} dt_n \theta(t_1-t_2)\theta(t_2-t_3)\ldots$$

$$\ldots\theta(t_{n-1}-t_n)H(t_1)H(t_2)\ldots H(t_n)$$

where the stepfunction $\theta(t) = 1$ for $t>0$ and $\theta(t) = 0$ for $t<0$.

In $I_n(t)$ the variables $t_1, t_2 \ldots t_n$ may be interchanged and subsequently the order of integration may be restored to the original order. Since we have $n!$ permutations of $t_1, t_2 \ldots t_n$ we may write

$$I_n(t) = \frac{1}{n!} \int_{t_o}^{t} dt_1 \int_{t_o}^{t} dt_2 \ldots \int_{t_o}^{t} dt_n \sum_P \theta(t_1-t_2)\theta(t_2-t_3)\ldots\theta(t_{n-1}-t_n)$$

$$H(t_1)H(t_2)\ldots H(t_n)$$

where $\sum_P$ denotes the summation over the permutations.

We define the time-ordering operator T_D^- in the following way

$$T_D^-(H(t_1)H(t_2)\ldots H(t_n)) = \sum_P \theta(t_1-t_2)\theta(t_2-t_3)\ldots\theta(t_{n-1}-t_n)$$

$$H(t_1)H(t_2)\ldots H(t_n) \quad \text{(I.13)}$$

This chronological operator was first introduced by Dyson[8] and the arrangement of the operators is by Blume[2] denoted as negative time-ordering. We, therefore, have

$$I_n(t) = \frac{1}{n!} \int_{t_o}^{t} dt_1 \int_{t_o}^{t} dt_2 \ldots \int_{t_o}^{t} dt_n T_D^-(H_1(t_1)H(t_2)\ldots H(t_n))$$

T_D^- arranges the time-labeled operators in the order of decreasing time-increments from the left to the right. For equal times the product is undefined. However, since $H(t)$ commutes with itself we write

$$T_D^-(H(t)H(t)) = H(t)H(t)$$

As an example, we have

$$T_D^-(H(t_1)H(t_2))=\theta(t_1-t_2)H(t_1)H(t_2)+\theta(t_2-t_1)H(t_2)H(t_1)$$

The solution to A(t) is now given by the series

$$A(t) = (1+\frac{1}{1!}(-i)\int_{t_o}^{t} H(t_1)dt_1+\frac{1}{2!}(-i)^2\int_{t_o}^{t} dt_1\int_{t_o}^{t} dt_2 T_D^-(H(t_1)H(t_2))\ldots$$

$$\ldots+\frac{1}{n!}(-i)^n\int_{t_o}^{t} dt_1\int_{t_o}^{t} dt_2\ldots\int_{t_o}^{t} dt_n T_D^-(H(t_1)H(t_2)\ldots H(t_n)))A(t_o)$$

which formally may be written

$$A(t) = T_D^-\exp(-i\int_{t_o}^{t} H(t')dt')A(t_o)=\exp_-(-i\int_{t_o}^{t} H(t')dt')A(t_o) \quad (I.14)$$

$\exp_-$ is Blume's notation for negative time-ordering[2].

As a check, we may differentiate the expression for A(t) to show that it is the solution to the initial differential equation.

In a similar way, we may find that the solution to the differential equation

$$i\,\frac{dA(t)}{dt} = A(t)H(t) \quad (I.15)$$

is given by the series

$$A(t)=A(t_o)\{1+\frac{1}{1!}(-i)\int_{t_o}^{t} H(t_1)dt_1+\frac{1}{2!}(-i)^2\int_{t_o}^{t} dt_1\int_{t_o}^{t} dt_2 T_D^+(H(t_1)H(t_2))\ldots$$

$$\ldots+\frac{1}{n!}(-i)^n\int_{t_o}^{t} dt_1\int_{t_o}^{t} dt_2\ldots\int_{t_o}^{t} dt_n T_D^+(H(t_1)H(t_2)\ldots H(t_n))\}$$

where T_D^+ is the chronological ordering operator for positive time-ordering defined by

$$T_D^+(H(t_1)H(t_2)\ldots H(t_n) =$$

$$\sum_P \theta(t_1-t_2)\theta(t_2-t_3)\ldots\theta(t_{n-1}-t_n)H(t_n)H(t_{n-1})\ldots H(t_2)H(t_1)$$

i.e. the time-labeled operators are arranged in the order of increasing time-increments from the left to the right.

The series expansion may formally be written

$$A(t)=A(t_o)T_D^+\exp(-i\int_{t_o}^{t} H(t')dt')=A(t_o)\exp_+(-i\int_{t_o}^{t} H(t')dt') \quad (I.16)$$

An important differential equation is

$$i\,\frac{dA(t)}{dt} = H^\times(t)A(t)=[H(t),\, A(t)] \quad (I.17)$$

The solution is given by

$$A(t)=\exp_-(-i\int_{t_o}^{t} H^\times(t')dt')A(t_o)=$$

$$\exp_-(-i\int_{t_o}^{t} H(t')dt')A(t_o)\exp_+(i\int_{t_o}^{t} H(t')dt') \quad (I.18)$$

which may be demonstrated by series expansion of A(t) or more conveniently by differentation of A(t) using the solutions (I.14) and (I.16) to Eqs.(I.12) and (I.15)

$$\frac{dA(t)}{dt} = -iH(t)A(t)+iA(t)H(t)=-iH^\times(t)A(t) \quad (I.19)$$

I.3. DENSITY OPERATORS

There are three main approaches to introduce the density operator. It was introduced by von Neumann[9] and by Dirac[10] to describe statistical concepts in quantum mechanics by the method of representative ensembles well-known in classical statistical mechanics. Later Dirac[11] and Husimi[12] introduced density matrices by averaging the product of the many-body wave function and its adjoint over all coordinates except those of a small number of electrons. This second method is usually known as the quantum mechanical approach, whereas the first method is called the statistical approach. We shall in this Chapter only briefly touch upon the second approach, which, however, has gained prominence within molecular orbital theory[13]. The third approach was used by Fano[14] and is known as the operational approach. In the later part of this Chapter, we shall follow the prominent points of view from Fano's exposition.

The elementary theory is outlined in several texts, e.g. Messiah[15] and Roman[16].

A quantum system characterized by the specification of a single ketvector, say $|m\rangle$, is said to be in a pure state. This is the case of maximum information. It is not possible to deduce any information about the behaviour of a quantum system which is not given implicitly by stating the ketvector $|m\rangle$.

For this system the expectation value of the observable A is given by

$$\langle A\rangle = \langle m|A|m\rangle$$

for normalized $|m\rangle$.

The mixed state or the case of less than maximum information obtains if we have to specify a whole set of possible normalized pure states or ketvectors, which are not necessarily different

$$|1\rangle, |2\rangle, |3\rangle \ldots |n\rangle$$

together with the normalized weights

$$p_1, p_2 \ldots p_n, p_m \geq 0$$

$$\text{where} \qquad \sum_{m=1}^{n} p_m = 1$$

The ensemble average or the average of the observable A for the mixed system is given by

$$\langle A\rangle = \sum_m p_m \langle A\rangle_m = \sum_m p_m \langle m|A|m\rangle \tag{I.20}$$

We emphasize at this point, that for mixed states we are considering two different concepts of averages. First, the familiar quantum mechanical average, or expectation value, and second, the ensemble average of these numbers with the weighting factors p_m. While the first averaging is inherent in the nature of quantization, the second is introduced for mixed states only because of our lack of maximum information.

We introduce now the density operator ρ defined by

$$\rho = \sum_m |m\rangle p_m \langle m| \tag{I.21}$$

which is seen to be hermitian.

Let us use as a base a complete set of discrete normalized ketvectors $|\zeta'\rangle$.

Since $\quad |m\rangle = \sum_{\zeta'} |\zeta'\rangle\langle\zeta'|m\rangle \quad$ and $\quad \langle m| = \sum_{\zeta'} \langle m|\zeta'\rangle\langle\zeta'|$

we have from equation (I.20)

$$\langle A\rangle = \sum_m p_m \langle m|A|m\rangle = \sum_{m,\zeta'} p_m \langle m|\zeta'\rangle\langle\zeta'|A|m\rangle = \sum_{m,\zeta'} \langle\zeta'|A|m\rangle p_m \langle m|\zeta'\rangle$$

$$= \sum_{\zeta'} \langle\zeta'|A\rho|\zeta'\rangle = \mathrm{Tr}(A\rho) = \mathrm{Tr}(\rho A) \tag{I.22}$$

For a continuous set of basis vectors we obtain in a similar fashion

$$\langle A\rangle = \int \langle\zeta'|\rho A|\zeta'\rangle d\zeta' = \mathrm{Tr}(\rho A) \tag{I.23}$$

A change in base to another complete set of ketvectors $|\mu'\rangle$ leaves the trace invariant. For this reason $\langle A\rangle$ has the same value in all representations. This invariance is of course also to be expected on physical grounds. A change in representation for the observable A is equivalent to a unitary transformation of the matrix A

$$A' = S^{-1}AS$$

Since $\langle A\rangle = Tr(A\rho) = Tr(S^{-1}A\rho S) = Tr(S^{-1}ASS^{-1}\rho S) = Tr(A'\rho')$

we find
$$\rho' = S^{-1}\rho S \qquad (I.24)$$

Accordingly, the density matrix follows the same transformation law as matrices representing observables. This observation is not trivial since ρ is related to state vectors rather than physical observables.

Suppose the system is a pure state $|k\rangle$ and accordingly $p_m = \delta_{mk}$. In this case the density operator ρ is the projection operator $\rho = |k\rangle\langle k|$.

The average value of the observable A is now the expectation value $\langle k|A|k\rangle$. However, it is still given by

$$\langle A\rangle = \sum_m (\rho A)_{mm} = \sum_m \langle m|k\rangle\langle k|A|m\rangle = \langle k|A|k\rangle$$

since $\langle m|k\rangle = \delta_{mk}$ for the pure state $|k\rangle$.

At this point we may follow Fano[14] and use the relation $\langle A\rangle = Tr(\rho A)$ as the definition of the density matrix, stating that "ρ is a hermitian operator characteristic of the system, whether mixed or pure, in the sense that the average value of any observable may be computed from the equation $\langle A\rangle = Tr(\rho A)$".

We now ascertain the general properties of the density operator.

a. ρ is hermitian as noted earlier.

b. $Tr\rho = 1$, since

$$Tr\rho = \sum_{\zeta'} \langle\zeta'|\rho|\zeta'\rangle = \sum_{\zeta',m} \langle\zeta'|m\rangle p_m \langle m|\zeta'\rangle = \sum_m p_m \sum_{\zeta'} |\langle m|\zeta'\rangle|^2 = \sum_m p_m = 1 \qquad (I.25)$$

follows from the completeness of the orthonormal set ζ'.

c. The diagonal elements in the representation with base $|\zeta'\rangle$ are non-negative and give the probability of finding the quantum system in the dynamical state represented by $|\zeta'\rangle$, since

$$\langle\zeta'|\rho|\zeta'\rangle=\sum_m\langle\zeta'|m\rangle p_m\langle m|\zeta'\rangle=\sum_m|\langle\zeta'|m\rangle|^2 p_m\geq 0 \qquad (I.26)$$

This statement is obviously valid in any base. We say that the density operator is positive definite. Thus ρ has no negative eigenvalues.

d. Consider a representation in which ρ is diagonal, i.e. $\rho_{nm}=\rho_n\delta_{nm}$. Then

$$\mathrm{Tr}(\rho^2) = \sum_n \rho_n{}^2 \leq \left(\sum_n \rho_n\right)^2 = (\mathrm{Tr}\rho)^2 = 1.$$

Since the trace is invariant under a similarity transformation, we have in general due to the hermiticity of ρ

$$\mathrm{Tr}(\rho^2) = \sum_{n,m} |\rho_{nm}|^2 \leq 1 \qquad (I.27)$$

We now wish to establish the additional properties ρ must have for a pure state.

e. The necessary and sufficient condition that ρ describes a pure state is $\rho^2 = \rho$, i.e. ρ is idempotent.

f. It follows that for a pure state

$$\mathrm{Tr}\rho^2 = 1 \qquad (I.28)$$

Comparing this equation with the general relation $\mathrm{Tr}\rho^2\leq 1$ we note that $\mathrm{Tr}\rho^2$ obtains its maximum value for a pure state.

g. All eigenvalues of ρ are 0 and 1, the latter occurring once only.

We leave the proofs of f-g as exercises.

Using the Schrödinger equation

$$i\,\frac{d|m(t)\rangle}{dt}=H|m(t)\rangle \quad \text{or} \quad -i\,\frac{d\langle m(t)|}{dt}=\langle m(t)|H \qquad (I.29)$$

the time-derivative of the density operator is given by

$$i\,\frac{d\rho(t)}{dt}=i\,\frac{d}{dt}\sum_m|m(t)\rangle p_m\langle m(t)|=i\sum_m\left(\frac{d|m(t)\rangle}{dt}p_m\langle m(t)|+|m(t)\rangle p_m\frac{d\langle m(t)|}{dt}\right)$$

$$=\sum_m(H|m(t)\rangle p_m\langle m(t)|-|m(t)\rangle p_m\langle m(t)|H)$$

$$i\,\frac{d\rho(t)}{dt} = [H,\rho(t)] = H^{\times}\rho(t) \qquad (I.30)$$

Eq.(I.30) is the quantum analogue of the classical equation of motion for the probability density $\rho(t)$

$$\frac{d\rho(t)}{dt} = \{H,\rho(t)\}$$

where { } denotes the Poisson bracket.

It is noteworthy that Eq.(I.30) has the same form, except for an opposite sign of i, as the equation of motion for an observable in the Heisenberg picture.

For the time-derivative of $<A>$ we deduce

$$i\,\frac{d<A>}{dt}=i\,\frac{d}{dt}\,Tr(\rho(t)A)=iTr(\frac{d\rho(t)}{dt}A)=Tr([H,\rho(t)]A)$$

$$=Tr(\rho(t)[A,H]) = <[A,H]> \qquad (I.31)$$

For time-independent H we may solve Eq.(I.30) by using Eq.(I.11), i.e.

$$\rho(t)=\exp(-i(t-t_o)H)\rho(t_o)\exp(i(t-t_o)H)$$

$$=\exp(-i(t-t_o)H^{\times})\rho(t_o)=T\rho(t_o)T^{-1} \qquad (I.32)$$

where

$$T = \exp(-i(t-t_o)H) \qquad (I.33)$$

It is frequently useful to expand Eq.(I.32) in the Hausdorff power series

$$\rho(t)=\sum_{n=0}^{\infty} \frac{(t-t_o)^n}{n!}\,(-i)^n[H,[H\ldots[H,\rho(t_o)]\ldots]]$$

So far we have made use of the Schrödinger picture with time-dependent state vectors and time-independent operators. In the Heisenberg picture the time-dependency is associated with the operators $A_H(t)$ rather than with the state vectors $|m>_H$.

We have

$$|m\rangle_H = T^{-1}|m(t)\rangle \text{ and } A_H(t) = T^{-1}AT$$

where T is given by Eq.(I.33).

For the density operator in the Heisenberg picture ρ_H follows from Eq.(I.32)

$$\rho_H = T^{-1}T\rho(t_o)T^{-1}T = \rho(t_o)$$

As expected, ρ_H is time-independent.

The hamiltonian consists frequently of a large time-independent interaction H_o and a small time-dependent term $H_1(t)$. The equation of motion for the density matrix is then

$$i\,\frac{d\rho(t)}{dt} = [H_o + H_1(t), \rho(t)] \tag{I.34}$$

Introducing the density operator $\rho^I(t)$ and the hamilton operator $H_1^I(t)$ in the interaction picture i.e.

$$\rho^I(t) = \exp(+iH_o t)\rho(t)\exp(-iH_o t) = \exp(+iH_o^{\times} t)\rho(t)$$

and $$H_1^I(t) = \exp(+iH_o^{\times} t)H_1(t)$$

we deduce for the equation of motion in the interaction picture

$$i\,\frac{d\rho^I(t)}{dt} = [H_1^I(t), \rho^I(t)] = (H_1^I(t))^{\times}\rho^I(t) \tag{I.35}$$

where $(H_1^I(t))^{\times}$ is the superoperator associated with $H_1^I(t)$.

For small $H_1(t)$ the variation in time of $\rho^I(t)$ is less pronounced than that of $\rho(t)$. This feature makes the interaction picture attractive in perturbation calculations.

The solution of the Eq.(I.35) follows Eq.(I.18), i.e.

$$\rho^I(t) = \exp_-(-i\int_{t_o}^{t} dt' H_1^I(t')^{\times})\rho^I(t_o) \tag{I.36}$$

The joint state of two interacting systems (1) and (2) may be represented by the density matrix $\rho^{(12)}$ given by

$$\rho^{(12)} = \sum_{m^{(1)},r^{(2)}} |m^{(1)}r^{(2)}>p_{mr}^{(12)}<m^{(1)}r^{(2)}|$$

where $|m^{(1)}>$ and $|r^{(2)}>$ are state vectors in (1) and (2) and $p_{mr}^{(12)}$ is the joint probability of state $m^{(1)}$ in (1) and $r^{(2)}$ in (2), i.e.

$$\sum_{m^{(1)},r^{(2)}} p_{mr}^{(12)} = 1.$$

An operator $Q^{(1)}$ of system (1) only may be treated as an operator of both systems when multiplied with the unit operator $I^{(2)}$ of system (2).

The average value of $Q^{(1)}$ is then given by

$$<Q^{(1)}> = \sum_{m^{(1)},r^{(2)}} p_{mr}^{(12)}<m^{(1)}r^{(2)}|Q^{(1)}I^{(2)}|m^{(1)}r^{(2)}>$$

As a base we introduce a complete set of discrete normalized ketvectors $|\xi^{(1)}>$ in system (1) and $|\xi^{(2)}>$ in system (2). Accordingly, we have

$$<Q^{(1)}> = \sum_{\substack{m^{(1)},r^{(2)} \\ \xi^{(1)},\xi^{(2)}}} p_{mr}^{(12)}<m^{(1)}r^{(2)}|\xi^{(1)}\xi^{(2)}><\xi^{(1)}\xi^{(2)}|Q^{(1)}I^{(2)}|m^{(1)}r^{(2)}>$$

$$= \sum_{\substack{m^{(1)},r^{(2)} \\ \xi^{(1)},\xi^{(2)}}} <\xi^{(1)}\xi^{(2)}|Q^{(1)}I^{(2)}|m^{(1)}r^{(2)}>p_{mr}^{(12)}<m^{(1)}r^{(2)}|\xi^{(1)}\xi^{(2)}>$$

$$= \sum_{\xi^{(1)},\xi^{(2)}} <\xi^{(1)}\xi^{(2)}|Q^{(1)}I^{(2)}\rho^{(12)}|\xi^{(1)}\xi^{(2)}> = \underset{(12)}{\mathrm{Tr}}(Q^{(1)}I^{(2)}\rho^{(12)}) \qquad (I.37)$$

Eq.(I.37) is an obvious extension of Eq.(I.22). Eq.(I.37) may be rewritten as follows

$$<Q^{(1)}> = \sum_{\xi^{(1)}} <\xi^{(1)}|Q^{(1)}\rho^{(1)}|\xi^{(1)}> = \underset{(1)}{\mathrm{Tr}}(Q^{(1)}\rho^{(1)}) \tag{I.38}$$

where the operator $\rho^{(1)}$ on system (1) is defined as

$$\rho^{(1)} = \sum_{\xi^{(2)}} <\xi^{(2)}|\rho^{(12)}|\xi^{(2)}> = \underset{(2)}{\mathrm{Tr}}\ \rho^{(12)} \tag{I.39}$$

i.e., by averaging the joint density operator $\rho^{(12)}$ over the coordinates irrelevant to $Q^{(1)}$.

The two systems are uncorrelated when $p_{mr}^{(12)} = p_m^{(1)} p_r^{(2)}$ and the density operator $\rho^{(12)}$ of the joint system is the direct product

$$\rho^{(12)} = \rho^{(1)} \times \rho^{(2)} \tag{I.40}$$

where $\rho^{(1)} = \sum_{m^{(1)}} |m^{(1)}>p_m^{(1)}<m^{(1)}|$ and $\rho^{(2)} = \sum_{r^{(2)}} |r^{(2)}>p_r^{(2)}<r^{(2)}|$

In this case $<Q^{(1)}Q^{(2)}> = <Q^{(1)}><Q^{(2)}>$.

Systems (1) and (2) may be the "system" and the "bath". They may consist of two different groups of particles or of different characteristics of the same particles, such as spin and orbital motion. Relative to the density operator $\rho^{(12)}$ of the joint system $\rho^{(1)}$ pertaining to system (1) is a reduced density operator.

Let us consider the density operator

$$\rho = \sum_m |m>p_m<m|$$

corresponding to a n-dimensional vector space, i.e. for any m

$$|m> = \sum_{r=1}^{n} |\xi_r><\xi_r|m>$$

where $|\xi_1>, |\xi_2> \ldots |\xi_n>$ is a complete set. The density operator then has n^2 matrix elements $<\xi_r|\rho|\xi_{r'}>$.

Assume furthermore that a hermitian operator A may be expanded into a set of n^2 base operators

$$A = \sum_{j=0}^{n^2-1} u_j^* U_j = \sum_{j=0}^{n^2-1} u_j U_j^\dagger \tag{I.41}$$

with the orthonormality condition

$$\mathrm{Tr}(U_i^\dagger U_j)=\mathrm{Tr}(U_i U_j^\dagger)=\delta_{ij}$$

The operators $U_o, U_1 \ldots U_{n^2-1}$ may be regarded as unit coordinate vectors in a n^2-dimensional vector space with the trace metric defining the scalar product of two operators

$$(A,B) = \mathrm{Tr}(A^\dagger B)$$

The coefficients in Eq.(I.41) are given by

$$u_j = \mathrm{Tr}(AU_j) \quad \text{and} \quad u_j^* = \mathrm{Tr}(AU_j^\dagger)$$

The coefficients a_j (or a_j^*) represent the operator A in the so-called Liouville representation.

The density operator ρ may be expanded in a similar fashion

$$\rho = \sum_{j=0}^{n^2-1} \rho_j^* U_j = \sum_{j=0}^{n^2-1} \rho_j U_j^\dagger \tag{I.42}$$

where the coefficients ρ_j are given by

$$\rho_j=\mathrm{Tr}(\rho U_j)=\langle U_j\rangle \quad \text{and} \quad \rho_j^*=\mathrm{Tr}(\rho U_j^\dagger)=\langle U_j^\dagger\rangle$$

Finally $\langle A\rangle$ may be written as follows

$$\langle A\rangle=\mathrm{Tr}(A\rho)= \sum_{i,j=0}^{n^2-1} u_j^*\rho_i \mathrm{Tr}(U_j U_i^\dagger)= \sum_{j=0}^{n^2-1} u_j^*\rho_j= \sum_{j=0}^{n^2-1} u_j\rho_j^*= \sum_{j=0}^{n^2-1} u_j^*\langle U_j\rangle= \sum_{j=0}^{n^2-1} u_j\langle A_j^\dagger\rangle \tag{I.43}$$

Eq.(I.43) shows that $\langle A\rangle$ is a linear combination of the average values $\langle U_j\rangle$ of the base operators. The expansion becomes particularly useful if we may choose the base operators in such a way that the average value of a number of base operators vanish. If preferable, the base operators may be chosen hermitian since non-hermitian operators U and $U^\dagger$ can be replaced by the hermitian combinations $(U+U^\dagger)$ and $i(U-U^\dagger)$. In this case, coefficiants ρ_j become real.

It is frequently advantageous to use the special Liouville representation for which the base operators include the normalized unit operator, i.e. $U_o = (\mathrm{Tr}I)^{-\frac{1}{2}}I$. The special Liouville representation is possible only if $\mathrm{Tr}(U_o^\dagger U_i) = (\mathrm{Tr}I)^{-\frac{1}{2}}\mathrm{Tr}U_i = 0$ for all $i \neq 0$, which means that traces must vanish for all base operators except the unit operator.

The equation of motion for the coefficients ρ_j may be derived by differentation of Eq.(I.42) and combination with Eq.(I.30), i.e.

$$\frac{d\rho}{dt} = \sum_{j=0}^{n^2-1} \frac{d\rho_j}{dt} U_j^\dagger = i[\rho,H]$$

Multiplication by U_r and taking the trace give

$$\frac{d\rho_r}{dt} = i\mathrm{Tr}(U_r[\rho,H]) = i\mathrm{Tr}(H[U_r,\rho])$$

By expansion of ρ according to Eq.(I.42) we obtain finally

$$\frac{d\rho_r}{dt} = i\sum_{j=0}^{n^2-1} \mathrm{Tr}(H[U_r,U_j^\dagger]\rho_j) = i\sum_{j=0}^{n^2-1} \mathrm{Tr}(H[U_r,U_j^\dagger])\langle U_j\rangle \qquad (I.44)$$

A simple and well-known example of the application of the special Liouville representation is the expansion of the density operator for spins with $I=\frac{1}{2}$ into the normalized unit operator and the three normalized Pauli operators[14].

In the statistical approach the density operator was defined in terms of a set of ketvectors $|m\rangle$ necessary to describe a quantum system adequately to allow calculation of the average values of certain operators. Characteristics irrelevant to these calculations were suppressed. By construction the density operator was therefore incapable of giving any information about average values of observables depending upon the suppressed characteristics. In the quantum-mechanical approach we start with a complete many-body state vector for an isolated system, define a projection operator, and by averaging over the characteristics of no interest to our particular problem we derive a reduced density operator with the relevant information.

Let us assume that an isolated system is described by the stationary state described by the many-body ketvector $|\psi(1,2...N)\rangle$. For the isolated system the density operator is the projection operator $|\psi\rangle\langle\psi|$. Consider now an operator which depends on the coordinates of the subsystem containing the first n particles, but not on the coordinates of the subsystem containing the last m particles. When the two subsystems interact, we cannot express $|\psi\rangle$ as a product of kets for the two subsystems.

For the expectation value of A we have

$$\langle A\rangle = \langle\psi|A|\psi\rangle$$

Expansion in the coordinate representation gives

$$\langle A\rangle = \int \langle\psi|\bar{q}_n'\bar{q}_m'\rangle\langle\bar{q}_n'\bar{q}_m'|A|\bar{q}_n''\bar{q}_m''\rangle\langle\bar{q}_n''\bar{q}_m''|\psi\rangle d\bar{q}_n' d\bar{q}_m' d\bar{q}_n'' d\bar{q}_m''$$

where

$$|\bar{q}_n'\bar{q}_m'\rangle = |\bar{q}_1', \bar{q}_2' \dots \bar{q}_n', \bar{q}_{n+1}' \dots \bar{q}_N'\rangle$$

with $n+m = N$.

Since A does not operate on the last m coordinates we obtain

$$\langle A\rangle = \int \langle\psi|\bar{q}_n'\bar{q}_m'\rangle\langle\bar{q}_n'|A|\bar{q}_n''\rangle\delta(\bar{q}_m'-\bar{q}_m'')\langle\bar{q}_n''\bar{q}_m''|\psi\rangle d\bar{q}_n' d\bar{q}_m' d\bar{q}_n'' d\bar{q}_m''$$

or

$$\langle A\rangle = \int \langle\bar{q}_n'|A|\bar{q}_n''\rangle\langle\bar{q}_n''|\rho^{(n)}|\bar{q}_n'\rangle d\bar{q}_n' d\bar{q}_n'' = \mathrm{Tr}(A\rho^{(n)}) \qquad (I.45)$$

where the reduced density operator for n particles is given by

$$\rho^{(n)} = \int \langle\bar{q}_m'|\psi\rangle\langle\psi|\bar{q}_m'\rangle d\bar{q}_m' = \int \langle\bar{q}_m'|\rho|\bar{q}_m'\rangle d\bar{q}_m' = \mathrm{Tr}_{(m)}\rho \qquad (I.46)$$

Eq.(I.45) is equivalent to Eq.(I.37) just as Eq.(I.46) is equivalent to Eq.(I.39). It should be noted that the averaging over the m last coordinates is equivalent to the averaging by means of the weighting factors p_m in the statistical approach.

Matrix elements of $\rho^{(n)}$ are

$$<\bar{q}''_n|\rho^{(n)}|\bar{q}'_n> = \int <\bar{q}''_n\bar{q}'_m|\psi><\psi|\bar{q}'_n\bar{q}'_m>d\bar{q}'_m = \int \psi(\bar{q}''_n\bar{q}'_m)\psi^*(\bar{q}'_n\bar{q}'_m)d\bar{q}'_m$$

The diagonal element $<\bar{q}'_n|\rho^{(n)}|q'_n>$ is a measure of the probability of finding in the subsystem particle 1 in volume element $d\bar{q}'_1$, particle 2 in volume element $d\bar{q}_2$ etc. On the other hand, no matrix element of $\rho^{(n)}$ can give information on the whereabouts of the last m particles.

For N indistinguishable particles the relevant information is the probability of finding N particles occupying N selected volume elements in configuration space regardless of the order.

For indistinguishable particles we have

$$P<\bar{q}_1,\bar{q}_2\ldots\bar{q}_N|\rho|\bar{q}_1,\bar{q}_2\ldots\bar{q}_N> = <\bar{q}_1,\bar{q}_2\ldots\bar{q}_N|\rho|\bar{q}_1,\bar{q}_2\ldots\bar{q}_N>$$

where the operator P permutes the particles. The number of permutations is N! and we may therefore conveniently introduce a new N-particle density operator ρ_N. A diagonal element in the coordinate representation is a measure of the probability of having N particles occupying N selected volume elements regardless of the order, i.e.

$$<\bar{q}_1,\bar{q}_2\ldots\bar{q}_N|\rho_N|\bar{q}_1,\bar{q}_2\ldots\bar{q}_N> = N!<\bar{q}_1,\bar{q}_2\ldots\bar{q}_N|\rho|\bar{q}_1,\bar{q}_2\ldots\bar{q}_N>$$

or $\rho_N = N!\rho$

We pursue this idea further and define now reduced density operators ρ_n with the property that diagonal elements in the coordinate representation is a measure of the probability of finding n of the N particles regardless of the order at n selected positions of configuration space, i.e.

$$\rho_n = N(N-1)\ldots(N-n+1)\int <\bar{q}_{n+1}\ldots\bar{q}_N|\rho|\bar{q}_{n+1}\ldots\bar{q}_N>d\bar{q}_{n+1}\ldots d\bar{q}_N$$

Successive reduced density operators are connected by the recursion formula

$$(N-n)\rho_n = \int \langle \bar{q}_{n+1} | \rho_{n+1} | \bar{q}_{n+1} \rangle d\bar{q}_{n+1}$$

Of particular interest are ρ_1 and ρ_2. ρ_1 integrates to N and denotes therefore a number of density.

The calculation of expectation values of operators involving all particles in a symmetrical fashion is straightforward once the relevant reduced density operators have been determined.

As an example, we may consider the hamiltonian operator

$$H(1\ldots n) = \sum_{i=1}^{N} h(i) + \frac{1}{2} \sum_{\substack{i,j=1 \\ i \neq j}}^{N} q(i,j)$$

where $$h(i) = -\frac{\hbar^2}{2m}\bar{\nabla}_i^2 + V(i), \qquad q_{(ij)} = \frac{e^2}{r_{ij}}$$

The expectation value of H in the stationary state ψ is then given by the brief expression

$$\langle \psi | H | \psi \rangle = E = Tr(h\rho_1) + \frac{1}{2} Tr(q\rho_2).$$

REFERENCES

1. J.A. Crawford, Nuovo Cim. 10, 698 (1958).
2. M. Blume, Phys.Rev. 174, 351 (1968).
3. R. Kubo in "Lectures in Theoretical Physics". W.E. Brittin and L.G. Dunham, Eds. Interscience Publishers, Inc., New York 1959, Vol.I, p. 139.
4. R.M. Wilcox, J. Math. Phys. 8, 962 (1967).
5. C.N. Banwell and H. Primas, Mol. Phys. 6, 225 (1963).
6. P. Roman, "Advanced Quantum Theory", Addison-Wesley Publishing Company, Inc. Reading, Mass. 1965, p. 308.
7. S.S. Schweber, "An Introduction to Relativistic Quantum Field Theory", Harper and Row, New York 1964, p. 330.
8. F.J. Dyson, Phys. Rev. 75, 486 (1949).
9. J. von Neumann, Nachr. Ges. Wiss. Göttingen, 1, 245,273 (1927).

10. P.A.M. Dirac, Proc. Camb. Phil. Soc. 25, 62 (1929); "Principles of Quantum Mechanics", Oxford University Press, 4.Ed. Oxford 1957, p. 130.

11. P.A.M. Dirac, Proc. Camb. Phil. Soc. 26, 361, 376 (1930).

12. K. Husimi, Proc. Phys. Math. Soc. Japan 22, 264 (1940).

13. R. McWeeny, Rev. Mod. Phys. 32, 335 (1960).

14. U. Fano, Rev. Mod. Phys. 29, 74 (1957); Contribution in E.R. Caianello, Ed., "Lectures on the Many-Body Problem", Vol. 2, Academic Press, New York 1964, p. 217.

15. A. Messiah, "Quantum Mechanics", North-Holland Publishing Company, Amsterdam, 1964, Vol. I, p. 331.

16. P. Roman, "Advanced Quantum Theory", Addison-Wesley Publishing Co., Inc. Reading, Mass. 1965, p. 90.

STOCHASTIC PROCESSES

J. Boiden Pedersen

Aarhus University

II.1. STOCHASTIC (RANDOM) VARIABLES AND PROBABILITY

The elementary theory of stochastic variables and probability is based on the idea of an experiment capable of being repeated. In an experiment we observe one or more outcomes, e.g. the outcomes may be the energy, the magnetization etc. of a system or it may be the number of eyes in a die experiment. We shall use the word "trial" to designate a single performance of a well-defined experiment in which a single outcome is observed. The outcome of a trial is called an event. The collection of all events is called the sample space; the points of which (the sample points) are the irreducible (simple) events. The physical sample space is the phase space.

Events which are not simple events are called reducible (compound) events. A reducible event is an aggregate of simple events (or a subset of the sample space). If in a trial the outcome is α, then we say that the event A occurred at this trial, if α is a subset of A. Thus the event A will occur at all trials whose outcomes are subsets of A. It is meaningful to talk about an event A only when it may be clearly stated for every outcome of the trial whether the event A has or has not occurred.

A stochastic (or random) variable X is a function which assigns a real number to any event. The value of X for an event may be natural in the sense that the physical system itself supplies us with a definite number for that event. This is the case when we measure the energy, the magnetization etc. In the case that the system does not supply us with a number for an event, we must choose the number. No restriction exists for this choice (indeed it may be done at random). An example is a toss of an unbiased coin. Here we can choose the values of X to be one for the occurrence of head and zero for the occurrence of tail.

To any event A we assign a probability measure $P\{A\}$ by the following conditions

a) $P\{A\} \geq = 0$

b) $P\{\Omega\} = 1$ where Ω is the total sample space

c) $P\{A_1 \cup A_2\} = P\{A_1\} + P\{A_2\}$ when $A_1 \cap A_2 = 0$

We may think of $P\{A\}$ as the fraction of trials with occurrence of A when the number of trials is large. The $P\{A\}$ then depends on the actual system we want to describe.
The probability measure $P\{A\}$ might be used to find the probability distribution F(x), describing the relative occurrence of the reducible event $\{X \leq x\}$.

$$F(x) \simeq N_x/N \quad \text{when N is large}$$

N_x is the number of trials with outcomes $X \leq x$ out of a total number of trials equal to N.
F(x) has the following properties (the range of X is most generally taken as $(-\infty,\infty)$)

$$\begin{aligned}
&F(x) = \text{Prob}\{X \leq x\} \\
&F(x) \text{ is a never decreasing function} \\
&F(x) \text{ is continuous from the right} \qquad \text{(II.1)} \\
&F(-\infty) = 0,\ F(+\infty) = 1 \\
&\text{Prob}\{a<X\leq b\} = F(b) - F(a)
\end{aligned}$$

We usually work with the probability density P(x) rather than the probability distribution F(x). P(x) is defined by

$$P(x) = \frac{dF(x)}{dx} \qquad \text{(II.2)}$$

and may contain delta peaks.

From the properties of F(x) it follows that

$$\begin{aligned} &P(x)dx = \text{Prob}\{x<X\leq x+dx\} \\ &P(x)\geq 0 \quad \text{for all } x \\ &\int_{-\infty}^{+\infty} P(x)dx = 1 \end{aligned} \qquad \text{(II.3)}$$

If the range of X is discrete, say $(x_1, x_2 \ldots)$, then the probability density P(x) is given by

$$P(x)\ \text{discrete} = \sum_i P_i \delta(x-x_i) \qquad \text{(II.4)}$$

where P_i is the discontinuous jump of F(x) at the point x_i.

$$P_i = \text{Prob}\{X = x_i\}$$

All the following equations will be written without specifying the range of X. They hold whether the range is continuous, discrete, or both. It is sometimes convenient to specify that the range is discrete. This may be done by the transformation

$$\int P(x)f(x)dx \rightarrow \sum_n P_n f_n \quad \text{where } f_n = f(x_n). \qquad \text{(II.5)}$$

From the definition of a stochastic variable it follows that

$$kX,\ X+k,\ X+Y,\ X-Y,\ X\cdot Y, \quad \text{and } g(X)$$

are also stochastic variables when X and Y are stochastic variables, k is a real constant, and g is a real function.

Mean value (average value or expectation value) of a stochastic variable g(X) is defined by

$$E\left\{g(X)\right\} = \int g(x)dF(x) = \int g(x)P(x)dx \qquad (II.6)$$

Common ways of writing the mean value are $E\left\{g(X)\right\}$, $\langle g(X)\rangle$, and $\overline{g(X)}$.

Special cases of mean values are the moments and the variance. It should be noticed that these may not exist for certain distribution functions.

$$m_n = E\left\{X^n\right\} \quad \text{the n'th moment} \qquad (II.7)$$

$$m = m_1 = E\left\{X\right\} \text{ is the mean of X.}$$

$$\mu_n = E\left\{(X-m_1)^n\right\} \text{ the n'th central moment} \qquad (II.8)$$

The variance of X is defined by

$$\text{var } X = \sigma^2 = E\left\{(X-EX)^2\right\} = E\left\{X^2\right\} - (E\left\{X\right\})^2 \qquad (II.9)$$

σ is known as the standard deviation.

It follows immediately that

$$E(k\,X) = k\,E(X),\ \text{var}(k\,X) = k^2\,\text{var}(X) \qquad (II.10)$$

$$E(X+k) = E(X)+k,\ \text{var}(X+k) = \text{var}(X) \qquad (II.11)$$

$$E(X+Y) = E(X)+E(Y) \qquad (II.12)$$

X and Y are stochastic variables, k is a real constant.

Sometimes it is convenient to use a stochastic variable with mean zero, variance one, or both, by defining new variables as

$$X_1 = X - m_1 \qquad \text{mean} = 0$$

$$X_2 = X/\sigma \qquad \text{variance} = 1$$

$$X_3 = (X-m_1)/\sigma \qquad \text{mean} = 0,\ \text{variance} = 1$$

X_3 is denoted the normalized variable.

It is sometimes necessary to use more than one stochastic variable. In this case each stochastic variable must be specified (i.e. its range and its probability density). We may define complex variables $Z = X_1 + iX_2$

vector variables $X = (X_1, \ldots X_n)$

The joint probability density of n stochastic variables $X_1 \ldots X_n$ is defined by

$$P(x_1 \ldots x_n)dx_1 \ldots dx_n = \text{Prob}\left\{x_1 < X_1 \leq x_1 + dx_1 \ldots x_n < X_n \leq x_n + dx_n\right\}$$

The joint probability has the properties

$$P_n(x_1 \ldots x_n) \geq 0 \tag{II.13}$$

$$\int P_n(x_1 \ldots x_{i-1}, x_i, x_{i+1} \ldots x_n)dx_i = P_{n-1}(x_1 \ldots x_{i-1}, x_{i+1} \ldots x_n) \tag{II.14}$$

$$\int \ldots \int P_n(x_1 \ldots x_n)dx_1 \ldots dx_n = 1 \tag{II.15}$$

We define conditional probability density by

$$P_{n,k}(x_1 \ldots x_k | x_{k+1} \ldots x_n) = P_n(x_1 \ldots x_n)/P_k(x_1 \ldots x_k) \tag{II.16}$$

$$P_{n,k}(x_1 \ldots x_k | x_{k+1} \ldots x_n)dx_{k+1} \ldots dx_n = \text{Prob}\left\{\text{given } X_1 = x_1 \ldots X_k = x_k \,\middle|\, \text{that } x_{k+1} < X_{k+1} \leq x_{k+1} + dx_{k+1} \ldots x_n < X_n \leq x_n + dx_n\right\}$$

We have

$$P_{n,k}(x_1 \ldots x_k | x_{k+1} \ldots x_n) \geq 0 \tag{II.17}$$

$$\int \ldots \int P_{n,k}(x_1 \ldots x_k | x_{k+1} \ldots x_n)dx_{k+1} \ldots dx_n = 1 \tag{II.18}$$

$$\int P_{n,k}(x_1 \ldots x_k | x_{k+1} \ldots x_n)dx_n = P_{n-1,k}(x_1 \ldots x_k | x_{k+1} \ldots x_{n-1}) \tag{II.19}$$

$P_{n,1}(x_1 \ldots x_{n-1}|x_n)$ will be written as $P_n(x_1 \ldots x_{n-1}|x_n)$

$P_{2,1}(x_1|x_2)$ will be written as $P(x_1|x_2)$.

It should be noted that several different ways of writing conditional probability appear in the literature.

A real or complex function $g(X_1 \ldots X_n)$ of the stochastic variables $X_1 \ldots X_n$ is a stochastic variable with mean value

$$E\{g(X_1 \ldots X_n)\} = \int \ldots \int g(x_1 \ldots x_n) P_n(x_1 \ldots x_n) dx_1 \ldots dx_n \qquad \text{(II.20)}$$

Special cases are the joint moments and the covariance. The joint moment is defined by

$$m_{k_1 \ldots k_n} = E\{X_1^{k_1} \ldots X_n^{k_n}\} \qquad \text{(II.21)}$$

The covariance of X and Y is defined by

$$\text{cov}(X,Y) = E\{(X-EX)^*(Y-EY)\} \qquad \text{(II.22)}$$

where X^* denotes the complex conjugate of X.

For a stochastic vector variable $X = (X_1 \ldots X_n)$ we define a n×n covariance matrix Γ with elements

$$\Gamma_{rs} = \text{cov}(X_r, X_s) \qquad \text{(II.23)}$$

We may prove that this matrix is hermitian and nonnegative definite. The correlation coefficient for X and Y is

$$\rho_{X,Y} = \text{cov}(X,Y)/[\text{var}X\text{var}Y]^{\frac{1}{2}} = \text{cov}(X,Y)/(\sigma_x \sigma_y). \qquad \text{(II.24)}$$

Let us show that $|\rho_{X,Y}| \leq 1$ (II.25)

Proof:

When $\text{var}(X+kY) = \text{var}X + k^2\text{var}Y + 2k\text{cov}(X,Y) \geq 0$

for all k, then the discriminant must be ≤ 0, i.e.

$\text{cov}(X,Y)^2 - \text{var}X\text{var}Y \leq 0$

or $\text{cov}(X,Y)^2/(\text{var}X\text{var}Y) = \rho^2_{X,Y} \leq 1$

X and Y are said to be strong correlated (positive or negative) if $\rho_{X,Y} \simeq \pm 1$

X and Y are weak correlated if $|\rho_{X,Y}| \ll 1$

X and Y are uncorrelated if $\rho_{X,Y} = 0 \Longleftrightarrow \text{cov}(X,Y) = 0 \Longleftrightarrow E(XY) = EXEY$

$X_1, X_2, \ldots X_n$ are independent random variables if

$$P_n(x_1 \ldots x_n) = P(x_1)P(x_2)\ldots P(x_n) \qquad \text{(II.26)}$$

for all values x_i.

For independent variables we have

$$E\{X \cdot Y\} = E\,X \cdot E\,Y \quad \text{and} \quad \text{var}(X+Y) = \text{var}X + \text{var}Y \qquad \text{(II.27)}$$

Independent variables are thus uncorrelated. The inverse statement is not true in general.

A stochastic variable X is a Gaussian variable if its probability density is a Gaussian distribution

$$p(x) = (2\pi b^2)^{-\frac{1}{2}} \exp\left[-(x-a)^2/2b^2\right] \qquad \text{(II.28)}$$

A Gaussian distribution is completely described by its mean and variance because of the following relations

$$E\{X\} = m = a, \text{var}X = \sigma^2 = b^2 \qquad \text{(II.29)}$$

$$\mu_{2n+1} = 0 \qquad \text{for } n = o,1,2,\ldots$$

$$\mu_{2n} = 1\ 3\ 5 \ldots (2n-1)(\sigma^2)^n \text{ for } n = 1,2,\ldots$$

Gaussian variables play an important role due to the existence of the following theorem.

The Central-Limit Theorem. The central-limit theorem says that the probability density P(x) of the sum $X = X_1 + \ldots + X_n$ of n independent stochastic variables X_i approaches a Gaussian (normal) distribution as n increases

$$P(x) \to (2\pi\sigma^2)^{-\frac{1}{2}} \exp\left[-(x-m)^2/2\sigma^2\right] \qquad \text{(II.30)}$$

as $n \to \infty$ (n equal to 10 is usually sufficient).

with $E\{X\} = m = m_1+\ldots+m_n$

and $\text{var}\{X\} = \sigma^2 = \sigma_1^2+\ldots\sigma_n^2$

if certain rather weak conditions[8] are satisfied.

The joint characteristic function of n stochastic variables $X_1\ldots X_n$ is defined as

$$\Phi(\omega_1\ldots\omega_n) = E\{\exp[i(\omega_1X_1+\ldots\omega_nX_n)]\} \tag{II.31}$$

i.e. Φ is the Fourier transform of P_n. There is thus a one to one correspondence between the probability density and the characteristic function. The joint moments may be obtained from the characteristic function as

$$(i)^r m_{k_1\ldots k_n} = \frac{\partial^r\Phi(0\ldots0)}{\partial\omega_1^{k_1}\ldots\partial\omega_n^{k_n}} \tag{II.32}$$

where $r = k_1+\ldots k_n$.

If $X_1\ldots X_n$ are independent variables then $\exp\{i\omega_1X_1\},\ldots\ldots$ $\exp\{i\omega_nX_n\}$ are also independent variables and

$$\Phi(\omega_1\ldots\omega_n) = \Phi_1(\omega_1)\ldots\Phi_n(\omega_n).$$

As an example of the usefulness of the characteristic function we shall derive the probability density of the sum $X=X_1+\ldots X_n$ of n independent variables.
We have

$$\Phi_X(\omega) = \Phi_1(\omega)\ldots\Phi_n(\omega) \tag{II.33}$$

By an inverse Fourier transformation we get that the density of X equals the convolution of the densities of $X_1\ldots X_n$.

$$P(x) = P_1(x)\circ P_2(x)\circ\ldots\circ P_n(x) \tag{II.34}$$

If n = 2 then P(x) is

$$P(x) = P_1(x) \circ P_2(x) = \int P_1(x-x')P_2(x')dx'.$$

We shall now give an example showing the usefulness of stochastic variables to magnetic relaxation problems.

Example. Spin-dephasing in a spin-echo experiment by diffusion.

Let us consider a spin-echo experiment with a linear field gradient. We want to calculate the dephasing of a precessing spin caused by the transport (diffusion) of the spin into a region where the applied field is slightly different.

We use a random walk model to describe the diffusion: A molecule remains at a given z position for exactly τ seconds when it abruptly jumps to a new position whose z coordinate differs from the previous one by $\Delta z a_i$, where Δz is a fixed distance and a_i is a random variable whose value is +1 or -1, with equal probability $P(\pm 1) = \frac{1}{2}$.

Let the gradient of H_z in the z direction be constant, and, in magnitude, G gauss/cm, and let $H_z(0)$ be the field in which a given spin finds itself at t=0.

At a later time $t=j\tau$ the spin will find itself in a field $H_z(j\tau)$ given by

$$H_z(j\tau) = H_z(0) + G\Delta z \sum_{i=1}^{j} a_i$$

After N steps, i.e. at time $t = N\tau$, the phase ϕ of the precessing spin will differ from the value ϕ_o, it would have had at this same instant if the spin had remained in the same place by an angle

$$\phi_D = \phi - \phi_o = \sum_{j=1}^{N} \gamma\tau \left[H_z(j\tau) - H_z(0) \right]$$

$$= G\Delta z \gamma\tau \sum_{j=1}^{N} \sum_{i=1}^{j} a_i$$

$$=G\Delta z\gamma\tau\sum_{j=1}^{N}(N+1-j)a_j$$

The last equality may be seen by the following argument. In the double summation we get an a_k each time $j\geq k$. This happens N+1-k times.

The stochastic variable ϕ_D is a sum of independent stochastic variables $\phi_D(j)$

$$\phi_D(j)=G\Delta z\gamma\tau(N+1-j)a_j$$

with $$E\{\phi_D(j)\}=G\Delta z\gamma\tau(N+1-j)E\{a_j\}=0$$

and $$\mathrm{var}\{\phi_D(j)\}=G^2\Delta z^2\gamma^2\tau^2(N+1-j)^2$$

We, therefore, have $E\{\phi_D\}=0$

and $$\mathrm{var}\{\phi_D\}=G^2\Delta z^2\gamma^2\tau^2\sum_{j=1}^{N}(N+1-j)^2$$

$$=G^2\Delta z^2\gamma^2\tau^2\sum_{j=1}^{N}\left[(N+1)^2+j^2-2j(N+L)\right]$$

$$=G^2\Delta z^2\gamma^2\tau^2\left[N(N+1)^2+N(N+1)(2N+1)/6-(N+1)N(N+1)\right]$$

As we let $N\to\infty$ we may drop the lower powers of N. Then

$$\mathrm{var}\{\phi_D\}=G^2\Delta z^2\gamma^2\tau^2N^3/3$$

By the central-limit theorem (II.30) we obtain

$$P(\theta_D)=(2\pi G^2\Delta z^2\gamma^2\tau^2N^3/3)^{-\frac{1}{2}}\exp(-3\phi_D^2/2G^2\Delta z^2\gamma^2N^3)$$

Substitution of $N=t/\tau$, taking the limits $\Delta z\to 0$, $\tau\to 0$, and assuming that

$$D=\lim_{\Delta z,\tau\to 0}\Delta z^2/2\tau$$

exists, gives the distribution-in-phase at time t

$$P(0|\theta_D,t)=(4\pi\gamma^2G^2Dt^3/3)^{-\frac{1}{2}}\exp(-3\phi_D^2/4G^2\gamma^2Dt^3)$$

This result was originally derived by Carr and Purcell[16] in their discussion of spin-echo experiments.

II.2. GENERAL REMARKS ON STOCHASTIC (RANDOM) PROCESSES

A stochastic process X(t) is a process in which the variable X does not depend in a definite way on the independent variable t(= time), as in a causal process; instead we get different functions $X_\alpha(t)$ in different observations. Let the allowed values of t be denoted by T. T may be continuous or discrete. In the latter case it is customary to call X(t) a stochastic chain. A specified function $X_\alpha(t), t\varepsilon T$, is called a realization of the process. Let it be denoted by α. In the language of physics $X_\alpha(t)$ is the trajectory in the phase space of a phase point representing a member of the initial ensemble. The α's constitute a set with a time-independent density ρ(α). The distribution ρ(α) is given by the construction of the ensemble.

The stochastic process is completely described by the following set of joint probability densities $P_n(x_1,t_1\ldots x_n,t_n)$ n=1,2,3...

$$P_n(x_1,t_1\ldots x_n,t_n)dx_1dx_2\ldots dx_n =$$

$$\mathrm{Prob}\left\{x_1<X(t_1)\le x_1+dx_1\ldots x_n<X(t_n)\le x_n+dx_n\right\}$$

Strictly speaking X(t) is a function of $\alpha, X_\alpha(t)$, thus

$$P_n(x_1,t_1\ldots x_n,t_n)dx_1\ldots dx_n = \int\limits_{\substack{x_1<X_\alpha(t_1)\le x_1+dx_1\\ \vdots \\ x_n<X_\alpha(t_n)\le x_n+dx_n}} \rho(\alpha)d\alpha$$

$$= dx_1 \ldots dx_n \int \delta(X_\alpha(t_1)-x_1)\ldots\delta(X_\alpha(t_n)-x_n)\rho(\alpha)d\alpha$$

Mainly we just write X(t) and do not specify that X(t) is a function of α. In fact we may think of X(t) as a family of stochastic variables depending on a parameter t (the time). This allows us to use the expressions from II.1 when we interpret X_1 as $X(t_1)$, X_2 as $X(t_2)$ etc.

P_n has the following properties

$$P_n \geq 0 \qquad \text{for all } n, x_i \tag{II.35}$$

$$\int P_n(x_1,t_1;\ldots x_n,t_n)dx_1\ldots dx_n = 1$$

$$\int P_n(x_1,t_1\ldots x_{i-1},t_{i-1},x_i,t_i,x_{i+1},t_{i+1}\ldots x_n,t_n)dx_i =$$

$$P_{n-1}(x_1,t_1\ldots x_{i-1},t_{i-1},x_{i+1},t_{i+1}\ldots x_n,t_n)$$

P_n is symmetrical in the set of variables $x_1t_1;x_2t_2\ldots x_nt_n$.

Mean of an arbitrary function $g(X(t_1),\ldots X(t_n))$ is

$$E\left\{g(X(t_1),\ldots X(t_n))\right\} = \int\ldots\int g(x_1,\ldots x_n)P_n(x_1t_1\ldots x_n,t_n)dx_1\ldots dx_n \tag{II.36}$$

Special mean value is that of the stochastic integral

$$E\left\{\int_0^t g(X(t)dt\right\} = \lim_{n\to\infty} E\left\{\sum_{k=1}^{n} g(X(k\Delta t))\Delta t\right\} \tag{II.36a}$$

We have devided the interval $[0,t]$ in n parts of equal length $\Delta t = t/n$.

The joint moments are defined by

$$m_{r_1 r_2 \dots r_n}(t_1 \dots t_n) = E\left\{X(t_1)^{r_1} X(t_2)^{r_2} \dots X(t_n)^{r_n}\right\} \qquad \text{(II.36b)}$$

$$m(t) = m_1(t) = E\left\{X(t)\right\} \text{ is the mean of } X(t) \qquad \text{(II.36c)}$$

$$\mathrm{var}X(t) = \sigma^2(t) = m_2(t) - \left[m(t)\right]^2 = E\left[X(t) - m(t)\right]^2 \qquad \text{(II.36d)}$$

The central-moments $\mu_n(t)$ are defined by

$$\mu_n(t) = E\left\{\left[X(t) - m(t)\right]^n\right\} \qquad \text{(II.36e)}$$

The characteristic function is

$$\phi(q,t) = E\left\{\exp\left[iqX(t)\right]\right\} \qquad \text{(II.36f)}$$

i.e. the Fourier transformed of the probability density.

The covariance function is defined

$$\Gamma(s,t) = E\left\{\left[X(s) - m(s)\right]\left[X(t) - m(t)\right]\right\} = m_{11}(s,t) - m(s)m(t) \qquad \text{(II.37)}$$

We shall always suppose that $X(t)$ is regular i.e. $E|X(t)|^2 < \infty$ for all $t \epsilon T$. This insures that the covariance function exists.

The auto-correlation function for the process is

$$\psi(s,t) = \Gamma(s,t) / \left[\sigma(s)\sigma(t)\right] \qquad \text{(II.38)}$$

For a complex process $Z(t) = X(t) + iY(t)$ we use the following definitions in analogy with those for stochastic variables

$$\begin{aligned} E\left\{Z(t)\right\} &= E\left\{X(t)\right\} + iE\left\{Y(t)\right\} \\ m_2(t) &= E|Z(t)|^2 = E\left\{Z(t)^* Z(t)\right\} \\ m_{11}(s,t) &= E\left\{Z(s)^* Z(t)\right\} \end{aligned} \qquad \text{(II.39)}$$

A stochastic process is said to be stationary in the strict sense if

$$P_n(x_1, t_1 + \tau \dots x_n, t_n + \tau) = P_n(x_1, t_1 \dots x_n, t_n) \qquad \text{(II.40)}$$

for all $n \geq 1, t_1, t_2 \ldots t_n, \tau \varepsilon T$

In particular $P_1(x,t) = P_1(x)$ and both $E\,X(t) = m(t) = m$ and $\mathrm{var}X(t) = \sigma^2(t) = \sigma^2$ are independent of t. We may, therefore, put m=0 and $\sigma^2=1$ without loss of generality. We shall do that in the following. The auto-correlation function has thus the following property

$$\psi(t_1,t_2) = \iint x_1^* x_2 P(x_1,t_1;x_2,t_2)dx_1dx_2$$

$$\iint x_1^* x_2 P(x_1,0;x_2,t_2-t_1)dx_1dx_2 \tag{II.41}$$

$$= \psi(0,t_2-t_1) \equiv \psi(t_2-t_1)$$

Using the symmetry properties of P_n we have

$$\psi(t_2-t_1) = \psi(t_1-t_2)^* \tag{II.42}$$

A stochastic process is said to be stationary in the wide sense if $m(t) = m, \sigma^2(t) = \sigma^2$, and $\psi(t_1,t_2) = \psi(t_2-t_1)$.

We notice that $\psi(0) = 1$ and $|\psi(t)| \leq 1$ for all $t \varepsilon T$. The inequality was shown for stochastic variables (II.22) and applies equally well for stochastic processes. It may also be derived from Schwarz's inequality. The value of $|\psi(t)|$ gives a measure of the degree of correlation. $|\psi(t) \simeq 1|$ means strong correlation. $|\psi(t) \simeq 0|$ means weak correlation. The degree of correlation means the extent to which the value of $X(t_1)$ affects the value of $X(t_2)$. For a system tending towards an equilibrium state we have

$$\psi(t) \to 0 \quad \text{as} \quad t \to \infty$$

The correlation time τ_c is defined by

$$\tau_c = \int_0^\infty \psi(t)dt \tag{II.43}$$

The feasibility of this definition may be seen by applying it to the following two correlation functions

$$\text{a)}\ \psi(t) = \begin{cases} 1 & \text{for } t \le T \\ 0 & \text{for } t > T \end{cases} \qquad \text{then } \tau_c = T$$

$$\text{b)}\ \psi(t) = \exp(-t/T) \qquad \text{then } \tau_c = T$$

This definition allows us to state that

$$\psi(t) \simeq \psi(0) = 1 \qquad \text{for } t \ll \tau_c$$

$$\psi(t) \simeq 0 \qquad \text{for } t \gg \tau_c$$

The auto-correlation tensor $\Psi(t)$ for a stationary vector process $X(t)$ with m components $X_i(t)$ and with mean zero and variance unity is a m×m matrix with elements

$$\psi_{rs}(t) = E\left\{X_r(0)^* X_s(t)\right\}$$

This matrix is hermitian i.e. $\psi_{rs}(t) = \psi_{sr}(t)^*$

The auto-correlation function is the trace

$$\psi(t) = E\left\{X(0)^* \cdot X(t)\right\} = \mathrm{Tr}\Psi(t) \tag{II.44}$$

The covariance function is important for the determination of the mathematical properties of a stochastic process. In the following we shall suppose the process is stationary unless otherwise stated. We can then use the auto-correlation function instead of the covariance function.

For the proofs of the following three theorems for stationary processes we refer to Prabhu[4].

Convergence in the mean square (m.s.) is defined as

$$X(t) \to X(\text{m.s.}) \qquad \text{as } t \to t_o$$

$$\text{if} \quad E|X(t)-X|^2 \to 0 \qquad \text{as } t \to t_o$$

1°. $X(t)$ is continuous (m.s.) at $t\varepsilon T$ if, and only if, $\psi(t)$ is continuous at $t = 0$. If this is the case then $\psi(t)$ is continuous at all $t\varepsilon T$.

2°. $X(t)$ has a derivative $X'(t)$ at $t\varepsilon T$ if, and only if, $\ddot{\psi}(t)$ exists at $t = 0$. If this is the case then $\ddot{\psi}(t)$ exists at all $t\varepsilon T$. Moreover the correlation function of $\dot{X}(t)$ is $-\ddot{\psi}(t)$ and

$$E\{\frac{dX(t)}{dt}\} = \frac{d}{dt} E\{X(t)\}$$

3°. $X(t)$ is Riemann-integrable over $[a,b]$ if, and only if, the integral $\int_a^b \psi(t)dt$ exists. If this is the case then

$$E\{\int_a^b X(0)^* X(t)dt\} = \int_a^b \psi(t)dt$$

and

$$E\{\int_a^b X(t)dt\} = \int_a^b E\{X(t)\}dt.$$

Let us consider a stationary process with mean zero, and variance one. The time average of $X(t)$ (written as m_T) is defined by

$$m_T = \lim_{t\to\infty} \frac{1}{2t} \int_{-t}^{t} X(t')dt' \qquad \text{(II.45)}$$

A special case is $\psi_T(\tau)$ defined as

$$\psi_T(\tau) = \lim_{t\to\infty} \frac{1}{2t} \int_{-t}^{t} X(t')^* X(t'+\tau)dt' \qquad \text{(II.46)}$$

The time averages are obviously stochastic variables. The possible values of m_T is $m_T(\alpha)$ defined by

$$m_T(\alpha) = \lim_{t\to\infty} 1/2t \int_{-t}^{t} X_\alpha(t')dt'$$

$m_T(\alpha)$ is the time average of the quantity X for a single realization (α)

X(t) is ergodic if time averages equal ensemble averages, i.e.

$$m = m_T, \psi(t) = \psi_T(t), \text{ etc.}$$

Ergodicity states that we may calculate ensemble averages as time averages of a single arbitrary realization $X_\alpha(t)$, i.e. the probability density of $m_T(\alpha)$ must be very narrow (its variance must equal zero).

In some cases we may prove whether a process is ergodic[8] but mostly ergodicity is taken as an assumption

It should be noted that we may have ergodicity at different levels, e.g. a process may be ergodic in the mean ($m=m_T$) but not in the variance ($\sigma^2 \neq \sigma_T^2$).

The spectral density (or power spectrum) $S(\omega)$ of a stochastic process X(t) is the Fourier transform of the auto-correlation function

$$S(\omega) = \frac{1}{2\pi} \int_{-\infty}^{\infty} \psi(t) \exp\left[-i\omega t\right] dt \tag{II.47}$$

As $\psi(-t) = \psi^*(t)$ we conclude that $S(\omega)$ is a real function.

By the Fourier inversion formula we have

$$\psi(t) = \int_{-\infty}^{\infty} S(\omega) \exp\left[i\omega t\right] d\omega \tag{II.48}$$

Setting t = 0 we observe that $S(\omega)$ is normalized to unity and equals the "average power" of the process X(t)

$$1 = \psi(0) = E\left\{|X(t)|^2\right\} = \int_{-\infty}^{\infty} S(\omega) d\omega$$

It may be proved[8] that $S(\omega)$ is non-negative for all ω.

If the process is real then $\psi(t)$ is real and even. $S(\omega)$ is therefore also even

$$S(-\omega) = S(\omega)$$

and the relationships between $\psi(t)$ and $S(\omega)$ reduce to

$$S(\omega) = \frac{1}{\pi}\int_0^\infty \psi(t)\cos(\omega t)dt \qquad \psi(t) = 2\int_0^\infty S(\omega)\cos(\omega t)d\omega$$

If the process is ergodic then the correlation function is

$$\psi(\tau) = \lim_{T\to\infty} \frac{1}{2T}\int_{-T}^{T} x^*(t)x(t+\tau)dt \tag{II.49a}$$

Substitution of (II.49a) in (II.47) gives

$$S(\omega) = \frac{1}{2\pi}\lim_{T\to\infty}\frac{1}{2T}\int_{-T}^{T}\int_{-T}^{T} x^*(t)x(t+\tau)\exp\left[-i\omega\tau\right]dtd\tau$$

We notice that

$$\exp\left[-i\omega\tau\right] = \exp\left[-i\omega(t+\tau)\right]\exp\left[i\omega t\right]$$

and change the integration variables to

$$t' = t \quad \text{and } \tau' = t+\tau$$

Then we have

$$S(\omega) = \frac{1}{2\pi}\lim_{T\to\infty}\frac{1}{2T}\int_{-T}^{T} x^*(t)\exp\left[i\omega t\right]\int_{-T+t}^{T+t} x(\tau)\exp\left[-i\omega\tau\right]d\tau dt$$

or

$$S(\omega) = \frac{1}{2\pi}\lim_{T\to\infty}\frac{1}{2T}\left|\int_{-T}^{T} x(t)\exp\left[-i\omega t\right]dt\right|^2 \tag{II.49b}$$

We observe that $S(\omega)\geq 0$ for all ω.

This expression is a familiar one in physics and it is used by Anderson[11]. In his article $x(t)$ is the element (for the line under consideration) of the radiating magnetic dipole moment. Anderson demonstrates by quantum mechanics without using stochastic arguments that the expression (II.49b) for $S(\omega)$ is the resonant intensity distribution (or resonance absorption spectrum).

The procedure originated by Anderson and extended by Kubo[12] is the prototype of much work on magnetic relaxation. We shall now define the relaxation function and then show the connection to the method of Anderson.

II.3. THE RELAXATION FUNCTION

Consider an oscillator with the equation of motion

$$\frac{dX}{dt} = i\omega(t)X \tag{II.50}$$

$\omega(t)$ is the frequency of a random modulation and, therefore, a stochastic process. In addition, we assume $\omega(t)$ to be stationary and ergodic. $X(t)$ is, therefore, also a stochastic process.

The time average of $\omega(t)$ is ω_o. We shall write $\omega(t)$ as

$$\omega(t) = \omega_o + \omega_1(t) \tag{II.51}$$

$\omega_1(t)$ represents the fluctuation in frequency with time average zero.

$$E\left\{\omega_1(t)\right\} = 0$$

A formal solution to (II.50) is

$$X(t)=X(0)\exp\left[i\int_o^t \omega(t')dt'\right]=X(0)\exp(i\omega_o t)\exp\left[i\int_o^t \omega_1(t')dt'\right]$$

The correlation function of $X(t)$ is

$$\begin{aligned}\psi(t) &= E\left\{X^*(0)X(t)\right\}/\sigma^2 \\ &= \exp(i\omega_o t)E\left\{\exp\left[i\int_o^t \omega_1(t')dt'\right]\right\}\end{aligned} \tag{II.52}$$

The relaxation function g(t) is defined by

$$g(t) = E\left\{\exp\left[i \int_{o}^{t} \omega_1(t')dt'\right]\right\} \tag{II.53}$$

The spectral density of X is thus

$$S(\omega) = \frac{1}{2\pi} \int_{-\infty}^{\infty} g(t)\exp\left[-i(\omega-\omega_o)t\right]dt$$

$$= \frac{1}{2\pi} \lim_{T\to\infty} \frac{1}{2T}\left| \int_{-T}^{T} x(t)\exp(-i\omega t)dt\right|^2 \tag{II.54}$$

We shall now follow Anderson's original procedure[11]. It can be shown in general that the spectrum of the radiation from any quantum mechanical system is given by

$$I(\omega) = \lim_{T\to\infty} \frac{1}{2T} \mathrm{Tr}\left| \int_{-T}^{T} \mu(t)\exp(-i\omega t)dt\right|^2 \tag{II.55}$$

where μ(t) is the radiating dipole moment matrix in the Heisenberg representation

$$i \frac{d\mu}{dt} = H\mu - \mu H$$

H is the complete hamiltonian which we may split up into three parts

$$H = H_o + H_p + H_m$$

H_o is the unperturbed hamiltonian which causes the energy differences which lead to the observed spectral lines we wish to study. H_o is usually the Zeeman hamiltonian. H_p is a perturbing hamiltonian, which does not commute with H_o and it, therefore, causes a broadening of the strong lines due to H_o or a fine structure. H_p is generally the dipolar interaction between the moments, although it may also involve other interactions, e.g. hyperfine splitting. H_m is the motional hamiltonian whose characteristic is that it commutes with both H_o and μ. Thus H_m can have no direct effect upon the radiation emitted or absorbed by the system. On the other hand H_m does not commute with H_p and can thus cause a time dependence of H_p by the relation

$$i\dot{H}_p = \left[H,H_p\right] = \left[H_o,H_p\right] + \left[H_m,H_p\right] \qquad \text{(II.56)}$$

It is this time-dependence which "narrows" the line-broadening which H_p otherwise would cause.

The assumptions for the following derivation is

a) Equation (II.55)

b) Equation (II.56)

c) $\left[H_m,\mu\right] = 0$

d) $\left[H_m,H_o\right] = 0$

e) H_p has no important matrix elements connecting different states $E_i^{(o)}$ and $E_j^{(o)}$ of the unperturbed hamiltonian H_o.

We shall only consider one spectral line corresponding to the transition $E_i^{(o)} \to E_j^{(o)}$. For this line

$$\underset{i\to j}{I(\omega)} = \lim_{T\to\infty} \frac{1}{2T} \left| \int_{-T}^{T} \mu_{ij}(t)\exp(-i\omega t)dt \right|^2 \qquad \text{(II.57)}$$

We transform $\mu_{ij}(t)$ to a new matrix element $\mu'_{ij}(t)$ which contains the time-dependence due to H_o explicitly

$$\mu_{ij}(t) = \mu'_{ij}(t)\exp(i\omega_{ij}^{(o)}t)$$

where $\omega_{ij}^{(o)} = E_i^{(o)} - E_j^{(o)}$

The equation of motion for $\mu'(t)$ is then

$$i\,\frac{d\mu'}{dt} = \left[H_p+H_m,\mu'\right]$$

$$= H_p\mu'-\mu' H_p$$

We have here used assumption c).

By the use of assumption e) we may write

$$i\,\frac{d\mu'_{ij}}{dt} = (H_p)_{ii}\mu'_{ij}-\mu'_{ij}(H_p)_{jj}$$

By defining ω'_{ij} as

$$\omega'_{ij}(t) = H_p(t)_{ii}-H_p(t)_{jj}$$

we have

$$\frac{d\mu'_{ij}(t)}{dt} = i\omega'_{ij}(t)\mu'_{ij}(t)$$

or

$$\frac{d\mu_{ij}(t)}{dt} = i\left[\omega_{ij}^{(o)}+\omega'_{ij}(t)\right]\mu_{ij}(t) \qquad \text{(II.58)}$$

Assuming that $\omega'_{ij}(t)$ is a stochastic process which is ergodic and stationary with mean zero then the equation of motion (II.58) has the form (II.50) and by using (II.57) and (II.54) we have

$$I_{i\to j}(\omega) = 1/2\pi\int_{-\infty}^{\infty} g(t)\exp\left[-i(\omega-\omega_{ij}^{(o)})t\right]dt \qquad \text{(II.59)}$$

where the relaxation function g(t) is given by (II.53)

$$g(t) = E\left\{\exp\left[i\int_{o}^{t}\omega'_{ij}(t)dt\right]\right\} \qquad \text{(II.60)}$$

Our main problem is, therefore, that of evaluating the relaxation function. It is likely, that this is hopeless unless the stochastic nature of the modulation $\omega'_{ij}(t)$ falls into one of the two simple cases, the Markoffian or the Gaussian which we shall now define.

II.4. GAUSSIAN AND MARKOV PROCESSES

We define conditional probability as for stochastic variables

$$P_n(x_1,t_1 \ldots x_{n-1},t_{n-1}|x_n,t_n)=\frac{P_n(x_1t_1\ldots x_nt_n)}{P_{n-1}(x_1t_1\ldots x_{n-1}t_{n-1})} \tag{II.61}$$

We take $t_1<t_2<\ldots<t_n$.

If $t_n = t_{n-1}$ then

$$P_n(x_1t_1\ldots x_{n-1}t_{n-1}|x_nt_n)=\delta(x_n-x_{n-1}) \tag{II.62}$$

We have also

$$\int P_n(x_1t_1\ldots x_{n-1}t_{n-1}|x_nt_n)dx_n=1 \tag{II.63}$$

$$P_n(x_1t_1\ldots x_{n-1}t_{n-1}|x_nt_n)\geq 0 \quad \text{for all } n,x_j.$$

A stochastic process X(t) is a Markov process if the conditional probability has the property

$$P_n(x_1t_1\ldots x_{n-1}t_{n-1}|x_nt_n)=P(x_{n-1}t_{n-1}|x_nt_n) \tag{II.64}$$

for all x_j and for all $t_1<t_2<\ldots<t_n \varepsilon T$.

It follows from the definition of conditional probability (II.61) that we have for a Markov process

$$P_n(x_1t_1\ldots x_nt_n) =$$

$$P(x_1,t_1)P(x_1t_1|x_2t_2)P(x_2t_2)|x_3t_3)\ldots P(x_{n-1}t_{n-1}|x_nt_n) \tag{II.65}$$

A Markov process is thus completely described by P(x,t) and $P(x_1t_1|x_2t_2)$.

For a stationary Markov process

$$P(x,t) = P(x) \text{ and } P(x_1t_1|x_2t_2)=P(x_1|x_2,t_2-t_1)$$

When the system tends to an equilibrium state we have

$$\lim_{t\to\infty} P(x_1|x_2t) = P(x_2)$$

In this case the $P(x_1,0;\ x_2,t)$ or $P(x_1|x_2,t)$ describes the process completely.

Besides these relations $P(x_1|x_2,t)$ must fulfill the so called S-C-K-equation to be derived later in this Section.

The relaxation function for a continuous and stationary Markov process may be evaluated by means of the functional integration technique of Kac (Sec. II.7).

A Gaussian process is a process $X(t)$ such that for $n\geq 1, t_1, t_2 \ldots \ldots t_n \varepsilon T$ the stochastic variables $X(t_1),\ldots,X(t_n)$ have a n-variate Gaussian distribution with

$$E\left\{X(t_p)\right\} = m(t_p)$$

$$\mathrm{var}\left\{X(t_p)\right\} = \Gamma(t_p,t_p)$$

$$\mathrm{covar}\left\{X(t_p),X(t_q)\right\} = \Gamma(t_p,t_q)$$

We assume that the covariance matrix Γ ($\Gamma_{pq} = \Gamma(t_p,t_q)$) is non-singular and denote the inverse matrix Γ^{-1}. The joint probability density of $X(t_1),\ldots X(t_n)$ is

$$P(x_1,t_1;\ldots;\ x_n,t_n) =$$

$$|\Gamma^{-1}|^{\frac{1}{2}}(2\pi)^{-n/2}\exp\left[-\tfrac{1}{2}\sum_{p,\ q=1}^{n}\sum \Gamma^{-1}_{pq}(x_p-m(t_p))(x_q-m(t_q))\right] \quad \text{(II.66)}$$

By the properties of a Gaussian distribution we see that a Gaussian process is stationary in the strict sense if it is stationary in the wide sense.

The characteristic functional of a process $X(t)$ is defined by

$$G[\xi] = E\left\{\exp\left[i\int_0^t X(t')\xi(t')dt'\right]\right\} \tag{II.67}$$

where $\xi(t)$ is an arbitrary function of t.

It is shown in appendix 1 that the characteristic functional of a stationary Gaussian process with mean zero is

$$G[\xi]=\exp\left[-\tfrac{1}{2}\sigma^2\int_0^t\int_0^t \psi(t_2-t_1)\xi(t_1)\xi(t_2)dt_1dt_2\right] \tag{II.68}$$

Taking $\xi(t) = 1$ for all t we obtain the following expression for the relaxation function which we shall use in the following discussion of the spectral density of a Gaussian modulated random process

$$g(t)=\exp\left[-\tfrac{1}{2}\sigma^2\int_0^t\int_0^t \psi(t_2-t_1)dt_1dt_2\right]$$

$$=\exp\left[-\sigma^2\int_0^t (t-\tau)\psi(\tau)d\tau\right] \tag{II.69}$$

Doob[18] has proved that a Gaussian process is Markoffian and stationary if, and only if, the correlation function has the form

$$\psi(t) = \exp(-\beta t) \qquad ,\beta>0 \tag{II.70}$$

The only process satisfying this requirement is the Ornstein-Uhlenbeck (O.U.) process with the conditional probability density

$$P(x_o|x,t)=\left[2\pi\sigma^2(t)\right]^{-\frac{1}{2}}\exp\left[-(x-m(t))^2/2\sigma^2(t)\right] \tag{II.71}$$

where $m(t)=x_o\exp(-\beta t)$ $\qquad ,\beta>0$

and $\sigma^2(t) = \left[1-\exp(-2\beta t)\right]D/\beta$ $\qquad ,D>0$

The O.U. process has the spectral density

$$S(\omega)\simeq 1/(\beta^2+\omega^2)$$

An interesting process is the Wiener-Einstein (W.E.) process which is Markoffian, stationary and Gaussian in the conditional probability. However, it has no joint probability. The conditional probability density is

$$P(x_o|x,t)=(4\pi Dt)^{-\frac{1}{2}}\exp\left[-(x-x_o)^2/4Dt\right] \quad ,D>0 \qquad \text{(II.72)}$$

The O.U. process and the W.E. process are described by two special Fokker-Planck equations, to be discussed later (Sec.II.6).

We shall now return to the behaviour of g(t) for two different regions of t using the expression (II.69) for g(t).

1) $t<<\tau_c$. As mentioned earlier (II.43) we may set $\psi(t)\simeq 1$ in this region. Then g(t) behaves as

$$g(t)\simeq\exp(-\tfrac{1}{2}\sigma^2t^2) \qquad |t|<<\tau_c \qquad \text{(II.73)}$$

2) $t>>\tau_c$. As $\psi(t)\simeq 0$ for $t>>\tau_c$ we may replace the upper limit of the integral by ∞. By integrating we get

$$\begin{aligned} g(t)&\simeq\exp(-\sigma^2\tau_c t+\text{const}) \\ &\simeq\exp(-\sigma^2\tau_c|t|) \\ &\equiv\exp(-|t|/\tau_r) \end{aligned} \qquad |t|>>\tau_c \qquad \text{(II.74)}$$

where the relaxation time τ_r is defined by

$$\tau_r=1/\sigma^2\tau_c$$

We shall now investigate the spectral density (II.54) for slow and fast modulation respectively.

a) Slow Modulation ($\sigma\tau_c>>1$).

g(t) may mainly be represented by the Gaussian type function (II.73). The spectral density becomes Gaussian

$$S(\omega)=\frac{1}{2\pi}\int_{-\infty}^{\infty}\exp(-\tfrac{1}{2}\sigma^2t^2)\exp(-i(\omega-\omega_o)t)dt$$

$$=(2\pi\sigma^2)^{-\frac{1}{2}}\exp(-(\omega-\omega_o)^2/2\sigma^2) \qquad \text{(II.73a)}$$

with the half-width σ. Thus the spectral density reflects directly the distribution of the modulation.

b) Fast Modulation ($\sigma\tau_c$<<1).

g(t) may mainly be represented by the exponential decay function (II.74). The spectral density becomes Lorentzian

$$S(\omega)=\frac{1}{2\pi}\int_{-\infty}^{\infty}\exp(-|t|/\tau_r)\exp(-i(\omega-\omega_o)t)dt$$

$$=\frac{1}{\pi}\frac{\gamma}{(\omega-\omega_o)^2+\gamma^2} \qquad \text{(II.74a)}$$

where the half-width γ is determined by $\gamma=\sigma^2\tau_c=1/\tau_r$.

If $\sigma\tau_c \ll 1$ then $\gamma \ll \sigma$ and the line is much narrowed relative to the slow modulation case. The condition $\sigma\tau_c \ll 1$ is, therefore, called the narrowing condition. This effect that the fluctuation is smoothed out and the spectral line becomes sharp around the centre is called motional narrowing.

We shall now investigate the properties of Markov processes and we shall show that Markov processes are characterized by the Smoluchowski-Chapman-Kolmogorov (S-C-K) equation. By some weak assumptions the Markov processes fulfill the Master Equation (M.E.) and the Fokker-Planck (F-P) equation also.

We shall derive the S-C-K equation for a non-stationary Markov process. By (II.35), (II.61), and (II.64) we have

$$P(x_1t_1;\ x_3t_3) = \int P_3(x_1t_1;\ x_2t_2;\ x_3t_3)dx_2$$

$$= \int P_3(x_1t_1;\ x_2t_2|x_3t_3)P(x_1t_1;\ x_2t_2)dx_2$$

$$= \int P(x_2t_2|x_3t_3)P(x_1t_1;\ x_2t_2)dx_2$$

$$P(x_1t_1;\ x_3t_3) = P(x_1t_1|x_3t_3)P(x_1,t_1)$$

$$P(x_1t_1;\ x_2t_2) = P(x_1t_1|x_2t_2)Px_1t_1)$$

Substitution of the last two equations into the first gives the S-C-K equation

$$P(x_1t_1|x_3t_3) = \int P(x_1t_1|x_2t_2)P(x_2t_2|x_3t_3)dx_2 \qquad \text{(II.75a)}$$

for all x and for all $t_1<t_2<t_3$

For stationary Markov processes the S-C-K equation is

$$P(x_1|x_3,t+\tau) = \int P(x_1|x_2,t)P(x_2|x_3,\tau)dx_2 \qquad \text{(II.75b)}$$

for all x and for all $t,\tau>0$

II.5. THE MASTER EQUATION (M.E.)

The M.E. is a differential form of the S-C-K equation. We shall derive the M.E. for a stationary Markov process only. We assume the existence of

$$\lim_{\tau\to 0}\left[P(x_2|x_3\tau) - \delta(x_2-x_3)\right]/\tau \equiv A(x_2|x_3)$$

As $P(x_2|x_3\tau) = \delta(x_2-x_3)$ for $\tau = 0$

We may use the following Taylor expansion about $\tau = 0$.

$$P(x_2|x_3\tau) = \delta(x_2-x_3)+\tau\ A(x_2,x_3)+0(\tau^2) \tag{II.76a}$$

Substitution into the S-C-K equation (II.75) leads to

$$P(x_1|x_3,t+\tau) = P(x_1|x_3t)+\tau\int P(x_1|x_2t)A(x_2,x_3)dx_2$$

Taking the limit $\tau\to 0$ gives

$$\frac{\partial}{\partial t} P(x_1|x_3t) = \int P(x_1|x_2t)A(x_2,x_3)dx_2 \tag{II.76b}$$

This expression is the M.E. $A(x_2,x_3)$ is the transition rate for a transition from x_2 to x_3. It follows from the definition of $A(x_2,x_3)$ that

$$A(x_2,x_3)\geq 0 \qquad \text{for } x_2 \neq x_3 \tag{II.77}$$

and

$$\int A(x_2,x_3)dx_3 = 0 \tag{II.78}$$

Therefore, $A(x,x)<0$. To indicate this point it is customary to write $A(x_2,x_3)$ as

$$A(x_2,x_3) = W(x_2|x_3) - w(x_2)\delta(x_2-x_3) \tag{II.79}$$

where both $W(x_2|x_3)$ and $w(x_2)$ are positive and

$$w(x_2) = \int W(x_2|x)dx$$

Inserting into (II.76) gives

$$\frac{\partial}{\partial t} P(x_1|x_3t) = \int \Big[P(x_1|x_2t)W(x_2|x_3)-P(x_1|x_3t)W(x_3|x_2)\Big]dx_2$$

which is usually written as

$$\frac{\partial}{\partial t} P(x,t) = \int \Big[P(x',t)W(x'|x) - P(x,t)W(x|x') \Big] dx' \qquad \text{(II.80)}$$

$P(x,t)$ means $P(x_o|x,t)$ i.e. (II.80) is solved subject to the initial condition $P(x,0) = \delta(x-x_o)$.

This is the common form of the M.E. and is readily interpreted as a rate equation for probability. $W(x_2|x_3)$ is the transition probability per unit time from x_2 to x_3 (the transition rate for a transition from x_2 to x_3).

The transition rates W for a process are generally chosen by physical intuition.

From equilibrium statistical mechanics we know that there may exist a time-independent solution $P^{eq}(x)$ to the M.E.

$$\frac{\delta}{\delta t} P^{eq}(x) = 0 = \int \Big[P^{eq}(x')W(x'|x) - P^{eq}(x)W(x|x') \Big] dx'$$

This solution is generally known from statistical mechanics and may be used in the evaluation of unknown parameters. We have

$$P(x_o|xt) \to P^{eq}(x) \qquad \text{as } t \to \infty.$$

Processes do not necessarily tend to an equilibrium state e.g. due to an external disturbance.

The S-C-K equation (II.75) in the discrete stationary case is

$$P_{jk}(t) = \sum_m P_{jm}(t-\tau)P_{mk}(\tau) \qquad \text{(II.81)}$$

where we have written $P_{jk}(t)$ for $P(j|k,t)$.

It is customary in the discrete case to expand $P_{jk}(\tau)$ in the following way

$$P_{jk}(\tau) = \begin{cases} 1-C_j\tau & \text{for } j = k \\ C_j p_{jk}\tau & \text{for } j \neq k \end{cases} \tag{II.82}$$

The probabilistic interpretation is: $C_j\tau$ is the probability that the system, known to be in state j at time t, changes to another state ($\neq$ j) during the time interval $[t,t+\tau]$. C_j may, therefore, be taken as the inverse of the lifetime of state j. p_{jk} can be interpreted as the conditional probability that if a change from state j occurs this change leads to state k.

Substitution of (II.82) into (II.81) and taking the limit $\tau\to 0$ gives the discrete M.E.

$$\frac{d}{dt} P_{jk}(t) = -P_{jk}(t)C_k + \sum_{m\neq k} P_{jm}(t)C_m p_{mk} \tag{II.83}$$

We may look upon the $P_{jk}(t)$'s as elements of the stochastic matrix P(t). By defining a matrix D with elements

$$D_{jj} = C_j \qquad D_{jk} = -C_j p_{jk} \qquad (j\neq k) \tag{II.84}$$

the M.E. (II.83) may be rewritten as

$$\dot{P} = -PD \tag{II.85}$$

with the formal solution

$$P(t) = P(0)\cdot\exp(-Dt) = \exp(-Dt) \tag{II.86}$$

$P(0) = 1$ (the unity matrix) because $P_{jk}(0)=\delta_{jk}$

The equilibrium probabilities P_j^{eq} may be arranged in a row vector ξ

$$\xi = (P_1^{eq},\ P_2^{eq},\ldots P_r^{eq}) \tag{II.87a}$$

which is the only solution of the equation

$$\xi\cdot D = 0 \tag{II.87b}$$

The condition (II.63) $\sum_k P_{jk}(t) = 1$ implies that

$$\sum_{k\neq j} P_{jk} = 1 \qquad \text{or} \qquad \sum_k D_{jk} = 0 \tag{II.87c}$$

which may be written as

$$D\cdot\eta = 0 \tag{II.87d}$$

where η is a column vector with all elements equal to one.

Let us assume that the random modulation $\omega(t)$ in Eq.(II.50) is a discrete stationary Markov process; then it is shown by Kubo[12] that $g(t)$ and $S(\omega)$ may be written as

$$g(t) = \xi\cdot\exp[(i\Omega-D)t]\cdot\eta \tag{II.87e}$$

$$S(\omega) = \frac{1}{\pi}\,\mathrm{Re}\{\xi\,\frac{1}{i(\omega-\Omega)+D}\,\eta\} \tag{II.87f}$$

Ω is a diagonal matrix with elements $\omega_1,\omega_2,\ldots\omega_r$ (the possible values of ω). This equation may be used for jump models.

The principle of detailed balance (d.b.) which is a physical principle states that

$$P^{eq}(x')W(x'|x) = P^{eq}(x)W(x|x') \tag{II.88a}$$

or more generally

$$P^{eq}(x')P(x'|x\tau) = P^{eq}(x)P(x|x'\tau) \tag{II.88b}$$

provided the microscopic equation of motion is reversible in time, that the macroscopic observables x are even functions of the particle velocities, and that there is no external magnetic field. If y are stochastic variables and odd functions of the particle velocities, and H is an external magnetic field then the principle of d.b. is

$$P^{eq}(x',y';H)W(x',y'|x,y;H)=P^{eq}(x,y;H)W(x,-y|x',-y';-H) \tag{II.88c}$$

A classical proof of d.b. is found in the book of de Groot and Mazur[2]. A quantum mechanical proof is given by van Kampen[3].

The M.E. implies that the H-theorem holds. Define

$$H(t) = -\int P(x_o|xt)\log\left[P(x_o|xt)/P^{eq}(x)\right]dx$$

After some algebraic manipulations[4] we obtain the H-theorem

$$\frac{dH}{dt} \geq 0$$

$$\frac{dH}{dt} = 0 \qquad \text{if, and only if,} \quad P(x_o|xt) = P^{eq}(x)$$

This implies that $P^{eq}(x)$ is the only time-independent solution and that $P(x_o|xt) \to P^{eq}(x)$ as $t \to \infty$.

As the M.E. (II.80) is valid only for Markov processes many authors investigate whether physical systems may be described by a M.E. There exist many attempts to derive a M.E. from first physical principles[4,5].

The fundamental assumption is that the process is Markoffian (called stosszahlansatz, molecular chaos, random phases, loss of information). This can be true only if we do choose neither too few nor too many variables for the macroscopic description. If we choose too few the process may be regarded as a projection of a Markov process.

If we choose to describe Brownian motion by the position of the Brownian particle only,then it is incorrect to assume that the process is a Markov process because the actual position of the Brownian particle must depend not only on the position at an earlier time but also on the velocity at that time. By including the velocity into the description the process is a Markov process. Indeed, the process is a Markov process if we describe it by the velocity only (the O.U. process). Further, it turns out that the process is a Markov process (the W.E. process), for a description by the position alone, for times $t \gg \beta^{-1}$ (β is the friction coefficient of the surrounding medium).

II.6. THE FOKKER-PLANCK (F-P) EQUATION

The F-P equation is an approximation to the M.E. The problem is to transform the integral operator in the M.E. into a differential operator. Many systematic expansions have been proposed[6,7]. We shall derive the F-P equation for a stationary Markov process only as a parabolic partial differential equation using a formal and intuitive procedure.

We start with the M.E. (II.80)

$$\frac{\partial}{\partial t} P(x,t) = \int P(x',t)W(x'|x)dx' - P(x,t)\int W(x|x')dx'$$

We define $\xi = x-x'$ as the jump length and write

$$W(x'|x) \quad \text{as } W(x'|\xi)$$

$W(x'|\xi)$ is thus the transition rate for a jump of length ξ starting at x'. $W(x'|\xi)$ is assumed to decrease rapidly with increasing $|\xi|$ but to vary slowly with x'. The M.E. then becomes

$$\frac{\partial}{\partial t} P(x,t) = \int P(x-\xi,t)W(x-\xi|\xi)d\xi - P(x,t)\int W(x|-\xi)d\xi$$

Noticing that

$$\int W(x|-\xi)d\xi = \int W(x|\xi)d\xi$$

and expanding both P and W in a Taylor series in ξ to second order leads to

$$\frac{\partial}{\partial t} P(x,t) = \int\left[P(x,t) - \xi\,\frac{\partial P(x,t)}{\partial x} + \frac{1}{2}\xi^2\,\frac{\partial^2 P(x,t)}{\partial x^2} + \ldots\right]$$

$$\times\left[W(x|\xi)-\xi\frac{\partial W(x|\xi)}{\partial x}+\frac{1}{2}\xi^2\frac{\partial^2 W(x|\xi)}{\partial x^2}+\ldots\right]$$

$$-P(x,t)\int W(x|-\xi)d\xi$$

$$=\int\left[\xi\left(-\frac{\partial}{\partial x}\right)P(x,t)W(x|\xi)\right.$$

$$\left.+\frac{1}{2}\xi^2\left(\frac{\partial^2}{\partial x^2}\right)P(x,t)W(x|\xi)\right]d\xi$$

We define the moment of the jump as

$$<\xi^n>(x)=\int\xi^n W(x|\xi)d\xi \qquad \text{(II.89)}$$

and remember that $P(x,t)$ is actually $P(x_o|x,t)$. We may then write

$$\frac{\partial}{\partial t}P(x_o|x,t)=-\frac{\partial}{\partial x}\left[<\xi>(x)P(x_o|x,t)\right]+\frac{1}{2}\frac{\partial^2}{\partial x^2}\left[<\xi^2>(x)P(x_o|x,t)\right] \qquad \text{(II.90a)}$$

This equation is known as the forward F-P equation.

The corresponding backward F-P equation, which is not used too often in physics, is[17]

$$\frac{\partial}{\partial t}P(x_o|x,t)=<\xi>(x_o)\frac{\partial P(x_o|x,t)}{\partial x_o}+\frac{1}{2}<\xi^2>(x_o)\frac{\partial^2 P(x_o|x,t)}{\partial x_o^2} \qquad \text{(II.90b)}$$

The generalization of the forward equation to higher dimension stochastic variables $x=(x_1,x_2\ldots x_n)$ is

$$\frac{\partial}{\partial t}P(x_o|x,t)=-\sum_{i=1}^{n}\frac{\partial}{\partial x_i}\left[<\xi_i>P(x_o|x,t)\right]+\frac{1}{2}\sum_{i,j=1}\frac{\partial^2}{\partial x_i\partial x_j}\left[<\xi_i\xi_j>P(x_o|x,t)\right]$$

(II.90c)

Lemma's concerning solutions to the F-P equation and some illustrative examples may be found in Ref.9 and Ref.10.

One well-known F-P equation is

$$a)\ \frac{\partial P}{\partial t} = D\,\frac{\partial^2 P}{\partial x^2} \qquad ,D>0 \tag{II.91}$$

This equation is normally used to describe the position of a Brownian particle. It is strictly valid only for times $t \gg \beta^{-1}$ where β is the friction coefficient of the surrounding medium. A solution to this equation is the W.E. process (II.72).

Another common F-P equation is

$$b)\ \frac{\partial P}{\partial t} = \beta\,\frac{\partial}{\partial x}(xP) + D\,\frac{\partial^2 P}{\partial x^2} \qquad \beta,D>0 \tag{II.92}$$

This equation is used to describe the velocity of a Brownian particle and it is identical with the Langevin equation[1,9,10]

$$\frac{dv}{dt} = -\,\beta v + f(t) \tag{II.93}$$

which expresses the total force on the Brownian particle as a sum of a friction force and a random force f(t). The random force f(t) is assumed to be independent of the friction force, and to be a stationary Gaussian process with

$$E\{f(t)\} = 0 \qquad ,E\{f(t_1)f(t_2)\} = 2D\delta(t_1-t_2) \tag{II.93a}$$

The random force f(t) is thus a very rapidly varying function compared to v(t) and it accounts for the individual molecular collisions. The friction term treats the liquid as a continuum.

Let us show that the Langevin equation (II.93) is identical to the F-P equation (II.92). By the use of (II.79), (II.76a), and (II.89) the moments of the jump may be written as

$$\begin{aligned} \langle \Delta v^n \rangle &= \lim_{\Delta t\to 0} 1/\Delta t \int \Delta v^n P(v_o|\Delta v,\Delta t)\,d\Delta v \\ &= \lim_{\Delta t\to 0} 1/\Delta t\ E\{\Delta v^n\}(\Delta t)_{v_o} \end{aligned} \tag{II.93b}$$

By integrating Eq.(II.93) over a time interval which is short compared to the time where v(t) changes appreciable but long compared to the time of change of f(t) we obtain

$$\Delta v = -\beta v \Delta t + \int f(t)dt \qquad \text{(II.93c)}$$

Taking the mean value of (II.93d) as defined in (II.93c) and using (II.93a) gives

$$E\{\Delta v\}(\Delta t)_v = -\beta v \Delta t \qquad \text{(II.93d)}$$

Taking the limit $\Delta t \to 0$ then gives the first moment

$$\langle \Delta v \rangle = -\beta v \qquad \text{(II.93e)}$$

The second moment may be found similarly. By squaring Eq.(II.93c) we obtain

$$\Delta v^2 = -\beta^2 v^2 \Delta t^2 - 2\beta v \Delta t \int_0^{\Delta t} f(t)dt + \int_0^{\Delta t} dt' \int_0^{\Delta t} dt f(t')f(t) \qquad \text{(II.93f)}$$

Taking the mean value of (II.93f), using (II.93a), and taking the limit $\Delta t \to 0$ gives the second moment

$$\langle \Delta v^2 \rangle = 2D \qquad \text{(II.93g)}$$

By the use of Eq.(II.90a) we then gets Eq.(II.92).

A solution to (II.92) or (II.93) is the O.U. process (II.71).

With no truncation of the Taylor expansion we would be left with the equation

$$\frac{\partial}{\partial t} P(x_o|x,t) = \sum_{n=1}^{\infty} \frac{1}{n!}\left(-\frac{\partial}{\partial x}\right)^n \left[\langle \xi^n \rangle P(x_o|x,t)\right]$$

This equation is known as the Kramers-Moyal expansion. If the F-P equation has to be equivalent with the M.E. then all the moments $\langle \xi^n \rangle$ have to be zero for $n \geq 3$. This is generally not the case except for a Gaussian process. However, use of the F-P equation may be convenient to obtain an approximate solution. A discussion of the different approximations to the M.E. may be found in Ref.6 and Ref.7.

II.7. THE FUNCTIONAL INTEGRATION TECHNIQUE OF KAC

One of the most powerful methods for the evaluation of the relaxation function g(t) is the one of Kac[13] which we shall now derive neglecting the mathematical problems of convergence.
We shall write g(t) more generally as

$$g(t) = E\left\{\exp\left[-\int_0^t V(X(t'))dt'\right]\right\} \tag{II.94}$$

V(X(t)) is a function of a stochastic peocess X(t) which we shall assume to be Markoffian and stationary. The dimension of X is not important for the results. The idea of the method is to introduce a function Q(x,t) which has a simple relationship to g(t). Q(x,t) may be found as the solution to an integral equation which may often be transformed to a differential equation.
First we shall evaluate the conditional mean subject to the initial condition $X = x_o$ for t = 0. We shall denote g(t) for that initial condition $g(x_o|t)$. Expanding the exponential in (II.94) we have

$$g(x_o|t) = \sum_{n=0}^{\infty}(-1)^n \mu_n(t)/n!$$

where the μ_n's are defined by

$$\mu_n(t) = E\left\{\left[\int_0^t V(X(t'))dt'\right]^n\right\}$$

$$= n!\int_0^t dt_n \int_0^{t_n} dt_{n-1}\cdots\int_0^{t_2} dt_1 \int dx_n \cdots \int dx_1$$

$$P(x_o|x_1,t_1)P(x_1|x_2,t_2-t_1)\ldots P(x_{n-1}|x_n,t_n-t_{n-1})V(x_1)\ldots V(x_n)$$

In obtaining the last equality we have used (II.65).

We define the functions $Q_n(x_o|x,t)$ as

$$Q_o(x_o|x,t) = P(x_o|x,t) \tag{II.95a}$$

$$Q_{n+1}(x_o|x,t) = \int_o^t dt' \int dx' Q_n(x_o|x',t')P(x'|x,t-t')V(x') \tag{II.95b}$$

By the use of (II.63) we observe that

$$\int Q_{n+1}(x_o|x,t)dx = \int_o^t dt' \int dx' Q_n(x_o|x',t')V(x')$$

and we now see that

$$\mu_n(t) = k! \int Q_n(x_o|x,t)dx$$

We define $$Q(x_o|x,t) = \sum_{n=0}^{\infty} (-1)^n Q_n(x_o|x,t)$$

and obtain $$g(x_o|t) = \int Q(x_o|x,t)dx \tag{II.96}$$

We also observe that $Q(x_o|x,t)$ satisfy the integral equation

$$Q(x_o|x,t) + \int_o^t dt' \int dx' Q(x_o|x',t')P(x'|x,t-t')V(x') = Q_o(x_o|x,t) \tag{II.97}$$

with the initial condition

$$Q(x_o|x,t) \to \delta(x-x_o) \qquad \text{as } t\to 0 \tag{II.97a}$$

(II.97) may be shown by multiplying (II.95b) with $(-1)^{n+1}$ and summing over $n=0,1,2...$

We may sometimes transform the integral equation (II.97) to a differential equation; e.g. we may assume that the stochastic process is described by the diffusion equation (II.91)

$$\frac{\partial P(x_o|x,t)}{\partial t} = D\nabla^2 P(x_o|x,t) \tag{II.98}$$

Differentiating (II.97) with respect to t gives

$$\frac{\partial Q(x_o|x,t)}{\partial t} + \int dx' Q(x_o|x',t)P(x'|x,0)V(x')$$

$$+ \int_o^t dt' \int dx' \frac{\partial P(x'|x,t-t')}{\partial t} Q(x_o|x',t')V(x') = \frac{\partial Q_o(x_o|x,t)}{\partial t}$$

By the use of (II.95), (II.97), (II.98) and $P(x'|x,0)=\delta(x-x')$ we get

$$\frac{\partial Q(x_o|x,t)}{\partial t} - D\nabla^2 Q(x_o|x,t) + V(x)Q(x_o|x,t) = 0 \tag{II.99}$$

The problem of evaluating $g(x_o|t)$ is now transformed to the problem of solving the differential equation (II.99). Fixman[14] and Sillescu and Kivelson[15] have used this technique to magnetic relaxation. They use the unconditional mean. The connection between this quantity and the conditional mean is

$$g(t) = \int P(x_o)g(x_o|t)dx_o \tag{II.100}$$

We define Q(x,t) as

$$Q(x,t) = \int P(X_o)Q(x_o|x,t)dx_o \tag{II.101}$$

then we have

$$g(t) = \int Q(x,t)dx \tag{II.102}$$

We obtain the integral equation for Q(x,t) by multiplying (II.97) with $P(x_o)$ and integrate over x_o. So

$$Q(x,t)+\int_o^t dt'\int dx'Q(x,t)P(x'|x,t-t')V(x')=P(x) \qquad \text{(II.103)}$$

We have here used the fact that

$$\int P(x_o)P(x_o|x,t)dx_o=P(x)$$

We may easily show that Q(x,t) also satisfies the differential equation (II.99) if (II.98) holds.

APPENDIX. PROOF OF EQ.(II.68)

To prove that the characteristic functional of a real stationary Gaussian process may be written in the form (II.68) we shall make use of the following two lemma's.

First we observe that the joint probability density (II.66) of n Gaussian variables $X_1,\dots X_n$ with mean zero may be written in tensor notation, which we shall use throughout this appendix, as

$$P(x)=|\Gamma^{-1}|^{\frac{1}{2}}(2\pi)^{-n/2}\exp\left[-\tfrac{1}{2}x\cdot\Gamma^{-1}\cdot x\right] \qquad \text{(A1)}$$

where x is a tensor of first order with elements x_i, and Γ^{-1} is the inverse of the covariance matrix (II.20).

Lemma 1. It is always possible by a linear transformation to transform n Gaussian variables $X_1,\dots X_n$ with covariance matrix Γ into n independent Gaussian variables $X_1',\dots X_n'$ with

$$P(x_i') = (2\pi\sigma_i'^2)^{-\frac{1}{2}}\exp(-x_i'^2/2\sigma_i'^2)$$

and where

$$\prod_{i=1}^{n} \sigma_i'^2 = |\Gamma|$$

Proof: As Γ is symmetric (II.20) then Γ^{-1} is symmetric and, therefore, Γ^{-1} may be diagonalized by a similarity transformation using a real orthogonal matrix $U(U^{-1}_{ij} = U_{ji})$. The transformed diagonal matrix is

$$\Lambda^{-1} = U^{-1}\cdot\Gamma^{-1}\cdot U \qquad ,\Lambda^{-1}_{ij} = (\sigma'^2_i)^{-1}\delta_{ij} \tag{A2}$$

We define new variables X' by

$$X' = X\cdot U \qquad \text{or identical } X = U\cdot X' \tag{A3}$$

and obtain

$$x\cdot\Gamma^{-1}\cdot x = x\cdot U\cdot U^{-1}\cdot\Gamma^{-1}\cdot U\cdot U^{-1}\cdot x = x'\cdot\Lambda^{-1}\cdot x' = \sum_{i=1}^{n}\sigma'^2_i x'^2_i \tag{A4}$$

The joint probability density of the variables $X'_1,\ldots X'_n$ may be found by

$$P(x'_1,\ldots x'_n) = P(x_1,\ldots x_n)\partial(x_1,\ldots x_n)/\partial(x'_1,\ldots x'_n)$$

As the Jacobian is equal to detU = 1 and

$$|\Gamma|^{-1} = |\Gamma^{-1}| = |\Lambda^{-1}| = 1/\prod_{i=1}^{n}\sigma'^2_i$$

we have

$$P(x'_1,\ldots x'_n) = \prod_{i=1}^{n}(2\pi\sigma'^2_i)^{-\frac{1}{2}}\exp(-x'^2_i/2\sigma'^2_i) \tag{A5}$$

which completes the proof.

Lemma 2. If $X_1,\ldots X_n$ is n Gaussian variables with mean zero then for arbitrary ξ'_is

$$E\left\{\exp\left[iX\cdot\xi\right]\right\} = \exp\left[-\tfrac{1}{2}\xi\cdot\Gamma\cdot\xi\right] \tag{A6}$$

Proof:

$$E\left\{\exp\left[iX\cdot\xi\right]\right\} = |\Gamma^{-1}|^{\frac{1}{2}}(2\pi)^{-n/2}\int\ldots\int dx_1\ldots dx_n \exp\left[-\tfrac{1}{2}x\cdot\Gamma^{-1}\cdot x + ix\cdot\xi\right]$$

Introduction of the independent variables X' gives

$$E\left\{\exp\left[iX\cdot\xi\right]\right\}=$$

$$|\Gamma^{-1}|^{\frac{1}{2}}(2\pi)^{-n/2}\int\cdots\int dx'_1\cdots dx'_n \exp\left[-\tfrac{1}{2}x'\cdot\Lambda^{-1}\cdot x'+i\xi\cdot U\cdot x'\right] \tag{A7}$$

Writing the exponential terms out in elements gives

$$-\tfrac{1}{2}x'\cdot\Lambda^{-1}\cdot x'+i\xi\cdot U\cdot x'$$

$$= -\tfrac{1}{2}\sum_i \Lambda_{ii}^{-1}\, x_i'^2+i\sum_j\sum_i \xi_j U_{ji} x_i'$$

$$= -\tfrac{1}{2}\sum_i \Lambda_{ii}^{-1}(x_i'^2-2ix_i'\sum_j \xi_j U_{ji}\Lambda_{ii})$$

$$= -\tfrac{1}{2}\sum_i \Lambda_{ii}^{-1}(x_i'-i\beta_i)^2-\tfrac{1}{2}\sum_i \Lambda_{ii}^{-1}\beta_i^2 \tag{A8}$$

where we have defined β_i by

$$\beta_i = \sum_j \xi_j U_{ji}\Lambda_{ii} \tag{A9}$$

Introducing (A8) into (A7), substitution of $x_i''=(x_i'-i\beta_i)$, and carrying out the integration gives

$$E\left\{\exp\left[iX\cdot\xi\right]\right\} = \exp\left[-\tfrac{1}{2}\sum_i \Lambda_{ii}^{-1}\beta_i^2\right] \tag{A10}$$

By inserting (A9) into (A10) we get

$$E\left\{\exp\left[iX\cdot\xi\right]\right\}$$

$$= \exp\left[-\tfrac{1}{2}\sum_i \Lambda_{ii}^{-1}\Lambda_{ii}^2\sum_j\sum_k \xi_k U_{ji} U_{ki}\right]$$

$$= \exp\left[-\tfrac{1}{2}\sum_j\sum_k \xi_j\xi_k \sum_i U_{ki}\Lambda_{ii}U^{-1}_{ij}\right]$$

$$= \exp\left[-\tfrac{1}{2}\sum_j\sum_k \xi_j\xi_k\Gamma_{kj}\right]$$

$$= \exp\left[-\tfrac{1}{2}\xi\cdot\Gamma\cdot\xi\right]$$

which proves lemma 2.

The characteristic functional (II.67) for a stationary Gaussian process with mean zero follows from lemma 2.

$$G[\xi] \equiv E\left\{\exp\left[i\int_0^t X(t')\xi(t')dt'\right]\right\} = \lim_{n\to\infty} E\left\{\exp\left[i\sum_{j=1}^n X(j\Delta t)\xi(j\Delta t)\Delta t\right]\right\}$$

We define $X_j = X(j\Delta t)$ and $\xi_j = \xi(j\Delta t)\Delta t$ and by use of lemma 2 (A6) we obtain

$$G[\xi] = \lim_{n\to\infty} \exp\left[-\tfrac{1}{2}\xi\cdot\Gamma\cdot\xi\right]$$

$$= \lim_{n\to\infty} \exp\left[-\tfrac{1}{2}\sum_{j=1}^n\sum_{k=1}^n \Gamma(j\Delta t,k\Delta t)\xi(j\Delta t)\xi(k\Delta t)\Delta t^2\right]$$

$$= \exp\left[-\tfrac{1}{2}\int_0^t\int_0^t \Gamma(t_1,t_2)\xi(t_1)\xi(t_2)dt_1dt_2\right]$$

$$= \exp\left[-\tfrac{1}{2}\sigma^2\int_0^t\int_0^t \psi(t_2-t_1)\xi(t_1)\xi(t_2)dt_1dt_2\right] \qquad \text{(A11)}$$

Equation (A11) is (II.68) which we have now proved.

REFERENCES

1) N.U. Prabhu, Stochastic Processes. The Macmillan Company N.Y. 1965.

2) S.R. de Groot and P. Mazur, Non-Equilibrium Thermodynamics. North-Holland Publ.Co.(1962).

3) N.G. van Kampen, Physica 20 (1954) 603.

4) N.G. van Kampen, Fundamental Problems in Statistical Mechanics, (p.173) North-Holland Publ.Co., Amsterdam (1962).

5) L. van Hove, Fundamental Problems in Statistical Mechanics (p.157) North-Holland Publ.Co., Amsterdam (1962).

6) I. Oppenheim, K.E. Shuler, G.H. Weis, Adv. in Molecular Relaxation Processes, 1 (1967) 13.

7) N.G. van Kampen, R.E. Burgess (Ed), in Fluctuation Phenomena in Solids, Academic Press, New York, 1965, (p.139).

8) A. Papaulis, Probability, Random Variables, and Stochastic Processes, McGraw-Hill, 1965.

9) S. Chandrasekkar, Rev.Mod.Phys. Vol.15, No.1, p.1 reprinted in: Selected papers on noise and stochastic processes. N.Wax (editor), Dover Publ.Inc., New York, 1954.

10) M.C. Wang and G.E. Uhlenbeck, Rev.Mod.Phys.,Vol.17, No.2 and 3, (p.113) reprinted in: Selected papers on noise and stochastic processes, N.Wax (editor), Dover Publ.Inc., New York 1954.

11) P.W. Anderson, J.Phys.Soc. of Japan, Vol.9, No.3,(p.316)(1954)

12) R. Kubo, Fluctuation, Relaxation, and Resonance in Magnetic Systems, (p.24), D. Ter Haar (editor) Oliver and Boyd, London, 1962.

13) M. Kac, Probability and Related Topics in Physical Sciences, (Interscience Publishers, Inc., New York, 1959), Chap.4.

14) M. Fixman, J.Chem.Phys., 48, 223 (1968).

15) H. Sillescu and D. Kivelson, J.Chem.Phys., 48, 3493 (1968).

16) H.Y. Carr and E.M. Purcell, Phys.Rev., 94, 630, (1954).

17) W. Feller, An Introduction to Probability Theory and Its Applications, John Wiley and Sons, Inc., (1961).

18) J.L. Doob, Annals of Math., Vol. 43, No. 2, p. 351 (1942). reprinted in:Selected papers on noise and stochastic processes, N. Wax (editor), Dover, New York 1954.

AN INTRODUCTION TO THE STOCHASTIC THEORY OF E.S.R. LINE SHAPES

S.J. Knak Jensen

Aarhus University

In this chapter I shall present an introduction to the effect of the stochastic motion of the lattice on the e.s.r. line shape of a radical in solution.

The stochastic Liouville equation[1-3] will be applied to a few simple and well-defined stochastic processes to study the line shape as a function of the correlation time of the stochastic process.

As the first example we consider the spectrum from a radical jumping between a discrete set of different sites. As the second example we treat a molecule undergoing isotropic rotational diffusion and we investigate the effect of changing the rotational diffusion constant from the fast diffusion limit (motional narrowing case) to the slow diffusion limit characteristic, e.g. for radicals in viscous solvents and for spin-labeled macromolecules. The stochastic Liouville equation implies that the stochastic process is a stationary Markov process and that the interaction between the spin system and the lattice does not depend on the spin state. For convenience we confine ourselves to the high temperature approximation and neglect of saturation effects.

III.1. THE JUMP MODEL

We consider the spectrum $I(\omega)$ from a spin system, which may jump between a discrete set of p states $(\lambda_1,\lambda_2...\lambda_p)$ where λ_i is a stochastic variable which specifies the i'th state. This model was first considered by P.W. Anderson[4] and Kubo[5]. Anderson restricted himself to the case where the random perturbation $H_1(\lambda)$ did commute with the static Zeeman interaction, H_o (i.e. the secular case). The model was later extended to include also nonsecular effects, f.ex. by Blume[6] and we treat these terms also in this presentation. The hamiltonian, $H(\lambda)$, is expressed in a basis where stochastic and spin variables are separable in order to allow the trace and averaging processes to be performed independently.

We introduce the stochastic Liouville equation for $S_x(\lambda,t)$ discussed by Kubo[1-2].

$$\frac{dS_x(\lambda,t)}{dt} = iH^{\times}(\lambda)S_x(\lambda,t) + \Gamma_\lambda S_x(\lambda,t) \qquad \text{(III.1)}$$

$S_x(\lambda,t)$ is an average of S_x corresponding to particular values for λ and t given in terms of the joint probability $P(\lambda,S_x,t)$ of finding λ and S_x at time t as

$$S_x(\lambda,t) = \int S_x P(\lambda,S_x,t)dS_x$$

The time-independent Markov operator, Γ_λ, for the jump model describes the jumping between the states and is given by the equation

$$\frac{dP(t)}{dt} = \Gamma_\lambda P(t) \qquad \text{(III.2)}$$

The matrix element of P(t) between stochastic states a and b is the probability that the system will be found in the state b at time t, when it was in the state a at time 0. The matrix elements of Γ_λ are

$$(a|\Gamma_\lambda|b) = \begin{cases} -c_b & a = b \\ c_b p_{ba} & a \neq b \end{cases}$$

where c_b is the reciprocal of the lifetime in state b and p_{ba} is the transition probability for the jump from b to a such that

$$\sum_{a(\neq b)} p_{ba} = 1.$$

That is we have

$$\sum_a (a|\Gamma_\lambda|b) = 0$$

For the high temperature approximation and no saturation we may write the spectrum as

$$I(\omega) = \frac{1}{2\pi}\int_{-\infty}^{\infty} \mathrm{Tr}(S_x\{S_x(t)\})e^{-i\omega t}dt \qquad \text{(III.3)}$$

where $\{\ldots\}$ means an average over lattice states.

$$\{S_x(t)\} = \sum_{a,b} P_a \langle a|S_x(\lambda,t)|b\rangle \qquad \text{(III.4)}$$

P_a is the equilibrium probability of finding the system in the state a.

Combining Eq.(III.3) and (III.4) leads to

$$I(\omega) = \frac{1}{2\pi}\,\mathrm{Tr}(S_x \int_{-\infty}^{\infty}\sum_{a,b} P_a\langle a|S_x(\lambda,t)|b\rangle\, e^{-i\omega t}dt)$$

$$= \frac{1}{2\pi}\,\mathrm{Tr}(S_x \sum_{a,b} P_a \langle a| \int_{-\infty}^{\infty} S_x(\lambda,t)e^{-i\omega t}dt\,|b\rangle)$$

$$= \frac{1}{\pi}\,\mathrm{Re}\,\mathrm{Tr}(S_x \sum_{a,b} P_a \langle a|S_x(\lambda,\omega)|b\rangle) \qquad \text{(III.5)}$$

where

$$S_x(\lambda,\omega) = \int_0^{\infty} S_x(\lambda,t)e^{-i\omega t}dt$$

We transform the stochastic Liouville equation to find $S_x(\lambda,\omega)$

$$\int_0^\infty \frac{d}{dt}S_x(\lambda,t)e^{-i\omega t}dt = \int_0^\infty iH^\times(\lambda)S_x(\lambda,t)e^{-i\omega t}dt + \int_0^\infty \Gamma_\lambda S_x(\lambda,t)e^{-i\omega t}dt \tag{III.6}$$

which gives

$$-S_x(0) + i\omega S_x(\lambda,\omega) = iH^\times(\lambda)S_x(\lambda,\omega) + \Gamma_\lambda S_x(\lambda,\omega)$$

or $$S_x(\lambda,\omega) = (i\omega - iH^\times(\lambda) - \Gamma_\lambda)^{-1}S_x(0) \tag{III.7}$$

Combining Eq.(III.5) and (III.7) leads to

$$I(\omega) = \frac{1}{\pi}\mathrm{Re}\sum_{\substack{n,m\\n',m'}} \langle n|S_x|m\rangle \sum_{a,b} P_a \langle m,n,a|\left[i\omega - iH^\times(\lambda) - \Gamma_\lambda\right]^{-1}|m',n',b\rangle\langle m'|S_x|n'\rangle \tag{III.8}$$

where we have used Eq.(I.4).

Accordingly, the line shape function may be obtained by inverting a non-Hermitian matrix corresponding to the operator $(i\omega - iH^\times - \Gamma_\lambda)$. The matrix is of dimensionality

$$\left\{(2S+1)\prod_j (2I_j+1)\right\}^2 p$$

where S is the spin of the radical and I_j is the nuclear spin of the j'th nucleus. The matrix representation of $(i\omega - H^\times(\lambda) - \Gamma_\lambda)$ is shown in Table I for the simplest possible case that of a system with $S=\frac{1}{2}$ and $I_j=0$ jumping between two states, -1 and 1. The hamiltonian, $H(\lambda)$, in Tabel I is

$$\begin{aligned} H(\lambda) &= g\beta BS_z + g\beta\bar{h}(\lambda)\bar{S} \\ &\equiv \omega_o S_z + a(\lambda)g\beta(\tfrac{1}{2}(h_x+ih_y)S_- + \tfrac{1}{2}(h_x-ih_y)S_+ + h_zS_z) \\ &\equiv \omega_o S_z + a(\lambda)(\omega_+S_- + \omega_-S_+ + \omega_zS_z) \end{aligned} \tag{III.9}$$

where B and h are the static and the fluctuating field. $a(\lambda)$ is 1

m,n,a \ m',n',b	½, ½, 1	½, ½,-1	½,-½, 1	½,-½,-1	-½, ½, 1	-½, ½,-1	-½,-½, 1	-½,-½,-1
½, ½, 1	$i\omega+\Omega$	$-\Omega$	$i\omega_+$		$-i\omega_-$			
½, ½,-1	$-\Omega$	$i\omega+\Omega$		$-i\omega_+$		$i\omega_-$		
½,-½, 1	$i\omega_-$		$i\omega+\Omega$ $-i(\omega_o+\omega_z)$	$-\Omega$			$-i\omega_-$	
½,-½,-1		$-i\omega_-$	$-\Omega$	$i\omega+\Omega$ $-i(\omega_o-\omega_z)$				$i\omega_-$
-½, ½, 1	$-i\omega_+$				$i\omega+\Omega$ $+i(\omega_o+\omega_z)$	$-\Omega$	$i\omega_+$	
-½, ½,-1		$i\omega_+$			$-\Omega$	$i\omega+\Omega$ $+i(\omega_o-\omega_z)$		$-i\omega_+$
-½,-½, 1			$-i\omega_+$		$i\omega_-$		$i\omega+\Omega$	$-\Omega$
-½,-½,-1				$i\omega_+$		$-i\omega_-$	$-\Omega$	$i\omega+\Omega$

TABLE I

Matrix Representation of the Operator $(i\omega-iH^{\times}(\lambda)-\Gamma_{\lambda})$ in Eqs. (III.8-10) in the state 1 and -1 in the state -1. The matrix elements of Γ_λ are

$$(m,n,a|\Gamma_\lambda|m',n',b)=\pm\delta_{mm'}\delta_{nn'}\Omega \tag{III.10}$$

where the upper sign applies if a and b are different, and the lower sign if they are identical.

In general, we have to invert the matrix by a numerical rather than an analytical procedure due to its complexity. However, we briefly consider the special situation, where the non-secular terms are zero ($\omega_+=\omega_-=0$) to find an analytical solution for the secular interactions.
In this case the matrix representation of $(i\omega-iH^\times(\lambda)-\Gamma_\lambda)$ reduces to a block diagonal matrix consisting of four 2 by 2 matrices. The inverse of a block diagonal matrix is another block diagonal matrix consisting of the inverted matrices, so the problem is substantially reduced. Furthermore, the first and fourth of the blocks do not contribute to $I(\omega)$ as they have to be multiplied by vanishing matrix elements $<\frac{1}{2}|S_x|\frac{1}{2}>$ and $<-\frac{1}{2}|S_x|-\frac{1}{2}>$.
We assume $P_1 = P_{-1} = \frac{1}{2}$ and find the contribution to $I(\omega)$ from the second block as

$$I(\omega) = \frac{1}{2\pi}\frac{\Omega\,\omega_z^2}{((\omega-\omega_o)^2-\omega_z^2)^2+4\Omega^2(\omega-\omega_o)^2} \tag{III.11}$$

The contribution from the third block is similar except $\omega-\omega_o$ is replaced by $\omega+\omega_o$. Accordingly, the third block will give the same spectrum as (III.11) but symmetrical with respect to $\omega=0$.

We can now use Eq.(III.11) to discuss the effect of changing Ω. If $\Omega<<\omega_z$ we have a slow jumping rate and we would, therefore, expect the spectrum to be pretty much the same as the unperturbed one (i.e. the rigid lattice spectrum). For $\Omega<<\omega_z$ $I(\omega)$ has two maxima for $\omega\simeq\omega_o\pm\omega_z$ and we find in the neighbourhood of $\omega\simeq\omega_o+\omega_z$

$$I(\omega) = \frac{1}{8\pi}\frac{\Omega}{(\omega-\omega_o-\omega_z)^2+\Omega^2} \tag{III.12}$$

that is a Lorentzian line with width Ω. In the limit $\Omega\to 0$ we obtain the rigid lattice spectrum

$$I(\omega) = \frac{1}{8}\,\delta(\omega-\omega_o-\omega_z) \tag{III.13}$$

where we have used[8]

$$\delta(\omega-\omega_o) = \frac{1}{\pi}\lim_{\varepsilon\to+0}\frac{\varepsilon}{(\omega-\omega_o)^2+\varepsilon^2} \tag{III.14}$$

Similar expressions are obtained for $\omega\simeq\omega_o-\omega_z$.

For the fast jumping case, $\Omega\gg\omega_z$, $I(\omega)$ reduces to

$$I(\omega) = \frac{1}{2\pi}\,\frac{\Omega\,\omega_z^2}{(\omega-\omega_o)^4+\omega_z^4+4\Omega^2(\omega-\omega_o)^2} \tag{III.15}$$

$I(\omega)$ has a maximum for $\omega=\omega_o$. In the vicinity of the maximum $I(\omega)$ is given by

$$I(\omega) = \frac{1}{2\pi}\,\frac{\Omega\,\omega_z^2}{4\Omega^2(\omega-\omega_o)^2+\omega_z^4} = \frac{1}{4\pi}\,\frac{\frac{\omega_z^2}{2\Omega}}{(\omega-\omega_o)^2+\left(\frac{\omega_z^2}{2\Omega}\right)^2} \tag{III.16}$$

Eq.(III.16) represents a Lorentzian line centered around ω_o with width

$$\frac{\omega_z^2}{2\Omega}$$

If Ω increases the line is sharpening up (motional narrowing) and in the limit $\Omega\to\infty$ we obtain

$$I(\omega) = \frac{1}{4}\,\delta(\omega-\omega_o) \tag{III.17}$$

$I(\omega)$ is displayed in Fig.1 for several values of ω_z/Ω. A jumping invariant linewidth T_2^{-1} was included in the construction of the spectra by letting $\omega_o\to\omega_o+iT_2^{-1}$.

The behavior shown in Fig.1 was observed in many cases by changing the temperature of the sample (i.e. varying Ω). References 9 - 11 give an extensive list of examples.

Before leaving the jump model it should be pointed out - as Gordon and McGinnis[12] did - that the computational work in finding $I(\omega)$ from Eq.(III.8) can be substantially reduced by a simple transformation. We introduce the notation

$$<n,m,a|S_x|n',m',b>=\delta_{n,n'}\delta_{m,m'}\delta_{a,b}<n|S_x|m> \qquad \text{(III.18)}$$

and the stochastic operator P

$$<n,m,a|P|n',m',b>=P_a\delta_{nn'}\delta_{mm'}\delta_{ab} \qquad \text{(III.19)}$$

Eq.(III.8) can then be rewritten as

$$I(\omega)=\frac{1}{\pi}\mathrm{Re}\sum <n,m,a|S_xP|n_o',m_o',a_o><n_o',m_o',a_o|(i\omega-iH^{\times}(\lambda)-\Gamma_\lambda)^{-1}|$$

$$n_o'',m_o'',b_o><n_o'',m_o'',b_o|S_x|n',m',b> \qquad \text{(III.20)}$$

where we are summing over all indicated states.

The equality of Eqs.(III.8) and (III.20) is seen by substituting Eqs.(III.18) and (III.19) into (III.20).

Using a single index α for $|n,m,a>$ we may express $I(\omega)$ as

$$I(\omega)=\frac{1}{\pi}\mathrm{Re}\sum_{\alpha,\beta} <\alpha|PS_xQ^{-1}Q(i\omega-iH^{\times}(\lambda)-\Gamma_\lambda)^{-1}Q^{-1}QS_x|\beta> \qquad \text{(III.21)}$$

where we have introduced the unit operator $Q^{-1}Q$.

$$I(\omega)=\frac{1}{\pi}\mathrm{Re}\sum_{\substack{\alpha,\beta,\\ \gamma,\delta}} <\alpha|PS_xQ^{-1}|\gamma><\gamma|\left[Q(i\omega-iH^{\times}(\lambda)-\Gamma_\lambda)Q^{-1}\right]^{-1}|\delta><\delta|QS_x|\beta> \text{(III.22)}$$

Now the idea is to choose Q so that

$$Q(i\omega-iH^{\times}(\lambda)-\Gamma_\lambda)Q^{-1} \equiv i\omega-A \qquad \text{(III.23)}$$

is diagonal. Then we may write $I(\omega)$ as

$$I(\omega)=\frac{1}{\pi}\mathrm{Re}\sum_{\gamma}<\gamma|(i\omega-A)^{-1}|\gamma>\sum_{\alpha,\beta}<\alpha|PS_xQ^{-1}|\gamma><\gamma|QS_x|\beta>$$

$$\equiv \frac{1}{\pi}\,\mathrm{Re}\sum_{\gamma}\frac{x_\gamma}{i\omega-A_\gamma} \qquad \text{(III.24)}$$

x_γ and A_γ do not involve the frequency ω. Therefore, $I(\omega)$ can be computed from Eq.(III.24) by one diagonalisation of a non-Hermitian matrix (e.g. by the QR transformation[13]) followed by a simple summation for each frequency. In contrast Eq.(III.8) requires a matrix inversion for each frequency.

III.2. THE ISOTROPIC ROTATIONAL DIFFUSION MODEL

This model has recently been applied by Fixman[14], Silescu and Kivelson[15] as well as Freed, Bruno and Polnaszek[3] to study magnetic relaxation in liquids. The equation which describes the stochastic motion of the radical is the isotropic rotational diffusion equation

$$\frac{\partial P(\Omega,t)}{\partial t} = D\nabla^2 P(\Omega,t). \qquad \text{(III.25)}$$

Ω is now a set of Eulerian angles which specify the orientation of a molecule fixed coordinate system relative to a space fixed coordinate system, and ∇^2 is the Laplacian operator.
$P(\Omega,t)$ is the probability of finding the radical in an orientation specified by Ω at time t and D is the isotropic rotational diffusion constant.
The hamiltonian consists of two parts: a static hamiltonian, H_o, and a perturbation, $H_1(\Omega)$ dependent upon the stochastic process

$$H = H_o + H_1(\Omega) \qquad \text{(III.26)}$$

We use again the stochastic Liouville equation, which for this stochastic process is

$$\frac{\partial S_x(\Omega,t)}{\partial t} = iH^{\times}(\Omega)S_x(\Omega,t) + D\nabla^2 S_x(\Omega,t) \qquad \text{(III.27)}$$

for the operator $S_x(\Omega,t)$.

The spectrum is given by

$$I(\omega) = \frac{1}{2\pi}\int_{-\infty}^{\infty}\mathrm{Tr}(S_x\{S_x(t)\})e^{-i\omega t}dt \tag{III.28}$$

where

$$\{S_x(t)\} = \frac{1}{4\pi}\int S_x(\Omega,t)d\Omega \tag{III.29}$$

is an average over all orientations.

We may rewrite $I(\omega)$ as

$$I(\omega) = \frac{1}{\pi}\,\mathrm{Re}\,\mathrm{Tr}(S_x\frac{1}{4\pi}\int S_x(\Omega,\omega)d\Omega) \tag{III.30}$$

where we have introduced

$$S_x(\Omega,\omega) = \int_{o}^{\infty} S_x(\Omega,t)e^{-i\omega t}dt$$

The idea now is to expand $S_x(\Omega,\omega)$ in an appropriate set of basis functions $\psi_n(\Omega)$ independent of spin operators as

$$S_x(\Omega,\omega) = \sum_n f_n(\omega)\psi_n(\Omega) \tag{III.31}$$

where $f_n(\omega)$ is a spin operator.

Combining Eqs.(III.30) and (III.31) leads to

$$I(\omega) = \frac{1}{\pi}\,\mathrm{Re}\,\mathrm{Tr}(S_x\sum_n f_n(\omega)\{\psi_n\}) \tag{III.32}$$

If $\psi_n(\Omega)$ is chosen so that

$\{\psi_n\} = k_o\delta_{n,o}$ then (III.32) reduces to

$$I(\omega) = \frac{1}{2\pi} \mathrm{Re}(k_o(<-\tfrac{1}{2}|f_o(\omega)|\tfrac{1}{2}>+<\tfrac{1}{2}|f_o(\omega)|-\tfrac{1}{2}>)) \tag{III.33}$$

The operator $f_o(\omega)$ can be determined from the stochastic Liouville equation as is demonstrated in the following simple example.

This type of analysis has recently been given in great detail by Freed, Bruno and Polnaszek[3], who made an expansion of the density matrix similar to the one in Eq.(III.31) using the Wigner rotation matrices $D^{\ell}_{nm}(\Omega)$ as the basis set.

We apply this method to a simple example to study the lineshape as a function of the rotational diffusion constant. We consider a radical ($S=\frac{1}{2}$) with axially symmetric g tensor and no hyperfine interaction.

The hamiltonian is

$$\begin{aligned} H(\Omega) &= g\beta BS_z + P_2(\cos\theta)\,\tfrac{2}{3}\beta B(g_{\parallel}-g_{\perp})S_z \\ &\equiv \omega_o S_z + P_2(s)QS_z \end{aligned} \tag{III.34}$$

where $P_2(\cos\theta) \equiv P_2(s)$ is the second Legendre polynomial, and θ is the angle between the static magnetic field B and the molecular z-axis.

We transform the stochastic Liouville equation to

$$\int_o^{\infty} \frac{dS_x(\theta,t)}{dt} e^{-i\omega t}dt = \int_o^{\infty} iH^{\times}(\theta)S_x(\theta,t)e^{-i\omega t}dt + \int_o^{\infty} D\nabla^2 S_x(\theta,t)e^{-i\omega t}dt \tag{III.35}$$

or

$$i\omega S_x(\theta,\omega)-S_x = iH^{\times}(\theta)S_x(\theta,\omega)+D\nabla^2 S_x(\theta,\omega) \tag{III.36}$$

We choose to expand $S_x(\theta,\omega)$ in Legendre polynomials, $P_{\ell}(s)$, because the hamiltonian is expressible in these functions and because they are eigenfunctions for ∇^2

$$\nabla^2 P_{\ell}(s) = -\ell(\ell+1)P_{\ell}(s) \tag{III.37}$$

$$S_x(\theta,\omega) \equiv S_x(s,\omega) = \sum_{\ell=0}^{\infty} f_\ell(\omega) P_{2\ell}(s) \qquad \text{(III.38)}$$

where we have used that $S_x(\theta,\omega)$ is even in θ.

Legendre polynomials have the property

$$\int_{-1}^{1} P_\ell(s)ds = 2\delta_{\ell,0} \qquad \text{(III.39)}$$

and the spectrum is found from Eq.(III.33) as

$$I(\omega) = \frac{1}{\pi} \text{Re}(\langle -\tfrac{1}{2}|f_o(\omega)|\tfrac{1}{2}\rangle + \langle \tfrac{1}{2}|f_o(\omega)|-\tfrac{1}{2}\rangle) \qquad \text{(III.40)}$$

We shall now find $f_o(\omega)$ from the stochastic Liouville equation. We substitute Eq.(III.38) into Eq.(III.36), multiply by $P_{2n}(s)$, integrate over $-1\leq s\leq 1$ (that is all space) and obtain

$$i\omega f_n(\omega) \frac{2}{4n+1} - \frac{2}{4n+1} \delta_{n,o} S_x = \frac{2}{4n+1} i\omega_o S_z^\times f_n(\omega)$$
$$+ iQS_z^\times \sum_\ell f_\ell(\omega) \int_{-1}^{1} P_{2n}(s)P_2(s)P_{2\ell}(s)ds - D2n(2n+1)\frac{2}{4n+1} f_n(\omega) \qquad \text{(III.41)}$$

where we have used

$$\int_{-1}^{1} P_{2n}(s)P_{2\ell}(s)ds = \frac{2}{4n+1} \delta_{n\ell} \qquad \text{(III.42)}$$

The integral on the right side of Eq.(III.41) is evaluated using the following property of the Legendre polynomials

$$sP_n(s) = \frac{(n+1)P_{n+1}(s)+nP_{n-1}(s)}{2n+1}, \qquad \text{(III.43)}$$

as well as $P_2(s) = (\frac{3}{2} s^2-\frac{1}{2})$. Eq.(III.41) is then transformed to

$$i\omega f_n(\omega)-\delta_{n,o}S_x = i\omega_o S_z^{\times} f_n(\omega)+iQS_z^{\times}(a_n f_{n-1}(\omega)$$

$$+ b_n f_n(\omega)+c_n f_{n+1}(\omega))-D2n(2n+1)f_n(\omega) \tag{III.44}$$

where

$$a_n = \frac{3}{2}\frac{2n(2n-1)}{(4n-3)(4n-1)}$$

$$b_n = \frac{(2n)(2n+1)}{(4n+3)(4n-1)}$$

$$c_n = \frac{3}{2}\frac{(2n+1)(2n+2)}{(4n+3)(4n+5)}$$

We take the matrix element of the above equation between the states $<-\frac{1}{2}|$ and $|\frac{1}{2}>$. For $n\geq 1$ this procedure leads to

$$i\omega F_n(\omega) = -i\omega_o F_n(\omega)-iQ(a_n F_{n-1}+b_n F_n(\omega)+c_n F_{n+1}(\omega))-D2n(2n+1)F_n(\omega) \tag{III.45}$$

where we have used

$$-<-\tfrac{1}{2}|S_z^{\times} f_n(\omega)|\tfrac{1}{2}> = <-\tfrac{1}{2}|f_n(\omega)|\tfrac{1}{2}>\equiv F_n(\omega) \tag{III.46}$$

Introducing $G_n(\omega) = F_n(\omega)/F_{n-1}(\omega)$ in Eq.(III.45) leads to

$$G_n(\omega) = -a_n\left[\frac{\omega}{Q} + \frac{\omega_o}{Q} + b_n+c_n G_{n+1} - i\frac{D}{Q}\,2n(2n+1)\right]^{-1} \tag{III.47}$$

For n=0 Eq.(III.44) reduces to

$$F_o(\omega) = -\frac{i}{2Q}\left[\frac{\omega}{Q} + \frac{\omega_o}{Q} + \frac{1}{5}G_1(\omega)\right]^{-1} \tag{III.48}$$

Similarily, we find when taking matrix elements of Eq.(III.44) between the states $<\frac{1}{2}|$ and $|-\frac{1}{2}>$

$$G'_n(\omega) = -a_n\left[-\frac{\omega}{Q} + \frac{\omega_o}{Q} + b_n+c_n G'_{n+1}(\omega)+i\frac{D}{Q}2n(2n+1)\right]^{-1} \quad (n>0) \tag{III.49}$$

$$F'_o(\omega) = -\frac{i}{2Q}\left[\frac{\omega}{Q} - \frac{\omega_o}{Q} - \frac{1}{5}G'_1(\omega)\right]^{-1} \quad (n=0) \tag{III.50}$$

where $G'_n(\omega) = F'_n(\omega)/F'_{n-1}(\omega)=<\frac{1}{2}|f_n(\omega)|-\frac{1}{2}>/<\frac{1}{2}|f_{n-1}(\omega)|-\frac{1}{2}>$

Eqs.(III.47) and (III.49) imply that

$$G'_n(\omega) = G^*_n(-\omega) \tag{III.51}$$

Our final result for $I(\omega)$ is obtained by combining Eqs.(III.40), (III.48) and (III.50)

$$I(\omega) = \frac{1}{\pi} \mathrm{Re}\left(-\frac{i}{2Q}\left(\left[\frac{\omega}{Q} + \frac{\omega_o}{Q} + \frac{1}{5} G_1(\omega)\right]^{-1} + \left[\frac{\omega}{Q} - \frac{\omega_o}{Q} - \frac{1}{5} G'_1(\omega)\right]^{-1}\right)\right) \tag{III.52}$$

This formula is too complex to handle analytically as it involves a continued fractional expansion through Eqs.(III.47) and (III.49). However, in the limit $|\frac{Q}{D}| << 1$ (fast diffusion) we may neglect $G_2(\omega)$ and $G'_2(\omega)$ in the expresion for $G_1(\omega)$ and $G'_1(\omega)$. In this limit we find two peaks centered around ω_o and $-\omega_o$. For $\omega \simeq \omega_o$ we obtain

$$I(\omega) = \frac{1}{2\pi} \frac{\frac{Q^2}{30D}}{(\omega-\omega_o)^2 + (\frac{Q^2}{30D})^2} \tag{III.53}$$

Near ω_o the line shape is Lorentzian with width $\frac{Q^2}{30D}$. If we increase D the line will sharpen up (motional narrowing) and in the limit $D \to \infty$ $I(\omega) \to \frac{1}{2}\delta(\omega-\omega_o)$. For other values of $\frac{Q}{D}$ we may compute $I(\omega)$ by a numerical procedure. We assume that $G_{n+1}(\omega)$ and $G'_{n+1}(\omega)$ are zero for a large n and use Eqs.(III.47) and (III.49) to find $G_1(\omega)$ and $G'_1(\omega)$. The computations show that for $\frac{Q}{D} = 1$ there is no observable change in $I(\omega)$ by increasing n beyond 3. $g_{\parallel}-g_{\perp}$ is taken to be positive.
Similarily, for $\frac{Q}{D} = 30000$ the computation need not be extended beyond n ∿ 35.

The absorption line shapes for several values of $\frac{Q}{D}$ are shown in Fig.2. A motionally invariant linewidth T_2^{-1} has been included in the calculation by letting $\omega_o \to \omega_o + iT_2^{-1}$.

It is seen that the first effect of increasing Q/D from the motional narrowing limit is to broaden the line. As Q/D is increased further ($1 < Q/D < \infty$) the line shifts downfield and becomes increasingly assym-

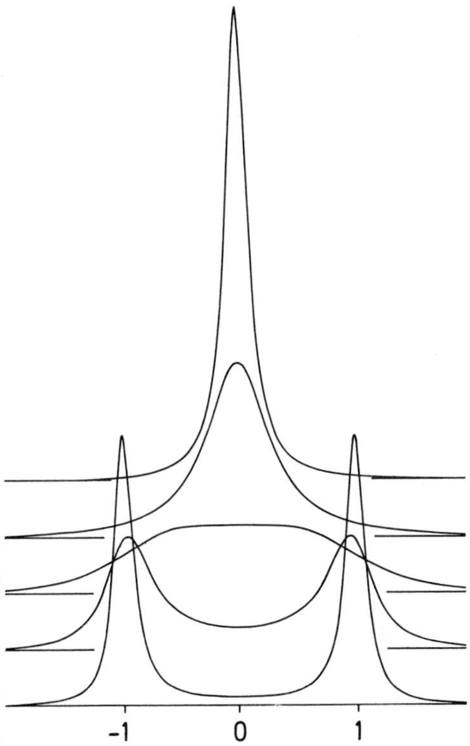

Fig.1. ESR absorption line shapes for a radical in a fluctuating field, h_z, parallel to the static field plotted against $(\omega-\omega_o)/\omega_z$.
The spectra (starting from the bottom) are characterized by ω_z/Ω = 15,4.7,1.5,0.54,0.16.

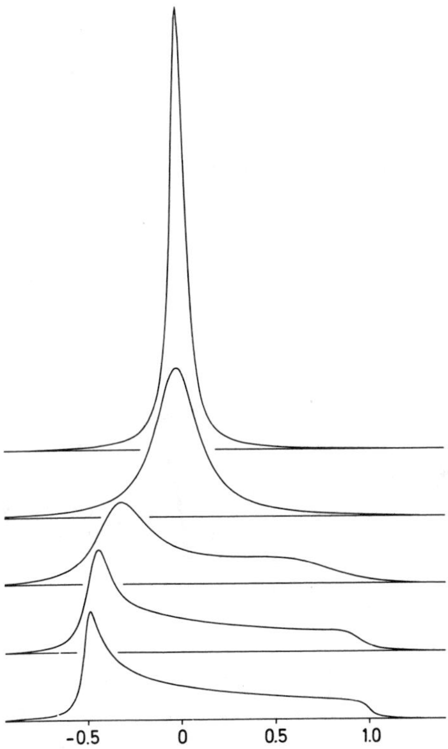

Fig.2. ESR absorption line shapes for a radical with axially symmetric g-tensor undergoing isotropic rotational diffusion. The spectra are plotted against $(\omega-\omega_o)/Q$ and are characterized (starting from the bottom) by
Q/D = 30000, 300, 30, 3, 0.3
($g_\parallel-g_\perp$ is taken to be positive).

metric and approaches eventually the rigid lattice spectrum derived in Appendix A. Although Fig.2 does not indicate it very clearly the computations show a small high field peak appearing for $10<Q/D<100$.

APPENDIX A

We consider the calculation of the rigid lattice line shape from a radical ($S=\frac{1}{2}$) with axially symmetric g-tensor. The hamiltonian is

$$H(\theta) = g\beta BS_z + (3\cos^2\theta-1)\frac{Q}{2}S_z$$

$$= \omega_o S_z+(3s^2-1)\frac{Q}{2}S_z \qquad \text{(III.A-1)}$$

where $s = \cos\theta$

$$I(\omega) = \frac{1}{2\pi}\int_{\infty}^{\infty} \mathrm{Tr}(S_x\{S_x(t)\})e^{-i\omega t}dt$$

where $$\{S_x(t)\} = \frac{1}{\pi}\int_{o}^{\pi} S_x(\theta,t)\sin\theta d\theta=\frac{1}{\pi}\int_{-1}^{1} S_x(s,t)ds \qquad \text{(III.A-2)}$$

$S_x(s,t)$ is obtained by integration of the Heisenberg equation[17]

$$\frac{dS_x(s,t)}{dt} = iH^{\times}S_x(s,t)$$

$$S_x(s,t) = \exp(i(\omega_o+(3s^2-1)\frac{Q}{2})tS_z)S_x\exp(-i(\omega_o+(3s^2-1)\frac{Q}{2})tS_z)$$

$$= \cos((\omega_o+(3s^2-1)\frac{Q}{2})t)S_x-\sin((\omega_o+(3s^2-1)\frac{Q}{2})t)S_y \qquad \text{(III.A-3)}$$

From Eq.(III.A-3) we find

$$\mathrm{Tr}(S_x\{S_x(t)\}) = \frac{1}{\pi}\int_{o}^{1}\cos((\omega_o+(3s^2-1)\frac{Q}{2})t)ds \qquad \text{(III.A-4)}$$

where we have taken the trace before averaging. The spectrum may now be written as

$$I(\omega) = \frac{1}{2\pi}\int_{-\infty}^{\infty}\left[\frac{1}{\pi}\int_{0}^{1}\cos((\omega_o+(3s^2-1)\tfrac{Q}{2})t)ds\right]e^{-i\omega t}dt \qquad \text{(III.A-5)}$$

Interchanging the order of integration and utilizing

$$\delta(\omega) = \frac{1}{2\pi}\int_{-\infty}^{\infty}\exp(-i\omega t)dt$$

leads to

$$I(\omega) = \frac{1}{2\pi}\int_{0}^{1}[\delta(\omega_o+(3s^2-1)\tfrac{Q}{2} -\omega)+\delta(\omega_o+(3s^2-1)\tfrac{Q}{2} +\omega)]ds \qquad \text{(III.A-6)}$$

This is integrated by noting

$$\int_{0}^{1}\delta(f(s))ds=\int_{f(s=0)}^{f(s=1)}\frac{\delta(z)}{\frac{\delta z}{\delta s}}dz=\int_{-\infty}^{\infty}\frac{\delta(z)g(z)}{\frac{\delta z}{\delta s}}dz=2\pi\sum_{i=1}^{N}\left.\frac{g(s_i)}{\frac{\delta z}{\delta s}}\right|_{s=s_i}$$

where $z = f(s)$ is monotonic, $g(z) = 1$ for $f(s=0)\leq z\leq f(s=1)$ (or $0\leq s\leq 1$) and zero otherwise. N is the number of values s_i, which makes z zero. The final expression for $I(\omega)$ is then

$$I(\omega) = \frac{1}{3Q}\frac{R\left(\frac{\omega-\omega_o+Q/2}{3Q/2}\right)}{\sqrt{\frac{\omega-\omega_o+Q/2}{3Q/2}}} + \frac{1}{3Q}\frac{R\left(\frac{-\omega-\omega_o+Q/2}{3Q/2}\right)}{\sqrt{\frac{-\omega-\omega_o+Q/2}{3Q/2}}} \qquad \text{(III.A-7)}$$

Where $R(k)$ is one if $0\leq k\leq 1$ and zero otherwise.

REFERENCES

1. R. Kubo, "Stochastic Processes in Chemical Physics", Advances in Chemical Physics Vol.XV, K.E. Shuler, Ed. (John Wiley and Sons, New York 1969) p.101.
2. R. Kubo, J.Phys.Soc. Japan 26, Suplement 1, (1969).
3. J.H. Freed, G.V. Bruno, C. Polnaszek (to be published).
4. P.W. Anderson, J.Phys.Soc. of Japan 9, 316 (1954).
5. a) R. Kubo, J.Phys.Soc. of Japan 9, 935 (1954).
   b) R. Kubo, Nuovo Cimento Suppl. 6, 1063 (1957).
   c) R. Kubo, "Fluctuation, Relaxation and Resonance in Magnetic Systems", D.ter Haar, Ed. (Oliver and Boyd, Edinburgh, (1962) p.23.
6. M. Blume, Phys.Rev. 174, 351 (1968).
7. J.H. Freed, J.Chem.Phys. 49, 376 (1968).
8. A. Messiah, Quantum Mechanics (North Holland Publishing Company, Amsterdam, 1965) p.469.
9. C.S. Johnson, Jr., Adv. in Magn. Res. Vol.1, J. Waugh, Ed. (Academic Press, New York, 1965) p.33.
10. P.D. Sullivan, J.R. Bolton, Adv. in Magn. Res. Vol.4, J. Waugh, Ed. (Academic Press, New York, 1970) p.39.
11. A. Hudson, G.R. Luckhurst, Chem.Rev. 69, 191 (1969).
12. R.G. Gordon, R.P. McGinnis, J.Chem.Phys. 49, 2455 (1968)
13. J.G.F. Francis, The Computer Journal, 4, 265 and 322 (1961).
14. M. Fixman, J.Chem.Phys. 48, 223 (1968).
15. H. Silescu, D. Kivelson, J.Chem.Phys. 48, 3493 (1968).
16. R. Courant, D. Hilbert, Methods of Mathematical Physics, (Interscience Publishers Inc., New York, 1953) p. 86.
17. C.P. Slichter, Principles of Magnetic Resonance, (Harper & Row, New York, 1963) p.26.

PROJECTION OPERATORS

O. Platz

Aarhus University

The application of projection operators within irreversible statistical mechanics has increased steadily during the last decennium. Provotorov, Argyres and Kelley, Terwiel and Mazur as well as Shimizu used projection operator procedures to problems within spin relaxation.

The application of projection operators offers several advantages. The interaction between the spin system and the bath may be handled in an orderly fashion and the perturbation may include higher order terms. In addition, the physical nature of the approximations needed to obtain a usable "Master Equation" is more transparent than in other types of perturbation calculations.

Following Zwanzig this chapter presents the general derivation of a "Non-Markoffian" Master Equation. A proper choice of the projection operator included in the M.E. leads to the procedure used by Argyres and Kelley. We shall confine the lectures to their method.

IV.1. "NON-MARKOFFIAN" MASTER EQUATION

The equation of motion for the density operator is

$$i \frac{\partial}{\partial t} \rho(t) = H^{\times}\rho(t) \qquad \text{(IV.1)}$$

where we put $\hbar = 1$. H is a time-independent hamilton operator. The projection operator P separates out the "relevant" part $\rho_1(t)$ and discards the "irrelevant" part $\rho_2(t)$ of the density operator, i.e.

$$\begin{aligned} \rho(t) &= \rho_1(t)+\rho_2(t) \\ P\rho(t) &= \rho_1(t) \\ (1-P)\rho(t) &= \rho_2(t) \end{aligned} \tag{IV.2}$$

Application of the time-independent projection operators P and (1-P) to Eq.(IV.1) gives

$$i \frac{\partial}{\partial t} \rho_1(t) = PH^{\times}(\rho_1(t)+\rho_2(t)) \tag{IV.3}$$

$$i \frac{\partial}{\partial t} \rho_2(t) = (1-P)H^{\times}(\rho_1(t)+\rho_2(t)) \tag{IV.4}$$

A formal solution to Eq.(IV.4) may be written as

$$\rho_2(t)=\exp[-it(1-P)H^{\times}]\rho_2(0)-i\int_0^t d\tau \exp[-i\tau(1-P)H^{\times}](1-P)H^{\times}\rho_1(t-\tau) \tag{IV.5}$$

Combination of Eqs.(IV.3) and (IV.5) now yields

$$\begin{aligned} i \frac{\partial}{\partial t} \rho_1(t) &= PH^{\times}\exp[-it(1-P)H^{\times}]\rho_2(0)+PH^{\times}\rho_1(t) \\ &-i \int_0^t d\tau PH^{\times}\exp[-i\tau(1-P)H^{\times}](1-P)H^{\times}\rho_1(t-\tau) \end{aligned} \tag{IV.6}$$

Eq.(IV.5) presents an exact equation for $\rho_1(t)$. The equation is "non-Markoffian" since the right-hand side includes a convolution integral containing ρ, with time-arguments different from t.

For H time-dependent the following equation with time-ordered exponentials obtains rather than Eqs.(IV.5) and (IV.6)

$$\begin{aligned} \rho_2(t) &= \exp\_[-i \int_0^t dt'(1-P)H^{\times}(t')]\rho_2(0) \\ &- \int_0^t d\tau \exp\_[-i \int_\tau^t dt'(1-P)H^{\times}(t')](1-P)H^{\times}(\tau)\rho_1(\tau) \end{aligned} \tag{IV.5a}$$

and

$$i \frac{\partial}{\partial t} \rho_1(t) = PH^\times(t)\exp_-[-i \int_0^t dt'(1-P)H^\times(t')]\rho_2(0)$$

$$+ PH^\times(t)\rho_1(t)$$

$$-i \int_0^t d\tau \exp_-[-i \int_\tau^t dt'(1-P)H^\times(t')](1-P)H^\times(\tau)\rho_1(\tau) \qquad \text{(IV.6a)}$$

IV.2. APPLICATION TO THE INTERACTION BETWEEN A SPIN SYSTEM AND A BATH

The hamiltonian for the spin system and bath combination may be written as

$$H = H_s + H_B + V = H_o + V$$

where H_s is the hamilton operator for the spin system in an external static magnetic field, H_B is the hamiltonian of the bath, and V represents the interaction between spin system and bath. It follows from these definitions that

$$[H_s, H_B] = 0$$

The Liouville equation of the total system, i.e. spin system and bath, is given by

$$i \frac{\partial}{\partial t} \rho_T = H^\times \rho_T(t) \qquad \text{(IV.7)}$$

where $\rho_T(t)$ is the density operator for the total system.

The density operator $\rho(t)$ for the spin system obtains after averaging or taking the trace of ρ_T over the bath variables, i.e.

$$\rho(t) = \underset{(B)}{\mathrm{Tr}}\, \rho_T(t) \qquad \text{(IV.8)}$$

We assume that ρ_T at time t = 0 or just prior to the start of the bath-spin system interaction is given by

$$\rho_T(0) = \rho_B \rho(0); \quad \rho_B = \frac{\exp(-\beta H_B)}{\underset{(B)}{\mathrm{Tr}}(\exp(-\beta H_B))}; \quad \beta = (kT)^{-1} \qquad \text{(IV.9)}$$

As projection operator it is convenient to choose $P = \rho_B \mathrm{Tr}_{(B)}$. We have now

$$P\rho_T(t) = \rho_1(t) = \rho_B\rho(t)$$

and

$$(1-P)\rho_T(t) = \rho_2(t)$$

$$\rho_T(t) = \rho_B\rho(t)+\rho_2(t) \tag{IV.10}$$

Combination of Eqs.(IV.10) with Eq.(IV.6) gives the expression

$$i\frac{\partial}{\partial t}\rho(t) = \mathrm{Tr}_{(B)}H^{\times}\exp(-it(1-\rho_B\mathrm{Tr}_{(B)})H^{\times})\rho_2(0)+\mathrm{Tr}_{(B)}H^{\times}\rho_B\rho(t)$$
$$-i\int_o^t d\tau \mathrm{Tr}_{(B)}H^{\times}\exp(-i\tau(1-\rho_B\mathrm{Tr}_{(B)})H^{\times})(1-\rho_B\mathrm{Tr}_{(B)})H^{\times}\rho_B\rho(t-\tau) \tag{IV.11}$$

This equation may undergo extensive simplification. We observe that

$$\mathrm{Tr}_{(B)}H^{\times}\rho_2(t) = \mathrm{Tr}_{(B)}H_s^{\times}\rho_2(t)+\mathrm{Tr}_{(B)}H_B^{\times}\rho_2(t)+\mathrm{Tr}_{(B)}V^{\times}\rho_2(t)$$

From Eqs.(IV.10) and (IV.8) we now have that

$$\mathrm{Tr}_{(B)}H_s^{\times}\rho_2(t) = H_s^{\times}\mathrm{Tr}_{(B)}\rho_2(t) = 0$$

$$\mathrm{Tr}_{(B)}H_B^{\times}\rho_2(t) = 0$$

Accordingly,

$$\mathrm{Tr}_{(B)}H^{\times}\rho_2(t) = \mathrm{Tr}_{(B)}V^{\times}\rho_2(t) \tag{IV.12}$$

Equation (IV.12) implies that the first $H^{\times}$ may be replaced by $V^{\times}$ in the first and in the last term of the right-hand side of Eq.(IV.11). Furthermore, Eqs.(IV.10) and (IV.9) show that $\rho_2(0) = 0$. For this reason, the first term on the right-hand side of Eq.(IV.11) vanishes.

Without loss of generality we assume that $\mathrm{Tr}_B V\rho_B = 0$, since any finite value might be taken care of by redefining H_s.

The second term on the right-hand side of Eq.(IV.11) may now be rewritten by observing that

$$\mathop{Tr}_{(B)}H^\times\rho_B\rho(t) = H_s^\times\rho(t)\mathop{Tr}_{(B)}\rho_B+\rho(t)\mathop{Tr}_{(B)}H_B^\times\rho_B+\mathop{Tr}_{(B)}V^\times\rho_B\rho(t) = H_s^\times\rho(t)$$

In addition,we have that

$$(1-\rho_B\mathop{Tr}_{(B)})H^\times\rho_B\rho(t-\tau) = V^\times\rho_B\rho(t-\tau)$$

and

$$[(1-\rho_B\mathop{Tr}_{(B)})H^\times]^n V^\times\rho_B\rho(t-\tau) = [H_o^\times+(1-\rho_B\mathop{Tr}_{(B)})V^\times]^n V^\times\rho_B\rho(t-\tau) \quad \text{(IV.12a)}$$

Equation (IV.12a) may be demonstrated by observing that for

$$A_{n+1} = (1-\rho_B\mathop{Tr}_{(B)})H^\times A_n$$

It follows that

$$\mathop{Tr}_{(B)}A_{n+1} = 0$$

Accordingly, we have

$$(1-\rho_B\mathop{Tr}_{(B)})H^\times A_{n+1} = (H_o^\times+(1-\rho_B\mathop{Tr}_{(B)})V^\times)A_{n+1}$$

In toto, we may therefore simplify Eq.(IV.11) as follows

$$i\frac{\partial}{\partial t}\rho(t) = H_s^\times\rho(t)-i\int_o^t d\tau\mathop{Tr}_{(B)}V^\times\exp(-i\tau(H_o^\times+(1-\rho_B\mathop{Tr}_{(B)})V^\times)V^\times\rho_B\rho(t-\tau) \quad \text{(IV.13)}$$

Eq.(IV.13) is an exact expression for the density operator of a spin system fulfilling the initial conditions in Eq.(IV.9). The first term on the right-hand side accounts for the unperturbed motion of the spin system, whereas the second term takes care of the effect of the bath on the time-development of the spin system. Eq.(IV.13) reduces to the Liouville equation in the rigid lattice approximation, i.e. $V = 0$.

In the following discussion the spin system bath interaction is included to lowest order only. The paper by Argyres and Kelley should be consulted for a discussion of the inclusion of higher order terms.

Eq.(IV.13) reduces to

$$i \frac{\partial}{\partial t} \rho(t) = H_s^\times \rho(t) - i \int_0^t d\tau \operatorname{Tr}_{(B)} V^\times \exp(-i\tau H_o^\times) V^\times \rho_B \rho(t-\tau) \qquad \text{(IV.14)}$$

We may now expand V into a complete set of operators $V = \sum_i v_i u_i$ where the v_i's pertain to the spin system only and the u_i's are bath operators exclusively.

We define now the following correlation functions for the Heisenberg operators of the bath $u_n(t)$

$$c_{nm}(\tau) = \operatorname{Tr}_{(B)} \rho_B u_n(\tau) u_m$$

Introduction of these correlation functions into Eq.(IV.14) gives

$$i \frac{\partial}{\partial t} \rho(t) = H_s^\times \rho(t) - i \sum_{nm} \int_0^t d\tau \{ c_{nm}(\tau) v_n^\times \exp(-i\tau H_s^\times) v_m \rho(t-\tau)$$

$$- c_{mn}(-\tau) v_n^\times \exp(-i\tau H_s^\times) \rho(t-\tau) v_m \} \qquad \text{(IV.15)}$$

We observe that all information about the bath is contained in the correlation functions.

The spin system-bath interaction may be taken as a stochastic process. For $\beta = 0$, i.e. the high temperature approximation, we have $c_{nm}(\tau) = c_{mn}(-\tau)$ and Eq.(IV.15) is now

$$i \frac{\partial}{\partial t} \rho(t) = H_s^\times \rho(t) - i \sum_{nm} \int_0^t d\tau c_{nm}(\tau) v_n^\times \exp(-i\tau H_s^\times) v_m^\times \rho(t-\tau) \qquad \text{(IV.16)}$$

Correlation functions $c_{nm}(\tau)$ are characteristics of the stochastic process. Denoting the correlation time for the bath τ_c we have the usual condition

$$c_{nm}(\tau) \to 0 \qquad \text{for } |\tau| > \tau_c .$$

For $t \gg \tau_c$ the integral in Eq.(IV.16) may accordingly be taken between limits 0 and ∞.

IV.3. THE REDFIELD APPROXIMATION

In the "non-Markoffian" Eq.(IV.16) we may make the approximation

$$i \frac{\partial}{\partial t} \rho(t) = H_s^{\times}\rho(t)$$

or $$\rho(t) = \exp[-iH_s^{\times}t]\rho(0)$$

or $$\rho(t-\tau) = \exp[iH_s^{\times}\tau]\rho(t)$$

Substitution of the expression for $\rho(t-\tau)$ into the last term of Eq.(IV.16) gives the Markoffian expression

$$i \frac{\partial}{\partial t} \rho(t) \simeq H_s^{\times}\rho(t) - i \sum_{nm} \int_0^t d\tau c_{nm}(\tau) v_n^{\times} \exp(-i\tau H_s^{\times}) v_m^{\times} \exp(i\tau H_s^{\times})\rho(t) \tag{IV.17}$$

For $t \gg \tau_c$ Eq.(IV.17) is equivalent to the Redfield equation. The equivalence may be ascertained by using the matrix elements of ρ in a representation diagonalizing H_s. However, it appears that the transition from the non-Markoffian Eq.(IV.16) to the Markoffian Eq.(IV.17) is not transparent.

IV.4. THE TWO JUMP MODELS

As a simple example let us apply Eq.(IV.16) to an ensemble of independent spins interacting with a bath. We assume that the bath is represented by a magnetic field $h_z(t)$ fluctuating in the direction of the z-axis.

The external magnetic field has two components: a static field h_o in the direction of the z-axis and a rotating field $\bar{h}_r(t) = h_1(\bar{i}\cos\omega t - \bar{j}\sin\omega t)$.

The rotating field may be taken care of in the relaxation equation (IV.16) by inclusion in the hamilton operator of the first term on the right-hand side.

For the rotating field the hamiltonian may be written

$H_r(t) = -\frac{1}{2}\omega_1(I^- \exp(-i\omega t) + I^+ \exp(i\omega t))$ and $H_s = -\omega_o I_z$, $V = -\gamma h_z(t) I_z$

where γ is the gyromagnetic ratio $\omega_1 = \gamma h_1$ and $\omega_o = \gamma h_o$.

Inserting into Eq.(IV.16) gives

$$i\frac{\partial}{\partial t}\rho(t) = (H_s^\times + H_r^\times(t))\rho(t) - i\int_o^t c_{zz} I_z^\times \exp(i\omega_o \tau I_z^\times) I_z^\times \rho(t-\tau)$$

where $c_{zz}(\tau) = \gamma^2 \langle h_z(0) h_z(\tau)\rangle$

Transformation to the coordinate system rotating in phase with $h_r(t)$ gives

$$\rho_r(t) = \exp(-i\omega t I_z^\times)\rho(t)$$

and

$$\frac{d}{dt}\rho_r(t) = -i((\omega-\omega_o)I_z^\times - \omega_1 I_x^\times)\rho_r(t)$$
$$- \int_o^t d\tau c_{zz}(\tau)\exp(i(\omega_o-\omega)\tau I_z^\times) I_z^\times I_z^\times \rho_r(t-\tau)$$

Assumption of a "steady state" in the sense that $\rho_r(t) \to \rho_r(\omega)$ for $t \to \infty$ leads to

$$0 = -i((\omega-\omega_o)I_z^\times - \omega_1 I_x^\times)\rho_r(\omega)$$
$$- \int_o^\infty d\tau c_{zz}(\tau)\exp(i(\omega_o-\omega)\tau I_z^\times) I_z^\times I_z^\times \rho_r(\omega)$$

The magnetization $\bar{M}^r$ in the rotating coordinate system is now given by

$$\bar{M}^r(\omega) = n\gamma \mathrm{Tr}(\bar{I}\rho_r(\omega))$$

n being the number density of spins.

Introducing the definition

$$j_z(\omega) = \int_o^\infty c_{zz}(\tau)\exp(i\omega\tau)d\tau$$

we obtain for the transverse components of the magnetization in the rotating coordinate system

$$0 = M_x^r \mathrm{Re} j_z(\omega-\omega_o) + M_y^r(\mathrm{Im} j_z(\omega-\omega_o)-(\omega-\omega_o))$$

$$0 = -M_x^r(\mathrm{Im} j_z(\omega-\omega_o)-(\omega-\omega_o)) + M_y^r \mathrm{Re} j_z(\omega-\omega_o) + M_z^r \omega_1$$

We notice that M_z^r differs from $M_z^{eq.}$ by terms which are of second order in h_1. Therefore, we take $M_z^r = M_z^{eq.}$ and obtain

$$M_y^r = \frac{-\omega_1 M_z \mathrm{Re} j_z(\omega-\omega_o)}{(\mathrm{Re} j_z(\omega-\omega_o))^2+((\omega_o-\omega)+\mathrm{Im} j_z(\omega-\omega_o))^2}$$

The energy absorption $I(\omega)$ is proportional to M_y^r.

We assume now that $h_z(t)$ is a magnetic field jumping at random between $+\frac{\delta}{\gamma}$ and $-\frac{\delta}{\gamma}$ with a probability per unit time given by $(2\tau_c)^{-1}$. Accordingly, we have

$$c_{zz}(\tau) = \delta^2 \exp(-|\tau|/\tau_c)$$

and

$$j_z(\omega) = \delta^2(\tau_c^{-1}+i\omega)(\tau_c^{-2}+\omega^2)^{-1}$$

Combination of these expressions with the equation for $I(\omega)$ yields now

$$I(\omega) = \frac{\delta^2\tau_c^{-1}}{(\omega-\omega_o)^4+\delta^4+(\tau_c^{-2}-2\delta^2)(\omega-\omega_o)^2}$$

This result is noteworty since it is identical to that from the Kubo-Anderson model. In the derivation above we introduced the correlation function for the stochastic process only whereas the derivation from the Kubo-Anderson model requires all the higher moments.

It is possible to include in the spin hamilton operator H_s a strong rotating field giving saturation phenomena. This inclusion invalidates the approximations in the simple example shown above. Argyres and Kelley included saturation phenomena. They used the same projection operator $P = \rho_B \mathrm{Tr}_{(B)}$. Since $H_s = H_s(t)$ is time-dependent they substituted the non-Markoffian equation (IV.6a) for the equation (IV.6).

REFERENCES

1. B.N. Provotorov, J.Exptl.Theoret.Phys. (U.S.S.R.) 41, 1582,(1961). (Soviet Phys. Jetp. 14, 1126 (1962)).

2. P.N. Argyres and P.L. Kelley, Phys.Rev. 134, A98 (1964).

3. R.H. Terwiel and P. Mazur, Physica 32, 1813 (1966).

4. T. Shimizu, J.Phys.Soc. Japan 29, 74 (1970).

5. R. Zwanzig, J.Chem.Phys. 33, 1338 (1960).

6. R. Zwanzig, "Lectures in Theoretical Physics III" p. 106 Interscience Publ., New York 1961.

CUMULANT EXPANSION

J. Aase Nielsen

Aarhus University

Earlier theories give an approximate solution to the equation of motion for the spin density operator. The solution is carried out to second order in the spin-perturbation operator. Formally, we may extend the calculation to higher orders. It is difficult, however, to group the higher order terms in convergent series.

By means of the cumulant rather than the moment expansion we shall in the following show how to obtain a general and formally correct solution to higher order.

First we shall introduce the cumulant expansion both for one- and multivariable distribution functions and make a comparison with the moments. Next a generalization is made to provide a solution to the equation of motion for the spin density operator. It will be shown that the relaxation matrix, R, can be found to high order and for finite times.

V.1. ONE DIMENSIONAL MOMENT AND CUMULANT EXPANSION

Let F(x) be a one-dimensional distribution function.

$$M(s) = \int_{-\infty}^{\infty} e^{sx}\, dF(x)$$

is then called the moment generating function for $F(x)$. Differentiation of $M(s)$ ν times with respect to s gives

$$dM^{\nu}(s)/ds^{\nu} = \int_{-\infty}^{\infty} x^{\nu} e^{sx} dF(x).$$

Accordingly, we obtain

$$dM^{\nu}(s)/ds^{\nu}\Big|_{s=0} = \int_{-\infty}^{\infty} x^{\nu} dF(x) \equiv m_{\nu}$$

where m_{ν} is the ν'th moment of $F(x)$. The moment generating function may now be written

$$M(s) = 1 + \sum_{1}^{k} \frac{m_{\nu}}{\nu!} s^{\nu} + o(s^{k})$$

where $o(s^k)$ is defined as

$$\lim_{s \to o} \frac{o(s^k)}{s^k} = 0.$$

Expansion of log $M(s)$ gives

$$\log M(s) = \sum_{1}^{k} \frac{H_{\nu}}{\nu!} s^{\nu} + o(s^{k}).$$

The coefficients H_{ν} are called the cumulants or semi-invariants of the distribution. We may find a relation between the cumulants and the moments using the identities

$$\log M(s) = \log(1+\sum \frac{m_{\nu}}{\nu!} s^{\nu}) = \sum \frac{H_{\nu}}{\nu!} s^{\nu}$$

e.g.

$$H_1 = m_1$$

$$H_2 = m_2 - m_1^2$$

$$\vdots$$

V.2. MULTIVARIABLE MOMENT AND CUMULANT EXPANSIONS

The above definitions can readily be extended to multivariable expansions.

For n random variables $x_1, x_2, \ldots, x_n$ the moments and the cumulants are defined by

$$M(s_1 \ldots s_n) \equiv \langle \exp \sum_{j=1}^{n} s_j x_j \rangle = \sum_{\nu_1 \ldots \nu_n = 0}^{\infty} \left(\prod_j \frac{s_j^{\nu_j}}{\nu_j!} \right) m(\nu_1 \ldots \nu_n) \tag{V.1a}$$

$$= \exp \sum_{\nu_1 \ldots \nu_n}{}' \left(\prod_j \frac{s_j^{\nu_j}}{\nu_j!} \right) H(\nu_1 \ldots \nu_n) \tag{V.1b}$$

$$= \exp \left[K(s_1 \ldots s_n) \right] \tag{V.1c}$$

M(s) and K(s) are called the moment generating function and the cumulant function.

$\sum'$ means summation over $\nu_1 \ldots \nu_n$ excluding $\nu_1 = \nu_2 = \ldots = \nu_n = 0$.

We further introduce the notation

$$H(\nu_1 \ldots \nu_n) \equiv \langle x_1^{\nu_1} \ldots x_n^{\nu_n} \rangle_c \tag{V.2}$$

$$m(\nu_1 \ldots \nu_n) \equiv \langle x_1^{\nu_1} \ldots x_n^{\nu_n} \rangle \tag{V.3}$$

The c in (V.2) indicates that the cumulant is a certain type of average of $x_1 \ldots x_n$.

Comparison of the coefficients in (V.1a) with those in (V.1b) gives the cumulants expressed by the moments, e.g.

$$\langle \exp \sum_{j=1}^{2} s_j x_j \rangle = \sum_{\nu} \frac{m(\nu_1 \nu_2)}{\prod_j \nu_j!} s_1^{\nu_1} s_2^{\nu_2}$$

$$= m(oo)+m(o1)s_2+m(1o)s_1$$

$$+ m(11)s_1s_2+ \ldots$$

$$= \exp\left\{\langle x_1\rangle_c s_1+\langle x_2\rangle_c s_2+\langle x_1x_2\rangle_c s_1s_2+\ldots\right\}$$

$$= 1+\langle x_1\rangle_c s_1+\langle x_2\rangle_c s_2+\langle x_1x_2\rangle_c s_1s_2+\tfrac{1}{2}\langle x_1\rangle_c^2 s_1^2$$

$$+ \ldots+\langle x_1\rangle_c s_1\langle x_2\rangle_c s_2+\ldots$$

i.e. $$m(1o) = \langle x_1\rangle = \langle x_1\rangle_c$$

$$m(11) = \langle x_1x_2\rangle = \langle x_1x_2\rangle_c+\langle x_1\rangle_c\langle x_2\rangle_c$$

or $$\langle x_1x_2\rangle_c = \langle x_1x_2\rangle-\langle x_1\rangle\langle x_2\rangle.$$

The cumulant has a very useful property, namely that the cumulant is zero if at least one of the random variables is uncorrelated with the other ones. Let us assume that x_i is uncorrelated with $x_1\ldots x_{i-1}x_{i+1}\ldots x_n$. Then

$$\langle \exp\sum_1^n s_jx_j\rangle = \langle \exp s_ix_i\rangle\langle \exp\sum_{k=i}^n s_kx_k\rangle,$$

i.e. there is no cumulant with x_i mixed with the other variables.

By replacing the discrete index j with the continuous parameter t we obtain the cumulant and moment expansion for a continuous stochastic variable x(t).

$$\sum_j s_jx_j \to \sum_j s(t_j)x(t_j)\delta t_j \to \int_o^t s(t')x(t')dt'$$

for $\delta t_j \to 0$. In addition, we note that only cumulants with $0 \leq \nu_r \leq 1$ are different from zero, i.e.

$$\lim_{\delta t_j \to 0} \sum_j \langle x(t_j)^2\rangle_c s(t_j)^2(\delta t_j)^2$$

$$= \lim_{\delta t_j \to o} \delta t_j \int_o^t \langle x(t')^2\rangle_c s(t')^2 dt' = 0$$

Accordingly, we may write the general expression

$$\langle \exp \int_o^t x(t')s(t')dt'\rangle$$

$$= \exp\left\{\sum_{n=1}^{\infty} \frac{1}{n!} \int_o^t dt_1 \ldots \int_o^t dt_n \langle x(t_1)\ldots x(t_n)\rangle_c s(t_1)\ldots s(t_n)\right\}$$

$$= \exp\left\{\sum_{n=1}^{\infty} \int_o^t dt_1 \int_o^{t_1} dt_2 \ldots \int_o^{t_{n-1}} dt_n \langle x(t_1)\ldots x(t_n)\rangle_c s(t_1)..s(t_n)\right\}$$

Replacement of s(t) by i gives the important equation

$$\langle \exp i \int_o^t x(t)dt\rangle$$

$$= \exp\left\{\sum_{1}^{\infty} i^n \int_o^t dt_1 \ldots \int_o^{t_{n-1}} dt_n \langle x(t_1)\ldots x(t_n)\rangle_c\right\}$$

$$= \exp K(t) \qquad \text{(V.4)}$$

V.3. GENERALIZATION

The cumulant expansion will be used to give a general solution to the spin-relaxation problem.

The system is described by the hamiltonian

$$H = H_o + H_1(t), \tag{V.5}$$

the time-independent part H_o and the perturbation $H_1(t)$ which is a random function of time.

The spectrum is given by

$$I(\omega) = \mathrm{Re}\,\frac{4}{\pi}\int_o^\infty G(t)e^{i\omega t}d\tau \tag{V.6}$$

where

$$G(t) = \mathrm{Tr}\Big[S_x(t)S_x\Big],$$

$$S_x(t) = \langle \exp(iHt)S_x\exp(-iHt)\rangle \tag{V.7}$$

$$= \langle \tilde{S}_x(t)\rangle$$

The ensemble average in (V.7) is taken over the ensemble defined by the probability distribution characteristic of $H_1(t)$.

The equation of motion for a single independent spin system in the interaction representation

$$S_x^I(t) = \exp(-iH_o t)S_x(t)\exp(iH_o t)$$

is given by

$$(d/dt)\tilde{S}_x^I(t) = iH_1^{Ix}(t)\tilde{S}_x^I \tag{V.8}$$

A formally correct solution to this is

$$\tilde{S}_x^I(t) = \exp_-\Big[i\int_o^t dt' H_1^{Ix}(t')\Big]S_x(o)$$

which on taking an ensemble average becomes

$$S_x^I(t) = \langle \exp_-\left[i\int_o^t dt' H_1^{Ix}(t')\right]\rangle S_x \qquad (V.9)$$

The exponent is defined as

$$\langle \exp_-\left[i\int_o^t dt' H_1^{Ix}(t')\right]\rangle = \sum_{n=o}^{\infty} M_n(t) \qquad (V.10)$$

$$M_o(t) = 1 \qquad (V.11)$$

$$M_n(t) = i^n \int_o^t dt_1 \ldots \int_o^{t_{n-1}} dt_n \langle H_1^{Ix}(t_1)..H_1^{Ix}(t_n)\rangle \text{for } n\geq 1 \qquad (V.12)$$

The expansion of (V.10) may be regarded as a generalized moment expansion.

The generalized cumulant, $K_n(t)$, is given by

$$\langle \exp_-\left[i\int_o^t dt' H_1^{Ix}(t')\right]\rangle = \exp_- K(t) \qquad (V.13)$$

$$K(t) = \sum_{n=1}^{\infty} K_n(t) \qquad (V.14)$$

The moment $m_n(t_1t_2\ldots t_n) = \langle H_1^{Ix}(t_1)\ldots H_1^{Ix}(t_n)\rangle$ and the cumulant $K_n(t)$ are operators.

By comparing (V.10) and (V.13) one can express K_n as sums of products of M_i with $i\leq n$.

Insertion of (V.4) into (V.9) gives on differentiation

$$(d/dt)S_x^I(t)=\sum_{n=1}^{\infty} i^n \int_o^t dt_1 \ldots \int_o^{t_{n-2}} t_{n-1} \langle OH_1^{Ix}(t)H_1^{Ix}(t_1)\ldots H_1^{Ix}(t_{n-1})\rangle_c S_x^I(t) \tag{V.15}$$

$$(d/dt)K_n(t)=i^n \int_o^t dt_1 \ldots \int_o^{t_{n-2}} dt_{n-1} \langle OH_1^{Ix}(t)H_1^{Ix}(t_1)\ldots H_1^{Ix}(t_{n-1})\rangle_c \tag{V.16}$$

O means time ordering.

$dK_n(t)/dt$ can, by letting $\tau_i = t_{i-1}-t_i$, be rewritten as

$$dK_n(t)/dt = i^n \int_o^t d\tau_1 \int_o^{t-\tau_1} d\tau_2 \ldots \int_o^{t-\sum_1^{n-2}\tau_i} d\tau_{n-1}$$

$$\times \langle OH_1^{Ix}(t)H_1^{Ix}(t-\tau_1)\ldots H_1^{Ix}(t-\sum_1^{n-1}\tau_j)\rangle_c \tag{V.17}$$

The cumulant average is zero, if at least one of the $H_1^{Ix}(t_m)$ is uncorrelated with the other ones. This lack of correlation occurs for $\tau_i > \tau_c$, where τ_c is the correlation time. For $t \gg \tau_c$ we can therefore, extend the upper limit in (V.17) to infinity without introduction of any serious error. Assuming that the average is of the form

$$\langle H_1^{Ix}(t)H_1^{Ix}(t-\tau)\rangle = \langle |H_1^{Ix}(t)|^2\rangle \exp(-|\tau|/\tau_c) \tag{V.18}$$

infers that $dK_n(t)/dt$ for $t \gg \tau_c$ may be written as

$$dK_n(t \gg \tau_c)/dt \simeq dK_n(t\to\infty)/dt \equiv \exp(-i\Omega t)R^{(n)*} \tag{V.19}$$

R^* is the complex conjugate of the familiar relaxation matrix. The successive terms in (V.17) differs by a factor of order $|H_1^{Ix}(t)|\tau_c$.

Therefore, we shall presumably get a convergent expansion for $|H_1^{Ix}(t)|\tau_c<1$.

The cumulants may also be determined at finite times.

Assume that $\left[H_1(t),H_o\right] = 0$

so that $$H_1^{Ix}(t) = H_1^x(t).$$

Furthermore, assume that $\langle H_1(t)\rangle = 0$ and that (V.18) is valid. The differentiated second order cumulant is then

$$dK_2(t)/dt = -\langle|H_1^x(t)|^2\rangle \int_o^t \exp(-|\tau|/\tau_c)d\tau \tag{V.20}$$

For the secular perturbation one has

$$\left[H_1^x(t)\right]_{\alpha\alpha'\beta\beta'} = \left[H_1(t)_{\alpha\alpha} - H_1(t)_{\alpha'\alpha'}\right] \delta_{\alpha\beta}\delta_{\alpha'\beta'}$$

so that

$$(d/dt)S^I_{x_{\alpha\alpha'}} = \sum_{\beta\beta'} (d/dt)K_2(t)_{\alpha\alpha'\beta\beta'}S^I_{x_{\beta\beta'}}$$

$$= -\langle|H_1(t)_{\alpha\alpha} - H_1(t)_{\alpha'\alpha'}|^2\rangle\tau_c\left[1-\exp(-|t|/\tau_c)\right]S^I_{x_{\alpha\alpha'}}$$

$$\equiv - \Delta^2\tau_c\left[1-\exp(-|t|/\tau_c)\right]S^I_{x_{\alpha\alpha'}}$$

We then derive

$$S^1_{x_{\alpha\alpha'}}(t) = \left[\exp K_2(t)\right]_{\alpha\alpha'\alpha\alpha'}S_{x_{\alpha\alpha'}}$$

By using (V.6) we obtain (to second order)

$$I(\omega) = \frac{\exp(\Delta^2\tau_c^2)}{\pi\Delta}\sum_{n=o}^{\infty}\frac{(-1)^n}{n!}(\Delta\tau_c)^{2n}$$

$$\times \frac{(n/\Delta\tau_c) + \Delta\tau_c}{\left[(n/\Delta\tau_c) + \Delta\tau_c\right]^2 + \left[(\omega-\omega_{\alpha\alpha'})^2/\Delta^2\right]} \tag{V.21}$$

For $\Delta^2\tau_c^2 \ll 1$ (V.21) becomes

$$I(\omega) \simeq \frac{\Delta^2\tau_c}{\pi}\left\{\left[(\Delta^2\tau_c)^2 + (\omega-\omega_o)^2\right]^{-1}\right.$$

$$\left. -\left[\tau_c^{-2} + (\omega-\omega_o)^2\right]^{-1}\right. \tag{V.22}$$

At resonance, the amplitude for the new Lorentzian line is $\Delta^4\tau_c^4$ times that of the main line. After integration of (V.22) we find that the ratio of these intensities is $\Delta^2\tau_c^2$, meaning that the new line is a second order correction to the main line.

If we include $K_3(t)$ assuming $m_3 = \Delta^3\exp\left[-(|\tau_1|+|\tau_2|)/\tau_c\right]$ we find a shift of the correction line by Δ.

V.4. CONCLUSION

In the above we have given a method to obtain a solution to the equation of motion for the spin angular momentum or the spin density operator, following

$$d\sigma/dt = -iH_1^\times(t)\sigma(t).$$

The solution has been found for finite times,and the calculation may be extended to higher orders.

REFERENCES

1. J. Freed, J.Chem.Phys. 49, 376 (1968)
2. R. Kubo, J.Phys.Soc. Japan 17, 1100 (1962)
3. R. Kubo, A Stochastic Theory of Line-Shape and Relaxation. Chapter in D. ter Haar, Editor, Fluctuation, Relaxation and Resonance in Magnetic Systems. Oliver and Boyd, London 1961.

LINEAR RESPONSE THEORY AND SPIN RELAXATION

Irwin Oppenheim

Massachusetts Institute of Technology

Cambridge, Massachusetts 02139

VI.1. INTRODUCTION

Linear response theory is a powerful tool for interpreting and predicting the results of macroscopic experiments which measure transport and relaxation properties of non-equilibrium systems. The application of linear response theory proceeds in several steps, each of which yields more useful and detailed information concerning the time dependence of the non-equilibrium system.

In the first step the response of the system to an externally applied time dependent force is obtained in the form

$$a(t) = \int_{-\infty}^{t} \phi(t-\tau)F(\tau)d\tau \tag{1}$$

where a(t) is the value of some property of the system at time t, $\phi(t-\tau)$ is the response function of the system which is independent of F and which can be written in terms of equilibrium correlation functions, and $F(\tau)$ is the external force at time τ. The linear relation between $\underline{a}$ and F depends on the fact that F is not too large so that the system is not too far from equilibrium. It also depends crucially on the choice of the variable $\underline{a}$. If the variable $\underline{a}$ depends sensitively on the dynamics of the N body system, Eq. (1) will, in general, not be valid no matter how small F is. We shall restrict our attention to $\underline{a}$ variables which vary slowly in space and time compared to molecular variables. We shall call $\underline{a}$ a macroscopic variable; it will be the average of some dynamical variable of the system.

Eq. (1) is useful because it relates the properties of non-equilibrium systems to the fluctuations in an equilibrium system described by the function ϕ. Thus the results of two types of macroscopic experiments are related.

Experiments of type 1 measure macroscopic quantities in perturbed systems which are close to equilibrium. Examples of this type of measurement are: mutual diffusion; nuclear magnetic resonance; electron spin resonance; ultrasonics; and conductivity measurements. In all of these experiments the system is perturbed either by an external field or agency or by setting up spatial gradients. Experiments of type 2 measure fluctuations in systems which are essentially at equilibrium. Examples of this type of measurement are: neutron scattering, light scattering; self diffusion using nuclear magnetic resonance; and line broadening. In these experiments, the measuring devices do not remove the system from its equilibrium state for all practical purposes.

In the second step of the application of linear response theory the value of the macroscopic variable at time t is related to its value at some previous time, t = 0, when the force $F(\tau)$ has the special time dependence

$$\begin{aligned} F(\tau) &= F\, e^{\varepsilon\tau} \qquad && \tau \leq 0 \qquad \varepsilon \to 0+ \\ F(\tau) &= 0 && \tau > 0 \end{aligned} \tag{2}$$

Under these conditions, we can write

$$a(t) = R(t)\, a(0) \tag{3}$$

where R(t) is a relaxation function which describes the decay of the macroscopic variable from its initial value a(0) to its equilibrium value a(∞). We shall choose variables $\underline{a}$ which have the property that they are zero in equilibrium. The quantity R(t) can also be written in terms of equilibrium time-dependent correlation functions. It is easy to obtain the equation for the time dependence of a(t) from Eq.(3) in the form

$$\dot{a}(t) = \dot{R}(t)\, R^{-1}(t)\, a(t) \tag{4}$$

Eq.(4) is a linear equation describing the time evolution of a(t).

We can now proceed in two ways. We can use the well-known phenomenological equations for macroscopic properties, such as the Bloch equations or the Navier Stokes equations, to obtain information concerning the behavior of the equilibrium correlation functions that appear in R(t). Or we can use our knowledge of equilibrium correlation functions to derive the macroscopic equations. We shall follow the latter procedure here.

All of the linearized macroscopic equations describing the evolution of macroscopic variables on time scales which are of interest can be written in the form

$$\dot{\underset{\sim}{a}}(t) = -\underset{\approx}{M}\cdot\underset{\sim}{a}(t) \tag{5}$$

where $\underset{\sim}{a}(t)$ is a vector whose components are the pertinent set of macroscopic variables under consideration and M is a time independent matrix of coefficients. Eq.(5) is simpler than Eq.(4) since the coefficients are time independent. This simplicity is obtained at the expense of considering a set of macroscopic variables rather than a single variable. Examples of pertinent sets of variables are the number density, the momentum density and the energy density for the hydrodynamic equations or the x,y, and z components of the total magnetization for the Bloch equations. Eq.(5) will be valid if the pertinent set of macroscopic variables relaxes much more slowly than all other dynamic variables in the system. Under these conditions, the relaxation matrix has the form

$$\underset{\approx}{R}(t) = \underset{\approx}{e}^{-Mt} \tag{6}$$

and

$$\underset{\sim}{a}(t) = \underset{\approx}{e}^{-Mt}\,\underset{\sim}{a}(0) \tag{7}$$

The time dependence of each macroscopic variable is then a sum of exponentials. We emphasize that in order to obtain Eq.(5) we must know the pertinent set of macroscopic variables. This set can be obtained either from phenomenological considerations or from detailed calculations of R(t).

In the next section, we shall illustrate the general considerations presented here by a detailed description of spin relaxation.

VI.2. SPIN RELAXATION

In this section we consider the application of the linear response formalism to spin relaxation. We wish to compute the linear response of the components of the total magnetization of the system to a small time-varying external magnetic field. We assume that all the spins in the system are identical and take the gyromagnetic ratio to be unity. The Hamiltonian for the system is

$$H_T(t) = H - M_\alpha H_\alpha(t) \tag{8}$$

where H is the Hamiltonian for the system in the absence of the small time varying uniform magnetic field $\underset{\sim}{H}(t)$ with components $H_\alpha(t)$. It has the form

$$H = H_z + H_\ell + \lambda H_1 \tag{9}$$

where H_z is the Hamiltonian for the spins in the presence of a uniform static magnetic field in the z direction, H_ℓ is the lattice Hamiltonian, and H_1 describes the interactions between the spins and the lattice which are characterized by a coupling constant λ. The quantity M_α is the αth component of the dynamical variable whose average is the αth component of the magnetization of the system. It is given by

$$M_\alpha = h \sum_i S_{i\alpha} \tag{10}$$

where $S_{i\alpha}$ is the αth component of the spin operator for the ith particle. In Eq. (8) and in the following there is an implied summation over repeated Greek indices.

We assume that at $t = -\infty$, the system is in equilibrium with respect to the Hamiltonian H and that the time varying field is turned on slowly.

The γth component of the macroscopic magnetization at time t, $m_\gamma(t)$ is given by

$$m_\gamma(t) = \mathrm{Tr}\{[\rho(t)]^T M_\gamma\} \tag{11}$$

where $[\rho(t)]^T$ the density matrix for the system is

$$[\rho(t)]^T = U(t)[\rho(-\infty)]^T U^\dagger(t) \tag{12}$$

$$U(t) = 1 - i/\hbar \int_{-\infty}^{t} H_T(\tau)U(\tau)d\tau \tag{13}$$

and

$$[\rho(-\infty)]^T = \rho_{eq} = e^{-\beta H}/\mathrm{Tr}\{e^{-\beta H}\} \tag{14}$$

Substitution of Eqs. (12) - (14) into Eq. (11) and expansion in powers of $\underset{\sim}{H}(t)$ yields to first order

$$\Delta m_\gamma(t) = \frac{1}{i\hbar} \int_{-\infty}^{t} \mathrm{Tr}\{[\rho_{eq}, M_\alpha(\tau - t)]M_\gamma\}H_\alpha(\tau)d\tau \tag{15}$$

where

$$\Delta m_\gamma(t) \equiv m_\gamma(t) - \langle M_\gamma \rangle \tag{16}$$

is the deviation of the γth component of the magnetization from its equilibrium value. The bracket notation implies an average with respect to ρ_{eq}, i.e.

$$\langle A \rangle \equiv \mathrm{Tr}\{\rho_{eq} A\} \tag{17}$$

and

$$M_\alpha(t) = e^{i/\hbar Ht} M_\alpha e^{-i/\hbar Ht}$$

Eq.(15) can now be written in the form of Eq.(1) by defining the response function

$$\phi_{\gamma\alpha}(t-\tau) \equiv \frac{1}{i\hbar} Tr\{[\rho_{eq}, M_\alpha(\tau-t)]M_\gamma\} \tag{18}$$

Substitution of Eq.(18) into Eq.(15) yields

$$\Delta m_\gamma(t) = \int_{-\infty}^{t} \phi_{\gamma\alpha}(t-\tau)H_\alpha(\tau)d\tau \tag{19}$$

The quantity $\phi_{\gamma\alpha}(t-\tau)$ describes the effect of the applied field at time τ, $H_\alpha(\tau)$, on the magnetization at time $t \geq \tau$, $m_\gamma(t)$. Eq. (18) can be written in a variety of useful ways for our future developments. Since Tr{AB} = Tr{BA}, Eq. (18) can be written

$$\phi_{\gamma\alpha}(t-\tau) = \frac{1}{i\hbar} <[M_\alpha(\tau-t), M_\gamma]> \tag{20}$$

$$= \frac{1}{i\hbar} <[M_\alpha, M_\gamma(t-\tau)]>$$

where we have used the translational time invariance of the equilibrium average for the second equality. Using an identity due to Kubo[1], Eq. (20) can be rewritten

$$\begin{aligned} \phi_{\gamma\alpha}(t-\tau) &= \int_0^\beta ds <\dot{M}_\alpha(-i\hbar s)M_\gamma(t-\tau)> \\ &= -\int_0^\beta ds <M_\alpha(-i\hbar s)\dot{M}_\gamma(t-\tau)> \end{aligned} \tag{21}$$

where

$$M_\alpha(-i\hbar s) \equiv e^{+sH} M_\alpha e^{-sH} \tag{22}$$

It is reasonable to assume that the response function approaches zero as $t-\tau \to \infty$ since the value of H_α in the infinite past should have no effect on the value of the magnetization at the present. Under this assumption, we take the Fourier transform of Eq. (19) from $-\infty$ to $+\infty$ to obtain

$$\Delta\hat{m}_\gamma(\omega) = \chi_{\gamma\alpha}(\omega)\hat{H}_\alpha(\omega) \tag{23}$$

where

$$\chi_{\gamma\alpha}(\omega) \equiv \int_0^\infty e^{i\omega t} \phi_{\gamma\alpha}(t) dt \tag{24}$$

is the susceptibility with a real part (dispersive) related to the
line shape and an imaginary part (absorptive) which is related to
the power absorbed by the sample.

Next we derive the analog of Eq.(3) when the external force
has the form of Eq.(2). We consider the special case where $H_\alpha(t)$
has the time dependence

$$H_\alpha(t) = H_\alpha e^{\varepsilon t} \qquad\qquad t \leq 0$$
$$\tag{25}$$
$$H_\alpha(t) = 0 \qquad\qquad t > 0$$

where ε is a small positive constant which approaches zero. Eq.(25)
corresponds to a step function disturbance where the magnetic field
$\underset{\sim}{H}(t)$ is finite until $t = 0$ and is then suddenly turned off. In this
situation we are interested in the relaxation behavior of $m_\gamma(t)$ for
positive times. We need compute only

$$\Delta \hat{m}_\gamma(\omega)_+ \equiv \int_0^\infty e^{i\omega t} \Delta m_\gamma(t) dt \tag{26}$$

since

$$\Delta m_\gamma(t) = \frac{1}{2\pi} \int_{-\infty}^\infty e^{-i\omega t} \Delta \hat{m}_\gamma(\omega)_+ d\omega \tag{27}$$

$$t > 0.$$

Eq.(27) follows from the fact that $\phi_{\gamma\alpha}(t) \to 0$ as $t \to \infty$ so that $\chi_{\gamma\alpha}(\omega)$
is analytic in the upper half of the complex ω plane. It is easy to
compute from Eqs.(26), (25) and (19) that

$$\Delta \hat{m}_\gamma(\omega)_+ = \lim_{\varepsilon \to 0+} \frac{\chi_{\gamma\alpha}(\omega) - \chi_{\gamma\alpha}(i\varepsilon)}{i\omega + \varepsilon} H_\alpha \tag{28}$$

$$\equiv \sigma_{\gamma\alpha}(\omega) H_\alpha$$

where $\sigma_{\gamma\alpha}(\omega)$ is defined by Eq.(28).

While we could proceed directly to the computation of $\Delta m_\gamma(t)$ for the step function disturbance using Eqs.(27) and (28) it is simpler and more instructive to return to Eq.(19) substitute Eq.(25) and obtain

$$\Delta m_\gamma(t) = \Phi_{\gamma\alpha}(t)H_\alpha \qquad t\geq 0 \tag{29}$$

where the relaxation function $\Phi_{\gamma\alpha}(t)$ is defined by

$$\Phi_{\gamma\alpha}(t) \equiv \int_t^\infty \phi_{\gamma\alpha}(\tau)d\tau \tag{30}$$

The analog of Eq.(3) follows directly in the form

$$\Delta \underset{\sim}{m}(t) = \underset{\approx}{\Phi}(t)\underset{\approx}{\Phi}^{-1}(0)\Delta\underset{\sim}{m}(0) \tag{31}$$

where $\Delta\underset{\sim}{m}(t)$ is a column vector with components $\Delta m_\alpha(t)$, $\underset{\approx}{\Phi}(t)$ is a matrix with elements $\Phi_{\gamma\alpha}(t)$ and $\underset{\approx}{\Phi}^{-1}(0)$ is the inverse matrix of $\underset{\approx}{\Phi}(0)$ with elements $\Phi_{\gamma\alpha}(0)$. In Fourier space, Eq.(31) becomes

$$\Delta\hat{\underset{\sim}{m}}(\omega)_+ = \hat{\underset{\approx}{\Phi}}(\omega)\underset{\approx}{\Phi}^{-1}(0)\Delta\underset{\sim}{m}(0) \tag{32}$$

Eqs.(31) and (32) are the fundamental equations for our future developments.

In the remainder of this section we shall obtain explicit forms for $\Phi_{\gamma\alpha}(t)$ and will derive the quantum-mechanical fluctuation-dissipation theorem. It follows from Eqs.(30) and (21) that

$$\Phi_{\gamma\alpha}(t) = \int_0^\beta ds[\langle M_\alpha(-i\hbar s)M_\gamma(t)\rangle - \lim_{t\to\infty}\langle M_\alpha(-i\hbar s)M_\gamma(t)\rangle] \tag{33}$$

$$= \int_0^\beta ds\langle \Delta M_\alpha(-i\hbar s)\Delta M_\gamma(t)\rangle \tag{34}$$

where

$$\Delta M_\gamma(t) = M_\gamma(t) - \langle M_\gamma(t)\rangle = M_\gamma(t) - \langle M_\gamma\rangle \tag{35}$$

where we have assumed that correlations are lost as $t \to \infty$ so that the second term on the right-hand side of Eq.(33) can be replaced by

$\langle M_\alpha\rangle\langle M_\gamma\rangle$. The result for $\Phi_{\gamma\alpha}(0)$ is of course

$$\Phi_{\gamma\alpha}(0) = \int_0^\beta ds \, \langle \Delta M_\alpha(-i\hbar s)\Delta M_\gamma\rangle \tag{36}$$

and for

$$\hat{\Phi}_{\gamma\alpha}(\omega) \equiv \int_0^\beta ds \int_0^\infty e^{i\omega t} dt \langle \Delta M_\alpha(-i\hbar s)\Delta M_\gamma(t)\rangle \tag{37}$$

$$= \int_0^\beta ds \, \langle \Delta M_\alpha(-i\hbar s)\Delta\hat{M}_\gamma(\omega)\rangle$$

where $\Delta\hat{M}_\gamma(\omega)$ is defined by Eq.(37).

Eqs.(31) - (37) contain the extremely interesting result that the relaxation of the macroscopic variables in a non-equilibrium system has the same time dependence as the regression of fluctuations in an equilibrium system. This is the quantum mechanical generalization of Onsager's hypothesis.

An even simpler way of deriving Eq.(31) uses the fact that the system subjected to an external force of the type given in Eq.(25) will be in equilibrium at t = 0 with respect to the Hamiltonian $H_T = H - M_\alpha H_\alpha$. Thus

$$m_\gamma(0) = Tr\{\rho(0)M_\gamma\} \tag{38}$$

and

$$m_\gamma(t) = Tr\{\rho(0)M_\gamma(t)\} \tag{39}$$

where

$$\rho(0) = e^{-\beta H_T}/Tr\{e^{-\beta H_T}\} \tag{40}$$

and

$$M_\gamma(t) = e^{i/\hbar Ht} M_\gamma \, e^{-i/\hbar Ht} \tag{41}$$

Expansion of $\rho(0)$ in Eqs.(38) and (39) to linear order in H_α immediately yields Eqs.(29), (31), and (34).

A more formal way of expressing this relationship is in terms of the quantum-mechanical fluctuation-dissipation theorem[2]. The relationship between the relaxation function $\Phi_{\gamma\alpha}(t)$ and the correlation function

$$g_{\gamma\alpha}(t) \equiv 1/2\langle[\Delta M_\gamma(t),\Delta M_\alpha]_+\rangle \tag{42}$$

is simply

$$\Phi_{\gamma\alpha}(t) = \int_{-\infty}^{\infty} B(t-\tau)g_{\gamma\alpha}(\tau)d\tau \tag{43}$$

where

$$B(t) = \frac{2}{\pi h}\log[\coth(\frac{\pi|t|}{2\beta\hbar})] \tag{44}$$

In the classical limit ($h \to 0$), $B(t) \to \beta\delta(t)$ and Eq. (43) becomes

$$\Phi_{\gamma\alpha}(t) = \beta g_{\gamma\alpha}(t) \tag{45}$$

This result is valid as long as the splitting between energy levels in the system is small compared to kT. This is almost always true for spin systems.

VI.3. DERIVATION OF THE BLOCH EQUATIONS

In this section we shall demonstrate how to reduce Eq. (31) to the analog of Eqs. (5) and (7).[3] We shall derive the Bloch equations which describe the time dependence of the magnetization of the system. In order to do this we shall study the properties of $\Phi_{\gamma\alpha}(\omega)$. If we can show that

$$\hat{\underset{\approx}{\Phi}}(\omega) = [-i\omega\Pi + \underset{\approx}{L}]^{-1}\underset{\approx}{\Phi}(0) \tag{46}$$

where Π is the unit tensor and where $\underset{\approx}{L}$ is independent of ω, then

$$\underset{\approx}{\Phi}(t) = \underset{\approx}{e}^{-Lt}\underset{\approx}{\Phi}(0) \tag{47}$$

and the Bloch equations will follow.

We start by defining the matrix $\underset{\approx}{L}(\omega)$ by the equation

$$\hat{\underset{\approx}{\Phi}}(\omega) \equiv [-i\omega\Pi + \underset{\approx}{L}(\omega)]^{-1}\underset{\approx}{\Phi}(0) \tag{48}$$

No assumption has been made in writing Eq. (48) since it is merely the definition of $\underset{\approx}{L}(\omega)$. The matrix $\underset{\approx}{L}(\omega)$ has 9 elements and in order to simplify its form we shall first investigate the symmetry properties of $\underset{\approx}{\Phi}$. Then we shall investigate the time dependences of the operators appearing in $\underset{\approx}{\Phi}$ to achieve further simplifications.

We recall that the time dependences and the density matrix needed to compute the average in Eq. (37) involve the Hamiltonian H of Eq. (9). Thus it is convenient to choose the components of $\underset{\sim}{M}$ as

$$M_o \equiv M_z$$

$$M_1 \equiv M_+ = M_x + iM_y \tag{49}$$

$$M_{-1} \equiv M_- = M_x - iM_y$$

and to denote these components by M_k, k = 0, +1, and -1. We consider the unitary rotation operators[2]

$$D(a,b,c) = e^{iaJ_x}\, e^{ibJ_y}\, e^{icJ_z} \tag{50}$$

where $J_\alpha = S_\alpha + L_\alpha, \alpha = x,y,z$ and S_α is the net α component of the spins and L_α is the net α component of the angular momentum of the internuclear axes in the system. The effects of the rotation operators on M_k are readily obtained as:

$$D(0,0,\pi)\, M_k D^{-1}(0,0,\pi) = M_k \cos(\pi k) \qquad k=0,\pm 1$$

$$D(0,0,\pi/2)\, M_k D^{-1}(0,0,\pi/2) = iM_k \sin(\tfrac{\pi}{2}k) \quad k=\pm 1$$

$$D(0,0,\pi/2)\, M_o D^{-1}(0,0,\pi/2) = M_o$$

$$D(0,\pi,0)\, M_o D^{-1}(0,\pi,0) = -M_o \tag{51}$$

and

$$D(0,\pi,0)\, M_k D^{-1}(0,\pi,0) = -M_{-k} \qquad k=\pm 1$$

The effects of these rotation operators on the Hamiltonian H are

$$D\, H\, D^{-1} = H \tag{52}$$

if it is understood that the operator $D(0,\pi,0)$ includes a reversal of the direction of the magnetic field in H_z. Application of these rotation operators to the elements of the $\underset{\approx}{\Phi}$ matrix yields the results that

$$\Phi_{ok} = \Phi_{ko} = \Phi_{kk} = 0 \qquad k = \pm 1$$

$$\Phi_{k-k,H_z} = \Phi_{-kk,-H_z} \tag{53}$$

where the subscript H_z indicates that the Zeeman field is parallel to z while the subscript $-H_z$ indicates that the Zeeman field is antiparallel to z. Use of Eq.(53) reduces the matrix equation (48) considerably. The result is

$$\hat{\Phi}_{oo}(\omega) = [-i\omega + L_{oo}(\omega)]^{-1} \quad \Phi_{oo}(0) \tag{54a}$$

$$\hat{\Phi}_{1-1}(\omega) = [-i\omega + L_{11}(\omega)]^{-1} \quad \Phi_{1-1}(0) \tag{54b}$$

$$\hat{\Phi}_{-11}(\omega) = [-i\omega + L_{-1-1}(\omega)]^{-1} \quad \Phi_{-11}(0) \tag{54c}$$

which will be used to determine the elements of $\underset{\approx}{L}(\omega)$. It is clear from Eqs.(53), (54b) and (54c) that $L_{11,H_z} = L_{-1-1,-H_z}$.

In order to simplify these equations further, we consider the commutator of the Hamiltonian

$$H_o \equiv H_z + H_\ell = H(\lambda = 0) \tag{55}$$

with M_k,

$$[H_o, M_k] = [H_z, M_k] = -\omega_o [M_o, M_k] \tag{56}$$

where $H_z = -\omega_o M_o$ and ω_o is the Larmor frequency. The results are

$$i/\hbar [H_o, M_k] = -ik\omega_o M_k \qquad k = 0, \pm 1 \tag{57}$$

or, equivalently,

$$M_k^{(o)}(t) \equiv e^{+i/\hbar H_o t} M_k e^{-i/\hbar H_o t} = M_k e^{-ik\omega_o t} \quad k=0,\pm 1 \tag{58}$$

We now define the operator $M_k^R(t)$ in the rotating frame by

$$M_k^R(t) \equiv e^{i/\hbar Ht}\, e^{-i/\hbar H_o t}\, M_k\, e^{i/\hbar H_o t}\, e^{-i/\hbar Ht}$$

$$= e^{ik\omega_o t}\, M_k(t) \tag{59}$$

The operator $M_k^R(t)$ has the convenient property that its time derivative is given by

$$\dot{M}_k^R(t) = i/\hbar\lambda\, e^{i/\hbar Ht}\, [H_1,M_k]\, e^{-i/\hbar Ht}\, e^{ik\omega_o t}$$

$$= i/\hbar\lambda\, [H_1(t),M_k^R(t)] \tag{60}$$

Thus, as $\lambda \to 0$, or equivalently as the spin-lattice interaction tends to zero, $\dot{M}_k^R(t) \to 0$. This fact is of extreme importance in the reduction of Eq.(31) to the analog of Eqs.(5) and (7) since it is only when the macroscopic variables change slowly that these equations are valid.

We now define the magnetization in the rotating frame, $m_k^R(t)$, by

$$m_k^R(t) \equiv e^{ik\omega_o t} m_k(t) \tag{61}$$

The analog of Eq.(32) is

$$\Delta\hat{\underset{\sim}{m}}^R(\omega)_+ = \hat{\underset{\approx}{\Phi}}^R(\omega)\underset{\approx}{\Phi}^{-1}(0)\Delta\underset{\sim}{m}(0) \tag{62}$$

where

$$\Phi_{kk'}^R(\omega) \equiv \int_0^\beta ds \int_0^\infty e^{i\omega t} dt \langle\Delta M_{k'}(-i\hbar s)\Delta M_k^R(t)\rangle$$

$$= \Phi_{kk'}(\omega+k\omega_o) \tag{63}$$

The matrix $\underset{\approx}{L}^R(\omega)$ is defined by the analog of Eq.(48) as

$$\underset{\approx}{\Phi}^R(\omega) \equiv [-i\omega\Pi + \underset{\approx}{L}^R(\omega)]^{-1}\underset{\approx}{\Phi}(0) \tag{64}$$

The symmetry properties in Eq.(53) still apply and Eqs.(54) remain true with a superscript R on $\hat{\Phi}_{kk'}(\omega)$ and on $L_{kk}(\omega)$.

We are now prepared to find an explicit molecular expression for $\underset{\approx}{L}^R$ from Eqs.(54) and (63). Our procedure involves developing $\underset{\approx}{L}^R$ in a power series in λ and finding the coefficients up to order $\tilde{\lambda}^2$. We note immediately from Eqs.(60) and (63) by integrating $\Phi^R_{kk'}(\omega) \to -\frac{\Phi_{kk'}(0)}{i\omega}$ as $\lambda \to 0$. Thus $\underset{\approx}{L}^R(\omega) \to 0$ as $\lambda \to 0$.

We shall illustrate the procedure by finding explicit expressions for $L_{\dot{o}o}(\omega)$. The expressions for $L^R_{11}(\omega)$ and $L^R_{-1-1}(\omega)$ can be found in a completely analogous fashion. We multiply Eq.(54a) by $-i\omega + L_{\dot{o}o}(\omega)$ and use the facts that

$$\Phi_{oo}(\omega) = \langle\langle \Delta M_o \Delta M_o(\omega)\rangle\rangle = -\frac{1}{i\omega}\langle\langle \Delta M_o \Delta M_o\rangle\rangle - \frac{1}{i\omega}\langle\langle \Delta M_o \Delta\dot{M}_o(\omega)\rangle\rangle \quad (65)$$

$$\langle\langle \Delta M_o \Delta\dot{M}_o(\omega)\rangle\rangle = -\frac{1}{i\omega}\langle\langle \Delta M_o \Delta\dot{M}_o\rangle\rangle + \frac{1}{i\omega}\langle\langle \Delta\dot{M}_o \Delta\dot{M}_o(\omega)\rangle\rangle \quad (66)$$

where the double bracket notation is defined by

$$\langle\langle AB\rangle\rangle \equiv \int_o^\beta ds \, \langle A(-i\hbar s)B\rangle \quad (67)$$

and we have integrated by parts with respect to t. The results are, after using Eq.(65),

$$\langle\langle \Delta M_o \Delta\dot{M}_o(\omega)\rangle\rangle + L_{oo}(\omega)\langle\langle \Delta M_o \Delta M_o(\omega)\rangle\rangle = 0 \quad (68)$$

and, after using Eqs.(65) and (66),

$$L_{oo}(\omega)\langle\langle \Delta M_o \Delta M_o\rangle\rangle = -L_{oo}(\omega)\langle\langle \Delta M_o \Delta\dot{M}_o(\omega)\rangle\rangle + \langle\langle \Delta\dot{M}_o \Delta\dot{M}_o(\omega)\rangle\rangle \quad (69)$$

In writing Eq.(69) we have used the fact that

$$\langle\langle \Delta M_o \Delta\dot{M}_o\rangle\rangle = \langle[\Delta M_o, \Delta M_o]\rangle = 0 \quad (70)$$

We now assume that the double bracketed quantities and $L_{oo}(\omega)$ can be expanded in powers of λ in the form

$$\begin{aligned} \langle\langle AB\rangle\rangle &= \langle\langle AB\rangle\rangle^{(0)} + \lambda\langle\langle AB\rangle\rangle^{(1)} + \lambda^2\langle\langle AB\rangle\rangle^{(2)} + \ldots \\ L_{oo}(\omega) &= L^{(0)}_{oo}(\omega) + \lambda L^{(1)}_{oo}(\omega) + \lambda^2 L^{(2)}_{oo}(\omega) + \ldots \end{aligned} \quad (71)$$

Substitution of Eq.(71) into (69), use of Eq.(60), and collecting coefficients of λ^o, λ and λ^2 yield:

$$L_{oo}^{(0)}(\omega)\langle\langle \Delta M_o \Delta M_o \rangle\rangle^{(0)} = 0 \tag{72}$$

which implies that $L_{oo}^{(0)}(\omega) = 0$,

$$L_{oo}^{(1)}(\omega)\langle\langle \Delta M_o \Delta M_o \rangle\rangle^{(0)} = 0 \tag{73}$$

which implies that $L_{oo}^{(1)}(\omega) = 0$; and

$$L_{oo}^{(2)}(\omega)\langle\langle \Delta M_o \Delta M_o \rangle\rangle^{(0)} = \langle\langle \Delta \dot{M}_o \Delta \dot{M}_o(\omega) \rangle\rangle^{(2)} \tag{74}$$

which implies that

$$L_{oo}^{(2)}(\omega) = \frac{\langle\langle \Delta \dot{M}_o \Delta \dot{M}_o(\omega) \rangle\rangle^{(2)}}{\langle\langle \Delta M_o \Delta M_o \rangle\rangle^{(0)}} \tag{75}$$

The explicit forms of the correlation functions appearing in Eq.(75) are easily obtained. Making use of the fact that H_o (Eq.55) commutes with ΔM_o we can write

$$\langle\langle \Delta M_o \Delta M_o \rangle\rangle^{(0)} = \int_o^\beta ds \frac{Tr\{e^{-\beta H_o} \Delta M_o \Delta M_o\}}{Tr\{e^{-\beta H_o}\}} \tag{76}$$

$$= \beta \langle \Delta M_o^2 \rangle_o$$

where the subscript $_o$ implies that the average is to be taken over a canonical ensemble with Hamiltonian H_o. From the definition, we can write

$$\langle\langle \Delta \dot{M}_o \Delta \dot{M}_o(\omega) \rangle\rangle =$$

$$\int_o^\beta ds \int_o^\infty e^{i\omega t} dt (\tfrac{i}{\hbar}\lambda)^2 \langle e^{sH}[H_1, \Delta M_o] e^{-sH} e^{i/\hbar Ht} [H_1, \Delta M_o] e^{-i/\hbar Ht} \rangle$$

$$\equiv (\tfrac{i}{\hbar}\lambda)^2 \langle\langle [H_1, \Delta M_o][H_1, \Delta M_o](\omega) \rangle\rangle \tag{77}$$

Thus

$$<<\Delta\dot{M}_o\Delta\dot{M}_o(\omega)>>^{(2)} = (\frac{i}{\hbar})^2<<[H_1,\Delta M_o][H_1,\Delta M_o](\omega)>>^{(0)}$$

$$= (\frac{i}{\hbar})^2 \int_o^\beta ds \int_o^\infty e^{i\omega t} dt<[H_1(-i\hbar s),\Delta M_o][H_1(t),\Delta M_o]>_o \quad (78)$$

Here the subscript $_o$ implies that the average is to be taken over a canonical ensemble with Hamiltonian H_o and that the time dependences of $H_1(-ihs)$ and $H_1(t)$ are also to be computed using the Hamiltonian H_o. In the classical limit where $\beta h\omega_o<<1$, Eq.(78) becomes

$$<<\Delta\dot{M}_o\Delta\dot{M}_o(\omega)>>^{(2)} = (\frac{i}{\hbar})^2\beta<[H_1,\Delta M_o][H_1(\omega),\Delta M_o]>_o \quad (79)$$

We shall now investigate the frequency dependence of $L_{oo}^{(2)}(\omega)$. It follows from symmetry that

$$<<\Delta M_{\pm 1}\Delta\dot{M}_o(\omega)>> = 0 \quad (80)$$

and from Eqs.(68), (72) and (73)

$$<<\Delta M_o\Delta\dot{M}_o(\omega)>>^{(1)} = 0 \quad (81)$$

Eq.(78) can be rewritten

$$<<\Delta\dot{M}_o\Delta\dot{M}_o(\omega >>^{(2)} = \frac{i}{\hbar} <<[H_1,\Delta M_o]\Delta\dot{M}_o(\omega)>>^{(1)} \quad (82)$$

Thus, to the appropriate order in λ, $\Delta\dot{M}_o(\omega)$ is orthogonal to the set of variables ΔM_k, $k = 0, \pm 1$. If these are the only pertinent slowly varying quantities in the system, the correlation function in Eq.(78) will go to zero on a short time scale which is independent of λ.

The pertinent time scale for the decay of $<<\Delta M_o\Delta M_o(t)>>$ or for $\Delta m_o(t)$ is proportional to λ^{-2} and the pertinent frequencies for describing this decay are proportional to λ^2. Under these conditions, the frequency dependence in Eq.(78) can be neglected, $L_{oo}^{(2)}$ becomes independent of frequency and we can rewrite Eqs.(54a) and (32) as

$$<<\Delta M_o\Delta M_o(\omega)>> = [-i\omega + \frac{1}{T_1}]^{-1}<<\Delta M_o\Delta M_o>> \quad (83)$$

and

$$\Delta\hat{m}_o(\omega)_+ = [-i\omega + \frac{1}{T_1}]^{-1}\Delta m_o(0) \tag{84}$$

where the spin lattice relaxation time T_1 is defined by

$$\frac{1}{T_1} \equiv \lambda^2 L_{oo}^{(2)}(0) = \frac{(\frac{i}{\hbar})^2 \lambda^2}{\beta\langle\Delta M_o^2\rangle_o} \int_o^\beta ds \int_o^\infty dt \langle[H_1(-i\hbar s),\Delta M_o][H_1(t),\Delta M_o]\rangle_o \tag{85}$$

Eq.(85) is the desired molecular expression for the transport coefficient $\frac{1}{T_1}$. The equation for the time dependence of $\Delta m_o(t)$ follows immediately from Eq.(84) in the form

$$\Delta\dot{m}_o(t) = -\frac{1}{T_1}\Delta m_o(t) \tag{86}$$

The Bloch equations for $\Delta m_\pm(t)$ can be derived in a completely analogous fashion.

We shall conclude this section by making some remarks concerning Eq.(85). It is immediately obvious from Eqs.(20), (21) and (30) that $\frac{1}{T_1}$ is real. More explicit forms for $\frac{1}{T_1}$ can be obtained for any desired system once H_1 is known. The correlation function in Eq.(85) can then be broken up into a sum of products of lattice correlation functions and spin correlation functions. We emphasize that the frequency independent form of Eq.(85) is valid only when the only slowly varying quantities in the system are ΔM_k. If there is another slowly varying quantity A for which $\langle\langle A\Delta\dot{M}_o(\omega)\rangle\rangle^{(1)} \neq 0$ $L_{oo}^{(2)}$ may be frequency dependent. In this case, in order to obtain linear macroscopic equations with frequency (or time) independent coefficients we would have to include A in our initial set of variables.

VI.4. CONCLUDING REMARKS

The procedures developed here can be used to derive all linear macroscopic equations. Thus, for example, the linearized Navier-Stokes equations of hydrodynamics can be obtained. The techniques are powerful enough to generalize these equations in a variety of ways. We could, for example, extend the Bloch equations to higher orders in λ if that were necessary. We can also include more variables than are usually considered. The usual hydrodynamic variables, the number density, the momentum density and the energy density, will have to be supplemented if there is slow exchange between the translational degrees of freedom and the vibrational or rotational molecular motions[4].

Finally, the procedure utilized here can be extended to include non-linear response[5]. This is necessary in order to obtain, for example, the non-linear hydrodynamic equations. The extension to non-linear response is also essential to check the range of validity of the linear response theory.

VI.5. ACKNOWLEDGEMENT

This work was supported in part by a grant from the National Science Foundation.

REFERENCES

1) R. Kubo, J. Phys. Soc. Japan 12, 570 (1957); see also
R. Kubo, "Some Aspects of the Statistical-Mechanical Theory of Irreversible Processes" in W.E. Brittin and L.G. Dunham, "Lectures in Theoretical Physics" Vol.I, Interscience Publishers, New York, 1959.

2) J.M. Deutch and I. Oppenheim, Adv. Mag. Resonance 3, 43 (1968).

3) The techniques described below have been utilized to derive the hydrodynamic equations by:

a) L.P. Kadanoff and P.C. Martin, Ann. Phys. 24, 419 (1963).
b) B.U. Felderhof and I. Oppenheim, Physica 31, 1441 (1965).
c) P.A. Selwyn and I. Oppenheim, Physica 54, 161 (1971).

4) I. Oppenheim and M. Weinberg, unpublished.

5) J. Weare and I. Oppenheim, unpublished.

TWO APPROACHES TO THE THEORY OF SPIN RELAXATION: I. THE REDFIELD LANGEVIN EQUATION; II. THE MULTIPLE TIME SCALE METHOD.

J.M. Deutch

Massachusetts Institute of Technology

Cambridge, Massachusetts 02139

VII.1. INTRODUCTION

Spin relaxation is one example of a general class of relaxation processes where the subsystem of interest is weakly coupled to a large bath or reservoir. The essential feature of such processes is the wide separation in the characteristic time scales of relaxation of the subsystem and the bath. If the coupling of the spin subsystem to the bath is 'weak' one finds that the spins relax on a 'slow' time scale relative to the lattice motion of the bath. One consequence of this separation of time scales is that the spin subsystem density matrix, to a certain degree of approximation, may be described by a master equation. To lowest order this equation of motion is only valid on the 'slow' time scale and is known as the Redfield equation. Another consequence is the existence of simple linear laws for the relaxation in the magnetization or other dynamical variables of interest. For many spin systems, in lowest order, the equations of motion for the macroscopic magnetization have the form of the Bloch equations which are also only valid on the slow spin time scale. A third consequence of the separation of time scales is that the relaxation times which appear in the Bloch or Redfield equations may be expressed as Fourier transforms of time correlation functions. To lowest order the time dependence of the correlation functions involves only the fast lattice motion.

In these lectures two developments are presented that emphasize the importance of the separation of time scales in spin relaxation. The first topic consists of a molecular derivation of a new equation called the Redfield-Langevin equation. Here a projection operator technique is employed to obtain a fundamental equation that simultaneously leads to the ordinary Redfield and Bloch equations upon suitable averaging. The equation is a microscopic equation which is valid on the slow spin time scale. On this slow time scale the lattice motion is extremely rapid and the effects of the lattice are shown to emerge as a fluctuating 'force' whose interpretation is similar to the fluctuating force appearing in the ordinary Langevin equation of Brownian motion. In the course of this section we will find it profitable to refer to and comment on other chapters in this volume.

The second section presents a derivation of the Redfield equation by the Multiple Time Scale (MTS) method. The MTS method provides an alternative procedure for obtaining relaxation equations for weakly coupled systems. The virtue of the MTS method is that it focuses attention, at the outset, on the existence of different time scales and consequently has advantage in displaying the physics of the relaxation process.

Neither the derivation of the Redfield-Langevin equation nor the MTS derivation is intended to provide a new, practical calculational procedure for ESR in fluids. Rather, we shall attempt to improve our understanding of relaxation processes by emphasizing the crucial consequences of the inherent existence of more than one time scale in the problem. Our development will, however, suggest systematic procedures for generalizing results beyond the lowest order. Possible generalizations and the problems these generalizations raise will be discussed in the course of the lectures.

For clarity, we explicitly consider here the case of a single spin relaxing via an intramolecular mechanism. Intermolecular mechanisms will be ignored which is frequently an adequate approximation in the interpretation of ESR experiments. The correct treatment of the entire N spin system when intermolecular and/or intramolecular relaxation mechanisms are present, is a more difficult task. The Hamiltonian for the system we consider is

$$H = H_o + \lambda H_1 = H_s + H_\ell + \lambda H_1 \tag{1.1}$$

where H_s is the Hamiltonian for the spin in the presence of a static magnetic field, H_ℓ is the Hamiltonian describing the lattice degrees of freedom, and H_1 is the interaction of the spin and the lattice characterized by the coupling constant λ. The Liouville equation describing the time evolution of the system density matrix $\rho(t)$ is $[\hbar = 1]$,

$$\frac{\partial\rho(t)}{\partial t} = -i[H,\rho(t)] \equiv -iL\rho(t) \tag{1.2}$$

where $L \equiv H^\times \equiv i[H,\ldots]$ is the Liouville operator of the system. The spin density matrix is obtained from $\rho(t)$ by taking a trace over lattice variables,

$$\sigma(t) = \mathrm{Tr}_\ell[\rho(t)] \tag{1.3}$$

and consequently satisfies the equation of motion

$$\frac{\partial\sigma(t)}{\partial t} = -i\mathrm{Tr}_\ell[L\rho(t)] \tag{1.4}$$

The equation of motion for an operator A is given by

$$\frac{\partial A(t)}{\partial t} = i[H,A(t)] = iLA(t) \tag{1.5}$$

with a formal solution

$$A(t) = e^{iHt}Ae^{-iHt} \equiv e^{iLt}A(0) \tag{1.6}$$

We refer to Chapter I for a complete discussion of the formal properties of these operators.

In ESR the observable of interest is the magnetization. The definition of the average macroscopic magnetization is

$$M_r(t) = \gamma \mathrm{Tr}\left[S_r(t)\rho(0)\right] = \gamma \mathrm{Tr}\left[S_r \rho(t)\right] \qquad r=x,y,z \tag{1.7}$$

where S_r is the spin operator, γ the gyromagnetic ratio, and the symbol Tr denotes a trace over both lattice and spin degrees of freedom. From Eqs. (1.3) and (1.7) it follows that

$$M_r(t) = \gamma \mathrm{Tr}_s\left[S_r \sigma(t)\right] \tag{1.8}$$

where Tr_s denotes a trace over the spin degrees of freedom. For convenience we shall set $\gamma = 1$. In problems of spin relaxation it is customary to define H_1 so that its average of the equilibrium lattice vanishes, i.e.

$$\langle H_1 \rangle = \mathrm{Tr}_\ell\left[\rho_\ell H_1\right] = 0 \tag{1.9}$$

In the following the angular bracket will denote an average over the equilibrium density matrix for the lattice,

$$\rho_\ell = \left\{\exp(-\beta H_\ell)/\mathrm{Tr}_\ell\left[\exp(-\beta H_\ell)\right]\right\} \tag{1.10}$$

VII.2. THE REDFIELD-LANGEVIN EQUATION[1]

A. Motivation

The Redfield equation[2] for the spin density matrix is an approximate equation of motion for the spin in weak interaction with a lattice. The equation is

$$\frac{\partial \langle\alpha|\sigma(t)|\alpha'\rangle}{\partial t} = -i\omega_{\alpha\alpha'}\langle\alpha|\sigma(t)|\alpha'\rangle + \sum_{\beta,\beta'} R_{\alpha\alpha'\beta\beta'}\langle\beta|\sigma(t)|\beta'\rangle \tag{2.1}$$

where $\omega_{\alpha\alpha'} = (E_\alpha - E_{\alpha'})$ and E_α is the energy of spin state $|\alpha\rangle$,

$$H_s|\alpha\rangle = E_\alpha|\alpha\rangle$$

The spin density matrix is an *averaged* quantity since it is obtained from the complete density matrix of the system by averaging over the lattice degrees of freedom. It is possible to obtain an equation of motion for the magnetization from the Redfield equation according to Eq.(1.8). The resulting equations of motion, which we shall refer to as Bloch equations, are, of course, equations for the *average* magnetization.

Our present purpose is to seek a more fundamental equation for a *microscopic* dynamical variable that leads to the ordinary Redfield and Bloch equations upon suitable averaging. We are led to consider the operator

$$G_{\alpha\alpha'}(t) = \exp(iHt)|\alpha'\rangle\langle\alpha|\exp(-iHt) = e^{iLt}|\alpha'\rangle\langle\alpha| \quad (2.2)$$

which is an operator in both the spin and lattice space of the system. We first wish to show that knowledge of the operator $G_{\alpha\alpha'}(t)$ is all that is required to obtain all quantities of interest in spin relaxation problems.

It follows immediately from the definitions in Eqs.(1.2), (1.4) and (2.2) that

$$\langle\alpha|\sigma(t)|\alpha'\rangle = \mathrm{Tr}[\rho(0)G_{\alpha\alpha'}(t)] \quad (2.3)$$

where $\rho(0)$ is the initial density matrix of the system. Furthermore, $S_r(t)$ may be expressed in terms of $G_{\alpha\alpha'}(t)$ according to

$$S_r(t) = \sum_{\alpha,\alpha'} G_{\alpha\alpha'}(t)\langle\alpha'|S_r|\alpha\rangle \quad (2.4)$$

We shall refer to the operator $S_r(t)$ when written in this way as $\mathcal{S}_r(t)$.

From Eq.(1.7) it follows that the average magnetization may be expressed as

$$M_r(t) = \mathrm{Tr}[\rho(0)\,\mathcal{S}_r(t)] \quad (2.5)$$

Under usual circumstances the initial density matrix may be approximated by

$$\rho(0) = \rho_{\ell}\sigma(0) \tag{2.6}$$

where $\sigma(0)$ is the initial non-equilibrium density matrix for the spin degrees of freedom. In this case Eqs.(2.3) and (2.5) simplify to

$$\langle\alpha|\sigma(t)|\alpha'\rangle = \mathrm{Tr}_s[\sigma(0)\langle G_{\alpha\alpha'}(t)\rangle] \tag{2.7}$$

and

$$M_r(t) = \mathrm{Tr}_s[\sigma(0)\langle \mathcal{S}_r(t)\rangle] \tag{2.8}$$

so that the equation of motion for the spin density matrix is the same as the equation for $\langle G_{\alpha\alpha'}(t)\rangle$ and the equation of motion for the magnetization is the same as the equation for $\langle \mathcal{S}_r(t)\rangle$. While the approximation of a factored initial density matrix, Eq.(2.6), is probably satisfactory for the interpretation of all usual experiments, it should be noted that a careful analysis of the consequences of initial correlations has not been presented for the spin problem.

Line shapes in ESR are directly proportional to the Fourier transforms of equilibrium correlation functions of the type

$$\phi_{rq}(t) = \langle S_r(t)S_q(0)\rangle_{av} \tag{2.9}$$

where the subscript av denotes an equilibrium average over both the lattice and spin degrees of freedom

$$\langle A\rangle_{av} = \mathrm{Tr}[\rho_o A] \tag{2.10}$$

Here

$$\rho_o = \{\exp(-\beta H_o)/\mathrm{Tr}[\exp(-\beta H_o)]\} = \rho_{\ell}\sigma^o \tag{2.11}$$

with σ^o the equilibrium spin density matrix,

$$\sigma^o = \{\exp(-\beta H_s)/\mathrm{Tr}_s[\exp(-\beta H_s)]\} \tag{2.12}$$

Note, once again, that correlations have been neglected in the characterization of the equilibrium system by the replacement of the complete Hamiltonian H by H_o. It follows from Eq.(2.4) that the correlation function $\phi_{rq}(t)$ may be expressed in terms of correlation functions of G,

$$\phi_{rq}(t) = \sum_{\substack{\alpha\alpha' \\ \gamma\gamma'}} [\langle G_{\alpha\alpha'}(t)G_{\gamma\gamma'}(0)\rangle_{av} \langle\alpha'|S_r|\alpha\rangle\langle\gamma'|S_q|\gamma\rangle] \tag{2.13}$$

Thus we see that all quantities of interest in ESR may be obtained once $G_{\alpha\alpha'}(t)$ is known. The equation of motion for $G_{\alpha\alpha'}(t)$ will be a microscopic equation in the sense that it includes the effects of the lattice. We shall assume that the spin system is in weak interaction with the lattice and obtain, by a projection operator method, an equation for $G_{\alpha\alpha'}(t)$ which is a valid description of the slow time scale on which the spins relax. We show that $G_{\alpha\alpha'}(t)$ satisfies an equation of motion of the Redfield form with the effects of the fast lattice motion appearing in an additional rapidly fluctuating term. In addition, we shall obtain an equation of motion for the magnetization operator $\mathcal{S}_r(t)$ directly from the equation of motion for $G_{\alpha\alpha'}(t)$. This equation is similar to the Bloch equation but an added fluctuating term is present which includes the effects of the lattice. When an average is performed over the lattice degrees of freedom the ordinary Redfield and Bloch equations are obtained. The stochastic properties of the fluctuating terms that appear in the equations for $\mathcal{S}_r(t)$ and $G_{\alpha\alpha'}(t)$ will be determined.

The point of our development is that the existence of a simple, linear relaxation equation for a macroscopic variable or the associated density matrix implies that the microscopic equations may also be expressed in a simple way on the slow time scale of the macroscopic variables. The resulting microscopic equations can be given an interpretation in terms of ordinary Brownian motion theory. This situation will hold for a variety of weakly coupled systems and has been discussed by Bixon and Zwanzig[3] in their treatment of the Boltzmann-Langevin equation for dilute gases and by Albers, Oppenheim and Deutch[4] in their treatment of Brownian motion. The trick is to find a dynamical variable which, upon averaging, leads to either an equation for the distribution function, i.e. density matrix, or the transport equations for the macroscopic variables. The analogy between the various situations is summarized below.

	Spins Ref.(1)	Dilute Gas Ref. (3)	Heavy Particle Brownian Motion Ref. (4)
Distribution Function Eq.	Redfield Eq.	Linearized Boltzmann Eq.	Fokker-Planck Eq.
Transport Equations	Bloch Eqs.	Linearized Hydrodynamic Eqs.	Momentum Relaxation
Generating Dynamical Variable	$G_{\alpha\alpha'}(t)$	$\phi(t)=\sum_{j=1}^{N}[\delta(\underset{\sim}{r}_j(t)-\underset{\sim}{r}) \times \delta(\underset{\sim}{v}_j(t)-\underset{\sim}{v})]$	$D(t)=\delta(\underset{\sim}{R}(t)-\hat{\underset{\sim}{R}}) \times \delta(\underset{\sim}{P}(t)-\hat{\underset{\sim}{P}})$
Generating Equation	Redfield-Langevin Eq.	Boltzmann-Langevin Eq.	Random Fokker Planck Eq.
Fluctuating Transport Eqs.	Bloch-Langevin Eq.	Fluctuating Hydrodynamic Eqs.	Langevin Eq. for $\underset{\sim}{P}(t)$

B. DERIVATION OF THE EQUATION OF MOTION FOR $G_{\alpha\alpha'}(t)$.

We seek an equation of motion for the operator $G_{\alpha\alpha'}(t)$ defined by Eq.(2.2). The operator identity

$$e^{i(A+B)t} = e^{iAt} + \int_0^t d\tau\, e^{i(A+B)(t-\tau)} iB\, e^{iA\tau} \tag{2.14}$$

is used to express the propagator exp(iLt) in terms of a modified propagator exp[i(1-P)Lt]. The modified propagator contains a projection operator P defined by

$$PA = \langle A\rangle = Tr_\ell[\rho_\ell A] \tag{2.15}$$

With the identification (A+B) = L and B = PL we obtain the operator identity

$$\exp(iLt) = \exp[i(1-P)Lt] + \int_0^t d\tau \exp[iL(t-\tau)] iPL \exp[i(1-P)L\tau] \quad (2.16)$$

If this operator identity acts on

$$(1-P)\dot{G}_{\alpha\alpha'}(0) = (1-P)iLG_{\alpha\alpha'}(0)$$

one obtains,

$$\dot{G}_{\alpha\alpha'}(t) = \exp(iLt)PiLG_{\alpha\alpha'}(0) + \exp[i(1-P)Lt](1-P)iLG_{\alpha\alpha'}(0)$$
$$+ \int_0^t d\tau \exp[iL(t-\tau)] iPL \exp[i(1-P)L\tau](1-P)iLG_{\alpha\alpha'}(0) \quad (2.17)$$

This equation may be considerably simplified. From the form of the Hamiltonian, Eq.(1.1), we see that the Liouville operator may be written as

$$L = L_s + L_\ell + \lambda L_1 = L_o + \lambda L_1 \quad (2.18)$$

where $L_o = [H_o, \ldots] = H_o^\times$ is the Liouville operator for the spins and lattice in the absence of any interaction and $L_1 = [H_1, \ldots] = H_1^\times$ is the interaction Liouville operator. It follows immediately from the definition of Eq.(2.15) that

$$PiL_\ell(\cdots) = 0 \quad (2.19)$$

and from the definition of $G_{\alpha\alpha'}(t)$, Eq.(2.2), that

$$PiL_s G_{\alpha\alpha'}(0) = -i\omega_{\alpha\alpha'} G_{\alpha\alpha'}(0) = iL_s G_{\alpha\alpha'}(0) \quad (2.20)$$

Furthermore from Eqs.(1.9), (2.2), and (2.15) it follows that

$$PiL_1 G_{\alpha\alpha'}(0) = 0 \quad (2.21)$$

so that one finds that

$$(1-P)iLG_{\alpha\alpha'}(0) = i\lambda L_1 G_{\alpha\alpha'}(0) \quad (2.22)$$

if use is made of Eqs.(2.19) - (2.21) and the fact $iL_\ell G_{\alpha\alpha'}(0) = 0$.

It is possible to rewrite Eq.(2.17) in the form

$$\dot{G}_{\alpha\alpha'}(t) = -i\omega_{\alpha\alpha'}G_{\alpha\alpha'}(t)+K_{\alpha\alpha'}(t) \tag{2.23}$$

$$+ \lambda^2 \int_o^t d\tau \exp[iL(t-\tau)]<iL_1\{\exp[i(1-P)L\tau]iL_1G_{\alpha\alpha'}(0)\}>$$

where we have introduced the random "force" term $K_{\alpha\alpha'}(t)$,

$$K_{\alpha\alpha'}(t) = \exp[i(1-P)Lt](1-P)iLG_{\alpha\alpha'}(0) = \exp[i(1-P)Lt]i\lambda L_1G_{\alpha\alpha'}(0) \tag{2.24}$$

In order to obtain Eq.(2.23) we have made use of Eqs.(2.20), (2.22), (2.15) and the eadily verifiable fact that

$$PK_{\alpha\alpha'}(t) = <K_{\alpha\alpha'}(t)> = 0 \tag{2.25}$$

If we take the ν,ν' spin space matrix elements of Eq.(2.23) we are led to the equation

$$<\nu|\dot{G}_{\alpha\alpha'}(t)|\nu'> = -i\omega_{\alpha\alpha'}<\nu|G_{\alpha\alpha'}(t)|\nu'>+<\nu|K_{\alpha\alpha'}(t,\lambda)|\nu'> \tag{2.26}$$

$$+\lambda^2 \sum_{\beta,\beta'} \int_o^t d\tau<\nu|\exp[iH(t-\tau)]|\beta'><\beta'|F_{\alpha\alpha'}(\tau,\lambda)|\beta><\beta|\exp[-iH(t-\tau)]|\nu'>$$

where $<\beta'|F_{\alpha\alpha'}(\tau,\lambda)|\beta>$ is defined by

$$<\beta'|F_{\alpha\alpha'}(\tau,\lambda)|\beta> = <\beta'|<iL_1K_{\alpha\alpha'}(\tau,\lambda)>|\beta>\lambda^{-1} = \tag{2.27}$$

$$<\beta'|<iL_1\exp[i(1-P)Lt]iL_1G_{\alpha\alpha'}(0)>|\beta>$$

In order to obtain Eq.(2.26) we have used the definition of exp[iLt] and introduced two complete sets of spin states β and β'. Since $<\beta'|F_{\alpha\alpha'}(\tau,\lambda)|\beta>$ is simply a function of time we may use the definition of $G_{\alpha\alpha'}(t)$ to arrive at the equation

$$\langle\nu|\dot{G}_{\alpha\alpha'}(t)|\nu'\rangle = -i\omega_{\alpha\alpha'}\langle\nu|G_{\alpha\alpha'}(t)|\nu'\rangle+\langle\nu|K_{\alpha\alpha'}(t,\lambda)|\nu'\rangle$$
$$+\lambda^2 \sum_{\beta,\beta'} \int_0^t d\tau\langle\beta'|F_{\alpha\alpha'}(\tau,\lambda)|\beta\rangle\langle\nu|G_{\beta\beta'}(t-\tau)|\nu'\rangle \tag{2.28}$$

or in operator form,

$$\dot{G}_{\alpha\alpha'}(t) = -i\omega_{\alpha\alpha'}G_{\alpha\alpha'}(t) + K_{\alpha\alpha'}(t,\lambda)$$
$$+ \lambda^2 \sum_{\beta,\beta'} \int_0^t d\tau\langle\beta'|F_{\alpha\alpha'}(\tau,\lambda)|\beta\rangle G_{\beta\beta'}(t-\tau). \tag{2.29}$$

This equation is the central formal result of our analysis. It has the form of a generalized Langevin equation with a random "force" $K_{\alpha\alpha'}(t)$ and a damping kernel given by the averaged quantity $\langle\beta'|F_{\alpha\alpha'}(t,\lambda)|\beta\rangle$. The equation is exact since, as yet, no approximations have been introduced. The effects of the bath motion enter through $K_{\alpha\alpha'}(t)$. The equation of motion for the average value of $G_{\alpha\alpha'}(t)$ is obtained by letting P operate on Eq.(2.29). The result is

$$\langle\dot{G}_{\alpha\alpha'}(t)\rangle = -i\omega_{\alpha\alpha'}\langle G_{\alpha\alpha'}(t)\rangle$$
$$+ \lambda^2 \sum_{\beta,\beta'} \int_0^t d\tau\langle\beta'|F_{\alpha\alpha'}(\tau,\lambda)|\beta\rangle\langle G_{\beta\beta'}(t-\tau)\rangle \tag{2.30}$$

where, of course, $\langle G_{\alpha\alpha'}(t)\rangle$ remains an operator in the spin space.

If we assume the customary initial condition, Eq.(2.6), then Eq.(2.7) holds and we may obtain an exact equation of motion for the spin density matrix from Eq.(2.30)

$$\langle\alpha|\dot{\sigma}(t)|\alpha'\rangle = -i\omega_{\alpha\alpha'}\langle\alpha|\sigma(t)|\alpha'\rangle$$
$$+ \lambda^2 \sum_{\beta,\beta'} \int_0^t d\tau\langle\beta'|F_{\alpha\alpha'}(\tau,\lambda)|\beta\rangle\langle\beta|\sigma(t-\tau)|\beta'\rangle \tag{2.31}$$

Of course there remains the crucial step of evaluating $\langle\beta'|F_{\alpha\alpha'}(t,\lambda)|\beta\rangle$. Before turning to this task we wish to compare our result for the exact equation of motion of the spin density matrix, Eq.(2.31), to the result obtained by the more traditional projection operator technique.

In the traditional projection operator techniques such as that of Argyres and Kelley[5], described in Chapter IV, one seeks an equation for $\sigma(t)$ directly. The projection operator employed[6] is

$$P^* = \rho_\ell \mathrm{Tr}_\ell(''') \tag{2.32}$$

One easily arrives, under the initial condition assumption in Eq. (2.6), at the operator equation[6]

$$\frac{\partial\sigma(t)}{\partial t} = -iL_s\sigma(t)+\lambda^2\int_0^t d\tau \mathrm{Tr}_\ell\{iL_1\exp[i(1-P^*)L\tau]iL_1\rho_\ell\sigma(t-\tau)\} \tag{2.33}$$

Note that this equation involves a different projection operator, but still is an exact equation and consequently completely equivalent to our equation, Eq.(2.31). The traditional result suffers from the disadvantage that in the exact form the modified propagator $\exp[i(1-P^*)L\tau]$ operates on all terms to the right. There is no simple way to directly extract the matrix elements of $\sigma(t)$ from the integral. In contrast, in Eq.(2.31) the kernel is not an operator on $\langle\beta|\sigma(t-\tau)|\beta'\rangle$. This leads to important advantages in generalizing the lowest order Redfield result and in comparing the resulting generalizations with those obtained by completely different means.

C. THE REDFIELD-LANGEVIN EQUATION - THE LOWEST ORDER RESULT

We note that in the exact equation for the rate of change of $G_{\alpha\alpha'}(t)$, Eq.(2.28), there are two terms proportional to G itself. Since we assume that λ is small it is useful to transform into an interaction representation in order to remove the Zeeman term. Thus we define

$$G^*_{\alpha\alpha'}(t) = e^{i\omega_{\alpha\alpha'}t}G_{\alpha\alpha'}(t) = e^{iLt}e^{-iL_ot}G_{\alpha\alpha'}(0) \tag{2.34}$$

and obtain the equation

$$\dot{G}^{*}_{\alpha\alpha'}(t) = K^{*}_{\alpha\alpha'}(t)$$

$$+\lambda^2 \sum_{\beta,\beta'} \exp(i\Delta t) \int_0^t d\tau \langle\beta'|F_{\alpha\alpha'}(\tau,\lambda)|\beta\rangle e^{+i\omega_{\beta\beta'}\tau} G^{*}_{\beta\beta'}(t-\tau) \qquad (2.35)$$

where $K^{*}_{\alpha\alpha'}(t)$ is the random "force" in the interaction representation,

$$K^{*}_{\alpha\alpha'}(t) = e^{i\omega_{\alpha\alpha'}t} K_{\alpha\alpha'}(t,\lambda) \qquad (2.36)$$

and Δ is the difference between the frequencies $\omega_{\alpha\alpha'}$ and $\omega_{\beta\beta'}$,

$$\Delta = \omega_{\alpha\alpha'} - \omega_{\beta\beta'} = (E_\alpha - E_{\alpha'} + E_{\beta'} - E_\beta). \qquad (2.37)$$

In order to proceed we must examine the function

$$F_{\alpha\alpha'\beta\beta'}(\tau,\lambda) = \langle\beta'|F_{\alpha\alpha'}(\tau,\lambda)|\beta\rangle e^{+i\omega_{\beta\beta'}\tau} \qquad (2.38)$$

From Eq.(2.27) we see that an evaluation of this function requires knowledge about $K_{\alpha\alpha'}(\tau,\lambda)$, Eq.(2.24), which involves the modified propagator $\exp[i(1-P)L\tau]$. The operator identity Eq.(2.14) permits one to express the modified propagator in terms of the propagator in the absence of interactions, $\exp(iL_o\tau)$. Let $(A+B) = L_o$ and $A = (1-P)L$ in Eq.(2.14). It follows that

$$-B = (1-P)L-L_o = \lambda(1-P)L_1-L_sP \qquad (2.39)$$

and hence

$$\exp[i(1-P)L\tau] = \exp(iL_o\tau)$$

$$+ \int_0^\tau d\tau' \exp[iL_o(\tau-\tau')][i\lambda(1-P)L_1-L_sP]\exp[i(1-P)L\tau'] \qquad (2.40)$$

If this identity operates on $K_{\alpha\alpha'}(0)$ we obtain

$$K_{\alpha\alpha'}(\tau,\lambda) = K^{o}_{\alpha\alpha'}(\tau)+\lambda \int_{o}^{\tau} d\tau' \exp[iL_{o}(\tau-\tau')]i(1-P)L_{1}K_{\alpha\alpha'}(\tau';\lambda) \qquad (2.41)$$

where use has been made of Eq.(2.25) and we have defined the random force to lowest order in $\lambda, K^{o}_{\alpha\alpha'}(\tau)$,

$$K^{o}_{\alpha\alpha'}(\tau) = \exp(iL_{o}\tau)K_{\alpha\alpha'}(0) = \exp(iL_{o}\tau)i\lambda L_{1}G_{\alpha\alpha'}(0) \qquad (2.42)$$

Note that $K^{o}_{\alpha\alpha'}(t)$ is proportional to λ and that $<K^{o}_{\alpha\alpha'}(t)> = 0$. Furthermore $K^{o}_{\alpha\alpha'}(t)$ will vary rapidly since L_1 contains lattice variables. Since the correction term in Eq.(2.41) is proportional to λ, a small quantity, we may expect that to lowest order one is justified in retaining only $K^{o}_{\alpha\alpha'}(\tau)$ in Eq.(2.41). With this basic assumption, $F_{\alpha\alpha'\beta\beta'}(\tau,\lambda)$ is replaced by

$$F_{\alpha\alpha'\beta\beta'}(\tau,0) = <\beta'|<iL_{1}K^{o}_{\alpha\alpha'}(\tau)>|\beta>\exp(+i\omega_{\beta\beta'}\tau) \qquad (2.43)$$

which is the lowest order result.

If the interaction Hamiltonian H_1 is of the form

$$H_{1} = \sum_{q} F_{q}S_{q}$$

where F_q is a lattice operator it is possible to evaluate $F_{\alpha\alpha'\beta\beta'}(\tau,0)$. The result is

$$\begin{aligned} F_{\alpha\alpha'\beta\beta'}(\tau,0) = \sum_{qq'} \{ & <\alpha|S_{q'}|\beta><\beta'|S_{q}|\alpha'>[g_{qq'}(\tau)e^{i\omega_{\beta\alpha}\tau} \\ & +g_{qq'}(-\tau)e^{-i\omega_{\beta'\alpha'}\tau}]-\delta_{\alpha'\beta'} \sum_{\gamma}<\alpha|S_{q}|\gamma><\gamma|S_{q'}|\beta>g_{qq'}(\tau)e^{i\omega_{\beta\gamma}\tau} \\ & -\delta_{\alpha\beta} \sum_{\gamma}<\beta'|S_{q}|\gamma><\gamma|S_{q'}|\alpha'>g_{qq'}(-\tau)e^{-i\omega_{\beta'\gamma}\tau}\} \end{aligned} \qquad (2.44)$$

where $g_{qq'}(t)$ are lattice correlation functions

$$g_{qq'}(t) = <F^{\ell}_{q}(t)F_{q'}(0)> \qquad (2.45)$$

with $F_q^{\ell}(t) = \exp(iL_{\ell}t)F_q^{\ell}(0)$.

If this expression for $F_{\alpha\alpha'\beta\beta'}(\tau,0)$ is replaced in the equation for $G^*_{\alpha\alpha'}(t)$, Eq.(2.34), we see that the rate of change of $G^*_{\alpha\alpha'}(t)$ is equal to a small term of order λ^2 which involves $G^*_{\beta\beta'}(t)$ plus a rapidly varying term which to lowest order may be taken to be

$$\tilde{K}^*_{\alpha\alpha'}(t) = K^o_{\alpha\alpha'}(t)\exp(i\omega_{\alpha\alpha'}t) \tag{2.46}$$

The lattice correlation functions involved in $F_{\alpha\alpha'\beta\beta'}(\tau,0)$ will decay in a time τ_c characteristic of the fluid. During this time $G^*_{\beta\beta'}(t-\tau)$ will vary according to the identity

$$G^*_{\beta\beta'}(t-\tau) = G^*_{\beta\beta'}(t) - \int_o^{\tau} \dot{G}^*_{\beta\beta'}(t-\sigma)d\sigma \tag{2.47}$$

Since $G^*_{\beta\beta'}$ is itself proportional to terms of order λ and λ^2 we may to lowest order replace $G^*_{\beta\beta'}(t-\tau)$ by $G^*_{\beta\beta'}(t)$ in Eq.(2.35). For time $t > \tau_c$ the upper limit on the integral may be extended to infinity and we obtain

$$\dot{G}^*_{\alpha\alpha'}(t) = \lambda^2 \sum_{\beta,\beta'} e^{i\Delta t} R_{\alpha\alpha'\beta\beta'} G^*_{\beta\beta'}(t) + \tilde{K}^*_{\alpha\alpha'}(t) \tag{2.48}$$

where

$$R_{\alpha\alpha'\beta\beta'} = \int_o^{\infty} d\tau F_{\alpha\alpha'\beta\beta'}(\tau,0) \tag{2.49}$$

It is an easy matter to show that the elements of $R_{\alpha\alpha'\beta\beta'}$ are identical to the usual Redfield tetradic. The major contribution to $G^*_{\alpha\alpha'}(t)$ arises from those terms for which $\Delta = 0$.

In the laboratory frame Eq.(2.48) takes the form

$$\dot{G}_{\alpha\alpha'}(t) = -i\omega_{\alpha\alpha'}G_{\alpha\alpha'}(t) + \lambda^2 \sum_{\beta,\beta'} R_{\alpha\alpha'\beta\beta'}G_{\beta\beta'}(t) + K^o_{\alpha\alpha'}(t) \tag{2.50}$$

This equation is the central result of our analysis. The form of the equation is identical to the Redfield equation except for the presence of the added fluctuating force $K^o_{\alpha\alpha'}(t)$ which describes the effect of the rapid lattice motion on $G_{\alpha\alpha'}(t)$. The approximations made in obtaining Eq.(2.48) indicate that the Redfield-Langevin equation, Eq.(2.50), is only valid if $(\lambda^2\tau_c) \ll 1$ and $t \gg \tau_c$. Since the spin relaxation times T_2^{-1} are proportional to λ^2 we see that the condition $(\lambda^2\tau_c) \ll 1$ is equivalent to the condition $(\tau_c/T_2) \ll 1$. Note that $\langle G_{\alpha\alpha'}(t)\rangle$ satisfies the Redfield equation, Eq.(2.1), since $\langle K_{\alpha\alpha'}(t)\rangle = 0$. Consequently, the Redfield-Langevin equation leads immediately to the Redfield equation for the spin density matrix according to Eq.(2.7), provided the initial condition is given by Eq.(2.6).

D. STOCHASTIC PROPERTIES OF THE REDFIELD-LANGEVIN EQUATION

In this section we shall discuss the properties of the Redfield-Langevin equation. In particular, we shall compute the correlation functions of the random quantities which appear in Eq.(2.51). The correlation function of $G_{\alpha\alpha'}(t)$ averaged over the equilibrium lattice motion is

$$\langle G_{\alpha\alpha'}(t)G_{\beta\beta'}(0)\rangle = \langle G_{\alpha\alpha'}(t)\rangle G_{\beta\beta'}(0) \tag{2.51}$$

The formal solution for $\langle G_{\alpha\alpha'}(t)\rangle$, obtained from Eq.(2.50), is

$$\langle G_{\alpha\alpha'}(t)\rangle = \sum_{\gamma,\gamma'} [\exp((-i\underline{\omega}+\underline{R})t)]_{\alpha\alpha'\gamma\gamma'} G_{\gamma\gamma'}(0) \tag{2.52}$$

where $\underline{\omega}$ is the frequency tetradic with elements

$$[\underline{\omega}]_{\alpha\alpha'\beta\beta'} = \omega_{\alpha\alpha'}\delta_{\alpha\beta}\delta_{\alpha'\beta'} \tag{2.53}$$

Thus the correlation function of $G_{\alpha\alpha'}(t)$ is determined from Eqs. (2.51) - (2.53). The <u>complete</u> equilibrium correlation function for $G_{\alpha\alpha'}(t)$ is defined according to Eq.(2.10),

$$\langle G_{\alpha\alpha'}(t)G_{\beta\beta'}(0)\rangle_{av} = Tr_s[\sigma^o \langle G_{\alpha\alpha'}(t)G_{\beta\beta'}(0)\rangle] \tag{2.54}$$

A direct calculation shows that

$$\langle G_{\alpha\alpha'}(t)G_{\beta\beta'}(0)\rangle_{av} = [\exp((-i\underline{\omega}+\underline{R})t)]_{\alpha\alpha'\beta'\beta}\sigma^{o}_{\beta\beta} \qquad (2.55)$$

where $\sigma^{o}_{\beta\beta}$ is the diagonal element of the equilibrium spin density matrix defined by Eq.(2.12). This result is of some importance since it permits direct calculation of the correlation function $\phi_{rq}(t)$, Eq.(2.13), that determines the line shape,

$$\phi_{rq}(t) = \sum_{\substack{\alpha\alpha' \\ \beta\beta'}} \{[\exp((-i\underline{\omega}+\underline{R})t)]_{\alpha\alpha'\beta'\beta}\sigma^{o}_{\beta\beta}\langle\alpha'|S_r|\alpha\rangle\langle\beta'|S_q|\beta\rangle\} \qquad (2.56)$$

The correlation function of the random "force" may also be computed. Note from Eqs.(2.10) and (2.42) that

$$\langle K^{o}_{\alpha\alpha'}(t_1)K^{o}_{\beta\beta'}(t_2)\rangle_{av} = \langle K^{o}_{\alpha\alpha'}(t_1-t_2)K^{o}_{\beta\beta'}(0)\rangle_{av.} \qquad (2.57)$$

Since the ramdom force is linear in the lattice variables we may expect that on the slow λ^2 time scale the time dependence of this correlation function will look like a delta function. Thus we write

$$\langle K^{o}_{\alpha\alpha'}(t_1)K^{o}_{\beta\beta'}(t_2)\rangle_{av} = B(\alpha,\alpha';\beta,\beta')\delta(t_1-t_2) \qquad (2.58)$$

and seek an expression for B. In order to determine B we rewrite Eq.(2.50) in the form

$$\delta\dot{G}_{\alpha\alpha'}(t) = i\omega_{\alpha\alpha'}\delta G_{\alpha\alpha'}(t)+\lambda^2\sum_{\beta,\beta'}R_{\alpha\alpha'\beta\beta'}\delta G_{\beta\beta'}(t)+K^{o}_{\alpha\alpha'}(t) \qquad (2.59)$$

where

$$\delta G_{\alpha\alpha'}(t) = G_{\alpha\alpha'}(t)-\sigma^{o}_{\alpha\alpha'} \qquad (2.60)$$

and $\sigma^{o}_{\alpha\alpha'}$ is the equilibrium spin density matrix defined in Eq.(2.12) with the necessary property[7] that

$$\sum_{\beta,\beta'}R_{\alpha\alpha'\beta\beta'}\sigma^{o}_{\beta\beta'} = 0 \qquad (2.61)$$

From Eq.(2.59) it follows that

$$\langle \delta G_{\alpha\alpha'}(t)\delta G_{\beta\beta'}(t)\rangle_{av} = T$$

$$+ \sum_{\substack{\gamma,\gamma' \\ \nu,\nu'}} \int_0^t d\tau_1 \int_0^t d\tau_2 [\exp((t-\tau_1)Q)]_{\alpha\alpha'\gamma\gamma'} [\exp((t-\tau_2)Q)]_{\beta,\beta',\nu,\nu'}$$

$$\times \langle K_{\gamma\gamma'}(\tau_1)K_{\nu\nu'}(\tau_2)\rangle_{av} \tag{2.62}$$

where $\underline{Q} = -i\underline{\omega} + \underline{R}$,

$$T = \langle\langle \delta G_{\alpha\alpha'}(t)\rangle\langle \delta G_{\beta\beta'}(t)\rangle\rangle_{av} \tag{2.63}$$

and use has been made of the fact that $\langle K^{o}_{\alpha\alpha'}(t)\rangle = 0$. We shall now take the long time limit of Eq.(2.62) and assume

$$\lim_{t\to\infty} \langle \delta G_{\alpha\alpha'}(t)\rangle = 0 \tag{2.64}$$

so that T may be taken to be zero. If $G_{\alpha\alpha'}(0)$ were a well-behaved matrix this property could be established since $\sigma^{o}_{\beta\beta'}$, the sole eigenmatrix of R with zero eigenvalue, is subtracted from $G_{\alpha\alpha'}(0)$. Strictly speaking, it is not possible to establish Eq.(2.64) without an examination of the subsequent spin average with which $\langle \delta G_{\alpha\alpha'}(t)\rangle$ is eventually to be associated. We may take T = 0 provided that we assume that the fluctuation formula will be employed only to compute a restricted class of averages for which Eq.(2.12) can be explicitly justified.

From the definition of the correlation function, Eq.(2.58), and Eq.(2.62), it is a straightforward but lengthy matter to show that

$$B(\alpha,\alpha',\beta,\beta') = -\sum_{\substack{\gamma,\gamma' \\ \eta,\eta'}} [Q_{\alpha\alpha'\gamma\gamma'} I_{\beta\beta'\eta\eta'} + I_{\alpha\alpha'\gamma\gamma'} Q_{\beta\beta'\eta\eta'}] \tag{2.65}$$

$$\times \langle \delta G_{\gamma\gamma'}(t)\delta G_{\eta\eta'}(t)\rangle_{av}; \quad I_{\alpha\alpha'\beta\beta'} = \delta_{\alpha\beta}\delta_{\alpha'\beta'}$$

To lowest order in the coupling constant λ, the equal time average on the right-hand side of this equation may be replaced by its initial value,

$$<\delta G_{\gamma\gamma'}(0)\delta G_{\eta\eta'}(0)>_{av} = \sigma^{o}_{\gamma'\gamma'}\delta_{\gamma'\eta}\delta_{\gamma\eta'}-\sigma^{o}_{\gamma\gamma'}\sigma^{o}_{\eta\eta'}$$

The result for B is

$$B(\alpha,\alpha';\beta\beta') = -[Q_{\alpha\alpha'\beta'\beta}\sigma^{o}_{\beta\beta}+Q_{\beta\beta'\alpha'\alpha}\sigma^{o}_{\alpha'\alpha'}] \tag{2.66}$$

$$= -[R_{\alpha\alpha'\beta'\beta}\sigma^{o}_{\beta\beta}+R_{\beta\beta'\alpha'\alpha}\sigma^{o}_{\alpha'\alpha}]$$

which determines the correlation function of the random "force" appearing in the Redfield-Langevin equation, Eq.(2.50).

E. THE BLOCH-LANGEVIN EQUATION

We now shall obtain stochastic equations of motion for the magnetization operator $\mathcal{M}_r(t)$, Eq.(2.4), from the Redfield-Langevin equation. We multiply Eq.(2.50) by $<\alpha'|S_r|\alpha>$ and sum over α and α'. This immediately leads to the equation

$$\dot{\mathcal{M}}_r(t) = -[\underset{\sim}{B}_o \times \underset{\sim}{\mathcal{M}}(t)]_r + \sum_{\substack{\alpha,\alpha' \\ \beta,\beta'}} R_{\alpha\alpha'\beta\beta'}G_{\beta\beta'}(t)<\alpha'|S_r|\alpha>+h_r(t) \tag{2.67}$$

where we have defined the random magnetic field operator

$$h_r(t) = \sum_{\alpha,\alpha'} K^{o}_{\alpha\alpha'}(t)<\alpha'|S_r|\alpha> \tag{2.68}$$

It is well known that in general the relaxation matrix is not always of a form which permits a reduction of Eq.(2.67). In fact, if the Redfield equation, Eq.(2.1), is valid and the average magnetization satisfies the usual Bloch equations

$$\dot{M}_r(t) = -[\underset{\sim}{B}_o \times \underset{\sim}{M}(t)]_r - \frac{1}{T_r}[M_r(t)-M^{o}_r] \qquad r = x,y,z \tag{2.69}$$

where $M^{o}_r = Tr_s[\sigma^{o}M_r]$, then a necessary condition for the simultaneous validity of the two equations is

$$\sum_{\alpha,\alpha'} [\langle\alpha'|S_r|\alpha\rangle R_{\alpha\alpha'\beta\beta'} + \frac{1}{T_r}\langle\alpha'|S_r|\alpha\rangle \delta_{\alpha\beta}\delta_{\alpha'\beta'}] = \delta_{\beta\beta'} \frac{M_r^o}{T_r} \qquad (2.70)$$

We shall assume that the form of the spin-lattice interaction, the multiplicity of the spin, and the temperature are such that Eq.(2.70) is valid. Systems for which Eq.(2.70) is valid are discussed in detail by Aleksandrov[8].

If Eq.(2.70) is used in Eq.(2.67) we arrive immediately at the Bloch-Langevin operator equation,

$$\dot{\mathscr{S}}_r(t) = -[\underset{\sim}{B}_o \times \underset{\sim}{\mathscr{S}}(t)]_r - \frac{1}{T_r}[\mathscr{S}_r(t) - \mathscr{S}_r^o] + h_r(t) \qquad (2.71)$$

where $\mathscr{S}_r^o = M_r^o$. This equation is an operator equation in both spin and lattice variables. If the average is performed over the lattice we obtain an operator equation that has the form of the Bloch equation

$$\langle \dot{\mathscr{S}}_r(t)\rangle = -[\underset{\sim}{B}_o \times \langle \mathscr{S}(t)\rangle]_r - \frac{1}{T_r}[\langle \mathscr{S}_r(t)\rangle - \mathscr{S}_r^o] \qquad (2.72)$$

From Eq.(2.8) we see that this equation immediately implies Eq. (2.69).

The significance of the Bloch-Langevin equation is that if the Redfield equation is valid and consistent with the Bloch equation for the average magnetization, then it necessarily follows that a magnetization type operator $\mathscr{S}_r(t)$ satisfies a Langevin equation. The operator $\mathscr{S}_r(t)$ is a useful description on the slow time scale and includes the effects of the fast lattice motion in a fluctuating term $h_r(t)$ whose average over the equilibrium lattice density matrix is zero.

The microscopic analogue of the Bloch equation[9] is obtained by averaging the microscopic Bloch-Langevin operator equation, Eq. (2.71), over an initial non-equilibrium spin density matrix $\sigma(0)$. Thus if we define

$$m_r(t) = Tr_s[\sigma(0)\, \mathscr{S}_r(t)] \qquad (2.73)$$

then

$$\dot{m}_r(t) = -[\underset{\sim}{B}_o \times m(t)]_r - \frac{1}{T_r}[m_r(t) - m_r^o] + b_r(t) \qquad (2.74)$$

where $m_r^o = M_r^o$ and $b_r(t)$ is a random magnetic field,

$$b_r(t) = Tr_s[\sigma(0)h_r(t)] \qquad (2.75)$$

This is an equation for the microscopic magnetization which depends upon lattice configurations. The observable magnetization is related to $m_r(t)$ by

$$M_r(t) = \langle m_r(t) \rangle \qquad (2.76)$$

Since $\langle b_r(t) \rangle = 0$ when Eq.(2.74) is averaged over the equilibrium lattice density matrix, one recovers the ordinary Bloch equations.

Correlation functions of the random quantities appearing in the Bloch-Langevin operator equation may also be computed. We rewrite Eq.(2.71) in the form

$$\delta\dot{\$}_q(t) = -L_q \delta\$_q(t) + h_q(t) \qquad q=0,\pm 1 \qquad (2.77)$$

where

$$\delta\$_o = \delta\$_z \qquad \delta\$_\pm = \delta\$_x \pm i\delta\$_y \qquad (2.78)$$

with a similar definition for h_q and

$$L_q = T_q^{-1} - iqB_o = T_q^{-1} + i\omega_q \qquad (2.79)$$

Here we have assumed that the external magnetic field lies along the z-axis so that $T_{\pm 1} = T_2$ and $T_o = T_1$.

It follows immediately that

$$\langle \delta\$_q(t)\delta\$_p(0) \rangle = \exp[-L_q t]\delta\$_q(0)\delta\$_p(0) \qquad p,q=0,\pm 1 \qquad (2.80)$$

and

$$\langle \delta\$_q(t)\delta\$_p(0) \rangle_{av} = \exp[-L_q t]\delta_{p-q}C(|q|) \qquad (2.81)$$

In Eq.(2.81) we have used the fact that

$$<\delta \not{S}_q \delta \not{S}_p>_{av} = \delta_{p-q} C(|q|) \qquad (2.82)$$

where C(|q|) denotes a simple spin average. Note that Eq.(2.81) is a statement of Onsager's assumption for the decay of equilibrium fluctuations on the slow spin time scale.

The correlation function of the random field operators may easily be computed from the definition of $h_r(t)$, Eq.(2.68),

$$<h_q(t)h_p(0)>_{av} = \sum_{\substack{\alpha,\alpha' \\ \beta,\beta'}} <K^o_{\alpha\alpha'}(t)K^o_{\beta\beta'}(0)>_{av} <\alpha'|S_q|\alpha><\beta'|S_p|\beta> \qquad (2.83)$$

and Eqs.(2.57), (2.66), and (2.70). The result is

$$<h_q(t)h_p(0)>_{av} = 2C(|q|)T_q^{-1}\delta_{q-p}\delta(t) \qquad (2.84)$$

Finally, the correlation function of the random force is simply related to the macroscopic relaxation times,

$$T_q^{-1} = C(|q|)^{-1} \int_o^\infty dt <h_q(t)h_{-q}(0)>_{av} \qquad (2.85)$$

which is a manifestation of the fluctuation-dissipation theorem.

F. FURTHER REMARKS

We conclude our discussion of the Redfield-Langevin equation with a few comments.

(1) Generalizations of the Lowest Order Result

The Redfield-Langevin equation to lowest order in λ^2 was obtained by a qualitative argument which led from the exact equation Eq.(2.35)

$$\dot{G}^*_{\alpha\alpha'}(t) = \lambda^2 \sum_{\beta,\beta'} \exp(i\Delta t) \int_o^t d\tau F_{\alpha\alpha'\beta\beta'}(\tau,\lambda) G^*_{\beta\beta'}(t-\tau) + K^*_{\alpha\alpha'}(t) \qquad (2.86)$$

to the approximate equation, valid to lowest order in λ and for times

$t >> \tau_c$, Eq.(2.48)

$$\dot{G}^*_{\alpha\alpha'}(t) = \lambda^2 \sum_{\beta,\beta'} \exp(i\Delta t)[\int_0^\infty d\tau F_{\alpha\alpha'\beta\beta'}(\tau,0)]G^*_{\beta\beta'}(t)+\tilde{K}^*_{\alpha\alpha'}(t) \quad (2.87)$$

Various generalizations of this lowest order result may be considered. For example, our reasoning suggests that a generalization of Eq.(2.87) valid to order λ^2 for short as well as long times, is of the form

$$\dot{G}^*_{\alpha\alpha'}(t) = \lambda^2 \sum_{\beta,\beta'} \exp(i\Delta t)[\int_0^t d\tau F_{\alpha\alpha'\beta\beta'}(\tau,0)]G^*_{\beta\beta'}(t)+\tilde{K}^*_{\alpha\alpha'}(t) \quad (2.88)$$

where the time integral is not extended to infinity.

If we assume that $F_{\alpha\alpha'\beta\beta'}(\tau,\lambda)$ decays to zero in a time τ_c for finite values of λ then we can obtain an equation for long times to all orders in λ. For time $t >> \tau_c$ an adequate approximation to the exact equation, Eq.(2.86), is

$$\dot{G}^*_{\alpha\alpha'}(t) = \lambda^2 \sum_{\beta,\beta'} \exp(i\Delta t)[\int_0^\infty F_{\alpha\alpha'\beta\beta'}(\tau,\lambda)d\tau]G^*_{\beta\beta'}(t)+K^*_{\alpha\alpha'}(t) \quad (2.89)$$

if $F_{\alpha\alpha'\beta\beta'}(\tau,\lambda)$ is sharply peaked. We may develop $F_{\alpha\alpha'\beta\beta'}(\tau,\lambda)$ in a power series in λ from the definition (see Eqs.(2.38) and (2.27))

$$F_{\alpha\alpha'\beta\beta'}(\tau,\lambda) = \langle\beta'|\langle iL_1 K_{\alpha\alpha'}(\tau,\lambda)\rangle|\beta\rangle e^{+i\omega_{\beta\beta'}\tau} \quad (2.90)$$

and iteration of the random force according to Eq.(2.41). Thus

$$F_{\alpha\alpha'\beta\beta'}(\tau,\lambda) = \sum_{n=0}^\infty F^{(n)}_{\alpha\alpha'\beta\beta'}(\tau)\lambda^n \;;\quad F^{(o)}_{\alpha\alpha'\beta\beta'}(\tau) = F_{\alpha\alpha'\beta\beta'}(\tau,0)$$
$$(2.91)$$

The resulting generalized Redfield equation obtained according to Eq.(2.7) with the initial condition assumption, Eq.(2.6), is

$$\frac{\partial\langle\alpha|\sigma(t)|\alpha'\rangle}{\partial t} = -i\omega_{\alpha\alpha'}\langle\alpha|\sigma(t)|\alpha'\rangle+\lambda^2 \sum_{n=0}^\infty \lambda^n R^{(n)}_{\alpha\alpha'\beta\beta'}\langle\beta|\sigma(t)|\beta'\rangle$$
$$(2.92)$$

with

$$R^{(n)}_{\alpha\alpha'\beta\beta'} = \int_0^\infty d\tau\ F^{(n)}_{\alpha\alpha'\beta\beta'}(\tau)d\tau \quad (2.93)$$

This generalized Redfield equation is of the same form as the generalized Redfield equation obtained by Freed[10] by application of the cumulant method of Kubo[11]. (This method is discussed in Chapter V). It is likely that the two procedures are completely equivalent and lead to identical results.

We see that the projection operator method leaves open the possibility of generalizing the lowest order result in various ways.

(2) Semiclassical Treatment

In a variety of applications it is an adequate approximation to consider the lattice motion to be classical. Under these circumstances it follows from Eq.(2.2) that

$$<\nu|<G_{\alpha\alpha'}(t)>|\nu'> = <\alpha|<G^{+}_{\nu\nu'}(t)>|\alpha'> \tag{2.94}$$

where + indicates complex conjugate. In this case the equation of motion for $S_r(t)$, Eq.(2.4), will be simplified. With use of Eq. (2.94), Eq.(2.4) may be rewritten as

$$<\nu|<S_r(t)>|\nu'> = \sum_{\alpha,\alpha'} <\alpha|<G^{+}_{\nu\nu'}(t)>|\alpha'><\alpha'|S_r|\alpha> \tag{2.95}$$

From the Redfield-Langevin equation, Eq.(2.50), we obtain directly the equation of motion

$$\frac{\partial}{\partial t}<\nu|<S_r(t)>|\nu'> = +i\omega_{\nu\nu'}<\nu|<S_r(t)>|\nu'> + \lambda^2 \sum_{\eta,\eta'} R^{+}_{\nu\nu'\eta\eta'}<\eta|<S_r(t)>|\eta'> \tag{2.96}$$

for a transition line or amplitude. In semiclassical treatments this equation may be employed to compute directly the correlation function $\phi_{rq}(t)$ and hence the line shape.

In the full quantum mechanical case treated here this equation of motion is not appropriate for computing the correlation function $\phi_{rq}(t)$, Eq.(2.9). In the quantum mechanical treatment $\phi_{rq}(t)$ is computed with use of the fluctuation theory result according to

Eq.(2.56). An alternative procedure is to compute the correlation function $\phi_{rq}(t)$ according to

$$\phi_{rq}(t) = Tr_s[S_r\bar{S}_q(-t)] \qquad (2.97)$$

where $\bar{S}_q(t)$ is an operator in spin space defined by

$$\begin{aligned} \langle\alpha|\bar{S}_q(t)|\alpha'\rangle &= \langle\alpha|Tr_\ell[(e^{iLt}S_q\rho_o)]|\alpha'\rangle \\ &= \sum_{\gamma,\gamma'} \langle\gamma'|\langle G^+_{\alpha\alpha'}(t)\rangle|\gamma\rangle\langle\gamma|S_q|\gamma'\rangle\sigma^o_{\gamma'\gamma'} \end{aligned} \qquad (2.98)$$

This <u>weighted</u> transition amplitude satisfies an equation similar to Eq.(2.96),

$$\frac{\partial}{\partial t}\langle\alpha|\bar{S}_q(t)|\alpha'\rangle=i\omega_{\alpha\alpha'}\langle\alpha|\bar{S}_q(t)|\alpha'\rangle+\sum_{\beta,\beta'}\lambda^2R^+_{\alpha\alpha'\beta\beta'}\langle\beta|\bar{S}_q(t)|\beta'\rangle \qquad (2.99)$$

Of course the result obtained for $\phi_{rq}(t)$ by this alternative procedure will be identical to Eq.(2.56) since

$$\langle G^+_{\alpha\alpha'}(-t)\rangle = \langle G_{\alpha\alpha'}(t)\rangle \qquad (2.100)$$

Another interesting class of semiclassical theories arises in the stochastic theory of line shapes where one introduces explicitly the idea of randomly varying Hamiltonians. The formalism developed here can profitably be used to discuss these theories. For example, one may investigate the connection between Modified Bloch Equations and a stochastic spin density matrix approach for describing problems of chemical exchange[12].

(3) Further Stochastic Interpretation of the Spin Problem

The Bloch-Langevin Equation for the magnetization, Eq.(2.74), contains a random magnetic field $b_r(t)$ whose mean is zero, $\langle b_r(t)\rangle=0$. It is tempting to suppose that under ordinary experimental conditions $b_q(t)$ may be treated as a Gaussian random process with second moment given by

$$\langle b_r(t)b_q(0)\rangle = 2C(|q|)T_q^{-1}\delta_{q-p}\delta(t) \tag{2.101}$$

It then follows from the theory of stochastic processes[13] (see Chapter II) that the probability of observing a given value of the magnetization at time t satisfies a Fokker-Planck equation. The appropriate Fokker-Planck equations are

$$\frac{\partial}{\partial t}\psi(m_z',t) = \frac{1}{T_1}\frac{\partial}{\partial m_z'}[m_z'-C(0)\frac{\partial}{\partial m_z'}]\psi(m_z',t); \quad m_z' = m_z-m_z^o \tag{2.102}$$

for the z-component of the magnetization and

$$\frac{\partial}{\partial t}\phi(m_x,m_y,t) = \omega_o(m_x\frac{\partial}{\partial m_y} - m_y\frac{\partial}{\partial m_x})\phi(m_x,m_y,t) \tag{2.103}$$

$$+\frac{1}{T_2}[\frac{\partial}{\partial m_x}(m_x+C(1)\frac{\partial}{\partial m_x})+\frac{\partial}{\partial m_y}(m_y+C(1)\frac{\partial}{\partial m_y})]\phi(m_x,m_y,t)$$

for the coupled x and y components of the magnetization. Note that as $t \to \infty$ the equilibrium distribution for each of the components is Gaussian,

$$\phi_{eq}(m_q) = (2\pi C(|q|))^{-\frac{1}{2}}\exp[-m_q^2/2C(|q|)] \tag{2.104}$$

The best way to give these Fokker-Planck equations a molecular justification is to derive directly an equation of motion for the probability of observing a particular value of the magnetization from the microscopic definition of this quantity,

$$P(\underset{\sim}{\delta m},t) = Tr\{\rho(0)e^{iLt}\delta[\underset{\sim}{S}-\underset{\sim}{S}^o-\underset{\sim}{\delta m}]\} \tag{2.105}$$

It is likely that this derivation can be economically accomplished by the projection operator method with reasonable assumptions about initial conditions.

VII.3. THE MULTIPLE TIME SCALE METHOD

A. Introduction

In this section we present the Multiple Time Scale (MTS) derivation of the Redfield equations. This method, originally developed to treat problems in non-linear mechanics, has recently been extensively employed to obtain kinetic equations for systems where there are two processes occurring on widely different time scales. Reference to applications of the MTS method may be found in the article by Cukier and Deutch[14].

The MTS method is designed to avoid the difficulties that arise in using standard perturbation theory to handle time dependent problems. The difficulties may be illustrated by considering, for the spin relaxation case, a perturbation solution to the Liouville equation, Eq.(1.2). In the interaction picture

$$\frac{\partial \rho^*}{\partial t} = -\lambda i L_1^*(t)\rho^*(t) \tag{3.1}$$

where

$$\rho^*(t) = e^{iL_o t}\rho(t) \quad \text{and} \quad L_1^*(t) = e^{iL_o t}L_1 \tag{3.2}$$

The spin density matrix in the interaction picture is

$$\sigma^*(t) = \mathrm{Tr}_\ell[\rho^*(t)] \tag{3.3}$$

If we seek a solution to Eq.(3.1) by an expansion of the form

$$\rho^*(t) = \rho_o^*(t)+\lambda\rho_1^*(t)+\ldots$$

with an analogous expansion for $\sigma^*(t)$ one finds in second order

$$\frac{\partial}{\partial t}\,\sigma_2^*(t) = -\mathrm{Tr}_\ell \int_o^t d\tau L_1^*(t)L_1^*(\tau)\rho_\ell\sigma(0) \tag{3.4}$$

In order to obtain this equation one assumes the validity of Eq.(1.9) and the initial condition

$$\rho_o^*(0) = \rho_\ell\sigma(0);\quad \rho_n^*(0) = 0,\quad n=1,\ldots$$

Since the right-hand side of Eq.(3.4) reaches a constant value after a time τ_c characteristic of the lattice motion, $\sigma_2^*(t)$ will grow linearly in time (secular behavior) for $t >> \tau_c$. This breakdown is encountered in all the standard perturbation treatments, e.g. Abragam[15] and is avoided by *ad hoc* procedures such as replacing $\sigma(0)$ by $\sigma_2^*(t)$ in Eq.(3.4). The difficulty arises because the standard perturbation theory, as employed, is only valid for short times.

The idea of the MTS method is that, when a parameter of smallness λ enters a differential equation such as Eqs.(1.2) and (1.4), the system will naturally evolve on characteristic time scales

$$\tau_n = \lambda^n t \tag{3.5}$$

In order to follow the system on the various time scales the MTS method introduces a set of time variables τ_o, τ_1, ..., τ_n each of which is treated as an independent variable in the place of the real time t. Thus

$$\rho(t) \to \tilde{\rho}(\tau_o, \ldots, \tau_n) \tag{3.6}$$

and

$$\sigma(t) \to \tilde{\sigma}(\tau_o, \ldots, \tau_n) \tag{3.7}$$

The MTS method deals explicitly with secular behavior by using the increased flexibility which accompanies the extended definition of the functions to eliminate secular behavior whenever it occurs.

We shall seek solutions to Eqs.(1.2) and (1.4) in the form

$$\tilde{\rho} = \tilde{\rho}_o(\tau_o,\ldots,\tau_n)+\lambda\tilde{\rho}_1(\tau_o,\ldots,\tau_n)+\lambda^2\tilde{\rho}_2(\tau_o,\ldots,\tau_n)+\ldots \tag{3.8}$$

and

$$\tilde{\sigma} = \tilde{\sigma}_o(\tau_o,\ldots,\tau_n)+\lambda\tilde{\sigma}_1(\tau_o,\ldots,\tau_n)+\lambda^2\tilde{\sigma}_2(\tau_o,\ldots,\tau_n)+\ldots \tag{3.9}$$

with the time derivatives replaced by

$$\frac{\partial}{\partial t} \to \frac{\partial}{\partial \tau_o} + \lambda \frac{\partial}{\partial \tau_1} + \lambda^2 \frac{\partial}{\partial \tau_2} + \ldots \tag{3.10}$$

Thus we shall solve a more complicated system of differential equations involving many time variables in a way that avoids secular behavior. When we restrict the extended functions to the physical time line according to Eq.(3.5) we recover $\rho(t)$ and $\sigma(t)$. Below we drop the tilde for economy.

B. DERIVATION OF THE EQUATIONS OF MOTION

If we substitute Eq.(3.8) and Eq.(3.10) into Eq.(1.2) we obtain, on equating powers of λ, the equations

$$\frac{\partial}{\partial\tau_o}\rho_o(\tau_o,\tau_1,\ldots) = -iL_o\rho_o(\tau_o,\tau_1,\ldots) \tag{3.11}$$

$$\frac{\partial}{\partial\tau_o}\rho_1(\tau_o,\tau_1,\ldots)+\frac{\partial}{\partial\tau_1}\rho_o(\tau_o,\tau_1,\ldots)=-iL_o\rho_1(\tau_o,\tau_1,\ldots)-iL_1\rho_o(\tau_o,\tau_1,\ldots) \tag{3.12}$$

$$\frac{\partial}{\partial\tau_o}\rho_2(\tau_o,\tau_1,\ldots)+\frac{\partial}{\partial\tau_1}\rho_1(\tau_o,\tau_1,\ldots)+\frac{\partial}{\partial\tau_2}\rho_o(\tau_o,\tau_1,\ldots)$$

$$= -iL_o\rho_2(\tau_o,\tau_1,\ldots)-iL_1\rho_1(\tau_o,\tau_1,\ldots) \tag{3.13}$$

When we replace the expansion for $\tilde{\sigma}$ and Eq.(3.10) in Eq.(1.4) and equate powers of λ, we obtain the set of equations

$$\frac{\partial}{\partial\tau_o}\sigma_o(\tau_o,\tau_1,\ldots) = -iTr_\ell[L_o\rho_o(\tau_o,\tau_1,\ldots)] \tag{3.14}$$

$$\frac{\partial}{\partial\tau_o}\sigma_1(\tau_o,\tau_1,\ldots)+\frac{\partial}{\partial\tau_1}\sigma_o(\tau_o,\tau_1,\ldots)$$

$$=-iTr_\ell[L_o\rho_1(\tau_o,\tau_1,\ldots)+L_1\rho_o(\tau_o,\tau_1,\ldots)] \tag{3.15}$$

$$\frac{\partial}{\partial\tau_o}\sigma_2+\frac{\partial}{\partial\tau_1}\sigma_1+\frac{\partial}{\partial\tau_2}\sigma_o = -iTr_\ell[L_o\rho_2+L_1\rho_1] \tag{3.16}$$

where for simplicity we have no longer explicitly indicated the dependence on $\{\tau_n\}$.

In order to proceed we must formulate appropriate initial conditions for the extended functions. We shall assume

$$\tilde{\rho}(0,\tau_1,\ldots,\tau_n) = \rho_\ell\tilde{\sigma}(0,\tau_1,\ldots,\tau_n) \tag{3.17}$$

which is an initial condition consistent with the customary one given by Eq.(2.6) at the origin of the physical time line given by Eq.(3.5). Note, however, that this condition only specifies the form of the function $\tilde{\rho}$ on the hypersurface at $\tau_o = 0$. It is possible to choose

$$\sigma_o(0,\tau_1,\ldots)=\tilde{\sigma}(0,\tau_1,\ldots) \quad \text{and} \quad \rho_o(0,\tau_1,\ldots)=\rho_\ell\tilde{\sigma}(0,\tau_1,\ldots,\tau_n) \tag{3.18}$$

so that

$$\sigma_n(0,\tau_1,\ldots) = \rho_n(0,\tau_1,\ldots) = 0 \qquad n \geq 1 \tag{3.19}$$

With these initial conditions the solution of Eq.(3.11) is

$$\rho_o(\tau_o,\tau_1,\ldots) = \rho_\ell e^{-iL_o\tau_o} \sigma_o(0,\tau_1,\ldots) \tag{3.20}$$

since $L_o\rho_\ell = 0$. This result may be introduced into Eq.(3.14) to obtain

$$\frac{\partial}{\partial\tau_o}\sigma_o = -iL_z \exp(-iL_z\tau_o)\sigma_o(0,\tau_1,\ldots) \tag{3.21}$$

where we have used the fact that $L_o = L_z + L_\ell$. The solution to Eq. (3.21) may be expressed as

$$\sigma_o(\tau_o,\tau_1,\ldots) = \exp(-iL_o\tau_o)\sigma_o(0,\tau_1,\ldots) \tag{3.22}$$

We now turn to Eq.(3.15). From Eqs.(3.20) and (1.9) it follows that $Tr_\ell[L_1\rho_o] = 0$. Thus we may write Eq.(3.15) as

$$\frac{\partial}{\partial\tau_1}\sigma_o = -Tr_\ell[iL_o\rho_1 + \frac{\partial}{\partial\tau_o}\rho_1] \tag{3.23}$$

which follows from the definition

$$\sigma_n = Tr_\ell\rho_n \tag{3.24}$$

Equation (3.23) may be rewritten as

$$\frac{\partial}{\partial\tau_1}\sigma_o = -e^{-iL_o\tau_o}\frac{\partial}{\partial\tau_o} Tr_\ell[e^{iL_o\tau_o}\rho_1] \tag{3.25}$$

where we have used the fact that

$$e^{iL_ot}Tr_\ell A = Tr_\ell e^{iL_ot}A \tag{3.26}$$

From Eqs.(3.22) and (3.25) it follows that

$$\frac{\partial}{\partial\tau_1}\sigma_o(0,\tau_1,\ldots) = -\frac{\partial}{\partial\tau_o}\mathrm{Tr}_\ell[e^{iL_o\tau_o}\rho_1(\tau_o,\ldots,\tau_n)] \tag{3.27}$$

The solution to this equation is

$$\tau_o\frac{\partial}{\partial\tau_1}\sigma_o(0,\tau_1,\ldots,\tau_n) = -\mathrm{Tr}_\ell[e^{iL_o\tau_o}\rho_1(\tau_o,\ldots,\tau_n)] \tag{3.28}$$

which according to Eqs.(3.26) and (3.22) may be rearranged to

$$\tau_o\frac{\partial}{\partial\tau_1}\sigma_o(\tau_o,\tau_1,\ldots) = -\sigma_1(\tau_o,\ldots,\tau_n) \tag{3.29}$$

The left-hand side of Eq.(3.28) will grow in time. In order to prevent this secular growth we set

$$\frac{\partial}{\partial\tau_1}\sigma_o(\tau_o,\tau_1,\ldots) = 0 \tag{3.30}$$

which implies σ_1 is identically zero. Consequently, $\sigma_o(\tau_o,\tau_2,\ldots,\tau_n)$ does not vary on the τ_1 time scale.

Next we consider Eq.(3.12) and integrate with respect to τ_o.

$$e^{iL_o\tau_o}\rho_1(\tau_o,\ldots,\tau_n) = -\int_o^{\tau_o}dx e^{iL_o x}[\frac{\partial}{\partial\tau_1}\rho_o(x,\tau_1,\ldots)+iL_1\rho_o(x,\tau_1,\ldots)] \tag{3.31}$$

The first term on the right-hand side is zero as a consequence of Eqs.(3.20) and (3.30). Consequently, we find

$$\rho_1(\tau_o,\tau_2,\ldots,\tau_n) = -i\int_o^{\tau_o}dx\ \exp[iL_o(x-\tau_o)]L_1\rho_\ell\sigma_o(x,\tau_2,\ldots)] \tag{3.32}$$

Note that $\sigma_1 = \mathrm{Tr}_\ell(\rho_1) = 0$ directly from this equation by use of Eq.(1.9).

The final step of the analysis is to integrate Eq.(3.16) with respect to τ_o,

$$\sigma_2(\tau_o,\tau_2,\ldots) = -\tau_o[\frac{\partial}{\partial\tau_2}\sigma_o(\tau_o,\tau_2,\ldots) + \frac{1}{\tau_o}\int_o^{\tau_o} dx \mathrm{Tr}_\ell\{\exp[iL_o(x-\tau_o)]iL_1\rho_1(x,\tau_2,\ldots)\}] \quad (3.33)$$

When Eq.(3.32) is introduced into this equation we obtain

$$\sigma_2(\tau_o,\tau_2,\ldots) = -\tau_o[\frac{\partial}{\partial\tau_2}\sigma_o(\tau_o,\tau_2,\ldots) - \frac{1}{\tau_o}\int_o^{\tau_o} dx \int_o^x dy \mathrm{Tr}_\ell\{\exp[iL_o(x-\tau_o)]iL_1\exp[iL_o(y-x)]iL_1\rho_\ell\}\sigma_o(y,\tau_2,\ldots)] \quad (3.34)$$

or alternatively with use of Eqs.(3.22) and (3.26)

$$e^{iL_o\tau_o}\sigma_2(\tau_o,\tau_2,\ldots) = -\tau_o[\frac{\partial}{\partial\tau_2}\sigma_o(0,\tau_2,\ldots) - \frac{1}{\tau_o}\int_o^{\tau_o} dx \int_o^x dy \mathrm{Tr}_\ell\{\exp(iL_o x)iL_1\exp[iL_o(y-x)]iL_1\rho_\ell\}\sigma_o(y,\tau_2,\ldots)] \quad (3.35)$$

If we introduce the definition $L_1(t) = [\exp(iL_o t)L_1]$ Eq.(3.35) becomes

$$e^{iL_o\tau_o}\sigma_2(\tau_o,\tau_2,\ldots) = -\tau_o[\frac{\partial}{\partial\tau_2}\sigma_o(0,\tau_2,\ldots) - \frac{1}{\tau_o}\int_o^{\tau_o} dx \int_o^x dy \mathrm{Tr}_\ell\{iL_1(x)iL_1(y)\rho_\ell\}\sigma_o(0,\tau_2,\ldots)]$$

Clearly there is a possibility of secular behavior in this equation. In order to determine the asymptotic time dependence of the right-hand side of this equation we take the $\alpha\alpha'$ matrix elements. After a good deal of computation one finds

$$e^{i\omega_{\alpha\alpha'}\tau_o}\langle\alpha|\sigma_2(\tau_o,\tau_2,\ldots)|\alpha'\rangle = -\tau_o[\frac{\partial}{\partial\tau_2}\langle\alpha|\sigma_o(0,\tau_2,\ldots)|\alpha'\rangle - \frac{1}{\tau_o}\sum_{\beta,\beta'}\int_o^{\tau_o} dx \int_o^x dy F_{\alpha\alpha'\beta\beta'}(x-y)\exp(i\Delta x)\langle\beta|\sigma_o(0,\tau_2,\ldots)|\beta'\rangle] \quad (3.36)$$

where we have introduced the intermediate states $\beta\beta'$. The kernel $F_{\alpha\alpha'\beta\beta'}(t)$ is given exactly by Eq.(2.44) and Δ is the difference in the frequencies defined in Eq.(2.37).

We consider the structure of the integral term I by inverting the order of integration,

$$I = \frac{1}{\tau_o} \sum_{\beta,\beta'} \{[\int_o^{\tau_o} ds \int_o^{\tau_o} dx - \int_o^{\tau_o} ds \int_o^{s} dx] F_{\alpha\alpha'\beta\beta'}(s) e^{i\Delta x} \langle\beta|\sigma_o(0,\tau_2,\ldots)|\beta'\rangle\} \tag{3.37}$$

Provided that $F_{\alpha\alpha'\beta\beta'}(t)$ decays rapidly on the fast time scale, the second term on the right-hand side of Eq.(3.37) may be neglected. The fast decay is expected because $F_{\alpha\alpha'\beta\beta'}(t)$ involves lattice correlation functions. Consequently, any secular behavior which is present must come from the first term. Tere are two cases of interest. The first case is when $\Delta \neq 0$. In this case as τ_o approaches infinity the x integration approaches zero. Consequently, the terms with $\Delta \neq 0$ do not give rise to secular behavior. In the case when $\Delta = 0$ the x integration gives rise to a term τ_o and we obtain asymptotically for the integral term as $t_o \to \infty$

$$I = - \sum_{\beta,\beta'}{}' R_{\alpha\alpha'\beta\beta'} \langle\beta|\sigma_o(0,\tau_2\ldots)|\beta'\rangle \tag{3.38}$$

where $R_{\alpha\alpha'\beta\beta'}$ is the Redfield relaxation tetradic defined by Eq. (2.49). The prime on the summation indicates that only those terms that satisfy the condition $\Delta = 0$ are to be included.

Consequently, in order to eliminate secular behavior in Eq.(3.36) we must set

$$\frac{\partial}{\partial\tau_2} \langle\alpha|\sigma_o(0,\tau_2,\ldots)|\alpha'\rangle = \sum_{\beta,\beta'}{}' R_{\alpha\alpha'\beta\beta'} \langle\beta|\sigma_o(0,\tau_2,\ldots)|\beta'\rangle \tag{3.39}$$

for those terms for which $\Delta = 0$. It follows that the time dependence of σ_2 will be determined by the equation

$$e^{i\omega_{\alpha\alpha'}\tau_o}\langle\alpha|\sigma_2(\tau_o,\ldots)|\alpha'\rangle$$

$$= \sum_{\beta,\beta'}'' \int_o^{\tau_o} dx \int_o^x dy F_{\alpha\alpha'\beta\beta'}(x-y) e^{i\Delta x} \langle\beta|\sigma_o(o,\tau_2,\ldots)|\beta'\rangle \quad (3.40)$$

where the double prime on the sum indicates that only those terms are retained for which $\Delta \neq 0$. From this equation[16] we can determine the equation of motion for σ_2 on the τ_o time scale,

$$\frac{\partial}{\partial\tau_o}\langle\alpha|\sigma_2(\tau_o,\ldots)|\alpha'\rangle + i\omega_{\alpha\alpha'}\langle\alpha|\sigma_2(\tau_o,\ldots)|\alpha'\rangle \quad (3.41)$$

$$= \sum_{\beta,\beta'}'' \int_o^{\tau_o} ds F_{\alpha\alpha'\beta\beta'}(s) e^{-i\omega_{\beta\beta'}\tau_o} \langle\beta|\sigma_o(0,\tau_2,\ldots)|\beta'\rangle$$

For large τ_o the upper limit of the integral may be extended to infinity and one obtains the equation

$$\frac{\partial}{\partial\tau_o}\langle\alpha|\sigma_2(\tau_o,\ldots)|\alpha'\rangle = -i\omega_{\alpha\alpha'}\langle\alpha|\sigma_2(\tau_o,\ldots)|\alpha'\rangle$$
$$+ \sum_{\beta,\beta'}'' R_{\alpha\alpha'\beta\beta'}\langle\beta|\sigma_o(\tau_o,\tau_2,\ldots)|\beta'\rangle \quad (3.42)$$

where we have used Eqs.(2.49) and (3.22).

We have determined the equation of motion for σ_o on the τ_o, τ_1, and τ_2 time scales according to Eqs.(2.22), (3.30), and (3.40). We have also determined that $\sigma_1 = 0$ and the equation of motion for σ_2 on the τ_o scale (Eq.(3.42)) for times $\tau_o >> \tau_c$. It follows from Eqs.(3.9) and (3.10) that to order λ^2

$$\frac{\partial}{\partial t}\sigma = \frac{\partial}{\partial\tau_o}\sigma_o + \lambda^2\frac{\partial}{\partial\tau_2}\sigma_o + \lambda^2\frac{\partial}{\partial\tau_o}\sigma_2 \quad (3.43)$$

or, using the results we have obtained in Eqs.(3.22), (3.40), and (3.42),

$$\frac{\partial}{\partial t}\langle\alpha|\sigma(t)|\alpha'\rangle = -i\omega_{\alpha\alpha'}[\langle\alpha|\sigma_o|\alpha'\rangle + \lambda^2\langle\alpha|\sigma_2|\alpha'\rangle] \quad (3.44)$$
$$+ \lambda^2 \sum_{\beta,\beta'} R_{\alpha\alpha'\beta\beta'}\langle\beta|\sigma_o|\beta'\rangle$$

where the sum is no longer restricted. Since $\sigma = \sigma_o + \lambda^2\sigma_2$, to order λ^2 we may replace σ_o by $\sigma_o + \lambda^2\sigma_2$ in the term on the right-hand side of Eq.(3.44),

$$\frac{\partial}{\partial t} <\alpha|\sigma|\alpha'> = -i\omega_{\alpha\alpha'}[<\alpha|[\sigma_o(\tau_o,\ldots)+\lambda^2\sigma_2(\tau_o,\ldots)]|\alpha'>] \qquad (3.45)$$

$$+ \lambda^2 \sum_{\beta,\beta'} R_{\alpha\alpha'\beta\beta'}<\beta|(\sigma_o(\tau_o,\ldots)+\lambda^2\sigma_2(\tau_o,\ldots))|\beta'>$$

This equation is valid to order λ^2 uniformly in time since all secular behavior has been eliminated by the MTS method. When we restrict $\sigma = \sigma_o + \lambda^2\sigma_2$ to the physical time line we recover the Redfield equation.

The MTS method provides a compact derivation of the Redfield equation. It focuses attention on the two time scales $\tau_o = t$ and $\tau_2 = \lambda^2 t$ which occur in the spin relaxation problem. Of course, it should be possible to use the systematic MTS procedure to obtain higher order corrections to Eq.(3.45). The corrections one might investigate are of three types: 1) The effect of a more general initial condition that includes some initial correlation between the spins and the lattice; 2) The short time corrections that arise when τ_o is not much greater than τ_c. Finally, one might wish to continue the analysis to higher order in λ. This would entail studying the behavior of σ_o on time scales slower than τ_2, σ_2 on time scales slower than τ_o, and the calculation of σ_3, σ_4, ... etc.

REFERENCES

1) The derivation of the Redfield-Langevin equation has been accomplished in collaboration with Dr. John Albers. In preparing these lectures I have profited from discussions with Miss Barbara Yoon.

2) See, for example, A.G. Refield in Advances in Magnetic Resonance Academic Press, Inc., New York and London,(1965), Vol.I, p.1 and references cited therein.

3) M. Bixon and R. Zwanzig, Phys. Rev. 187, 267 (1969).

4) J. Albers, J.M. Deutch and I. Oppenheim, J. Chem. Phys. 55,2613 (1971).

5) P.N. Argyres and P.L. Kelley, Phys. Rev. 134A, 98 (1964). See also R.I. Cukier, Ph.D. Thesis, Princeton University (1969), unpublished.

6) The type of projection operator employed by Mori in his generalized theory of Brownian motion [Prog. Theor. Phys. 33, 423 (1965)] is suitable for obtaining equations of motion for macroscopic variables not distribution functions. For magnetic relaxation the Mori projection operator is

$$P_M(\ldots) = (\ldots, \underset{\sim}{S}) \cdot (\underset{\sim}{S},\underset{\sim}{S})^{-1} \cdot \underset{\sim}{S}$$

where (A,B) is an appropriately defined equilibrium average.

7) The statement that σ^o is an eigenmatrix of R with zero eigenvalue is universally made. Explicit verification of this fact in a complete quantum mechanical treatment is quite difficult. Any difference which might exist in special cases between σ^o and the correct form of the equilibrium eigenmatrix of R would, under normal conditions, not be experimentally observable.

8) I.V. Aleksandrov, The Theory of Nuclear Magnetic Resonance (Academic Press, Inc., New York and London , 1966), pp. 65-74.

9) J.M. Deutch and Irwin Oppenheim in Advances in Magnetic Resonance, ed. J.S. Waugh (Academic Press, Inc., New York and London, 1968), Vol. 3, p. 43.

10) J. Freed, J. Chem. Phys. 49, 376 (1968). It should be noted that Freed's development explicitly assumes that the lattice motion may be treated in a stochastic manner. In contrast, in the development presented here, the Hamiltonian is entirely mechanical.

11) R. Kubo in Fluctuation Relaxation and Resonance in Magnetic Systems, ed. D. Ter Haar (Oliver and Boyd, London, 1962), p. 23.

12) A detailed review of these various procedures may be found in the excellent review article by C.S. Johnson, Jr. in Advances in Magnetic Resonances, ed. J.S. Waugh (Academic Press, Inc., New York and London, 1965), Vol. I, p. 33.

13) M.C. Wang and G.E. Uhlenbeck, Rev. Mod. Phys. 17, 323 (1945).

14) R.I. Cukier and J.M. Deutch, J. Chem. Phys. 50, 36 (1969).

15) A. Abragam, The Principles of Nuclear Magnetism (Oxford University Press, London, 1961), Chapter 8, p. 276 et seq.

16) The analysis presented in reference 14 includes only the $\Delta = 0$ terms. Here we extend the MTS treatment to include the $\Delta \neq 0$ contributions.

ESR RELAXATION AND LINESHAPES FROM THE GENERALIZED CUMULANT AND RELAXATION MATRIX VIEWPOINT

Jack H. Freed

Department of Chemistry, Cornell University

Ithaca, New York 14850

VIII.1. GENERAL APPROACH[1]

We start with the time rate of change of the spin density matrix for a single spin system:

$$\dot{\sigma}(t) = -i[\mathcal{H},\sigma] \equiv -i\mathcal{H}^{\times}\sigma \tag{1}$$

where $\mathcal{H} = \mathcal{H}_0 + \mathcal{H}_1(t)$, and we use the Kubo[2,3] notation for superoperators: $A^{\times}$, such that $A^{\times}B = [A,B]$. We define an interaction representation by:

$$\sigma^{\ddagger}(t) = e^{i\mathcal{H}_0^{\times}t}\sigma = e^{i\mathcal{H}_0 t}\sigma e^{-i\mathcal{H}_0 t} \tag{3}$$

and

$$\mathcal{H}_1^{\ddagger}(t) = e^{i\mathcal{H}_0^{\times}t}\mathcal{H}_1(t) = e^{i\mathcal{H}_0 t}\mathcal{H}_1(t)e^{-i\mathcal{H}_0 t} \quad . \tag{4}$$

Then, eq. (1) becomes in the interaction representation:

$$\dot{\sigma}^{\ddagger}(t) = -i[\mathcal{H}_1^{\ddagger}(t),\sigma^{\ddagger}] \quad . \tag{5}$$

We are, however, interested in an ensemble averaged $\sigma(t)$ which we denote by $\langle\sigma(t)\rangle$. One method of solution of eq. 5 is to iterate and then take ensemble averages.[4,5] But it is better to first write the formal solution of eq. 5 in terms of an ordered exponent, designated by the subscript 0 (cf. Ch. I by Muus on time ordering):

$$\langle\sigma^{\ddagger}(t)\rangle = \langle \exp_0[-i\int_0^t dt'\mathcal{H}_1^{\ddagger}(t')^{\times}]\rangle\sigma(0) \tag{6}$$

where $\sigma(0)$ is an arbitrary initial value for the ensemble and the operator exponential is defined in terms of its infinite series expansion:

$$\langle \exp_0 [-i\int_0^t dt' \mathcal{H}_1^{\ddagger}(t')^{\times}] \rangle = \sum_{n=0}^{\infty} M_n(t) \tag{7}$$

where $M_n(t) = \frac{(-i)^n}{n!} \langle 0 [\int_0^t dt' \mathcal{H}^{\ddagger}(t)]^n \rangle = (-i)^n \int_0^t dt_1 \int_0^{t_1} dt_2 \ldots$

$$\ldots \int_0^{t_{n-1}} dt_n m_n(t_1, t_2 \ldots t_n) \text{ for } n \geq 1 \tag{8}$$

and $M_0(t) = 1$ (8a)

Here

$$m_n(t_1, t_2 \ldots t_n) = \langle \mathcal{H}_1^{\ddagger}(t_1)^{\times} \mathcal{H}_1^{\ddagger}(t_2)^{\times} \ldots \mathcal{H}_1^{\ddagger}(t_n)^{\times} \rangle \tag{9}$$

is a generalized n^{th} order time correlation function of the random operator, $\mathcal{H}_1^{\ddagger}(t)$. It may also be regarded as a generalized moment.

In the cumulant method,[1-3] we replace the expansion eq. by the ordered exponent

$$\langle \exp_0 [-i\int_0^t dt' \mathcal{H}_1^{\ddagger}(t')^{\times}] \rangle = \exp_0 K(t) \tag{10}$$

where $K(t) = \sum_{n=1}^{\infty} K_n(t)$. (11)

Each K_n is still an operator and is of nth order in $\mathcal{H}_1^{\ddagger}(t)^{\times}$. The precise definition of the nth order cumulant is given by the infinite series expansion

$$\exp_0 [\sum_{n=1}^{\infty} K_n(t)] \equiv \sum_{p=0}^{\infty} (p!)^{-1} 0 [\sum_{n=1}^{\infty} K_n(t)]^p \tag{12}$$

where we must preserve the time ordering in the cumulants. The K_n are then obtained by equating the terms in the two expansion eqs. 7 and 12 of the same order in $\mathcal{H}_1^{\ddagger}(t)^{\times}$. Thus to fourth order one has:

$$K_1 = M_1 \tag{13a}$$

$$K_2 = M_2 - \tfrac{1}{2}0(M_1{}^2) \tag{13b}$$

$$K_3 = M_3 - \tfrac{1}{2}[0(M_1 M_2) + 0(M_2 M_1)] + \tfrac{1}{3}0(M_1{}^3) \tag{13c}$$

$$\begin{aligned} K_4 = M_4 &- \tfrac{1}{2}0(M_2{}^2) - \tfrac{1}{2}[0(M_1 M_3) + 0(M_3 M_1)] \\ &+ \tfrac{1}{3}[0(M_1{}^2 M_2) + 0(M_1 M_2 M_1) + 0(M_2 M_1{}^2)] \\ &- \tfrac{1}{4}0(M_1{}^4). \end{aligned} \tag{13d}$$

The nature of the ordering prescription is obtained directly from eq. 8. Thus for example:

$$O(M_1{}^2) = O[-i\int_0^t dt_1 m_1(t_1)]^2$$

$$= -2\int_0^t dt_1 \int_0^{t_1} dt_2 m_1(t_1) m_1(t_2). \tag{14}$$

Thus, $K_2 = -\int_0^t dt_1 \int_0^{t_1} dt_2 [m_2(t_1, t_2) - m_1(t_1)m_1(t_2)] =$

$$-\int_0^t dt_1 \int_0^{t_1} dt_2 \kappa_2(t_1, t_2) \tag{14a}$$

and is the dominant term which gives the Redfield theory.[6] In general, Kubo[2] shows that

$$K_n(t) = (-i)^n \int_0^t dt_1 \int_0^{t_1} dt_2 \ldots \int_0^{t_{n-1}} dt_n \kappa_n(t_1, t_2 \ldots t_n) \tag{15}$$

where the κ_n are the appropriate collections of the m_n:

$$\kappa_n(t, t_2 \ldots t_n) \equiv \langle O\mathcal{H}_1^\ddagger(t_1)^\times \mathcal{H}_1^\ddagger(t_2)^\times \ldots \mathcal{H}_1^\ddagger(t_n)^\times \rangle_c \tag{15a}$$

The most important properties of the cumulants κ_n are the following:

1) They are zero if any one of the (random) variables contained in them are uncorrelated with the others. For example, if $\mathcal{H}_1(t_i)$ is uncorrelated between two times t_1 and t_2, then we have:

$$m_2(t_1, t_2) = m_1(t_1) m_1(t_2) \tag{16a}$$

so $K_2(t) = 0.$ (16b)

This property automatically removes potential divergences in calculating the $M_n(t)$ as $t \to \infty$ by replacing them with the $K_n(t)$.

2) As opposed to a moment expansion, the cumulant expansion maintains at each level of the approximation a "generalized exponential-decay" solution for the equations of motion, thus bearing a closer relation to what is expected physically. This removes the problems and ambiguities of the earlier formulations[5,6,7]

which have to force a low order moment expansion into exponential form.

3) The approach is valid for all times t, whereas the earlier perturbation approaches require a coarse-graining in time solution, i.e. they are valid for $t \gg \tau_c$, where τ_c is a characteristic correlation time.

4) As long as the moment expansion is valid, the cumulant expansion represents a complete solution to the problem, which is valid for any stochastic process. It is usually assumed that: a) the stochastic process is stationary and b) it is ergodic. (a) means that the random process generating fluctuations in $\mathcal{H}_1(t)$ always remains at equilibrium, while (b) assumes that all the spin systems are able to experience the same range of effects from $\mathcal{H}_1(t)$. However these assumptions are not, in principle necessary. [In fact instead of a stochastic approximation, one can average over a canonical distribution of the ensemble, i.e. $\langle A\rangle_\beta = \mathrm{Tr}Ae^{-\beta H}/\mathrm{Tr}e^{-\beta H}$, $\beta = \hbar/kT$.]

To more precisely demonstrate how the cumulant expansion theory generalizes the Redfield-type perturbation theory, we proceed by rewriting the solution eq. 6 for $\sigma^\ddagger(t)$. We now drop the explicit averaging notation, and utilize eq. 10 to obtain:

$$\dot{\sigma}^\ddagger(t) = \dot{K}(t)\sigma^\ddagger(t)\ . \tag{17}$$

When the substitution $\tau_i = t_{i-1} - t_i$ is made, we have

$$\dot{K}_n(t) = (-i)^n \int_0^t d\tau_1 \int_0^{t-\tau_1} d\tau_2 \ldots . \int_0^{(t-\sum_{i=1}^{n-2}\tau_i)} d\tau_{n-1}\, K_n(t, t-\tau_1 \ldots . t-\sum_{j=1}^{n-1}\tau_j). \tag{18}$$

VIII.2. RELAXATION MATRIX AND SPECTRAL LINESHAPES

We recover the time-dependent perturbation theory by solving for $\dot{K}(t)$ for times $t \gg \tau_c$. Since the K_n vanish if any of the $\mathcal{H}_1^\ddagger(t)^\times$ in eq. 15a are uncorrelated, the only non-vanishing contributions to eq. 18 come from times $\tau_i \lesssim \tau_c$. Thus, a negligible error is introduced into eq. 18 by letting all the upper limits tend to infinity. We also note that any correlations in K_n which decay with time (e.g. an exponential decay) will go to zero. Thus the $\dot{K}(t)$ approaches an asymptotic steady-state value $\dot{K}(\infty)$ independent of t except for sinusoidal-type oscillations. (For stationary random processes the correlations depend only on the time differences τ_i, so we automatically obtain non-negligible steady-state values.) Thus we have

$$\dot{K}_n(t \gg \tau_c) \cong \dot{K}_n(t\to\infty) \equiv \exp(i\Omega_n t)R^{(n)} , \tag{19}$$

where $R^{(n)}$ and Ω_n are time-independent operators. [Note that it is also possible to obtain the limiting value from $\lim_{t\to\infty} \dot{K}_n(t) = \lim_{s\to 0} s\tilde{\dot{K}}_n(s)$, where $\tilde{\dot{K}}_n(s)$ is the Laplace-transform.] One can readily show that $\Omega_n = \mathcal{H}_0^\times$, for all n.[1] Thus, the long-time approximation to eq. 17, is, in matrix elements of $\mathcal{H}_0$:

$$\dot{\sigma}^{\ddagger}_{\alpha\alpha'}(t) = \sum_{\beta\beta'} \exp[i(\omega_{\alpha\alpha'}-\omega_{\beta\beta'})t]R_{\alpha\alpha'\beta\beta'}\sigma^{\ddagger}_{\beta\beta'}(t) \tag{20a}$$

or

$$\dot{\sigma}_{\alpha\alpha'}(t) = -i\omega_{\alpha\alpha'}\sigma_{\alpha\alpha'} + \sum_{\beta\beta'} R_{\alpha\alpha'\beta\beta'}\,\sigma_{\beta\beta'} \tag{20b}$$

where $R_{\alpha\alpha'\beta\beta'}$ is the time-independent relaxation matrix given to all orders by:

$$R_{\alpha\alpha'\beta\beta'} = \sum_{n=1}^{\infty} R^{(n)}_{\alpha\alpha'\beta\beta'} . \tag{21}$$

The cumulants then provide a precise prescription for generating all the $R^{(n)}$.

We note from eqs. 18 and 15a that $R^{(n)}$ is of order $\langle|\mathcal{H}_1^{\ddagger}(t)|^n\rangle\tau_c^{n-1}$, so convergent expansions are expected only for $|\mathcal{H}_1^{\ddagger}(t)^2|^{\frac{1}{2}}\tau_c < 1$. When this condition is not fulfilled the relaxation-matrix approach involving times $t \gg \tau_c$ is no longer appropriate, and one must investigate $\dot{K}(t)$ more carefully for finite t.

Note that, in general, the matrix elements $R_{\alpha\alpha'\beta\beta'}$ are complex where $\mathrm{Re}R_{\alpha\alpha'\beta\beta'}$ gives the relaxation effects and therefore must be intrinsically negative; while $\mathrm{Im}R_{\alpha\alpha',\beta\beta'}$ are the dynamic frequency shifts which must be added to the first term on the rhs of eq. 20b.

We note that the above method also applies to the solution of ensemble averaged operators, such as $S_x(t)$. Here we obtain:

$$\dot{S}_x^{\ddagger}(t) \overset{t\gg\tau_c}{=} e^{-it\mathcal{H}_0^{\times}} R^* S_x^{\ddagger}(t) \tag{22}$$

where R^* is the complex conjugate of R, or

$$\dot{S}_{x_{\alpha\alpha'}}(t) = i\omega_{\alpha\alpha'} S_{x_{\alpha\alpha'}} + \sum_{\beta\beta'} R^*_{\alpha\alpha'\beta\beta'} S_{x_{\beta\beta'}} \; . \tag{23}$$

Eq. 23 expresses the fact that we can get coupled relaxation of different transitions corresponding to different matrix elements $S_{x_{\alpha\alpha'}}$ and $S_{x_{\beta\beta'}}$. One must therefore diagonalize the non-Hermitian matrix with elements

$$M_{\alpha\alpha',\beta\beta'} = [i\omega_{\alpha\alpha'}\delta_{\alpha\alpha',\beta\beta'} + R^*_{\alpha\alpha',\beta\beta'}] \; . \tag{24}$$

Clearly, if

$$|R^*_{\alpha\alpha',\beta\beta'}| \ll |\omega_{\alpha\alpha'} - \omega_{\beta\beta'}| \tag{25}$$

the $\alpha\alpha'$ transition is "decoupled" from the $\beta\beta'$ transition and we can treat them independently. The most common case where we must consider coupled relaxation is when there are degenerate transitions, e.g.

$$\omega_{\alpha_i\alpha_i'} = \omega_{\alpha_j\alpha_j'} \equiv \omega_0^K \text{ for all } i, j = 1 \ldots. N \; . \tag{26}$$

Then in this N-fold degenerate transition subspace, we must diagonalize the N-fold matrix $R^K_{i,j}$.

At this stage we discuss line-shapes from the point of view of linear response theory, so we can write a normalized line-shape function[8] (see also Ch. X by Kivelson):

$$I(\omega) = \frac{4}{\pi}\int_0^\infty G(t)\cos\omega t\,dt \tag{27}$$

where $G(t) = \mathrm{Tr}_s[S_x(t)S_x]$, i.e. a trace over spin states. (28)

a) Simple line - Suppose $S_{x_{\alpha\alpha'}}$ is uncoupled to any other transition. Then we have from eqs. 27, 28 and 23

$$G(t) = 2\cos(\omega_0 t)\exp(-t/T_2) \tag{29a}$$

$$\text{and } I(\omega) = \frac{4}{\pi} \frac{T_2}{1+T_2^2(\omega-\omega_0)^2} \tag{29b}$$

$$\text{where } \omega_0 = \omega_{\alpha\alpha'} - \text{Im}R_{\alpha\alpha'\alpha\alpha'} \tag{30a}$$

$$\text{and } T_2^{-1} = -\text{Re}R_{\alpha\alpha'\alpha\alpha'} \tag{30b}$$

i.e. a Lorentzian with transverse relaxation time T_2^{-1} and resonant frequency ω_0.

b) Multiple (Degenerate line)[8] - We assume eq. 26 applies. We let the $S_{x_{\alpha_i\alpha_i'}} \equiv S_{x,i}$ be the components $X_i^{(K)}$ of an N-dimensional vector $\vec{X}^{(K)}$. Let $U^{(K)}$ be the transformation that diagonalizes $R^{(K)}$; (we assume for now it is unitary, but see below):

$$\left(U^{(K)-1}R^{(K)}U^{(K)}\right)_{ij} = \lambda_i^{(K)}\delta_{ij} \tag{31}$$

$$\text{Then one has } I_K(\omega) = \frac{4}{\pi}\sum_{i=1}^{N} |Y_i^{(K)}|^2 \frac{T_{2,i}^{(K)}}{1+[T_{2,i}^{K}]^2(\omega-\omega_i^{(K)})^2} \tag{32}$$

where we have set

$$[-\text{Re}\lambda_i^{(K)}]^{-1} = T_{2,i}^{(K)} \tag{33a}$$

$$\omega_i^{(K)} = \omega_0^{(K)} - \text{Im}\lambda_i^{(K)} \tag{33b}$$

$$\text{and } Y_i^{(K)} = \sum_{j=1}^{N} [U^{(K)}]^{-1}_{ij} X_j^{(K)} . \tag{33c}$$

Thus we have a superposition of Lorentzians. Of course, if the $T_{2,i}$ for different i do not differ greatly, one can adequately describe the line as a single Lorentzian with an average T_2.[8] We note, at this stage, that it is only the matrix elements of $\mathcal{H}_1(t)$ which are off-diagonal in eigenstates of $\mathcal{H}_0$, that can lead to off-diagonal elements $R_{i,j}$. Physically, the non-zero $R_{i,j}$ for $i \neq j$ mean that the random perturbation is mixing up eigenlevels and/or transitions in a way that reflects its random time dependence. Thus, there is uncertainty on the individual molecular level, as to which are the correct zero-order transitions that are induced by the very weak perturbing rf field. When the correct zero-order eigenstates remain time-independent despite fluctuations of $\mathcal{H}_1(t)$, and when each eigen level is involved in no more than one transition obeying eq. 25, then it will be possible to render the R matrix automatically diagonal by the proper choice of the zero-order eigenstates.

The case of multiple lines[8] represented by a superposition of Lorentzians becomes important when 1) there is hyperfine structure resulting from equivalent nuclei or 2) when $S > \frac{1}{2}$, so there are degenerate ESR transitions. In the case of equivalent nuclei, it is important to distinguish between cases where the fluctuating hyperfine and dipolar parameters of all equivalent nuclei are the same at all times, in which case the nuclei are said to be completely equivalent, or where only their time average hyperfine terms are equal, i.e. equivalent (but not completely equivalent) nuclei. Modulation of $a_i(t)$ for equivalent nuclei can lead to alternating line widths, wherein the components of a multiple hf line are affected very differently. Completely equivalent nuclei are best treated in the coupled representation, i.e. nuclear spin eigenfunctions of J and J_z where $J = \sum_i I_i$ the total spin of the completely equivalent group of nuclei. Then each component line belonging to a particular set of values of J and J_z will behave as a distinct and independent line in its relaxation properties.[8,9]

The question now arises as to when the spectrum eq. 27 is determined by the asymptotic form eq. 23 valid for $t \gg \tau_c$. One knows from Fourier transform theory that $I(\omega-\omega_0)$, where ω_0 is the center frequency, is determined mainly by the behavior of G(t) around the region $t \sim |\omega-\omega_0|^{-1}$. Now we have seen that -Re(R) gives the line widths for the spectrum. Thus $I(\omega-\omega_0)$ is non-negligible only for $|\omega-\omega_0| \sim |\mathrm{Re}(R)|$, so only times $t \gtrsim |\mathrm{Re}(R)|^{-1}$ contribute to the main portion of the line. Thus if

$$|\tau_c \mathrm{Re}(R)| \ll 1 \tag{34a}$$

it follows that only times $t \gg \tau_c$ contribute to the main portion of the line. Also, since Im(R) gives the dynamic frequency shift, which shifts the resonant frequency from ω_0 to $[\omega_0 - \mathrm{Im}(R)]$, these arguments also require:

$$|\tau_c \mathrm{Im}(R)| \ll 1 \tag{34b}$$

Eqs. 34 are independent of the order to which R has been calculated.

A. Properties of the Relaxation Matrix

It is possible to generate all the $R^{(n)}$ utilizing the prescription given above. We wish to examine some of the properties of the $R^{(n)}$ now. We mainly examine $R^{(2)}$, although higher order terms are given elsewhere.[1] Thus,

$$R^{(2)}_{\alpha\alpha'\beta\beta'} = L_{\alpha\beta,\beta'\alpha'}(\omega_{\alpha\beta}) + L_{\alpha\beta,\beta'\alpha'}(\omega_{\beta'\alpha'})$$
$$-\delta_{\alpha'\beta'}\sum_{\gamma} L_{\alpha\gamma,\gamma\beta}(\omega_{\gamma\beta}) - \delta_{\alpha\beta}\sum_{\gamma} L_{\beta'\gamma,\gamma\alpha'}(\omega_{\beta'\gamma}). \tag{35}$$

Here the spectral densities $L(\omega)$ are given as one-sided Fourier transforms of correlation functions, which may be written in terms of sine and cosine Fourier transforms:

$$L_{\alpha\beta,\beta'\alpha'} = J_{\alpha\beta,\beta'\alpha'}(\omega) - iK_{\alpha\beta,\beta'\alpha'}(\omega)$$
$$= J_{\alpha\beta,\beta'\alpha'}(-\omega) + iK_{\alpha\beta,\beta'\alpha'}(-\omega) \tag{36}$$

where

$$J_{\alpha\beta,\beta'\alpha'} = \int_0^\infty [\langle \mathcal{H}_1(t)_{\alpha\beta}\mathcal{H}_1(t+\tau)_{\beta'\alpha'}\rangle -$$
$$\langle \mathcal{H}_1(t)_{\alpha\beta}\rangle\langle \mathcal{H}_1(t)_{\beta\alpha}\rangle]\cos\omega\tau d\tau \tag{37a}$$

$$K_{\alpha\beta,\beta'\alpha'} = \int_0^\infty [\langle \mathcal{H}_1(t)_{\alpha\beta}\mathcal{H}_1(t+\tau)_{\beta'\alpha'}\rangle -$$
$$\langle \mathcal{H}_1(t)_{\alpha\beta}\rangle\langle \mathcal{H}_1(t)_{\beta\alpha}\rangle]\sin\omega\tau d\tau \tag{37b}$$

It follows from the properties of stationary random functions, as well as from the Hermitian character of $\mathcal{H}_1(t)$, that the spectral densities obey the relations:

$$J_{\alpha\beta,\beta'\alpha'}(\omega) = J_{\beta'\alpha'\alpha\beta}(\omega) = J_{\beta\alpha\alpha'\beta'}{}^*(\omega) \tag{38a}$$

$$K_{\alpha\beta,\beta'\alpha'}(\omega) = K_{\beta'\alpha'\alpha\beta}(\omega) = K_{\beta\alpha\alpha'\beta'}{}^*(\omega) \tag{38b}$$

It thus follows that

$$\mathrm{Re}R^{(2)}_{\alpha\alpha',\beta\beta'} = \mathrm{Re}R^{(2)}_{\beta'\beta,\alpha'\alpha} = \mathrm{Re}R^{(2)*}_{\beta\beta',\alpha\alpha'} \tag{39a}$$

$$\mathrm{Im}R^{(2)}_{\alpha\alpha',\beta\beta'} = \mathrm{Im}R^{(2)}_{\beta'\beta,\alpha'\alpha} = \mathrm{Im}R^{(2)*}_{\beta\beta',\alpha\alpha'} \tag{39b}$$

where by Re and Im we more precisely mean the $J(\omega)$ and $K(\omega)$ contributions respectively. Thus, both $\mathrm{Re}R^{(2)}_{\alpha\alpha',\beta\beta'}$ and $\mathrm{Im}R^{(2)}_{\alpha\alpha',\beta\beta'}$ are Hermitian matrices. We find in all examples of interest that they are also real, so they are real symmetric matrices. Thus the complete $R^{(2)}$ matrix is symmetric, but is neither Hermitian nor real. It may be diagonalized by a complex orthogonal matrix.

One often rewrites $\mathcal{H}_1(t)$ in the form:

$$\mathcal{H}_1(t) = \sum_q F_q(t)A_q \tag{40}$$

where $F_q(t)$ is a function of spatial variables and is thus a randomly varying classical function of the time, and A_q contains only the spin operators. Then

$$L_{\alpha\beta,\beta'\alpha'}(\omega_{\beta'\alpha'}) = \sum_{q,r} A_{q\ \alpha\beta}A_{r\beta'\alpha'}\,\ell(q,r;\omega_{\beta'\alpha'}) \tag{41a}$$

$$L_{\alpha\beta,\beta'\alpha'}(\omega_{\alpha'\beta'}) = \sum_{q,r} A_{q\ \alpha\beta}A_{r\beta'\alpha'}\,\ell(r,q;\omega_{\alpha\beta}) \tag{41b}$$

where the classical spectral densities $\ell(\omega)$ are Fourier transforms of correlation functions $g(\tau)$

$$\ell_2(q,r;\omega_{\alpha\beta}) = \int_0^\infty g_2(q,r;\tau)\exp(-i\omega_{\alpha\beta}\tau)d\tau \tag{42a}$$

$$\text{where } g_2(q,r;\tau) = \langle F_q(t)F_r(t-\tau)\rangle - \langle F_q(t)\rangle\langle F_r(t-\tau)\rangle. \tag{42b}$$

In an analogous manner to eq. 36, we can separate ℓ_2 into even and odd parts with respect to ω. Thus,

$$\ell_2(q,r;\omega) = j(q,r;\omega) - ik(q,r;\omega). \tag{43}$$

The higher order cumulants are found to involve one-sided Fourier transforms of higher-order time correlation functions, e.g. for $\langle F_q(t)\rangle = 0$ we get:

$$g_3(q,r,s;\tau_1,\tau_2) = \langle F_q(t)F_r(t-\tau_1)F_s(t-\tau_1-\tau_2)\rangle \tag{44}$$

and

$$g_4(q,r,s,u;\tau_1,\tau_2,\tau_3) =$$
$$\langle F_q(t)F_r(t-\tau_1)F_s(t-\tau_1-\tau_2)F_u(t-\tau_1-\tau_2-\tau_3)\rangle$$
$$- \langle F_q(t)F_r(t-\tau_1)\rangle\langle F_s(t-\tau_1-\tau_2)F_u(t-\tau_1-\tau_2-\tau_3)\rangle. \tag{45}$$

For a simple line uncoupled to other lines we have to consider only the diagonal line width term. Thus from $R^{(2)}$ we have:

$$T_2{}^{-1}{}_{a,b} = -\mathrm{Re}R_{ab,ab} = [J_{aa,aa}(0)+J_{bb,bb}(0)-2J_{aa,bb}(0)]$$
$$+\sum_{\gamma\neq a} J_{a\gamma,\gamma a}(\omega_{\gamma a}) + \sum_{\gamma\neq b} J_{b\gamma,\gamma b}(\omega_{b\gamma})$$
$$= \int_0^\infty \langle\omega(t)\omega(t-\tau)\rangle d\tau + \tfrac{1}{2}\left(\sum_{\gamma\neq a} W_{a\gamma} + \sum_{\gamma\neq b} W_{b\gamma}\right) \tag{46}$$

where we have let:

$$\omega(t) = [\mathcal{H}_1(t)_{aa} - \langle\mathcal{H}_1(t)\rangle_{aa}] - [\mathcal{H}_1(t)_{bb} - \langle\mathcal{H}_1(t)\rangle_{bb}] \tag{47}$$

The term in $\omega(t)$ in eq. 46 is thus seen to be a secular line width contribution, i.e. fluctuations in the energy difference between the two states a and b. The terms in eq. 46 of type $W_{a\gamma}$, $W_{b\gamma}$ give the mean of all the transitions away from states a and b. These are the non-secular terms yielding line-broadening due to the Heisenberg uncertainty in lifetime effect. Thus, as we shall see later, the transition probability between states a→b or $W_{a\rightarrow b}$ is given by

$$W_{a\rightarrow b} = 2J_{ba,ab}(\omega_{ab}) \tag{48}$$

so we have $W_{b\rightarrow a} = 2J_{ab,ba}(\omega_{ba}) = W_{a\rightarrow b}$ by eq. 37a. This is the usual microscopic reversibility, which is, however, not in general true in higher order, although $W_{Kb\rightarrow Ka} = W_{a\rightarrow b}$ where K is time reversal operator, if the Hamiltonian is invariant under time-reversal.[10] The equivalent second-order time-dependent transition probability is, for non-resonant intermediate states, (i.e. $\omega_{a\gamma}$, $\omega_{b\gamma}$, $\omega_{a\gamma'}$, $\omega_{b\gamma'}$, very large) and for $\mathcal{H}_1(t)_{ab} = 0$:

$$W^{(2)}_{a\to b} = -\sum_{\gamma,\gamma'} [J_{a\gamma,\gamma b,b\gamma',\gamma' a}(\omega_{\gamma a},\omega_{ba},\omega_{b\gamma'})$$

$$+ J_{a\gamma,\gamma b,b\gamma',\gamma' a}(\omega_{a\gamma'}\omega_{ab},\omega_{\gamma b})] \tag{49}$$

where $J_{a\gamma,\gamma b,b\gamma',\gamma' a}(\omega_i,\omega_j,\omega_k) =$

$$\mathrm{Re}\sum_{q,r,s,u} A_{q\ \alpha\gamma}A_{r\ \gamma\beta}A_{s\beta'\gamma'}A_{u\gamma'\alpha'}\ell_4(q,r,u,s;\omega_i,\omega_j,\omega_k) \tag{50}$$

$$\text{with } \ell_4(\ldots) = \int_0^\infty d\tau_1 e^{-i\omega_i\tau_1}\int_0^\infty d\tau_2 e^{-i\omega_j\tau_2}\int_0^\infty d\tau_3 e^{-i\omega_k\tau_3} \times$$

$$\times\ g_4(q,r,s,u;\tau_1,\tau_2,\tau_3) \tag{51}$$

But now we must be careful in defining $W_{a\to b}$. Specifically we let

$$\dot\sigma_{aa} = +\sum_b R_{aa'bb}\sigma_{bb} = W_{b\to a}\sigma_{bb} - \sum_{b\neq a} W_{a\to b}\sigma_{aa} \tag{52}$$

One then finds that:

$$W^{(2)}_{b\to a} = \sum_{\gamma,\gamma'} J_{a\gamma,\gamma b,b\gamma',\gamma' a}(\omega_{\gamma a},\omega_{ba},\omega_{\gamma' a})$$

$$+ J_{a\gamma,\gamma b,b\gamma',\gamma' a}(\omega_{a\gamma'},\omega_{ab},\omega_{a\gamma}). \tag{53}$$

Given that $\omega_{a\gamma}$, $\omega_{b\gamma}$, $\omega_{a\gamma'}$ and $\omega_{b\gamma'}$ are very large, the terms in $W_{b\to a}$ are proportional to $(\omega_{\gamma a}\omega_{\gamma' a})^{-1}$, while the equivalent terms in $W_{b\to a}$ (i.e. having the same matrix elements), are proportional to $(\omega_{\gamma a}\omega_{\gamma' b})^{-1}$ or $(\omega_{\gamma' a}\omega_{\gamma b})^{-1}$. Thus $W_{b\to a}$ and $W_{a\to b}$ are not quite equal if $E_a \neq E_b$.

A particular example of such a second-order lattice-induced transition in liquids is the g-tensor mechanism. In general, it is the spin-orbit (SO) and orbit-field (O-F) interactions $\lambda\vec{L}\cdot\vec{S}$ and $B_e\vec{L}\cdot\vec{B}_o$ which are random functions, since $\vec{L}$ is quantized in the molecular frame, while S for a polyatomic molecule and large values of B_o is coupled to the laboratory frame. From this point of view, g-tensor relaxation effects come from $R^{(4)}$, since they are quadratic in both S-O and O-F terms. (One must now consider matrix elements in combined spin and electronic space.) Furthermore, the g-shift is naturally found to be the dynamic frequency shift of $R^{(2)}$ arising from these terms. There is an associated rotational spin-orbit (RSO) relaxation mechanism from $ReR^{(2)}$, which should be unimportant. Various Orbach-type processes (cf. Kivelson Ch. X) illustrated by:

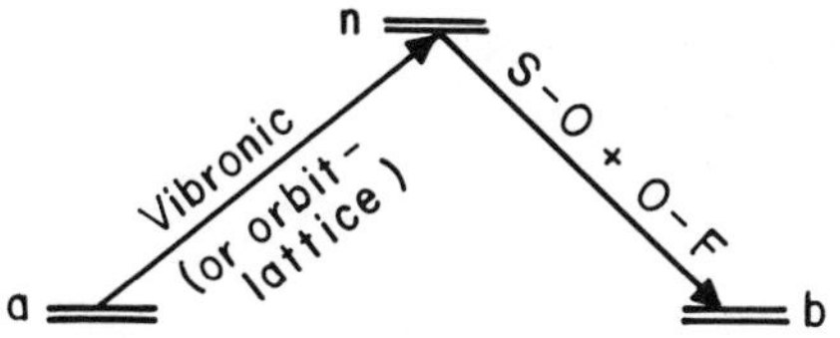

can be calculated in this way even when the orbit-lattice or vibronic modulation is approximated by a stochastic process that is independent of the rotational motion modulating the S-O process.[11] The relaxation terms are quadratic in both vibronic (or O-L) and in S-O, while the associated dynamic g-shift-type terms are quadratic in vibronic (or O-L) terms but linear in both S-O and O-F terms.

Note that when $\omega_{na} \sim \omega_{nb} \gg \omega_{ba}$, the results outlined above for $W^{(2)}_{b\rightarrow a}$ are well treated by utilizing a "quasi-solid" approximation and then only $R^{(2)}$. The "quasi-solid" approximation in the case of the g-tensor is just the usual approach of first calculating the solid g-tensor from S-O and O-F terms <u>and then</u> introducing rotational modulation into the resulting spin Hamiltonian.

Note that one usually finds $T_1 = T_2$ for the spin-orbit induced relaxation mechanisms in liquids.[11] The other mechanisms such as spin orbit pulse (SOP) and spin-orbit tunneling (SOT) are discussed elsewhere.[11]

VIII.3. NON-ASYMPTOTIC SOLUTIONS

The general solutions for $\dot{K}_n(t)$ instead of $R^{(n)}$, are obtained by replacing the infinite upper limits of the time integrals in the spectral densities by the appropriate finite values obtained from eq. 15. Let us examine the $\dot{K}_2(t)$ term (see also Nielsen, Ch. V). If we consider only a secular perturbation with

$$\langle \mathcal{H}_1(t) \rangle = 0 \tag{54}$$

and $$\Delta^2 \equiv \langle |\mathcal{H}_1(t)_{aa} - \mathcal{H}_1(t)_{bb}|^2 \rangle. \tag{55}$$

Then $$\dot{K}_2(t)_{ab,ab} = -\Delta^2 \int_0^t \bar{g}_2(\tau)d\tau \tag{56}$$

$$\text{with } \bar{g}_2(\tau) \equiv \frac{\langle \mathcal{H}_1(t)^\times \mathcal{H}_1(t-\tau)^\times \rangle_{ab,ab}}{\langle |\mathcal{H}_1(t)^\times|^2 \rangle_{ab,ab}} = \frac{\langle \mathcal{H}_1(t)^\times \mathcal{H}_1(t-\tau)^\times \rangle_{ab,ab}}{\Delta^2} . \tag{57}$$

If we assume a simple exponential decay:

$$\bar{g}_2(\tau) = \exp(-|\tau|/\tau_c) \tag{58}$$

$$\text{then } K_2(t) = -\Delta^2\tau_c^2[t/\tau_c - 1 + \exp(-|t|/\tau_c)] \tag{59}$$

$$\text{and } S^{\ddagger}_{x\alpha\alpha'}{}^{(2)}(t) = [\exp K_2(t)]_{\alpha\alpha',\alpha\alpha'} S_{x\alpha\alpha'}(0) \tag{60}$$

$$\text{Now } I^{(2)}(\omega) = \frac{1}{\pi}\mathrm{Re}\int_0^\infty e^{-i\omega t} S_x^{(2)}(t)$$

$$= \frac{\tau_c}{\pi} e^{\Delta^2\tau_c^2} \mathrm{Re}[(\Delta\tau_c)^{-2z}\gamma(z, \Delta^2\tau_c^2)] \tag{61}$$

where $\gamma(z,a)$ is the incomplete gamma function and $z = i(\omega-\omega_0)\tau_c + \Delta^2\tau_c^2$. A series expansion gives:

$$I^{(2)}(\omega) = \frac{e^{\Delta^2\tau_c^2}}{\pi\Delta} \sum_{n=0}^{\infty} \frac{(-)^n}{n!} (\Delta\tau_c)^{2n} \frac{(n/\Delta\tau_c)+\Delta\tau_c}{[(n/\Delta\tau_c)+\Delta\tau_c]^2 + [(\omega-\omega_0)^2/\Delta^2]} \tag{62}$$

which is a superposition of Lorentzians of width $n/\tau_c+\Delta^2\tau_c$ and resonance frequency ω_0. But the intensity of each such component is proportional to $(\Delta^2\tau_c^2)^n$, i.e. this is an expansion in powers of $(\Delta\tau_c)^2$. Thus, within the validity of utilizing only $K^{(2)}(t)$ we retain only the n=0 and 1 terms to get

$$I^{(2)}(\omega) \cong \frac{\Delta^2\tau_c}{\pi}\left[\frac{1}{(\Delta^2\tau_c)^2+(\omega-\omega_0)^2} - \frac{1}{\tau_c^{-2}+(\omega-\omega_0)^2}\right] \tag{63}$$

for $\Delta^2\tau_c^2 \ll 1$. There is thus a subsidary line, opposite in intensity to the main line, which has at resonance an amplitude $\Delta^4\tau_c^4$ times the subsidary line. The negative sign may be understood on simple physical grounds. It is well known that the Lorentzian line shape is not a very good approximation to the true line in the far wings, because its intensity does not decrease fast enough to yield convergent moments. The very broad subsidiary line will, however, be most effective in subtracting out some intensity from the wings of the main Lorentzian component, so that the resulting composite line does decay more rapidly. It is shown elsewhere[1] by considering $K_3(t)$, that the subsidiary line is shifted from ω_0 to $\omega_0+\Delta$.

In general, one must be more precise about specifying the stochastic process than by just writing eq. 58. We now discuss stochastic averaging.

VIII.4. STOCHASTIC AVERAGING (cf. Ch. II by Pedersen)

A. Gaussian Processes

A Gaussian random process which modulates $F_q(t)$ is characterized by the vanishing of all correlation functions $g_n(\tau_1 \ldots \tau_n)$ for $n>2$.[3] This leads to the result $K_n=0$ for $n>2$, so that the spectrum is entirely described by K_2, (where it is assumed that $\mathcal{H}_0$ has been defined so that $K_1=0$). Gaussian processes apply to physical situations, where each spin is weakly affected by many perturbers. Such an example is the exchange coupling in solids. Thus we have

$$K(t)_{\alpha\alpha',\beta\beta'} = \int_0^t \dot{K}(t')^{(2)}_{\alpha\alpha',\beta\beta'} dt' \tag{64}$$

where $\dot{K}(t')^{(2)}$ is obtained from $R^{(2)}$ by utilizing the time-dependent spectral density:

$$\ell_2(t,q,r;\alpha\beta) = \int_0^t g_2(q,r,\tau) e^{-i\omega_{\alpha\beta}\tau} d\tau \tag{65}$$

Once a proper choice is made for $g_2(q,r,\tau)$, it is, in principal, possible to determine the spectrum over the whole range from slow to fast modulation. To illustrate, we again assume a simple exponential decay, and a secular perturbation. Such a correlation function can only be an approximate choice, since it is discontinuous at $\tau=0$, but it is useful for illustrative purposes. In this case, the expansion eq. 62 yields <u>all</u> the higher order terms as an expansion in Lorentzians. For $\Delta\tau_c \ll 1$ only a simple Lorentzian of width $\Delta^2\tau_c$ is retained in the central portion. Gaussian random processes are known to lead to Gaussian widths in the limit of no modulation. This result is obtained in the present case by rewriting $K_2(t)$ with $\Delta\tau_c \equiv x \to \infty$ as:

$$K_2(t, x\to\infty) = \lim_{x\to 0}\{-x^2[\exp(-t\Delta/x)-1+t\Delta/x]\} = -\tfrac{1}{2}(\Delta t)^2 \tag{66}$$

so that $$e^{Kt} \xrightarrow[x \to \infty]{} \exp[-\tfrac{1}{2}(\Delta t)^2] \qquad (67)$$

and the Fourier transform of this result leads to a Gaussian line-shape with a second moment of Δ^2.* In general we note that a line is Lorentzian for $|\omega-\omega_0| \ll \tau_c^{-1}$ and Gaussian for $|\omega-\omega_0| \gg \tau_c^{-1}$.

Kubo and Tomita[5] find that the half-half width T_2^{-1} obtained from numerical solutions of eq. 62 is given well by:

$$T_2^{-2} \cong \frac{4\ln 2}{\pi} \Delta^2 \tan^{-1}\left\{\frac{\pi\tau_c}{(4\ln 2)T_2}\right\} \qquad (68)$$

Also the Lorentzian approximation is valid for $|\omega-\omega_0| = K\Delta$ where K obeys the inequality

$$\Delta\tau_c \leqq K\left[\tan\frac{\pi}{2}\left(1 - \frac{4}{9K^2}\right)\right]^{-1} \qquad (69)$$

B. Markov Processes

A stationary Markov process y(t) is completely determined by specifying the a priori probability $W(y_2)dy_2$ of finding y_2 in the range (y_2, y_2+dy_2) and the conditional probability $P(y_1|y_2,\tau)dy_2$ that given y_1 at an initial time, one finds y in the range (y_2, y_2+dy_2) at a time τ later. For Brownian motion problems the two are related by $W(y_2) = \lim_{\tau\to\infty} P(y_1|y_2,\tau)$. The joint probability

---

*A Gaussian correlation function of form $\bar{g}(\tau) = e^{-a^2\tau^2}$, which is continuous at τ=0, is often used in analyzing Gaussian random processes. For example in the simple theory of exchange narrowing, $a^2 = \frac{1}{4}\pi\omega_e^2$ where ω_e is the "exchange frequency". This choice of correlation function leads to:

$$K_2(t) = -\Delta^2/a[t\,\mathrm{erf}(at) + (1/2a)(e^{-a^2t^2}-1)]$$
$$= -\Delta^2/a[t - t\,\mathrm{erfc}(at) + \tfrac{1}{2a}(e^{-a^2t^2}-1)]$$

where $\mathrm{erf}(y) = 1 - \mathrm{erfc}(y) = \int_0^y e^{-z^2}dz$. The first expression is useful for expanding about the zero modulation region, $\Delta/a\to\infty$ (where $K_2(t) \cong -\Delta^2t^2$), while the latter is for $\Delta/a\to 0$ (where $K_2(t) \cong \Delta^2 t/a$). In neither case are simple Lorentzian expansions like that of eq. 62 obtained.

density of finding y in the range (y_2, y_2+dy_2) at any time t_2 and in the range (y_1, y_1+dy_1) at a later time $t_1 = t_2 + \tau$ is given by

$$W_2(y_2; y_1\tau) = W(y_2)P(y_2|y_1, \tau) \tag{70}$$

This may be generalized to an nth order joint probability density:

$$W_n(y_n, t_n; y_{n-1}\, t_{n-1}; \ldots y_2 t_2, y_1 t_1) = W_n(y_n; y_{n-1}\tau_{n-1}; \ldots y_2\tau_2, y_1\tau_1) =$$

$$= W(y_n) \prod_{i=1}^{n-1} P(y_{n-i+1}|y_{n-i}\tau_{n-i}) \tag{71}$$

where $t_1 > t_2 \ldots > t_{n-1} > t_n$.

Alternatively from the symmetry between the past and future we have:

$$W_n = W(y_1) \prod_{i=1}^{n-1} P(y_1|y_{i+1}, \tau_i) \tag{71a}$$

Then, for the nth order time correlation of the random function $F_q(y)$:

$$\langle F_{q_1}(t)F_{q_2}(t-\tau_1) \ldots F_{q_n}(t - \sum_{i=1}^{n-1}\tau_n)\rangle$$

$$= \int dy_n W(y_n)F_{q_n}(y_n)\iint\ldots\int dy_{n-1}\ldots dy_2 dy_1 \prod_{i=1}^{n-1} P(y_{n-i+1}|y_{n-i}, \tau_{n-i})$$

$$F_{q_{n-1}}\ldots F_{q_2}F_{q_1}. \tag{72}$$

The integrals in eq. 72 may be changed to summations when the variable y_i takes on a discrete set of values.

We shall consider internal rotations and (anisotropic) rotations below.

VIII.5. DIFFUSION MODELS

A. Internal Rotations

Very often the paramagnetic molecule will have an internal rotor, e.g. a methyl group or hydroxyl group. When one uses classical models for these motions, we have limiting cases of free rotation and torsional oscillations. We now consider the former.

We assume a rotor characterized by its moment of inertia I and friction constant $\beta' = \beta I$. By analogy with translational Brownian motion[12] we have a Langevin equation in the angle of rotation θ:[13]

$$\frac{d^2\theta}{dt^2} + \beta \frac{d\theta}{dt} = A(t) \tag{73}$$

where $A(t)$ is the random rotational acceleration. Now the probability of finding the rotor at any angle θ_0, $0 \le \theta_0 \le 2\pi$ is independent of θ_0 and is given by:

$$W(\theta_0)d\theta_0 = \frac{1}{2\pi}d\theta_0 \tag{74}$$

while the conditional probability that the rotor is at the angle θ at the time $t \geq 0$ if it was at θ_0 at the time $t=0$ is

$$P(\theta_0|\theta,t)d\theta = (4\pi Dt)^{-\frac{1}{2}}\exp[-(\theta-\theta_0)^2/4Dt]d\theta \tag{75}$$

This equation holds for long times, i.e. $t \gg \beta^{-1}$ and $D=kT/\beta I$. However, a real internal rotor may be better approximated by considering the possibility that free rotation occurs, in which the orientational changes are still partly determined by the angular velocity, which is not completely damped out for short enough times $t \lesssim \beta^{-1}$. We then use the more general solution:[12]

$$P(\theta_0,\dot{\theta}_0|\theta,t) = \left(\frac{Q}{\pi}\right)^{\frac{1}{2}}\exp\{-Q|\theta-\theta_0-\dot{\theta}_0(1-e^{-\beta t})/\beta|^2\} \tag{76}$$

where

$$Q = \frac{\beta}{2D}[2\beta t - 3 + 4e^{-\beta t} - e^{-2\beta t}]^{-1} \tag{77}$$

which has explicit dependence on the initial angular velocity $\dot{\theta}_0$, but becomes eq. 75 for $\beta t \gg 1$. I.e. for $\beta t \lesssim 1$ the process is a two dimensional Markoff process in $\theta, \dot{\theta}$. We now integrate eq. 76 over an initial Boltzmann distribution in velocity: $W(\dot{\theta}_0) = \left(\frac{I}{2\pi kT}\right)^{\frac{1}{2}} e^{(-I\dot{\theta}_0^2/2kT)}$ to obtain:

$$P(\theta_0|\theta,t) = \left(\frac{S}{\pi}\right)^{\frac{1}{2}}\exp-[(\theta-\theta_0)^2 S] \tag{78}$$

$$S = \frac{\beta}{4D}[\beta t - 1 + e^{-\beta t}]^{-1} \tag{79}$$

the appropriate generalization of eq. 75, including short times. Note that since the rotating group can, in principle, make many complete revolutions during the time t, the angle θ in eqs. 75-77 can be anywhere in the range $-\infty \le (\theta-\theta_0) \le \infty$. An equivalent eigenfunction expansion for eq. 77 is obtained as a Fourier Series expansion:

$$P(\theta_0 \,|\theta, t) = \frac{1}{2\pi}\sum_K e^{iK(\theta-\theta_0)}\exp[-K^2/4S] \tag{80}$$

where now the periodicity in θ has been accounted for so $0 \le \theta \le 2\pi$.[14] The time dependence for each term in K is thus seen to be identical to that for the Gaussian random process, and we may analyze the spectral densities similarly. Thus, for example consider:

$$g(\tau) = \langle e^{in\theta(\tau)}e^{in'\theta(0)}\rangle = e^{-n^2/4S}\delta_{n,-n'} \tag{81}$$

Then $j(\omega) = \mathrm{Re}\int_0^\infty e^{-i\omega\tau}g(\tau)d\tau = \beta^{-1}e^p\mathrm{Re}[p^{-z}\gamma(z,p)]$

$$= \frac{e^p}{\beta}\sum_{m=0}^{\infty}\frac{(-)^m}{m!}p^m\frac{m+p}{(m+p)^2+(\omega/\beta)^2} \tag{82}$$

where $p = n^2D/\beta$ and $z = i\omega+p$. Thus $j(\omega)$ is essentially a Lorentzian if $p \ll 1$, while it is essentially a Gaussian if $p \gg 1$, i.e.

$$j(\omega) \xrightarrow{t\to\infty} \frac{p\beta}{p^2\beta^2+\omega^2} \tag{83a}$$

$$j(\omega) \xrightarrow{t\to 0} \frac{1}{2}\sqrt{\frac{2\pi}{p\beta^2}}\;\exp\left[-\frac{1}{2}\frac{\omega^2}{\beta^2 p}\,t\right] \tag{83b}$$

Intermediate behavior consists of Lorentzian character near $\omega \sim 0$ and Gaussian character for asymptotically large ω.

Internal rotations are important mechanisms in the modulation of hyperfine interactions. That is, for an internal rotation with an n-fold symmetry axis, we may expand the instantaneous hyperfine interaction a(t) in a Fourier series as:

$$a(t) = \sum_{n=0}^{\infty}B_n\cos n\theta(t) = B_0 + \frac{1}{2}\sum_{n=1}^{\infty}B_n(e^{in\theta}+e^{-in\theta}) \tag{84}$$

Usually only the leading terms $n = 0, \pm 1$ are kept. Note from eqs. 80 and 81 we have

$$\langle(a(t)-B_0)(a(0)-B_0)\rangle = \frac{1}{2}\sum_{n=1}^{\infty}B_n^2 e^{-n^2/4S}. \tag{85}$$

Other, more detailed classical models are discussed elsewhere including analogous discussions of torsional oscillator correlation functions and spectral densities.[13]

B. Anisotropic Rotational Diffusion[14,15]

Suppose we assume that the rotation of a molecule can be compared to that of a rigid sphere of radius a in a medium of viscosity η, and that it can be described by a diffusion equation in the probability $P(\Omega, t)$

$$\frac{\partial P(\Omega,t)}{\partial t} = R\nabla_S^2 P(\Omega,t) \tag{86}$$

Here ∇_S^2 is the Laplacian operator on the surface of a sphere and we may use R, the rotational diffusion coefficient in the Stokes-Einstein form:

$$R = \frac{kT}{8\pi a^3 \eta}. \tag{87}$$

The probability $P(\Omega, t)$ gives the probability that the Euler angles between molecule fixed axes and an appropriate laboratory co-ordinate system have the value Ω at time t. We note that there is an exact analogy between eq. 86 and the time-dependent Schrodinger equation for a spherical top

$$i\hbar\frac{\partial\Psi(\Omega,t)}{\partial t} = -H\Psi = -\frac{\hbar^2}{2I}\nabla_S^2\,\Psi(\Omega,t)\ . \tag{88}$$

Thus the eigenfunctions of both eqs. 86 and 88 are the spherical harmonics $Y_m^1(\theta, \varphi)$.

When we have a non-spherical particle in an otherwise isotropic medium, then it is not surprising that the diffusion equation obeyed by $P(\Omega,t)$ is just the analogue of the Schrodinger equation for an asymmetric top. Thus we have

$$\frac{\partial P(\Omega,t)}{\partial t} = \Lambda P(\Omega,t) \tag{89}$$

where Λ is the quantum mechanical Hamiltonian for an asymmetric top, but now we replace the rotational constants for the rigid rotor in units of Planck's constant by the principal values of a diffusion tensor: R_1, R_2, and R_3 along molecular axes x', y', and z'.

The eigenfunctions and eigenvalues of Λ become those of the rigid rotor, i.e.

$$\Lambda\varphi_n(\Omega) = E_n\varphi_n(\Omega) \tag{90}$$

where the $\varphi_n(\Omega)$ are a complete orthonormal set of rigid rotor wave functions with eigenvalue E_n. Now we may write the general solution of eq. 89 as an eigenfunction expansion:

$$P(\Omega,t) = \sum_{n=0}^{\infty} C_n(t)\varphi_n(\Omega) \ . \tag{91}$$

Then from eqs. 89-91 and the orthonormality of the $\varphi_n(\Omega)$ we have:

$$\dot{C}_n(t) = -E_n C_n(t) \tag{91a}$$

with the solution:

$$C_n(t) = C_n(0)e^{-E_n t} \tag{91b}$$

If it is assumed that at t=0 the Euler angles are given by Ω_0, then $P(\Omega,\tau=0)$ is just the delta function $\delta(\Omega-\Omega_0)$, which may be expanded as:

$$\delta(\Omega-\Omega_0) = \sum_{n=0}^{\infty} \varphi_n^*(\Omega_0)\varphi_n(\Omega). \tag{92}$$

Thus $C_n(0) = \varphi_n^*(\Omega)$ (92a)

and we obtain an expression for the conditional probability $P(\Omega_0|\Omega,\tau)$:

$$P(\Omega_0|\Omega,\tau) = \sum_{n=0}^{\infty} \varphi_n^*(\Omega_0)\varphi_n(\Omega)e^{-E_n\tau} \tag{93}$$

Axially Symmetric Rotational Diffusion. When $R_1 = R_2 = R_\perp$ and $R_3 = R_{||}$, eq. 89 reduces to the Hamiltonian of the symmetric rotor, whose symmetry axis corresponds to the z′ axis. The φ_n are then the well-known symmetric rotor wave functions which may be classified in terms of the quantum numbers L, K, and M, i.e. $\varphi_n \rightarrow \varphi_{KM}^L(\Omega)$. It is useful to identify the wave functions with the Wigner rotation matrices. Thus

$$\varphi_{KM}^L(\Omega) = \left(\frac{2L+1}{8\pi^2}\right)^{\frac{1}{2}} \mathcal{D}_{KM}^L(\Omega) \tag{94}$$

Eq. 93 becomes

$$P(\Omega_0|\Omega,t) = \sum_{L,K,M} \left(\frac{2L+1}{8\pi^2}\right)\mathcal{D}_{KM}^{L*}(\Omega_0)\mathcal{D}_{KM}^L(\Omega)\exp(-E_{L,K}t) \tag{95}$$

where $E_{L,K} = R_\perp L(L+1) + (R_{||}-R_\perp)K^2$ (96)

Asymmetric Rotational Diffusion. When $R_1 \neq R_2 \neq R_3$, then the eigenfunctions of eq. 93 become rather complex, but they may be expressed as linear combinations of the symmetric rotor eigenfunctions. The nature of the problem is seen by writing the operator Λ as:

$$\Lambda = R_+\vec{\mathfrak{M}}^2 + (R_3 - R_+)\mathfrak{M}_{z'}^2 + \tfrac{1}{2}R_-(\mathfrak{M}_+^2 + \mathfrak{M}_-^2) \tag{97}$$

where $\vec{\mathfrak{M}}$ is an "angular momentum" operator such that $\mathfrak{M}^2\varphi_{KM}^L(\Omega) = L(L+1)\varphi_{KM}^L(\Omega)$ and $\mathfrak{M}_\pm\varphi_{KM}^L(\Omega) = [(L\mp K)(L\pm K+1)]^{\frac{1}{2}}\varphi_{K\pm1,M}^L(\Omega)$. Also $R_\pm = \frac{1}{2}(R_1 \pm R_2)$ and $\mathfrak{M}_\pm = \mathfrak{M}_{x'} \pm i\mathfrak{M}_{y'}$. The last term in eq. 97 couples only symmetric rotor functions for which L=L′, K=K′±2, and M=M′. The eigenfunctions corresponding to L=2 are given in ref. 15. When $R_- = 0$, the asymmetric rotor functions reduce to symmetric rotor functions (or simple linear combinations of the degenerate functions).

Eq. 93 may now be written for L=2 as:

$$P^{(2)}(\Omega_0|\Omega,t) = \sum_{K',M} \Phi_{K'M}^{(2)}(\Omega_0)\Phi_{K'M}^{(2)}(\Omega)e^{-E_{2,K'}t} \tag{98}$$

A simplification for the asymmetric diffusion case occurs when $R_3 \gg R_\pm$. This corresponds to rotational relaxation about the molecular z′-axis being much more rapid than about the other two axes. The solutions for asymmetric diffusion are then approximated by the axially symmetric diffusion solutions with $E_{2,\pm2} \cong 4R_3$ and $E_{2,0} \cong R_+$.

We may note that eq. 87 becomes for ellipsoids $R_i = kT/\beta_i$, where β_i are the principal values of the friction tensor. For an axially symmetric ellipsoid in the Stokes-Einstein approximation, letting $a_2 = a_3$, then the β_i become:

$$\beta_2 = \beta_3 = 3\pi\eta(a_1^4 - a_2^4)/3[(2a_1^2 - a_2^2)S - 2a_1] \tag{99a}$$

$$\beta_1 = 32\pi\eta a_2^2(a_1^2 - a_2^2)/3[2a_1 - a_2^2 S] \tag{99b}$$

where a_1, a_2, and a_3 are the lengths of the x′, y′ and z′ semi-axes. When $a_1 > a_2$

$$S = 2(a_1^2 - a_2^2)^{-\frac{1}{2}}\ln\{[a_1 + (a_1^2 - a_2^2)^{\frac{1}{2}}]/a_2\} \tag{100a}$$

When $a_1 < a_2$

$$S = 2(a_2^2 - a_1^2)^{-\frac{1}{2}}\tan^{-1}[(a_2^2 - a_1^2)^{\frac{1}{2}}/a_1] \tag{100b}$$

But on a more microscopic scale one would want to relate the β_i to the anisotropic intermolecular potential of a molecule with the surrounding molecules.

One can also attempt to generalize these results to include short time free rotational effects. The general problem is complex,[16-17] but for spherical symmetry a rough approximation is eq. 95 with now $E_L t = RL(L+1)/4S$ with S given by eq. 79 and R instead of D.

Another important diffusional model is jump diffusion where the molecule reorients through large angles as a result of strong collisions.[18]

Anisotropic rotational diffusion can, in principle, manifest itself through line width effects from the intramolecular anisotropic interactions: g-tensor, electron-nuclear dipolar, and quadrupolar.[15] These terms, which are 2nd rank tensors may be written as irreducible tensor components in the form of eq. 40 as:

$$\mathcal{H}_1(t) = \sum_{q,m,m'} F'^{(2,m)}_q \mathcal{D}^2_{-m,m'}(\Omega) A^{(2,m')}_q \tag{101}$$

where the prime on $F'^{(2,m)}_q$ indicates it is written in molecule-fixed axes, while unprimed $A^{(2,m')}_q$ is in laboratory axes. The $\mathcal{D}^2_{-m,m'}(\Omega)$ are the rotation matrix elements (with Ω the Euler angles) for the transformation between the two axis systems. One then needs correlation functions of form:

$$g(\tau) = \langle \mathcal{D}^L_{-m,q}(t) \mathcal{D}^{L'\,*}_{-m',q'}(t+\tau) \rangle_\Omega \tag{102}$$

which for axially-symmetric rotational diffusion is, from Eqs. 72, 95 and 96 and the orthonormality of the $\varphi^L_{KM}(\Omega)$ of Eq. 94:

$$g(\tau) = \frac{1}{2L+1} \exp[-E_{L,m}\tau] \delta_{L,L'} \delta_{m,m'} \delta_{qq'} \tag{103}$$

Then the spectral density terms, $J(\omega)$ of Eq. 37a become for L = 2

$$J(\omega) = \sum_{q,q',m} A^{(2,m)}_q A^{*\,(2,m)}_{q'} j^{(qq',2)}(\omega) \tag{104}$$

with $j^{(qq',2)}(\omega) = \frac{1}{2L+1} \sum_M F'^{(2,M)}_q F'^{*(2,M)}_{q'} [E_{2,M}^{-1}/(1+E_{2,M}^{-2}\omega^2)]$. (105)

This means that each irreducible tensor component $F'^{(2,M)}_q$ is "relaxed" with its own characteristic time $E_{L,|M|}^{-1}$, and it illustrates the great advantages in using irreducible tensor formalism.

VIII.6. SUMMATION OF THE GENERALIZED MOMENTS FOR A MARKOFF PROCESS: STOCHASTIC LIOUVILLE EQUATION

We have for a Markoff process:

$$\frac{\partial}{\partial t} P(\Omega, t) = -\Gamma_\Omega P(\Omega,t) \tag{106}$$

where $P(\Omega, t)$ is the probability of finding Ω at the particular state at time t. The process is assumed to be stationary, so that Γ is a time-independent Markoff operator, (e.g. it is $-\Lambda$ in eq. 89) and also that the process has a unique equilibrium distribution, $P_0(\Omega)$, characterized by:

$$\Gamma P_0(\Omega) = 0 \tag{107}$$

A formal solution to eq. 107 given that Ω initially had value Ω_0 is given by the conditional probability $P(\Omega_0|\Omega,t)$ which obeys:

$$P(\Omega_0|\Omega,t) = e^{-\Gamma_\Omega t}\delta(\Omega-\Omega_0) \tag{108}$$

An alternate form of eq. 108, in terms of orthogonal eigenfunctions G_m of the operator Γ_Ω is

$$\begin{aligned} P(\Omega_0|\Omega;\tau) &= \sum_m G_m^*(\Omega_0)e^{-\Gamma_\Omega \tau}G_m(\Omega) = \sum_m G_m^*(\Omega_0)e^{-E_m\tau}G_m(\Omega) \\ &= \sum_m e^{-\Gamma_\Omega \tau}|G_m(\Omega)\rangle\langle G_m(\Omega_0)| \end{aligned} \tag{109}$$

where $|G_m(\Omega)\rangle$ and $\langle G_m(\Omega_0)|$ are ket and bra vectors in the Hilbert space defined by the variable Ω, with Γ Hermitian for convenience.

It is then possible to show that

$$\begin{aligned} e^{-i\mathcal{H}_0^\times t} m_n(t_1..t_n) &= \langle P_0(\Omega)|e^{-(t-t_1)(i\mathcal{H}_0^\times+\Gamma)}\mathcal{H}_1^\times(\Omega)e^{-(t_1-t_2)(i\mathcal{H}_0^\times+\Gamma)} \times \\ &\mathcal{H}_1^\times(\Omega)e^{-(t_2-t_3)(i\mathcal{H}_0^\times+\Gamma)} \ldots\ldots\ldots \mathcal{H}_1^\times e^{-(t_{n-1}-t_n)(i\mathcal{H}_0^\times+\Gamma)} \times \\ &\mathcal{H}_1^\times e^{-t_n(i\mathcal{H}_0^\times+\Gamma)}|G_0\rangle \end{aligned} \tag{110}$$

where $\langle P_0(\Omega)|Q(\Omega)|G_0\rangle \equiv \int d\Omega P_0^*(\Omega)Q(\Omega)G_0(\Omega)$ (110a)

is a "matrix-element" of operator Q, and $|G_0(\Omega)\rangle \propto |P_0(\Omega)\rangle$.

In order to do this we first note that

$$[\mathcal{H}_0, \Gamma_\Omega] = 0 \tag{111a}$$

since $\mathcal{H}_0$ is taken to be independent of Ω, but in general,

$$[\mathcal{H}_0, \mathcal{H}_1] \neq 0 \tag{111b}$$

$$[\mathcal{H}_1, \Gamma] \neq 0 \ . \tag{111c}$$

Also, it is easy to see that the commutators of type $[\mathcal{H}_1^{\ddagger}, B]$ obey:

$$[\mathcal{H}_1^{\ddagger}, B] \equiv \mathcal{H}_1^{\ddagger\times} B = (e^{it\overset{\times}{\mathcal{H}}_0} \overset{\times}{\mathcal{H}}_1 e^{-it\overset{\times}{\mathcal{H}}_0})B \ . \tag{112}$$

Thus $e^{-(i\overset{\times}{\mathcal{H}}_0+\Gamma)t}|G_0\rangle = e^{-i\overset{\times}{\mathcal{H}}_0 t} e^{-\Gamma_\Omega t}|G_0\rangle = e^{-i\overset{\times}{\mathcal{H}}_0 t}|G_0\rangle$; similarly

$$\langle P_0(\Omega)|e^{-(i\overset{\times}{\mathcal{H}}_0+\Gamma)(t-t_1)} = e^{-i\overset{\times}{\mathcal{H}}_0 t}\langle P_0(\Omega)|e^{+i\overset{\times}{\mathcal{H}}_0 t_1}. \tag{113}$$

Thus from eq. 110 (and eqs. 111-113)

$$\begin{aligned} m_n(t_1..t_n) &= \langle P_0(\Omega)|e^{it_1\overset{\times}{\mathcal{H}}_0}\, \overset{\times}{\mathcal{H}}_1 e^{-(t_1-t_2)(i\overset{\times}{\mathcal{H}}_0+\Gamma)}\, \overset{\times}{\mathcal{H}}_1 e^{-(t_2-t_3)(i\overset{\times}{\mathcal{H}}_0+\Gamma)} \\ &\quad \ldots.\overset{\times}{\mathcal{H}}_1 e^{-(t_{n-1}-t_n)(i\overset{\times}{\mathcal{H}}_0+\Gamma)}\, \overset{\times}{\mathcal{H}}_1 e^{-it_n\overset{\times}{\mathcal{H}}_0}|G_0\rangle \\ &= \langle P_0(\Omega)|\mathcal{H}_1^{\ddagger}(t_1)^{\times} e^{-\Gamma(t_1-t_2)}\mathcal{H}_1^{\ddagger}(t_2)^{\times} e^{-\Gamma(t_2-t_3)}\ldots. \\ &\quad \ldots.\mathcal{H}_1^{\ddagger}(t_{n-1})^{\times} e^{-\Gamma(t_{n-1}-t_n)}\mathcal{H}_1^{\ddagger}(t_n)^{\times}|G_0\rangle \end{aligned} \tag{114}$$

We now note from eq. 108 that

$$\int P(\Omega|\Omega_i, t)Q(\Omega_i)d\Omega_i = e^{-\Gamma_\Omega t}Q(\Omega) \tag{115}$$

Eq. 115 is then used systematically from the right on the last form of eq. 114 to replace the operators $e^{\Gamma_\Omega(t_{i-1}-t_i)}$, finally yielding the usual Markovian correlation expression,[1] (cf. eq. 72, but with eq. 71a utilized and with $\mathcal{H}_1^{\ddagger}(\Omega_i, t_i)^{\times}$ instead of just $F(y_i)$). This completes the proof of eq. 110.

Now eq. 110 along with eqs. 6, 7 and 8 provides a solution for $\sigma(t)$

$$\sigma(t) = e^{-i\overset{\times}{\mathcal{H}}_0 t}\sum_{n=0}^{\infty} M_n(t) \tag{116}$$

The Laplace transform of $e^{-i\overset{\times}{\mathcal{H}}_0 t}M_n(t)$ or $\mathfrak{M}_n(s)$ is shown to be

$$\mathfrak{M}_n(s) = \langle P_0 | \left(\frac{-i}{s+i\mathcal{H}_0^\times+\Gamma_\Omega} \mathcal{H}_1(\Omega)^\times \right)^\eta \frac{1}{s+i\mathcal{H}_0^\times+\Gamma_\Omega} | G_0 \rangle \tag{117}$$

since the Laplace transform of the (multiple) convolution in eq. 8 is just the product of Laplace transforms. Then the Laplace transform of $\sigma(t)$ or $\tilde{\sigma}(s)$ is just:

$$\frac{\tilde{\sigma}(s)}{\sigma(0)} = \sum_{n=0}^{\infty} \mathfrak{M}_n(s) = \langle P_0 | \frac{1}{1+i\mathcal{H}_1^\times (s+i\mathcal{H}_0^\times+\Gamma_\Omega)^{-1}} \frac{1}{s+i\mathcal{H}_0^\times+\Gamma_\Omega} | G_0 \rangle$$

$$= \langle P_0 | [s + i\mathcal{H}^\times + \Gamma_\Omega]^{-1} | G_0 \rangle \tag{118}$$

This is equivalent to the stochastic Liouville result of Kubo,[19] (cf. Jensen Ch. III). [An equivalent result but with $i \rightarrow -i$ is obtained for $\tilde{S}(s)$.] Then we have

$$\sigma(t) = \langle P_0 | \exp[(-t)(i\mathcal{H}^\times+\Gamma)] | G_0 \rangle \sigma(0) \tag{119}$$

It follows from the definition, eq. 10 of K(t) that:

$$K(t) = \ln \langle P_0 | e^{it\mathcal{H}_0^\times} e^{(-i\mathcal{H}^\times-\Gamma)t} | G_0 \rangle \tag{120}$$

and, if $[\mathcal{H}_0, \mathcal{H}_1] = 0$ then $K(t) = \ln\langle P_0 | e^{-i\mathcal{H}_1^\times-\Gamma)t} | G_0 \rangle$. Also,

$$\dot{K}(t) = -e^{it\mathcal{H}_0^\times} \langle P_0 | (i\mathcal{H}_1-\Gamma_\Omega) e^{-(i\mathcal{H}^\times+\Gamma)t} | G_0 \rangle \left[\langle P_0 | e^{-(i\mathcal{H}^\times+\Gamma)t} | G_0 \rangle \right]^{-1} \times e^{-it\mathcal{H}_0^\times}, \tag{121}$$

which does not appear to be very useful for calculation purposes, although we shall see eq. 118 is very useful for nonperturbative solutions (cf. Ch. XIV).

VIII.7. ACKNOWLEDGEMENT

This work was supported in part by the Advanced Research Projects Agency and by the National Science Foundation.

References

1. J. H. Freed, J. Chem. Phys. 49, 376 (1968).
2. R. Kubo, J. Phys. Soc. Japan 17, 1100 (1962).
3. R. Kubo, in "Fluctuation, Relaxation, and Resonance in Magnetic Systems", D. ter Haar, Ed. (Oliver and Boyd, London, 1962), p. 23; J. Math. Phys. 4, 174 (1963).
4. A. Abragam, "The Principles of Nuclear Magnetism" (Oxford University Press, London, 1961).
5. R. Kubo and K. Tomita, J. Phys. Soc. Japan 9, 888 (1954).
6. A. G. Redfield, IBM J. Res. Develop. 1, 19 (1957).
7. F. Bloch, Phys. Rev. 102, 104 (1956).
8. J. H. Freed and G. K. Fraenkel, J. Chem. Phys. 39, 326 (1963); G. K. Fraenkel, J. Phys. Chem 71, 139 (1967).
9. J. H. Freed, J. Chem. Phys. 43, 2312 (1965).
10. A. Messiah, "Quantum Mechanics" (John Wiley and Sons, New York, 1962), p. 727.
11. J. H. Freed and R. G. Kooser, J. Chem. Phys. 49, 4715 (1968); J. H. Freed (to be published).
12. S. Chandrasekhar, Rev. Mod. Phys. 15, 1 (1943).
13. J. H. Freed and G. K. Fraenkel, J. Chem. Phys. 41, 3623 (1964).
14. L. D. Favro, in "Fluctuation Phenomena in Solids", R. E. Burgess, Ed. (Academic Press, New York, 1965), p. 79.
15. J. H. Freed, J. Chem. Phys. 41, 2077 (1964) and references cited therein.
16. R. A. Sack, Proc. Phys. Soc. (London) 70, 402, 414 (1957).
17. W. A. Steele, J. Chem. Phys. 38, 2404, 2411 (1963).
18. E. N. Ivanov, Sov. Phys. JETP 18, 1041 (1964).
19. R. Kubo, J. Phys. Soc. Japan 26, Supplement, 1 (1969).

SPIN RELAXATION VIA QUANTUM MOLECULAR SYSTEMS

Jack H. Freed

Department of Chemistry, Cornell University

Ithaca, New York 14850

Very often one has to consider the quantum nature of the molecular systems whose modulation induces spin relaxation. We first consider a "gas-like" model wherein strong collisions randomize the molecular degrees of freedom, more specifically the rotational states. Then we generalize the results to cover more general descriptions of the way that the molecular degrees of freedom relax through thermal contact.

IX.1. STRONG COLLISONAL RELAXATION[1]

We start with the density matrix ρ which refers to the combined spin-molecular degrees of freedom:

$$\dot{\rho} = -i\mathcal{H}^{\times}\rho \tag{1}$$

where $\hbar\mathcal{H}$ is the Hamiltonian for a combined system and is given by

$$\mathcal{H} = \mathcal{H}_M + \mathcal{H}_S + V = \mathcal{H}_0 + V \tag{2}$$

Here $\hbar\mathcal{H}_M$ and $\hbar\mathcal{H}_S$ are the unperturbed Hamiltonians of the molecular and spin systems, respectively, while V is the coupling term operating on both rotational and spin variables, and $\mathcal{H}$ is time independent. We often only include the rotational part of $\mathcal{H}_M$ designated by $\mathcal{H}_R$. Let us define V and $\mathcal{H}_S$ such that

$$\langle V \rangle \equiv \mathrm{Tr}_R\{B(R)V\} = 0 \tag{3}$$

where B(R) is the Boltzmann distribution in rotational states:

$$B(R) = Z(R)e^{-\hbar\mathcal{H}_R/kT} \; ; \; Z(R)^{-1} = \mathrm{Tr}_R\left\{e^{-\hbar\mathcal{H}_R/kT}\right\} . \tag{3a}$$

In the interaction representation with

$$\rho^{\ddagger}(t) = e^{i(\mathcal{H}_R+\mathcal{H}_S)t}\rho(t)e^{-i(\mathcal{H}_R+\mathcal{H}_S)t} \tag{4a}$$

and

$$V^{\ddagger}(t) = e^{i(\mathcal{H}_R+\mathcal{H}_S)t}Ve^{-i(\mathcal{H}_R+\mathcal{H}_S)t} \tag{4b}$$

one has

$$\frac{1}{i}\frac{d\rho^{\ddagger}(t)}{dt} = -[V^{\ddagger}(t), \rho^{\ddagger}(t)]. \tag{5}$$

Equation 5 may be solved by integrating to successive approximations obtaining a series expansion for the dependence of $\rho^{\ddagger}(t)$ on its value at some earlier time $\rho^{\ddagger}(t_0)$ (or better, by the cumulant expansion approach).

To proceed further, the following assumptions will be introduced.

1) $\rho(t)$ is approximately factorable into the product $\lambda(t)\sigma(t)$ where $\lambda(t)$ and $\sigma(t)$ are reduced density matrices depending only on the rotational and spin degrees of freedom respectively.

2) Collisions affect only the rotational degrees of freedom (i.e., they are perturbations with no matrix elements between spin states).

3) The collision takes place over an interval of time which is short enough that $\sigma(t)$ remains essentially constant.

4) The collision is strong in the sense that the distribution of rotational states just after collision is given by a Boltzmann distribution at the kinetic temperature of the molecules and is independent of their distribution just before collision, so that

$$\rho(t_0) = B(R)\sigma(t_0). \tag{6}$$

However, in order to be consistent with assumption (2), collisions must not change the spin symmetry. Thus, the symmetry of the rotational states will be unchanged and B(R) will be understood as normalized only over states R having the symmetry of interest.

5) The collisions are random with a mean time interval τ. The basic nature of this model is that collisions represent a very strong perturbation which rapidly restores the rotational states to equilibrium, while the spin-rotational interaction, V is a much weaker perturbation that slowly tends to bring the spins to equilibrium.

Assuming that a collision occurred at t_o, expanding Eq. 5 to second order, and taking Tr_R (i.e. a trace over rotational states)gives:

$$\frac{\partial\sigma^{\ddagger}(t,t_o)}{\partial t} = -iTr_R\left\{[V^{\ddagger}(t),\ B(R)\sigma^{\ddagger}(t_o)]\right\} - -Tr_R\left\{\int_{t_o}^{t} dt'[V^{\ddagger}(t),[V^{\ddagger}(t'),B(R)\sigma^{\ddagger}(t_o)]]\right\} + \ldots \quad (7)$$

If V does not connect states of different nuclear spin symmetry, and any change of nuclear spin symmetry resulting from collisions is neglected, then all molecules of a particular spin symmetry may be treated as a separate subensemble represented by a separate Eq. 7. It immediately follows from Eqs. 3 and 4 that the first term on the left of Eq. 7 vanishes.

To obtain $\sigma^{\ddagger}(t)$ from Eq. 7, $\sigma^{\ddagger}(t,t_o)$ must be integrated over all values of $t_o = t-\theta$. Thus

$$\sigma^{\ddagger}(t) = \int_0^{\infty} \sigma^{\ddagger}(t,t-\theta)\tau^{-1}e^{-\theta/\tau}\,d\theta. \quad (8)$$

Differentiating partially with respect to time gives

$$\frac{\partial}{\partial t}\sigma^{\ddagger}(t) = \int_0^{\infty}\left[\frac{\partial}{\partial t}\sigma^{\ddagger}(t,t_o)\right]_{t_o=t-\theta}\tau^{-1}e^{-\theta/\tau}d\theta - \int_0^{\infty}\frac{\partial}{\partial\theta}\sigma^{\ddagger}(t,t-\theta)\tau^{-1}e^{-\theta/\tau}d\theta. \quad (9)$$

The term $\left[\frac{\partial}{\partial\theta}\sigma^{\ddagger}(t,t_o)\right]_{t_o=t-\theta}$ in Eq. 9 is just that given by Eq. 7.

The second term in Eq. 9 is shown to be zero by first integrating it by parts giving $\tau^{-1}[\sigma^{\ddagger}(t)-\sigma^{\ddagger}(t,t_0=t)]$ and then by utilizing assumptions (3) and (5), which permit Eq. 8 to be written where $\sigma^{\ddagger}(t)$ on the LHS is replaced by $\sigma^{\ddagger}(t,t_0=t)$, and $t-\theta$ on the RHS is the time of the collision previous to the one at $t=t_0$. This leaves the following expression:

$$\frac{\partial\sigma^{\ddagger}(t)}{\partial t} = -Tr_R \int_0^{\infty} \tau^{-1} e^{-\theta/\tau} d\theta \int_0^{\theta} dx\, [V^{\ddagger}(t),$$

$$[V^{\ddagger}(t-x),\ B(R)\sigma^{\ddagger}(t-\theta)]] + \ldots \tag{10}$$

Equation 10 may be solved using the approximations: a) Replace $\sigma^{\ddagger}(t-\theta)$ on the right by $\sigma^{\ddagger}(t)$, and b) neglect higher order terms in the expansion. These approximations require $V^2 \ll \tau^{-2}$, which will lead to $T_1, T_2 \gg \tau$, i.e., the relaxation effects of the perturbation V are much weaker than those of the collisions.

Equation 10 is evaluated in a basis diagonal in H_R and H_S. That is, if ψ_r and φ_s are respectively complete sets of eigenfunctions of H_R and H_S having the correct symmetry for the spin species of interest, then the appropriate basis would be the set of products $\psi_r\varphi_s$. The final results (neglecting the second order frequency shifts) may be expressed in the usual relaxation matrix form:

$$\frac{\partial\sigma^{\ddagger}(t)_{\alpha\alpha'}}{\partial t} = \sum_{\beta\beta'} R_{\alpha\alpha'\beta\beta'}\sigma^{\ddagger}(t)_{\beta\beta'} \tag{11}$$

where for $V = \sum_q K^{(q)}F^{(q)}$ with $K^{(q)}$ a spin operator and $F^{(q)}$ a rotation operator:

$$R_{\alpha\alpha'\beta\beta'} = \sum_{q,q'}\Big[K_{\alpha\beta}^{(q)}K_{\beta'\alpha'}^{(q')}2j_{qq'}(\alpha-\beta)$$

$$-\sum_{\gamma}\delta_{\alpha\beta}K_{\beta'\gamma}^{(q')}K_{\gamma\alpha'}^{(q)}j_{qq'}(\gamma-\beta')$$

$$-\sum_{\gamma}\delta_{\alpha'\beta'}K_{\alpha\gamma}^{(q')}K_{\gamma\beta}^{(q)}j_{qq'}(\gamma-\beta)\Big]. \tag{12}$$

Here α, β, etc. label eigenstates of $\mathcal{H}_S$. Equation 12 is formally very similar to the Redfield Eq. (cf. eq. VIII-35) but the spectral densities $j(\alpha-\beta)$ are different:

$$j_{qq'}(\alpha-\beta) = Z(R) \sum_{r,r'} \left[e^{-E_{r'}/kT} F_{rr'}^{(q)} F_{r'r}^{(q')} \frac{\tau}{1+(\omega_{rr'}+\omega_{\alpha\beta})^2\tau^2} \right]$$

$$- \langle F^{(q)} \rangle \langle F^{(q')} \rangle \frac{\tau}{1+\omega_{\alpha\beta}^2 \tau^2}, \qquad (13)$$

where r, r′, etc. label eigenstates of $\mathcal{H}_R$ and $\hbar\omega_{rr'} = E_r - E_{r'}$, etc. Eq. 13 includes the correction for non-zero $\langle F^{(q)} \rangle$ in case Eq. 3 is not fulfilled.

Equations 12 and 13 require the usual energy restriction that

$$E_\alpha - E_\beta = E_{\alpha'} - E_{\beta'} \, . \qquad (14)$$

As is usually done, the ad hoc assumption that $\sigma^{\ddagger}(t)$ relaxes to

$$\sigma^T = Z(S) e^{-\hbar\mathcal{H}_S/kT} \; ; \; Z(S)^{-1} = Tr_S \left\{ e^{-\hbar\mathcal{H}_S/kT} \right\} \qquad (15)$$

is introduced by replacing $\sigma^{\ddagger}(t)$ with $\sigma^{\ddagger}(t) - \sigma^T$ in Eq. 11.

If, however, we now allow the possibility that V may connect states of different nuclear spin symmetry even though the collisions do not, we may utilize assumption (1) to write $\rho^i(t) = \lambda^i(t)\sigma^i(t)$ as that portion of the density matrix which only includes all states corresponding to the ith nuclear spin symmetry.[2] Then we may write ρ as the partitioned matrix:

$$\rho = \begin{pmatrix} \lambda^i \sigma^i & \lambda^{ij}\sigma^{ij} & \cdots \\ \lambda^{ji}\sigma^{ji} & \lambda^j \sigma^j & \cdots \\ \vdots & \vdots & \end{pmatrix} \quad i \neq j \qquad (16)$$

where submatrices $\lambda^{ij}\sigma^{ij}$ include all off-diagonal elements between states belonging to the ith and j^{th} symmetry classes. Terms such as $\lambda^{ji}\sigma^{ij}$ or $\lambda^{ij}\sigma^{i}$ for $i \neq j$ are not allowed, since they violate the Exclusion Principle. Note that the mixed superscripts may not be simply permuted. However, the Hermitian property of the density matrix and the separability of the two reduced matrices leads to

$$(\sigma^{ji}\lambda^{ji})^{\dagger} = \sigma^{ij}\lambda^{ij} \tag{17}$$

where † indicates the Hermitian conjugate. The normalization of the density matrices is taken to be

$$\mathrm{Tr}_S(\sigma) = 1 \tag{18a}$$

and

$$\mathrm{Tr}_{R_i}(\lambda^i) = \mathrm{Tr}_{R_j}(\lambda^j) = 1 \tag{18b}$$

so that any differences in population of states of different nuclear spin symmetries are contained in σ. The subscripts S_i and R_i limit the trace operations to spin states and to rotational states of ith symmetry, respectively.

The equation of motion for ρ may be written in terms of each submatrix. Thus for example

$$\frac{d}{dt}(\sigma^j\lambda^j) = -i\,[\mathcal{H}, \rho(t)]^{jj} \tag{19}$$

Note that, while the matrix elements of the commutator itself are restricted to states of symmetry j in eq. 19, both $\mathcal{H}$ and $\rho(t)$ within the commutator could have matrix elements involving states of other symmetry.

After a strong collision we have:

$$\mathrm{Tr}_R\left\{\rho^{\ddagger}(t_o)\right\} = \mathrm{Tr}_R\left\{B(R) \times \sigma(t_o)\right\} = \begin{pmatrix} \sigma^{i\ddagger}(t_o) & 0 & ---- \\ 0 & \sigma^{j\ddagger}(t_o) & ---- \\ \vdots & \vdots & \\ \vdots & \vdots & \end{pmatrix} \tag{20}$$

where B(R) may be partitioned so that

$$B^i(R) = Z^i(R)\,[\exp(-\hbar\mathcal{H}_R/kT)]^i \tag{21}$$

and

$$[Z^i(R)]^{-1} = \mathrm{Tr}_{R_i}\,[\exp(-\hbar\mathcal{H}_R/kT)]. \tag{22}$$

$[Z^i(R)]^{-1}$ is the rotational partition function normalized for the ith symmetry states. Thus the strong collision is assumed to restore each set of rotational states belonging to a particular spin symmetry to its respective Boltzmann distribution, while also having no effect on $\sigma(t_0)$, which includes the relative populations of states of different spin symmetry. The disappearance of off-diagonal submatrices $\lambda^{ij}(t_0)$ in Eq. 20 after a strong collision does not necessarily require that important nuclear-spin dependent intermolecular forces exist. These submatrices contain only off-diagonal elements between rotational eigenstates, so they are relaxed by secular mechanisms which broaden each of the coupled rotational states differently and by all nonsecular processes involving these levels. A derivation similar to that given above yields:

$$\dot{\sigma}^{\ddagger}(t)_{\alpha_i\alpha'_i} = \sum_{\beta_j\beta'_j} R_{\alpha_i\alpha'_i\beta_j\beta'_j}\left[\sigma^{\ddagger}(t)_{\beta_j\beta'_j} - \sigma_{o\beta_j\beta'_j}\right] \tag{23}$$

where $R_{\alpha_i\alpha'_i\beta_j\beta'_j}$ is again given by Eq. 12 after letting $\alpha\rightarrow\alpha_i$, $\beta\rightarrow\beta_j$, $\gamma\rightarrow\gamma_k$, etc. and

$$j_{qq'}(\alpha_i-\beta_j) =$$

$$\left[Z^j(R)\sum_{r_i,r'_j}\exp(-E_{r'_j}/kT)\,F^{(q)}_{r_ir'_j}F^{(q')}_{r'_jr_i}\frac{\tau}{1+(\omega_{r_ir'_j}+\omega_{\alpha_i\beta_j})^2\tau^2}\right.$$

$$-\left.\langle F^{(q)}\rangle_i\langle F^{(q')}\rangle_i\frac{\tau}{1+\omega^2_{\alpha_i\beta_j}\tau^2}\,\delta_{ij}\right] \tag{24}$$

We have introduced the ad hoc assumption that $\sigma^{\ddagger j}(t)$ relaxes to a Boltzmann distribution given by:

$$\sigma_o^{\,j} = f_j Z(S)\exp(-\hbar\mathcal{H}_S^{\,j}/kT) \tag{25}$$

where

$$[Z(S)]^{-1} = \mathrm{Tr}_S\left\{e^{-\hbar\mathcal{H}_S/kT}\right\} \tag{25a}$$

and

$$f_j = \sum_i Z^i(R)/Z^j(R) \tag{26}$$

which measures the fractional population of rotational levels of jth symmetry at thermal equilibrium, is introduced to account for the normalization of $B^i(R)$, $\sigma^i(t)$ and $\lambda^i(t)$ given respectively by Eqs. 22 and 18. Note that $\delta_{\alpha_i\beta_j}$ also implies δ_{ij}. The above result assumes $\langle V\rangle_j$ is the same for all j. When this is not so, the result is a little more complicated and is given in ref. 2.

IX.2. GENERAL FORMULATION[3]

It is now assumed that the equation of motion for $\rho(t)$ may be written as

$$i(d\rho/dt) = [\mathcal{H}, \rho] - i\Gamma\rho\ , \tag{27}$$

The term $\Gamma\rho$ has been introduced phenomenologically into Eq. 27 to describe in a general way the relaxation of the combined system as a result of its thermal contact. The fact that only the molecular systems (and not the spins) are assumed to be directly coupled to the thermal modes means that the "relaxation" matrix Γ will only affect the molecular systems directly, and the spin relaxation is achieved indirectly via the spin-molecular coupling term. Equation 27 is clearly valid in the limit V=0, since the molecular spin systems are uncoupled, and it is proper to treat the relaxation of the molecular degrees of freedom as independent of the spins. As long as

$$|\Gamma| \gg V, \tag{28}$$

then even as V is introduced, the effects of V will be negligible upon the molecular states when compared to their lattice-induced widths and relaxation transitions contained in Γ.

The relaxation transitions of the molecular-spin systems are described by

$$(\Gamma\rho)_{nn} = \sum_{n'} (W_{nn'}\rho_{n'n'} - W_{n'n}\rho_{nn}), \tag{29}$$

where $W_{nn'}$ is the transition probability from state n' to n and ρ_{nn} is the diagonal density-matrix element for the nth molecular state but is still an operator on spin states. The assumption of detailed balance yields

$$W_{nn'} = W_{n'n}\exp(\hbar\omega_{nn'}/kT). \tag{30}$$

However, no restriction is placed on the nature of the intermolecular interactions leading to the $W_{nn'}$. The n diagonal elements ρ_{nn} will, in general, relax in a coupled fashion requiring a normal modes solution for the eigenvalues. The off-diagonal density-matrix elements, whose relaxation is associated with the transition linewidths, are assumed to obey

$$(\Gamma\rho)_{nn'} = -\sum_{m,m'} \Gamma_{nn'mm'}\rho_{mm'} , \tag{31}$$

where

$$(\omega_{nn'} - \omega_{mm'})\Gamma^{-1} \ll 1 . \tag{31a}$$

Equation 31 implies that the off-diagonal elements may be coupled, but the "adiabatic assumption" which allows only off-diagonal elements between pairs of states with nearly the same energy differences to be coupled, is introduced by Eq. 31a. Any coupling via Γ between diagonal and off-diagonal elements of $\rho_{nn'}$ is being neglected. In the absence of any couplings of $\rho_{nn'}$ Eq. 31 becomes:

$$(\Gamma\rho_{nn'}) = -\Gamma_{nn'nn'}\rho_{nn'} \equiv -\Gamma_{nn'}\rho_{nn'} , \tag{32}$$

where $\Gamma_{nn'}$ is the "linewidth" for the $n \leftrightarrow n'$ transition. It is, in general, composed of secular and non-secular effects.

It is useful to obtain a basis for Γ corresponding to the normal modes of relaxation of the molecular part of $\rho(t)$. First $\rho(t)$ is partitioned into the distinct non-coupling components each distinguished by a different value for the subscript λ. The distinction between the normal modes for matrix elements of $\rho(t)$ which are diagonal and off-diagonal in molecular states is represented by $\lambda \rightarrow \delta$

and $\lambda \rightarrow \nu$, respectively. It is further useful to distinguish the normal modes in terms of the sets of molecular states whose diagonal density-matrix elements relax independently of one another (e.g., states of different spin symmetry). This leads to density-matrix components such as: $\rho_{\delta_i}(t) \equiv \rho_i(t)$ — diagonal in the ith set of molecular states, $\rho_{\nu_i}(t)$ — off-diagonal, involving only the ith set, $\rho_{\nu_{ij}}(t)$ — off-diagonal, involving both the ith and j^{th} set. Each such component is generally written as $\rho_\lambda(t)$.

Now Γ may be partitioned in the same manner to give the different component Γ_λ. Letting T_λ be the similarity transformation which diagonalizes Γ_λ, one has

$$(T_\lambda^{-1}\Gamma_\lambda T_\lambda)_{ab} = \Gamma'_{\lambda_a}\delta_{ab}, \tag{32a}$$

and

$$\chi_\lambda(t) = T_\lambda^{-1}\rho_\lambda(t), \tag{33}$$

where Γ'_{λ_a} is the eigenvalue for the λ_ath normal mode represented by $\chi_{\lambda_a}(t)$. Thus the relaxation of the diagonal density-matrix elements of the ith set of molecular states, represented by Γ_i, is given by Eq. 29 replacing n and n′ by n_i and n_i'. For the ith group of states there will be a zero root, $\Gamma'_{i1} = 0$, corresponding to the conservation of probability in such states.

In the interaction representation Eq. 27 becomes

$$\dot{\rho}^{\ddagger}(t) = -\Gamma\rho^{\ddagger}(t) - i[V^{\ddagger}(t), \rho^{\ddagger}(t)]. \tag{34}$$

We now look at the evolution of Eq. 34 for times of the order of t such that

$$|\Gamma|t \gg 1 \gg |V|t; \tag{35}$$

that is, for times long compared to the damping time of the molecular systems given formally by $|\Gamma|^{-1}$ but short enough that the effect of V is small enough to be expanded as a perturbation. Over this time domain an iterative expansion given by

$$\dot{\rho}_\lambda^{\ddagger(0)}(t) = -\Gamma_\lambda\rho_\lambda^{\ddagger(0)}(t), \tag{36a}$$

$$\dot{\rho}_\lambda^{\ddagger(1)}(t) = -\Gamma_\lambda \rho_\lambda^{\ddagger(1)} - i[V^\ddagger(t), \rho^{\ddagger(0)}(t)]_\lambda, \tag{36b}$$

$$\dot{\rho}_\lambda^{\ddagger(n)}(t) = -\Gamma_\lambda \rho_\lambda^{\ddagger(n)}(t) - i[V^\ddagger(t), \rho^{\ddagger(n-1)}(t)]_\lambda \tag{36c}$$

is employed. The commutator on the right corresponds to the λ th normal mode although $\rho^\ddagger(t)$ within need not. One obtains the relaxation matrix expression Eq. 23 where now

$$\begin{aligned} R_{\alpha_i\alpha'_i\beta_j\beta'_j} = \sum_{q,q'} \Big\{ & K_{\alpha_i\beta_j}^{(q)} K_{\beta'_j\alpha'_i}^{(q')} [k_{qq'}(\beta'_j-\alpha'_i)+\hat{k}_{qq'}(\alpha_i-\beta_j)] \\ & - \sum_{\gamma,k} K_{\alpha_i\gamma_k}^{(q')} K_{\gamma_k\beta_i}^{(q)} \delta_{\beta'_j\alpha'_i} \hat{k}_{qq'}(\gamma_k-\beta_j) - \\ & - \sum_{\gamma,k} K_{\beta'_j\gamma_k}^{(q')} K_{\gamma_k\alpha'_i}^{(q)} \delta_{\alpha_i\beta_j} k_{qq'}(\beta'_j-\gamma_k) \Big\} \end{aligned} \tag{37}$$

with

$$\begin{aligned} k_{qq'}(\beta'_j-\alpha'_i) = \sum_{\lambda,n,n',m,m'} & (F_{\lambda\dagger}^{(q)})_{n_jn'_j} (F_\lambda^{(q')})_{m'_jm_i} B_{m'_j}(M) \\ & \times [\Delta_\lambda(\omega_{\beta'_j\alpha'_i} + \omega_{m'_jm_i})_{n'_jn_im'_jm_i}] \end{aligned} \tag{38a}$$

and

$$\begin{aligned} \hat{k}_{qq'}(\alpha_i-\beta_j) = \sum_{\lambda,n,n',m,m'} & (F_\lambda^{(q)})_{m_im'_j} (F_{\lambda\dagger}^{(q')})_{n'_jn_i} B_{m'_j}(M) \\ & \times [\Delta_\lambda(\omega_{\alpha_j\beta_j} + \omega_{m_im_j'})_{n_in'_jm_im'_j}] \end{aligned} \tag{38b}$$

$$\Delta_\nu(\omega)_{n'_jn_im'_jm_i} = [(\Gamma_\nu + i\omega\mathbb{1})^{-1}]_{n'_jn_im_im'_j} \tag{39}$$

$$\Delta_i(\omega)_{nn,mm} = [(\Gamma_i + \omega\mathbb{1})^{-1}]_{nnmm} - [B_i(M)_{nn}/i\omega] \tag{40}$$

Note that m_i and m'_j must constitute one of the λth set of transitions, while in the $\lambda\dagger$th set the pairs of states are transposed. Also $\mathbb{1}$ is the unit matrix. A particularly simple case exists when

$$\Gamma'_{\lambda_a} = \tau_c^{-1}, \quad \lambda_a \neq i_1, \quad \Gamma_{i_1} = 0 \tag{41}$$

otherwise independent of λ_a. Then, one obtains the "strong-collision" approximation of Eq. 24.

When Γ_ν is just a 1$\times$1 matrix (uncoupled molecular width), Eq. 39 shows that the relaxation effects of off-diagonal matrix elements of $(F_\nu^{(q)})$ involve the lattice-induced molecular linewidths for the νth transition. Equation 40 shows that the relaxation effects of the diagonal-matrix elements of $(F_i^{(q)})$ involve just the lattice-induced transition probabilities amongst the ith set of molecular states.

IX.3. APPLICATIONS

A. Gas-Phase Relaxation

In applying the formalism to gas-phase relaxation, the rotational wavefunctions ψ_r must be reasonably well known so that the spectral density, Eq. 13 may be calculated. In cases where there are internal rotational degrees of freedom, ψ_r should include them as well. In this context it is important to recognize a difference that exists between the semiclassical theory of relaxation (cf. Ch. VIII) and the basically quantum mechanical formulation of the theory presented here. Internal and over-all rotations, as treated classically, will tend to average out anisotropic terms such as dipole-dipole interactions and will lead to spin relaxation effects which are dependent in part on the effective reorientation rates of the motions. However, in the quantum mechanical model, when collisions and related interactions are neglected, the molecule will be in a definite rotational quantum state and no significant spin relaxation is to be expected. Significant relaxation is introduced by the effect of collisions which themselves change the molecular rotational states and also broaden them so that V can be effective in energy transfers. Thus, in the present formulation the relaxation will depend directly upon the effective collision times causing reorientation and only indirectly on the quantum mechanical rotational frequencies.

One is free to make various assumptions about the reorientational collision times, or more precisely the Γ matrix. The simplest is, as we have seen, a strong-collision assumption. The strong collision assumption can be modified somewhat by assigning a separate "effective collision time" τ for each sub-ensemble of molecules which can be approximately treated as isolated from the rotational states accessible to the rest of the ensemble. Such a separation into sub-ensembles is appropriate for molecules of different nuclear-spin symmetry provided V has no matrix elements between them (e.g. ortho-hydrogen[1]). It is also possible, as has been discussed for ortho-hydrogen, to assume that for small molecules, because of the large energy differences between J levels, that collisions primarily redistribute molecules among the m_J magnetic substates of a given J and V is too weak to couple states of different J. Then Eq. 13 should be rewritten for each set of J states as:

$$j_{qq}^{J}(\alpha\text{-}\beta) = \frac{1}{2J+1} \sum_{m_J, m_J'} |F_{m_J m_J'}(q)|^2 \frac{\tau_J}{1+(\omega_{m_J m_J'}+\omega_{\alpha\beta})^2 \tau_J^2} \qquad (42)$$

and a separate relaxation equation (12 and 14) may be written for each sub-ensemble of molecules differing in their J value. One can then calculate the macroscopic magnetization $M_J(t)$ for each J state and appropriate $T_{1,J}$ and $T_{2,J}$ values are obtained. Now, provided collisional transition rates between J levels ($\tau_{J,J'}^{-1}$) are much smaller than $T_{1,J}^{-1}$ and $T_{2,J}^{-1}$ one can introduce such transitions into a form reminiscent of the modified Bloch equations for chemical exchange, except that the differences in Larmor frequencies of molecules in states having different values of J may be negligible,[1] the only differences being in their relaxation properties. One finds that if $\tau_{J,J'}^{-1}$ is much larger than differences in $T_{1,J}^{-1}$ and $T_{2,J}^{-1}$ between J levels, one still obtains a macroscopic magnetization characterized by a single $T_i^{-1} = \sum_J \frac{B_J}{T_{i,J}}$,[1] where B_J is the Boltzmann factor for the Jth rotational state.

However, even a modified strong collision approach cannot deal with selection rules for changes in magnetic sublevels m_J. This would require specifying the detailed $W_{(J,m)\rightarrow(J,m')}$ in Γ.[3] For heavier molecules, characterized by a closer spacing of the non-degenerate rotational energy levels, one would expect that collisions will be more effective in causing transitions between them, thus allowing for some significant changes in J. This may be treated in the strong collision approximation, or by introducing $W_{(J,m)\rightarrow(J',m')}$ terms into Γ.

A detailed discussion for the case of NMR of ortho-hydrogen is given elsewhere.[1] In general for NMR, the perturbation V should consist primarily of the dipole-dipole interactions of the nuclear spins, nuclear quadrupole interactions, and the nuclear spin-rotational magnetic coupling.

In the case of ESR, the electron spin-rotational magnetic coupling, the unquenched spin-orbit coupling, and electron-nuclear dipolar interactions should be the dominant terms to which the theory applies. One should, of course, also add to a relaxation equation (cf. Eq. 11) a Heisenberg spin-exchange term such as Eq. XVIII-105. Now, however, ω_{HE} of Eq. XVIII-106 should be calculated from the appropriate gas-phase collision theory rather than from liquid diffusion theory as is done in Ch. XVIII. The validity of uncorrelated R-matrix and ω_{HE} terms as employed in Ch. XVIII would probably require an experimental situation in which there is a dominant buffer gas present such that most collisions of radicals are with diamagnetic buffer molecules. Unless the pressure is high enough, it will not necessarily be true that the condition for the validity of the present theory (i.e. Eq. 28) is fulfilled; then an approach more like that of Ch. XIV would be required, i.e. Eq. 27 is solved explicitly without perturbation theory.

B. Quantum Effects of Methyl Group Tunneling[2]

This is an interesting case where V couples states of different nuclear-spin symmetry, so that a relaxation equation like Eqs. 23

and 24 is required. We illustrate with the ESR case where the dominant term in V usually (but not always) is the isotropic hyperfine interactions of the three methyl protons:

$$\hbar\mathcal{H}_{HF} = (\tfrac{8}{3})\,\pi\hbar^2\gamma_e\gamma_p\sum_{i=1}^{3}\delta(r-r_i)\vec{S}\cdot\vec{I}_i \tag{43}$$

One calculates the hyperfine interaction by assuming hyper-conjugative mixing of hydrogenic 1s orbitals and the methyl carbon orbitals with the unpaired electron in the carbon $2p_z$ orbital to which the methyl group is attached. Only the π-type symmetry linear combination of hydrogenic orbitals has the proper symmetry. One then obtains the hyperfine term as a function of the angle φ, which is the angle of rotation of the methyl group. The result is:

$$\mathcal{H}_{HF} \cong \Big(a\vec{I} + \frac{a}{2}\,e^{i2\varphi}[\vec{I}_1+\epsilon^*\vec{I}_2+\epsilon\vec{I}_3] + \frac{a}{2}\,e^{-i2\varphi}[\vec{I}_1+\epsilon\vec{I}_2+\epsilon^*\vec{I}_3]\Big)\cdot S \tag{44}$$

where $\vec{I} = \sum_{i=1}^{3}\vec{I}_i$, $\epsilon = \exp(2\pi i/3)$. In a classical average, only the first term of Eq. 44 remains. Quantum mechanically one must consider the permutation symmetry of the three protons, but the subgroup: E, (123), and (321) is sufficient. It is isomorphous with point group C_3 with irreducible representations A, E_a, and E_b. The last two terms in Eq. 44 are of symmetry E_a and E_b in the rotational ($e^{\pm 2i\varphi}$) and nuclear spin parts, although they are of overall A symmetry. Thus they lead to transitions between internal rotational (or tunneling states) of different nuclear-spin symmetry.

Interesting effects may be observed at low temperatures in solids when only the lowest torsional level is appreciably populated.

In the limit when $|\Gamma|\gg a$, the first term of Eq. 44 yields a four line hyperfine pattern of intensity ratio: 1:3:3:1. Each of the 3-fold degenerate lines are coupled to one another by the last two terms in Eq. 44, and if the torsional splitting Δ between ground-state A and E sub-levels obeys $|\Delta/\Gamma|\ll 1$ then the two central lines are predicted to broaden. However, for $|\Delta/\Gamma| \gg 1$, the A and E sub-levels are well enough separated, so that the A component is not broadened, and one predicts four sharp lines in the ratio 1:1:1:1 with doubly degenerate components of the two center lines broadened out. In the static limit for $\Delta \gg a \gg \Gamma$ the E lines reappear between the A lines leading to a 7 line spectrum with intensity ratio: 1:1:1:2:1:1:1, which has been observed. If, however, $a \gg \Delta, \Gamma$, the usual four line spectrum is observed. Detailed line-shape calculations have been made using a strong collision approximation,[2] but such features as differences of the widths and relaxation of the A and E symmetry torsional sub-levels would require the more general approach, Eq. 27.

C. Spin-Relaxation via Vibronic Relaxation

Another application of Eq. 27 is to the problem of spin-relaxation via excited electronic states. Various spin-orbit mechanisms in liquids were outlined in Ch. VIII which involved calculations from the appropriate terms in $R^{(4)}$. Orbach-type processes involving combined orbit-lattic (O-L) and spin-orbit (S-O) mechanisms can readily be calculated in the manner outlined for classical models of the dynamics of the O-L interaction. However, recent experimental work on degenerate ground-state free radicals[4] has strongly suggested that, while the anomalously large relaxation is spin-orbit in nature, it is essentially an intramolecular process. This suggests a vibronic-spin-orbit mechanism which takes advantage of the molecular vibronic relaxation. Thus Γ in Eq. 27 must include the vibronic relaxation, and one may deduce the spin relaxation from steady-state solutions to Eq. 27 utilizing a generalized form of time-independent perturbation theory (cf. Ch. XIV, discussion leading to Eqs. XIV-49 and 50).[5]

D. Non-Resonant Effects

It has been pointed out that the theory given here is for resonance-type effects.[3] That is, spin relaxation occurs by a resonant (within the appropriate linewidth of the molecular states) transfer of energy to the molecular systems. These widths are represented as Lorentzians by Eqs. 27-32. When, however, energy differences are large in the spectral densities of Eqs. 38 and 39, e.g. for

$$|\omega_{\beta'_j\alpha'_j} + \omega_{m'_j m_i}| \gg \Gamma_{m'_j m_i}$$

where $\Gamma_{m'_j m_i}$ is assumed to be a simple width, one may be looking too far into the "wings" of the line shape of the molecular transitions for the Lorentzian approximation to be valid. That is, the short time behavior of the intermolecular interactions, which determine the molecular line shapes in the wings may now be important. This is really no different than the short-time non-diffusive effects discussed in Ch. VIII with regard to classical Brownian-type diffusive reorientation. Quantum mechanically, we can speak of non-resonant or higher-order processes in this limit. And, as we shall note, the spectral densities of Eq. 38 and 39 given in this limit by terms of type

$$\frac{|F_{mm'}|^2 K_{\alpha\beta} K_{\beta\alpha}}{(\omega_{\beta\alpha} + \omega_{m'm})^2} \Gamma_{m'm} \equiv \alpha^2 \Gamma_{m'm} \quad , \tag{45}$$

are essentially of this non-resonant form. That is one reduces the molecular relaxation parameter $\Gamma_{m'm}$ by the mixing coefficient squared or α^2 which measures the degree to which the perturbation mixes molecular and spin states. However, one must exercise some caution in the proper interpretation of $\Gamma_{m'm}$ as we have just noted.

It should be possible to introduce the idea of short time behavior of Γ by analogy to the generalized Langevin equation[6] such that Eq. 27 is generalized to:

$$i\dot{\rho} = \mathcal{H}^{\times}\rho - i\int_{t_0}^{t} \Gamma(t-t')\rho(t')dt' \tag{46}$$

where, as in Eq. 27 we have not included the random force (see Deutch Ch. VII). A proper analysis of Eq. 46, including the random force, may be written as a hierarchy of continued fractions, following Mori.[7] That is, we write the ESR spectrum as being given by:

$$\chi(\omega) \propto \frac{1}{(\omega-\omega_{ab})+i\Gamma_1(\omega)-\Delta_1(\omega)} \tag{47}$$

where $\Gamma_1(\omega)$ and $\Delta_1(\omega)$ are respectively the "frequency-dependent" width and shift of the ESR line. These widths and shifts come from the spectral densities of Eqs. 38 and 39. That is, next in the hierarchy of continued fractions we have

$$\Delta_1(\omega)+i\Gamma_1(\omega) \propto \frac{|F_{mm'}|^2\ |K|^2}{(\omega-\omega_{mm'})+i\Gamma_2(\omega)-\Delta_2(\omega)} \tag{48}$$

where the terms $\Gamma_2(\omega)$ and $\Delta_2(\omega)$ give the real and imaginary parts of the one-sided Fourier transform of $\Gamma(t)$ in eq. 46. The frequency shift $\Delta_2(0)$ due to intermolecular interactions has been implicitly included in the zero-order molecular energy states.

The next stage in the hierarchy would be to calculate $\Gamma_2(\omega)$ in terms of the intermolecular interactions, utilizing a form like Eq. 48. However, one can terminate the hierarchy to a good approximation, whenever $\Gamma_{n+1} \gg \Gamma_n$. This was the basis of our analysis utilizing Eq. 28 such that $\Gamma_1 \ll \Gamma_2(0)$, and frequency-independent ESR widths and shifts are obtained in the central portion of the spectrum

The form Eq. 48 suggests that the non-resonant terms may be reasonably approximated by replacing the Γ_λ in Eqs. 39 and 40 by $\Gamma(\omega)$ [$=\Gamma_2(\omega)$]. We illustrate this with a simple example.

Suppose we have a spin S (spin system) which relaxes by its coupling to another spin I (considered here as the molecular system) due to a simple perturbation of type

$$V = AI_z[S_+ + S_-] \tag{49}$$

with I rapidly relaxing according to

$$(2T_1)^{-1} = W_I = \frac{|B|^2 \tau_c}{1+\omega_I^2\tau_c^2} \tag{49a}$$

where B is an appropriate perturbation matrix element. W_I in Eq. 49a belongs in Γ_2 and τ_c is, in principle, calculable from the intermolecular interactions (i.e. Γ_3). The usual, or resonant-type approach would yield:

$$W_S = \frac{A^2 W_I}{\omega_S^2+4W_I^2} \tag{50}$$

where W_S belongs in Γ_1.

In the non-resonant-type calculation, one first uses Eq. 49 to correct the zero order wave-functions, so that to first order

$$|\pm,\pm\rangle' = |\pm,\pm\rangle + \alpha|\mp,\pm\rangle \tag{51a}$$

and

$$|\pm,\mp\rangle' = |\pm,\mp\rangle - \alpha|\mp\,\mp\rangle \tag{51b}$$

$\alpha = \frac{A}{2\omega_S}$, and where the notation is $|M_S, M_I\rangle$, and primes indicate corrected wave functions. Now the unspecified calculation which lead to Eq. 49a, (i.e., a transition probability calculation from $R^{(2)}$ between states $|+,+\rangle \leftrightarrow |+,-\rangle$ and $|-,+\rangle \leftrightarrow |-,-\rangle$) is repeated for corrected states: $|+,+\rangle' \leftrightarrow |-,-\rangle'$ and $|+,-\rangle \leftrightarrow |-,+\rangle'$ [for matrix elements of type $B(t)(I_+ + I_-)$] and when averaged over a high temperature Boltzmann distribution in I_z, yields:

$$W_S = \frac{\frac{1}{2}A^2}{\omega_S^2}\,[W_I(\omega_S) + W_I(-\omega_S)] \tag{52}$$

where $W_I(\pm\omega_S)$ is just W_I of Eq. 49a, but with $\omega_I \rightarrow \omega_I \pm \omega_S$.

If $W_I \gg \omega_S$, then only the resonant form, Eq. 50 is appropriate. If, however, $W_I \ll \omega_S$ then Eq. 50 becomes:

$$W_S = \frac{A^2}{\omega_S^2}\,W_I \tag{50a}$$

Eq. 53 predicts the same result as Eq. 50a only when $(\omega_I \pm \omega_S)^2 \tau_c^2 \sim \omega_I^2 \tau_c^2$. If $\omega_S \gtrsim \omega_I$ then this requires $\omega_S^2 \tau_c^2 \ll 1$ or very short correlation times; or else if $\omega_S \ll \omega_I$, this is fulfilled.

When, however, $W_I \ll \omega_S$ and $|\omega_I \pm \omega_S|^2 \tau_c^2 \neq \omega_I^2 \tau_c^2$, then the non-resonant expression, Eq. 53, should be utilized. That is, the nature of Γ for Eqs. 38 and 39 has been changed.

IX.4. ACKNOWLEDGEMENT

This work was supported in part by the Advanced Research Projects Agency and by the National Science Foundation.

References

1. J. H. Freed, J. Chem. Phys. 41, 7 (1964).
2. J. H. Freed, J. Chem. Phys. 43, 1710 (1965).
3. J. H. Freed, J. Chem. Phys. 45, 1251 (1966).
4. M. R. Das, S. B. Wagner, and J. H. Freed, J. Chem. Phys. 52, 5404 (1970).
5. J. H. Freed (to be published).
6. R. Kubo, Rep. Prog. Phys. 29 (Part I), 255 (1966).
7. H. Mori, Prog. Theor. Phys. (Japan) 34, 399 (1965).

ELECTRON SPIN RELAXATION IN LIQUIDS. SELECTED TOPICS.

Daniel Kivelson

University of California, Los Angeles, California

90024

X.1. INTRODUCTION

In this chapter we shall discuss the principles associated with the study of molecular motions in liquids by means of electron spin relaxation. In particular, we shall concentrate on molecular reorientations although other motions will also be discussed. In fact, it should be noted that a relatively unique application of magnetic resonance in liquid studies arises from the fact that although a variety of experimental techniques can be used to study molecular reorientation, only magnetic resonance and rotational raman studies can yield direct information concerning molecular angular momentum in liquids, and this information is obtained not by studying reorientation but by studying spin-rotational effects which are discussed in Chapter XI.

Electron spins can be affected only by magnetic interactions. In the presence of a D.C. magnetic field the spins "precess" about the applied field, and if an rf magnetic field is applied at right angles to the D.C. field, it can induce transitions between spin levels. Furthermore, within the system of spins there are a host of magnetic interactions which give rise to local magnetic fields and these local fields can profoundly affect the electron spins. Some of these interactions are constant, i.e. time independent, and some are time dependent; the time dependence arises because the magnetic interactions are modulated by molecular motions. Time dependent magnetic interactions give rise to time dependent spin behavior; thus the time dependent behavior of the electron spins is determined not only by the magnetic interactions but also by the molecular motions that modulate these interactions, and these motions are, of course, dependent upon the intermolecular, and to some extent intramolecular, forces which depend largely on electrostatic intermolecular potentials. In liquids we can thus obtain information concerning intermolecular forces and molecular dynamics by studying the time dependent characteristics of the electron spin. One should remember in studying liquids that the magnetic interactions are weak and the spin system described by a "pure" spin hamiltonian ($\mathcal{H}_S$), is weakly coupled by means of a spin-lattice hamiltonian ($\mathcal{H}_{SL}$) to the molecular motions represented by the lattice hamiltonian ($\mathcal{H}_L$); thus the spin system is almost conserved except for this weak coupling to the rest of the system (the lattice). It follows that we obtain rather "second hand information" concerning the motions. However, the magnetic interactions can be well identified and we needn't worry about local field effects which complicate non-magnetic techniques, e.g. light scattering and dielectric relaxation; this feature is an asset in the study of molecular motions by means of magnetic resonance.

The time behavior of electron spins can be studied either by analyzing esr line widths, or even better, line shapes, or by observing directly the time dependent behavior or relaxation of the spins. We will say more about this below. The width of an esr line is related to the uncertainty in the energy of the spin states, and by means of the uncertainty principle this can be related to the spin relaxation times. We shall not pursue this approach but it is interesting to note that relaxation and linewidth data yield similar information.

In order to obtain information concerning liquids from magnetic resonance experiments, one must first identify the various sources of local magnetic fields and then consider the molecular motions that modulate them. Local static magnetic interactions merely add or subtract on to the effect of the applied D.C. field ($\vec{B}$); since the absorption frequency associated with an esr spectrum is dependent upon the applied static field, the local fields cause a frequency or field "shift" in the observed spectrum. If the static local fields can take on a finite number of distinct values, different spins are subjected to different net fields and their spectral lines are "frequency-shifted" different amounts, thus giving rise to a multiplet structure in the spectrum. If the local static environment differs from particle to particle in a continuous fashion, the "shifts" are not distinct and the spectral lines are <u>inhomogeneously broadened</u>, i.e. the apparent <u>linewidth</u> consists of many closely spaced, unresolved lines. If, on the other hand, the local magnetic interactions are time dependent, perhaps because of the time dependence of the separation between or of the orientation of the interacting magnetic moments, then the inhomogeneous effects discussed above tend to average out, the averaging becoming more effective as the rate of motion increases. The linewidths decrease and this effect is called <u>motional narrowing</u>. The line broadening arising from time dependent local fields is called <u>homogeneous broadening</u>; this broadening is not described by a distribution of unresolved lines, each corresponding to a given spin in a given environment, but each spin sees all possible local environments. This is called <u>secular</u> broadening. Over and above this effect one should note that components of the local magnetic field that have frequencies equal to the absorption frequencies (<u>Larmor frequency</u>), or other characteristic frequencies of the spins, can induce transitions between spin states, thereby providing a "spin-relaxation" mechanism, i.e. limiting the lifetime of a spin in a given state. This lifetime limitation also affects the linewidth because if the mean lifetime of a spin state is T_1, the corresponding Heisenberg uncertainty in angular frequency, and hence in linewidth (in units of sec^{-1}), is of order T_1^{-1}. This phenomenon is called nonsecular homogeneous broadening or spin-lattice relaxation; it involves energy transfer between spins and lattice.

We shall be interested in examining electron spins in a D.C. applied magnetic field and in studying the time-dependent local magnetic fields which act upon them. We will characterize these interactions in the next three sections. We will usually be concerned with a dilute paramagnetic species in a diamagnetic solvent, i.e. a species with electron spin $S > 0$ in a solvent with $S = 0$, although cases do arise in which all molecules are paramagnetic. (We restrict the term paramagnetic to refer to electronic paramagnetism, and discount the nuclear spin paramagnetism.) If the solutions, solid or liquid, are not dilute in paramagnetic species, the intermolecular spin-spin interactions must be included; in this study we wish to minimize this effect but in concentrated solutions spin exchange can be used to provide information concerning translational motions of molecules.

These lectures are specifically concerned with electron spin relaxation in liquids and, of course, this implies that we are interested in esr as a tool to be used in the study of molecular motions in liquids. It might be useful at this point to give a general overview of the material to be presented in order to give motivation to some of the calculational procedures and to give relevance to the various specialized discussions. In magnetic resonance experiments, the quantity measured is directly related to a component of the magnetization, $\vec{M}(t)$, a macroscopic quantity. We are especially interested in the time dependence of $\vec{M}(t)$ which can be described by

$$M_Z(t) = \sum_a M_Z^{(a)}(0)\exp(-t/T_{1a}) \qquad (1)$$

$$M_{\pm}(t) = \sum_a M_{\pm}^{(a)}(0)\exp(-i\omega_a t - t/T_{2a}) \qquad (2)$$

where $M_{\pm} = M_x \pm iM_y$, the initial value of $\vec{M}(t)$ is $\vec{M}(0)$, the frequencies ω_a are characteristic of the spin larmor frequencies, the relaxation times T_1 and T_2 are called spin-lattice (or longitudinal) and spin-spin (or transverse) relaxation times, respectively. Of special interest is the experimental fact that one can almost always describe $\vec{M}(t)$ by a small number (small sum over "a") of such exponentials. As we shall show, the absorption lines can be described by the real part of the Fourier transform of $M_{\pm}(t)$, and since the Fourier transform of an exponential is a Lorentzian frequency spectrum, the esr absorption spectra consist of small numbers of Lorentzian lines. We shall be concerned with rapidly tumbling molecules in liquids, i.e. relatively small molecules in solvents of low viscosity (the so-called Redfield or motionally narrowed limit); we shall always assume that the temperatures are high so that all $\hbar\omega_a \ll k_BT$, where k_BT is the thermal energy; and we shall assume that the lattice system ($\mathcal{H}_L$) has such a large heat capacity that it remains at equilibrium even though the spin system ($\mathcal{H}_S$) does not. The quantity $\vec{M}(t)$ is discussed in Sections 6 and 7.

We are next interested in relating the observed macroscopic quantity, $\vec{M}(t)$, to molecular quantities. This is done in Sections 6 and 7 by relating $\vec{M}(t)$ to the spin autocorrelation function, $\langle\vec{S}(t)\vec{S}\rangle$, where $\vec{S}$ is the electron spin and the brackets indicate an average over an equilibrium ensemble. The theory in § 5 is valid only for small perturbations of $\vec{M}(t)$ from its time independent equilibrium value; the statistical linear response theory of Kubo,[1] more fully described in Chapter VI, is used. The time dependence of $\langle\vec{S}(t)\vec{S}\rangle$ parallels that of $\vec{M}(t)$.

The time dependence of the spins, i.e. that of $\langle\vec{S}(t)\vec{S}\rangle$, depends intimately upon the local magnetic interactions ($\mathcal{H}_{SL}$) and, more importantly, upon the modulations of these interactions by molecular motions ($\underline{\mathcal{H}}_L$). Our particular interest in molecular motions, therefore, suggests that we derive interrelations between the spin correlation functions which are obtained, as described above, rather directly from the experimental measurements and correlation functions dependent more explicitly upon molecular quantities not connected with spin. In § 7 we make use of linear response theory to relate $<\underline{\vec{S}}(t)\underline{\vec{S}}>$ to the integral $\int_0^\infty <\underline{\mathcal{H}}_{SL}(\tau)\underline{\mathcal{H}}_{SL}>d\tau$, where $<\underline{\mathcal{H}}_{SL}(\tau)\underline{\mathcal{H}}_{SL}>$ is an autocorrelation function dependent both upon spin and lattice coordinates such as molecular orientation angles. It is the experimental determination of this integral that yields the totality of information, available from magnetic resonance, concerning molecular motions in liquids. It contains a spin dependent part, which we usually can specify quite completely, and a lattice part whose dynamics can sometimes be estimated by various theories but which cannot be exactly determined unless the problem of 10^{23} interacting particles is solved exactly. Useful comparisons can be made between theoretical predictions and the measured value of the integral. Of course, the integral is a single quantity and can be applied meaningfully to the study of liquid motions only if it is determined under a variety of different conditions, i.e. as a function of pressure, temperature, applied magnetic field, etc. As contrasted with numerous other techniques such as dielectric relaxation and light scattering, the information concerning liquids obtained from magnetic resonance is not obtained directly from the measured relaxation times but indirectl by relating the measured spin relaxation times to molecular motions. However, as we shall show, in magnetic resonance we obtain information concerning single particle correlations in liquids, information not readily obtainable from many of the other techniques. In the remaining sections, various applications and refinements are discussed, but spin rotational effects are included in Chapter XI and spin-spin exchange in Chapter XVIII.

The references in this article are not intended to be comprehensive but merely to give some contact with the literature.

X.2. HAMILTONIAN: TERMS WHICH DETERMINE FREQUENCIES

ESR spectra arise because of the interaction of electron spins with an applied D.C. magnetic field (B). Associated with an electron spin, $\vec{\underline{S}}$, is a magnetic moment, $g\mu_o\vec{\underline{S}}$, where μ_o is the Bohr magneton ($e\hbar/2mc$), g is the gyro-magnetic ratio (2.0023 for a free electron), and $\underline{S}_Z$ can assume the values ±1/2. In a D.C. magnetic field, $\vec{B}$, applied along the laboratory Z-axis, the hamiltonian, $\underline{\mathcal{H}}_Z$, for the spin on a free electron (in radians per second) is

$$\underline{\mathcal{H}}_Z = -(\mu_o g/\hbar)\underline{S}_Z B \ . \tag{3}$$

We shall call $\underline{\mathcal{H}}_Z$ the Zeeman hamiltonian, and it can readily be seen that its eigenvalues (in radians per second) are $\pm\ \gamma B/2$, where

$$\gamma = \mu_o g/\hbar \ . \tag{4}$$

Thus absorption of electromagnetic radiation can take place at the so-called Larmor precessional frequency, ω_o,

$$\omega_o = \gamma B \ . \tag{5}$$

In order to induce transitions in the spin, one ordinarily applies an A.C. (or rf) magnetic field at right angles to the D.C. field $\vec{B}$, e.g. the rf field may be along the laboratory Z-axis. The rf energy of interaction, $\underline{\mathcal{H}}_{rf}$ (in radians per sec.), is

$$\underline{\mathcal{H}}_{rf} = -\ \gamma\underline{S}_x b_x \cos\omega t \tag{6}$$

where b_x is the magnitude of the rf field.

In a slightly more complicated situation the electron spin, $\vec{\underline{S}}$, may interact with a nuclear spin, $\vec{\underline{I}}$, on the same atom, (e.g. in a hydrogen atom $S = 1/2$, $S_Z = \pm 1/2$, $I = 1/2$, $I_Z = \pm 1/2$). The nuclear spin has a magnetic moment, $(e\hbar/2M_pc)g_I\vec{\underline{I}}$, where M_p is the proton mass and g_I the nuclear gyromagnetic ratio. In atoms, the magnetic moment of the nucleus gives rise to a local field at the electron and the energy of interaction, $\underline{\mathcal{H}}_I$, is

$$\underline{\mathcal{H}}_I = a \, \vec{\underline{I}} \cdot \vec{\underline{S}} \tag{7}$$

where a is called the Fermi contact or the isotropic hyperfine interaction constant.[2] In a strong D.C. magnetic field the Zeeman term, $\underline{\mathcal{H}}_Z$, given in Eq. (3), dominates and the major contribution from the hyperfine interaction is that part of $\underline{\mathcal{H}}_I$ which is diagonal in the $|S_Z>$ representation, i.e.

$$\underline{\mathcal{H}}_I^{(o)} = a\underline{I}_Z\underline{S}_Z \ . \tag{8}$$

If one neglects the off-diagonal contributions of $\underline{\mathcal{H}}_I$ and takes account of the selection rules for electro-magnetic transitions, $(\Delta S_Z = \pm 1, \ \Delta I_Z = 0)$, then absorption is predicted at frequencies

$$\omega_o = \gamma B + a \ I_Z \ . \tag{9}$$

If, as for Cu^{65}, $I = 3/2$, then $I_Z = \pm 3/2, \pm 1/2$, and the spectrum consists of four hyperfine lines, with separation frequency, a. If the constant, a, is quite large, i.e. $a/\gamma B$ is not negligible, a second order calculation for the hamiltonian, $\underline{\mathcal{H}}_Z + \underline{\mathcal{H}}_I$, in the basis, $|I_Z S_Z>$, yields

$$\omega_o = \gamma B + aI_Z + \frac{a^2}{2\gamma B}\,[I(I+1) - I_Z^2] \ . \tag{10}$$

We will assume that B is always large enough for this equation to hold and, in fact, unless specifically stated, we will assume that Eq. (9) is valid.

If the unpaired electron has orbital angular moment, $\vec{\underline{\ell}}$, then the magnetic moment, $(\mu_o/\hbar)\vec{\ell}$, can interact with the applied D.C. magnetic field:

$$\underline{\mathcal{H}}_\ell = -(\mu_o/\hbar)\vec{\underline{\ell}} \cdot \vec{B} \tag{11}$$

The orbital magnetic moment can also interact directly with the spin-magnetic moment,

$$\underline{\mathcal{H}}_{so} = \zeta \vec{\underline{S}} \cdot \vec{\underline{\ell}} \ , \tag{12}$$

where ζ is called the spin-orbit coupling constant. In molecules where the symmetry is not spherical, the orbital angular momentum is quenched by the electrical interactions and the expectation value of $\vec{\underline{\ell}}$ in the ground electronic state is zero. In most of the discussion below, we shall be interested in the behavior of spins in the ground electronic state; to first order we can then average the hamiltonian in this electronic state and we then obtain what is called the first order spin-Hamiltonian, a hamiltonian dependent upon spins (and possibly nuclear positions) but not upon spatial electronic coordinates. In the first order spin-hamiltonian, $\underline{\mathcal{H}}_\ell$ and $\underline{\mathcal{H}}_{so}$ both vanish, but in the second order spin-hamiltonian, as discussed below, they can make a contribution.

Next we turn to the modifications required of Eq. (3) in an asymmetric molecule. The g-value must be replaced by a tensor, the anisotropic g-tensor, $\underline{\underline{g}}$:

$$\underline{\mathcal{H}}_Z = -(\mu_o/\hbar)\vec{\underline{S}}\cdot\underline{\underline{g}}\cdot\vec{B} \quad . \tag{13}$$

The tensor, $\underline{\underline{g}}$, can be diagonalized in a molecular coordinate system, and it represents the anisotropy in the interaction between the electron spin and the applied field, $\vec{B}$. This anisotropy arises because the applied field interacts with the incompletely quenched orbital angular momentum (averaged out in first order but not in second order if excited electronic states are not too energetic), which in turn interacts with the spin through spin-orbit interactions. Although for simplicity we will assume that the spin-orbit interaction is isotropic (scalar), the residual orbital motions reflect the anisotropy of the molecular electronic wave functions, and this anisotropy is mirrored in the orbital magnetic moment and transmitted in this way to the local field interacting with the electron spin. A brief second order calculation which combines $\underline{\mathcal{H}}_\ell$ and $\underline{\mathcal{H}}_{so}$ in Eqs. (11) and (12), yields

$$-2(\zeta\mu_o/\hbar)\sum_n \frac{\langle o|\underline{\ell}_\alpha\underline{S}_\alpha|n\rangle\langle n|\underline{\ell}_\alpha B_\alpha|o\rangle}{(E_o - E_n)} \quad , \tag{14}$$

where only terms linear in B have been kept, the subscript α represents a particular molecular axis and similar expressions hold for the other two axes, and $|o\rangle$ and $|n\rangle$ represent the ground and excited electronic states, respectively. Thus it can be seen that

$$g_{\alpha\alpha} = 2.0023 - 2\zeta \sum_n \frac{|\langle o|\ell_\alpha|n\rangle|^2}{E_o - E_n} ; \qquad (15)$$

and the deviation of $g_{\alpha\alpha}$ from 2.0023 is proportional to the spin-orbit coupling (ζ) and inversely proportional to the relevant energy gaps ($E_o - E_n$). The anisotropy arises because different components of $\vec{\ell}$ connect different electronic states, $|n\rangle$, which have different geometric properties. Note that the evaluation of $g_{\alpha\alpha}$ in the ground electronic state is consistent with our description of a spin-hamiltonian.

Now let us consider the anisotropic hamiltonian in Eq. (13) in more detail. We can separate out the trace g,

$$g = \frac{1}{3} (g_{xx} + g_{yy} + g_{zz}) , \qquad (16)$$

from the rest of the $\underline{\underline{g}}$ tensor, $\underline{\underline{G}}$:

$$\mathcal{H}_Z = -(\mu_o/\hbar) g \vec{S}\cdot\vec{B} - (\mu_o/\hbar)\vec{S}\cdot\underline{\underline{G}}\cdot\vec{B} \qquad (17)$$

where

$$\underline{\underline{G}} = \begin{pmatrix} g_{xx}-g & 0 & 0 \\ 0 & g_{yy}-g & 0 \\ 0 & 0 & g_{zz}-g \end{pmatrix} , \qquad (18)$$

and the subscripts refer to molecular axes in which $\underline{\underline{G}}$ has been diagonalized. The scalar g may differ from 2.0023 and the traceless tensor, $\underline{\underline{G}}$, indicates the anisotropy of the magnetic interaction. If the applied field, $\vec{B}$, is strong, the spin $\vec{S}$ is quantized along the laboratory Z-axis, and if the molecular z-axis is oriented along the laboratory Z-axis, Eq. (13) becomes

$$\text{(z-axis)} \qquad \underline{\mathcal{H}}_Z = -(\mu_o/\hbar) g_{zz} \underline{S}_Z B \; ; \tag{19a}$$

whereas if the molecular x-axis is oriented along the laboratory Z-axis,

$$\text{(x-axis)} \qquad \underline{\mathcal{H}}_Z = -(\mu_o/\hbar) g_{xx} \underline{S}_X B \; . \tag{19b}$$

The corresponding Zeeman frequencies are $(\mu_o/\hbar)g_{zz}B$ and $(\mu_o/\hbar)g_{xx}B$, respectively. In a crystal the molecules can be so oriented and g_{xx}, g_{yy} and g_{zz} determined; in a glass these values can also be determined.[3,4] In a liquid the situation is somewhat different; the molecule is rotating rapidly so that the orientation of the molecular coordinates relative to the laboratory axes is constantly changing. The first term in Eq. (17), the isotropic part of $\underline{\mathcal{H}}_Z$, is not altered by rotations (the trace is unaffected by rotations or unitary transformations). The components of $\underline{\underline{G}}$ in the molecular system, $\underline{\underline{G}}_{molec}$, consist of two molecular constants, e.g. g_{xx} and g_{yy} with g_{zz} determined by the traceless character of $\underline{\underline{G}}$; thus in the molecular system, $\underline{\underline{G}}_{molec}$ is independent of the orientation of the molecule. However, since $\vec{B}$ is directed along a laboratory axis, and we have also decided to quantize $\vec{\underline{S}}$ along a laboratory axis, we wish to transform from $\vec{\underline{S}}_{molec}$, $\vec{B}_{molec}$, $\underline{\underline{G}}_{molec}$, to $\vec{\underline{S}}_{lab}$, $\vec{B}_{lab}$, $\underline{\underline{G}}_{lab}$. In the laboratory system, $\underline{\underline{G}}_{lab}$ depends upon molecular orientation. Since $\vec{\underline{S}} \cdot \underline{\underline{G}} \cdot \vec{B}$ must be invariant to changes of coordinate systems, we write

$$\vec{\underline{S}}_{molec} \cdot \underline{\underline{G}}_{molec} \cdot \vec{B}_{molec} = \vec{\underline{S}}_{lab} \cdot \underline{\underline{G}}_{lab} \cdot \vec{B}_{lab} \tag{20}$$

where

$$\vec{B}_{lab} = \underline{\underline{D}} \cdot \vec{B}_{molec} \tag{21a}$$

$$\underline{\underline{G}}_{lab} = \underline{\underline{D}} \, \underline{\underline{G}}_{molec} \cdot \underline{\underline{D}}^{-1} \tag{21b}$$

$$\vec{\underline{S}}_{lab} = \vec{\underline{S}}_{molec} \cdot \underline{\underline{D}}^{-1} \tag{21c}$$

and $\underline{\underline{D}}$ is a unitary first order rotational transformation relating the molecular to the laboratory axes by means of Eulerian angles.[5]

The traceless tensor $\underline{\underline{G}}$ is transformed by rotations and $\underline{\underline{G}}_{lab}$ depends upon the Eulerian angles specifying the orientation of the molecule relative to the laboratory system; this dependence enters because $\underline{\underline{D}}$ depends upon the Eulerian angles. If the molecule rotates rapidly enough, $\underline{\underline{G}}_{lab}$ averages to zero, and the anisotropic term in Eq. (17) can be neglected; the spectrum for a molecule is then identical with that in Eq. (5) except that γ has a value different than its free electron value, (i.e. the trace of the $\underline{\underline{g}}$-tensor), whereas in a solid, where rotations are quenched, all three components can be measured. Thus in a liquid only the trace of $\underline{\underline{g}}$ can be determined. To measure all components in a solid, the solid must be dilute in the paramagnetic species or else the magnetic moments of spins on neighboring molecules will interfere with the measurements. We shall, of course, return to $\underline{\underline{G}}_{lab}$ in liquids since it plays a significant role in spin-relaxation; we shall assume that the values of the molecular parameters, g_{xx}, g_{yy}, g_{zz} determined in solids, is essentially unchanged by changes in temperature and by the solid-liquid phase change.

There is also an anisotropic hyperfine interaction term in the spin-hamiltonian of molecules:

$$\mathcal{H}_I = \vec{\underline{I}} \circ \underline{\underline{a}} \circ \vec{\underline{S}} \tag{22}$$

where $\underline{\underline{a}}$ is the <u>anisotropic hyperfine tensor</u>. The tensors $\underline{\underline{a}}$ and $\underline{\underline{g}}$ need not necessarily be diagonizable in the same molecular coordinate system, but we will assume that they can be simultaneously diagonalized. The anisotropy of $\underline{\underline{a}}$ arises in large part from the magnetic dipole-dipole interaction between the electron and nuclear spin on the same molecule; it also has a contribution that is second order and proportional to ζ in analogy with the results in Eq. (14). We can separate the hamiltonian, $\mathcal{H}_I$, into an isotropic and an anisotropic part:

$$\mathcal{H}_I = a\vec{\underline{I}}\cdot\vec{\underline{S}} + \vec{\underline{I}}\circ\underline{\underline{A}}\circ\vec{\underline{S}} \tag{23}$$

where a is the trace,

$$a = \frac{1}{3}(a_{xx} + a_{yy} + a_{zz}) \ . \tag{24}$$

$\underline{\underline{A}}$ is the traceless anisotropic hyperfine tensor,

$$\underline{\underline{A}} = \begin{pmatrix} a_{xx}-a & 0 & 0 \\ 0 & a_{yy}-a & 0 \\ 0 & 0 & a_{zz}-a \end{pmatrix} \tag{25}$$

and the subscripts refer to molecular axes. By completely analogous procedures to those used in discussing the $\underline{\underline{g}}$-tensor above, we can measure a_{xx}, a_{yy} and a_{zz} in paramagnetically dilute crystals or glasses.[4] In liquids, the first or isotropic term in Eq. (23), is not altered by rotations, but the second or anisotropic term is. The traceless tensor in the second term transforms as

$$\underline{\underline{A}}_{lab} = \underline{\underline{D}}\,\underline{\underline{A}}_{molec}\,\underline{\underline{D}}^{-1} \tag{26}$$

where $\underline{\underline{A}}_{molec}$ is a constant tensor, but the $\underline{\underline{D}}$'s are functions of the Euler angles. (See Eq. 21.) If the molecule rotates rapidly enough, the anisotropic term averages out and the esr spectrum is once again given by Eq. (9).

A few points of interest can be made concerning $\mathcal{H}_Z$ and $\mathcal{H}_I$. If $\underline{\underline{g}}$ and $\underline{\underline{a}}$ cannot be simultaneously diagonalized, it is a simple matter to extend the theory, requiring nothing but infinite patience and the expectation that the experimental data are accurate enough to warrant the effort. The Hamiltonians discussed above are valid for a single molecule with one electron and one interacting nucleus; subscripts should be added to indicate which electrons and which nuclei reside on which molecules. We will consider these matters below. Finally, it should be noted that analogous descriptions apply for nmr; the electron spin is replaced by a nuclear spin $\vec{I}_a$, the Bohr magneton is replaced by the nuclear magneton, the appropriate g-factor is substituted, the g-tensor is replaced by the chemical shift tensor, and the hyperfine interaction is replaced by the spin-spin interaction between nuclear spins $\vec{I}_a$ and $\vec{I}_b$.

X.3. HAMILTONIAN: TERMS WHICH DETERMINE RELAXATION

Time dependent terms in the spin-hamiltonian are spin-lattice interactions because they depend upon the spin ($\vec{\underline{S}}$) and are modulated by the motions of the molecules; we will call everything other than the spin, the lattice. Thus molecular positions and orientations, as well as internuclear coordinates, are lattice coordinates. In liquids the anisotropic terms in the spin hamiltonian developed in § 2 are time dependent because of molecular tumbling. We can rewrite ($\underline{\mathcal{H}}_Z + \underline{\mathcal{H}}_I$) as the sum of an isotropic term or pure spin term ($\underline{\mathcal{H}}_S$) and an anisotropic term ($\underline{\mathcal{H}}_{ANIS}$) where

$$\underline{\mathcal{H}}_S = \gamma B \underline{S}_Z + a\vec{\underline{I}}\cdot\vec{\underline{S}} \tag{27}$$

$$\underline{\mathcal{H}}_{ANIS} = (\mu_o/\hbar)\vec{\underline{S}}\cdot\underline{\underline{G}}\cdot\vec{B} + \vec{\underline{I}}\cdot\underline{\underline{A}}\cdot\vec{\underline{S}} \ . \tag{28}$$

As we have seen, if $\vec{\underline{S}}$ and $\vec{\underline{I}}$ are quantized along the laboratory Z-axis, i.e. along $\vec{B}$, then $\underline{\underline{G}}$ and $\underline{\underline{A}}$ are functions of Eulerian angles, and since the molecules tumble, $\underline{\underline{G}}$ and $\underline{\underline{A}}$, and hence $\underline{\mathcal{H}}_{ANIS}$, are time dependent. Thus $\underline{\mathcal{H}}_{ANIS}$ gives rise to a time dependent magnetic interaction with the electron spin and is, consequently, a spin-lattice term. It is convenient to rewrite the anisotropic tensors, $\underline{\underline{G}}$ and $\underline{\underline{A}}$, in terms of the anisotropic parameters:

$$\Delta g = g_{zz} - \frac{1}{2}(g_{xx} + g_{yy}) \tag{29a}$$

$$\delta g = \frac{1}{2}(g_{xx} - g_{yy}) \tag{29b}$$

$$\Delta a = a_{zz} - \frac{1}{2}(a_{xx} - a_{yy}) \tag{29c}$$

$$\delta a = \frac{1}{2}(a_{xx} - a_{yy}) \ , \tag{29d}$$

where the subscripts refer to the molecular axes. If the molecule has axial symmetry, $\delta g = \delta a = 0$.

A third major contributor to the spin-lattice interactions is the spin-rotational interaction,

$$\mathcal{H}_{SR} = \hbar\, \vec{\underline{J}}\cdot\underline{\underline{C}}\circ\vec{\underline{S}}\ , \tag{30}$$

where $\underline{\underline{C}}$ is the spin rotational interaction tensor and $\vec{\underline{J}}$ is the rotational angular momentum of the molecule. This term arises from the interaction of the spin with the rotational magnetic moment of the molecule. We will assume that $\underline{\underline{C}}$ can be diagonalized in the same molecular frame as was $\underline{\underline{G}}$ and $\underline{\underline{A}}$; in this frame, $\underline{\underline{C}}$ can then be described by three molecular spin-rotational constants. Since $\vec{\underline{S}}$ is quantized in the laboratory frame, but $\vec{\underline{J}}$ is quantized in a molecular frame, (i.e. we can readily express the rotational energy in terms of constant moments of inertia in this frame), we are interested in $\underline{\underline{C}}_{\substack{semi\\lab}}$ where

$$\underline{\underline{C}}_{\substack{semi\\lab}} = \underline{\underline{C}}_{molec}\cdot\underline{\underline{D}}^{-1}\ . \tag{31}$$

$\underline{\underline{C}}_{semi}$ is clearly dependent upon molecular orientation. Furthermore, the angular momentum, $\vec{\underline{J}}$, is altered by the presence of intermolecular torques which in turn depend upon intermolecular positions and orientations. Thus both $\underline{\underline{J}}$ and $\underline{\underline{C}}$ are lattice variables and can be time dependent. Spin-rotational interactions are discussed more fully in Chapter XI.

Another contribution to the spin-lattice interaction is the term describing fluctuations in the isotropic g and a values. This term can be expressed as

$$\mathcal{H}_{FL} = (\mu_o/\hbar)[g - \langle g\rangle]\vec{\underline{S}}\circ B + \hbar[a - \langle a\rangle]\vec{\underline{I}}\cdot\vec{\underline{S}} \tag{32}$$

where g and a are the instantaneous values of the isotropic g-value and Fermi constant, respectively. $\langle g\rangle$ and $\langle a\rangle$ are the average values of g and a, respectively. Both g and a fluctuate with changes in nuclear or electronic distributions; such changes can be brought about by intermolecular collisions and interactions or by intramolecular reorientations. Thus g and a are dependent upon lattice variables, and $\mathcal{H}_{FL}$ is a spin-lattice interaction which can be time-dependent. We will, however, neglect this term since in most cases it is probably not very significant. (In Eq. (27), γ and a should be replaced by $<\gamma>$ and $<a>$ but we shall not do this.) Similar fluctuations in $\underline{\underline{G}}$ and $\underline{\underline{A}}$ could be considered.

All the spin-lattice terms above involve <u>intramolecular magnetic interactions</u>; we shall now list a number of spin-lattice terms which involve <u>intermolecular magnetic interactions</u>. First we will mention the <u>intermolecular electron spin-spin interactions</u>,

$$\mathcal{H}_{SS} = \sum_{i \neq j} \mathcal{J}_{ij} \vec{S}_i \cdot \vec{S}_j + \sum_{i \neq j} \vec{S}_i \cdot \underline{\underline{\mathcal{D}}}_{ij} \cdot \vec{S}_j \qquad (33)$$

where the first term involves the exchange integral, $\mathcal{J}_{ij}$, i.e. the exchange integral for the unpaired electrons on the ith and j^{th} molecules, and the second term represents the intermolecular magnetic dipolar interaction between the electron spins. The first term is known as the <u>exchange term</u> and the second one is the <u>electron spin-spin dipolar term</u>. $\underline{\underline{\mathcal{D}}}_{ij}$ represents the interaction of the two dipoles averaged over the electronic coordinates; however, $\underline{\underline{\mathcal{D}}}_{ij}$ still depends upon the intermolecular distances and relative orientations; it is traceless, depends on lattice variables, and is time dependent if the molecules move. $\mathcal{J}_{ij}$ has less angular dependence, is not traceless and also depends upon the intermolecular distances and hence upon lattice parameters and time; it probably has a finite value only at very small molecular separations where it is probably much larger than $\underline{\underline{\mathcal{D}}}_{ij}$. The term $\mathcal{H}_{SS}$ is important for concentrated solutions of paramagnetic molecules where the average intermolecular distance is small but negligible in very dilute solutions. In these lectures we shall consider dilute solutions in which only intramolecular magnetic interactions are important; in Chapter XVIII a discussion of electron spin-spin exchange is presented; spin exchange is a phenomenon which depends upon the translational diffusion of the paramagnetic solute molecules since they can undergo spin exchange only when they come up very close to each other.

The intermolecular magnetic dipolar interactions between the electron spin and the nuclear spins on neighboring molecules are usually not important in esr experiments though of major significance in NMR experiments. In the measurement of ESR linewidths, the linewidths seem to be unaltered if a solvent containing hydrogens is deuterated, thus confirming the statement above. Exceptions occur when ion-pairs are formed and the unpaired electron overlaps appreciably with a nucleus in the non-paramagnetic partner of the pair; in this case the unresolved hyperfine structure arising from the overlap can contribute to the linewidth. This phenomenon has been observed in the benzene negative ion paired to Na^+ ions.[6] We shall, however, neglect these interactions.

In §15, we discuss Orbach operators. The Orbach terms arise when the spin-hamiltonian approximation is no longer valid and spin transitions are accompanied by electronic (orbital) transitions; they may be important in liquids containing species with degenerate or nearly degenerate ground electronic (orbital) states, species such as $C_6H_6^-$. We shall not include intramolecular spin-spin interactions characteristic of electron spin triplets; in such molecules the intramolecular spin hamiltonian, in the absence of nuclear hyperfine interactions, is

$$\underline{\mathcal{H}}_{triplet} = g(\mu_o/\hbar)\underline{\vec{S}}\cdot\vec{B} + \underline{\vec{S}}\cdot\underline{\underline{\mathcal{D}}}\cdot\underline{\vec{S}} \tag{34}$$

where the spin quantum numbers are S=1 and $S_z = \pm 1,0$. $\underline{\underline{\mathcal{D}}}$ represents the magnetic dipolar interactions between the two electrons and is often called the zero field tensor; it is a constant tensor in a molecular frame but since $\underline{\vec{S}}$ is quantized in a laboratory frame, $\underline{\underline{\mathcal{D}}}_{lab} = \underline{\underline{D}}\underline{\underline{\mathcal{D}}}_{molec}\underline{\underline{D}}^{-1}$, where the time-dependent transformations, $\underline{\underline{D}}$, have already been discussed. Thus $\underline{\vec{S}}\cdot\underline{\underline{\mathcal{D}}}\cdot\underline{\vec{S}}$ becomes a spin-lattice interaction with the same basic form as $\underline{\mathcal{H}}_{ANIS}$.[7]

We have neglected the "pure" nuclear spin terms; they do not contribute significantly in esr but we will assume throughout that all nuclear terms are expressed in a representation $|I_Z>$.

We will also neglect nuclear quadrupole interactions of the form $\underline{I}\cdot\underline{\underline{Q}}\cdot\underline{I}$ where $\underline{\underline{Q}}$ is a second rank tensor (nuclear quadrupole interaction tensor) which has transformation properties similar to those of $\underline{\underline{G}}$ and $\underline{\underline{A}}$. It is not often important in esr relaxation of liquids but there are exceptions to this.[8] In NMR experiments, quadrupole interactions, if not spherically symmetric, usually provide the principle relaxation mechanism for nuclei with quadrupole moments, i.e. with $I \geq 1$. In these cases the effect of quadrupole relaxation is quite analogous to that of triplet zero field interactions in ESR.

In NMR relaxation the anisotropic chemical shift is analogous to the anisotropic g-tensor, the anisotropic spin-spin interaction is analogous to the anisotropic hyperfine interaction, and there exists an important nuclear spin-rotation interaction of the form, $\underline{I}\cdot\underline{\underline{C}}_N\cdot\vec{\underline{J}}$.

A detailed discussion of both spin hamiltonians[8,9] and nuclear spin hamiltonians[10,11] are given in various books.

X.4. COMPLETE HAMILTONIAN AND DENSITY MATRIX

The dynamics of the system as a whole, in the absence of perturbing rf or pulsed fields, can be determined by a Hamiltonian, $\underline{\mathcal{H}}$, which is composed of three quite different terms:

$$\underline{\mathcal{H}} = \underline{\mathcal{H}}_S + \underline{\mathcal{H}}_L + \underline{\mathcal{H}}_{SL} , \qquad (35)$$

where the "pure" spin Hamiltonian, $\underline{\mathcal{H}}_S$, is a function only of electron $(\vec{\underline{S}})$ and nuclear spins $(\vec{\underline{I}})$, molecular constants and the applied magnetic field $(\vec{B})$; the lattice Hamiltonian, $\underline{\mathcal{H}}_L$, is a function only of intra- and inter-molecular nuclear coordinates; and the spin-lattice Hamiltonian, $\underline{\mathcal{H}}_{SL}$, is a function of both spins and the various nuclear coordinates. $\underline{\mathcal{H}}_{SL}$ corresponds to the local magnetic interactions described in the last section. In this division of terms, the lattice corresponds to all terms not dependent upon spin;

it is assumed to be a constant temperature heat bath characterized by an infinite heat capacity, i.e. the properties of the lattice are assumed to be the same both in the presence or absence of spins. This latter property is safeguarded by the fact that the "spin-system" described by $\mathcal{H}_S$ has a relatively low heat capacity and the coupling, $\mathcal{H}_{SL}$, between the spin and lattice systems is small. (At very low temperatures the lattice and spin systems cannot be so separated.) The spin and spin lattice terms together, $\mathcal{H}_S + \mathcal{H}_{SL}$, constitute the "spin Hamiltonian", a Hamiltonian in which all spatial electronic coordinates have been averaged out to second order.

The pure spin term, $\mathcal{H}_S$, is given by Eq. (27); the spin lattice term, $\mathcal{H}_{SL}$, is the sum of the interactions in Eqs. (28), (30), (32), (33) and possibly others. The lattice terms, $\mathcal{H}_L$, are the molecular kinetic energies, intermolecular electrostatic interactions, and intramolecular nuclear energies; although we will make use of a quantum mechanical development, eventually we will treat $\mathcal{H}_L$ classically.

It can readily be seen that

$$[\mathcal{H}_S,\mathcal{H}_L] = 0 \ , \qquad (36)$$

but that $\mathcal{H}_{SL}$ does not commute with either $\mathcal{H}_S$ or $\mathcal{H}_L$. Thus the lattice motions ($\mathcal{H}_L$) affect the spins ($\mathcal{H}_S$) by way of the spin-lattice interactions.

The density matrix,[12,13] $\underline{\rho}$, plays an important role in the development of the theory. The average value of a molecular quantity, $\underline{A}$, can be expressed as

$$\text{Av. of } \underline{A}(t) = \text{Tr } \underline{\rho}(t)\underline{A} \ . \qquad (37)$$

In particular, we shall have occasion to consider the magnetization, $\vec{M}(t)$, which is a macroscopic quantity obtained by averaging the electron spin magnetic moments, $\sum_j \underline{\underline{g}}_j \cdot \vec{\underline{S}}_j(\mu_o/\hbar)$, where j represents the jth spin.

$$\vec{M}(t) = \text{Tr } \underline{\rho}(t) \sum_j \underline{\underline{g}}_j \cdot \vec{\underline{S}}_j(\mu_o/\hbar) \qquad (38)$$

where $\underline{\rho}(t)$ is the instantaneous density matrix which involves sums over all spins. Since we will not consider intermolecular spin interactions, we replace $\sum \underline{\underline{g}}_j \cdot \vec{\underline{S}}_j$ by $\underline{\underline{g}} \cdot \vec{\underline{S}}(N/V)$ where (N/V) is the spin number density:

$$\vec{M}(t) = \mathrm{Tr}\ \rho(t)\underline{\underline{g}} \cdot \vec{\underline{S}}(\mu_o/\hbar)(N/V) \ . \tag{39}$$

In a magnetic resonance experiment the quantity actually measured is a component of $\vec{M}(t)$.

The density matrix, $\underline{\rho}(eq)$, of an equilibrium system enters frequently into the theory:

$$\underline{\rho}(eq) = e^{-\beta\underline{\mathcal{H}}}/\mathrm{Tr}\ e^{-\beta\underline{\mathcal{H}}} \tag{40}$$

where $\underline{\mathcal{H}}$ is the time independent hamiltonian of the system and $\beta = 1/k_B T$. The equilibrium average of an operator, $\underline{A}$, which we shall write as $\langle\underline{A}\rangle$, is

$$\langle\underline{A}\rangle = \mathrm{Tr}\ \underline{\rho}(eq)\underline{A} \ . \tag{41}$$

Since we will be interested in expansions in powers of $\underline{\mathcal{H}}_{SL}$, we can make use of the theorem in Appendix A to expand both the numerator and denominator of $\underline{\rho}(eq)$ in powers of the small spin-lattice interaction, $\underline{\mathcal{H}}_{SL}$:

$$\underline{\rho}(eq) = \underline{\rho}_o[\underline{1} - \int_o^\beta d\lambda\, e^{\lambda(\underline{\mathcal{H}}_S+\underline{\mathcal{H}}_L)}\{\underline{\mathcal{H}}_{SL} - \langle\underline{\mathcal{H}}_{SL}\rangle_o\}\, e^{-\lambda(\underline{\mathcal{H}}_S+\underline{\mathcal{H}}_L)} \tag{42}$$
$$+ \ldots\ldots$$

where $\underline{\rho}_o$ is

$$\underline{\rho}_o = e^{-\beta(\underline{\mathcal{H}}_S+\underline{\mathcal{H}}_L)}/\mathrm{Tr}\ e^{-\beta(\underline{\mathcal{H}}_S+\underline{\mathcal{H}}_L)} \tag{43}$$

and the average, $\langle\underline{A}\rangle_o$, is defined as

$$\langle\underline{A}\rangle_o = \mathrm{Tr}\ \underline{\rho}_o\underline{A} \ . \tag{44}$$

In the high temperature limit, Eq. (42) becomes

$$\underline{\rho}(eq) = \underline{\rho}_o[1 - \beta\{\underline{\mathcal{H}}_{SL} - \langle\underline{\mathcal{H}}_{SL}\rangle_o\}] \ , \tag{45}$$

and at extremely high temperature

$$\underline{\rho}(eq) \approx \underline{\rho}_o \ . \tag{46}$$

This latter is the approximation we shall use throughout although Eq. (45) can readily be employed, without much profit, at the cost of additional labor and confusion.

X.5. TIME-DEPENDENT PERTURBATION EXPANSIONS

In the development of the theory of spin-relaxation, we will have need of time dependent perturbation theory because the exact time evolution of the density matrix, $\underline{\rho}(t)$, and the electron spin operator, $\vec{\underline{S}}(t)$, cannot, in general, be obtained. In this section we will discuss a time dependent perturbation procedure which makes use of an interaction representation and is useful when working in a Heisenberg representation.

Let us consider a total hamiltonian, $\underline{\mathcal{H}}_T$, composed of two terms,

$$\underline{\mathcal{H}}_T = \underline{\mathcal{H}}_1 + \lambda\underline{\mathcal{H}}_2(t) \tag{47}$$

where $\underline{\mathcal{H}}_1$ is the principal time independent contribution, and $\lambda\underline{\mathcal{H}}_2(t)$ is a small perturbation which in some cases may be explicitly time dependent, e.g. it may represent the interaction of a spin system with an oscillating external field (cf. Eq. 6). The equation of motion for a spin operator, $\vec{\underline{S}}$, subject to the hamiltonian, $\underline{\mathcal{H}}_T$, is[13]

$$\dot{\vec{\underline{S}}}(t) = i[\underline{\mathcal{H}}_T(t),\vec{\underline{S}}(t)] \tag{48}$$

and the corresponding equation of motion for the density matrix, $\underline{\rho}$, is[12,13]

$$\dot{\underline{\rho}}(t) = -i[\underline{\mathcal{H}}_T(t),\underline{\rho}(t)] \ . \tag{49}$$

The difference in sign in these two equations of motion is carried over in classical mechanics where $\vec{\underline{S}}(t)$ could represent a dynamical variable, $\underline{\rho}(t)$ the distribution function, and the commutator must be replaced by the poisson bracket divided by i.

We wish to obtain perturbation expansions for $\vec{\underline{S}}(t)$ and $\underline{\rho}(t)$ in powers of $\lambda\underline{\mathcal{H}}_2$. The principal (rapid) time dependence of both $\vec{\underline{S}}(t)$ and $\underline{\rho}(t)$ arises from the hamiltonian, $\underline{\mathcal{H}}_T$, and we assume that the contribution of the small term, $\lambda\underline{\mathcal{H}}_2$, is manifested in an additional slow time dependence. In particular, we shall be concerned with $\vec{\underline{S}}(t)$ when $\underline{\mathcal{H}}_1$ represents the uncoupled spin (Larmor precession) and lattice motions, both of which may be rapid, and $\lambda\underline{\mathcal{H}}_2$ represents the spin-lattice interactions ($\underline{\mathcal{H}}_{SL}$) which couple the spins weakly to the lattice and give rise to relatively slow spin relaxation. The traditional procedure for handling this problem is to transform into a rotating coordinate system where the coordinate axes rotate at the Larmor frequency, and in which only motions over and above the Larmor precession contribute to the time dependence of the spins. These additional motions are usually much slower than the Larmor motions and constitute the spin relaxation processes. In our perturbation analysis of this problem, we transform to an interaction representation which is analogous to a rotating coordinate system, and we look at the time dependence of $\vec{\underline{S}}(t)$ which arises exclusively because of the presence of the weak interaction term, $\underline{\mathcal{H}}_{SL}$. As a second example, we shall consider $\underline{\rho}(t)$ when $\underline{\mathcal{H}}_1$ represents the total spin hamiltonian in the absence of an external rf field and $\lambda\underline{\mathcal{H}}_2$ represents a weakly coupled external rf interaction. Here too we move into an interaction representation in which the time dependence is slow and arises exclusively because of the presence of the weak rf interaction.

We will carry out the expansion for $\vec{\underline{S}}(t)$ and note that we need only replace i by -i to obtain the corresponding result for $\underline{\rho}(t)$. The spin, $\vec{\underline{S}}(t)$ in the interaction representation is, $\vec{\underline{S}}^I(t)$,

$$\vec{\underline{S}}(t)^I = e^{-i\underline{\mathcal{H}}_1 t}\vec{\underline{S}}(t)e^{i\underline{\mathcal{H}}_1 t} \tag{50}$$

and this can readily be seen, with the aid of Eq. (48), to correspond to the equation of motion;

$$\dot{\vec{\underline{S}}}(t)^I = i[\lambda\underline{\mathcal{H}}_2^I(t),\vec{\underline{S}}^I(t)] \tag{51}$$

where

$$\underline{\mathcal{H}}_2^I(t) = e^{-i\underline{\mathcal{H}}_1 t}\underline{\mathcal{H}}_2(t)e^{i\underline{\mathcal{H}}_1 t} \quad . \tag{52}$$

A careful examination of either Eq. (50) or (51) indicates that in the absence of $\lambda\underline{\mathcal{H}}_2$, $\vec{\underline{S}}(t)^I$ has no time dependence; therefore, we expect $\vec{\underline{S}}(t)^I$ to evolve slowly in time, whereas $\vec{\underline{S}}(t)$ may evolve rapidly in time because of the contribution of the $\underline{\mathcal{H}}_1$ term. (Note, however, that although $\vec{\underline{S}}^I(t)$ is time independent if $\lambda\underline{\mathcal{H}}_2 = 0$, in the presence of $\lambda\underline{\mathcal{H}}_2$ its time dependence is also affected by $\underline{\mathcal{H}}_1$ because $\exp(-i\underline{\mathcal{H}}_1 t)\exp(-i\underline{\mathcal{H}}_1 t - i\lambda\underline{\mathcal{H}}_2 t) \neq \exp(-i\lambda\underline{\mathcal{H}}_2 t)$ when $\underline{\mathcal{H}}_1$ and $\underline{\mathcal{H}}_2$ are non-commuting operators.) Eq. (51) can be integrated:

$$\vec{\underline{S}}(t)^I = \vec{\underline{S}}(t_o)^I + i\int_{t_o}^{t} dt'[\lambda\underline{\mathcal{H}}_2^I(t'),\underline{S}^I(t')] \quad . \tag{53}$$

This is an integral equation and as such does not represent a useful result. However, we can substitute this expression for $\vec{\underline{S}}(t)^I$ back into the integrand of Eq. (53):

$$\vec{\underline{S}}^I(t) = \vec{\underline{S}}^I(t_o) + i\lambda\int_{t_o}^{t} dt'[\underline{\mathcal{H}}_2^I(t'),\vec{\underline{S}}^I(t_o)] - \lambda^2\int_{t_o}^{t} dt'\int_{t_o}^{t'} dt''[\underline{\mathcal{H}}_2^I(t'),[\underline{\mathcal{H}}_2^I(t''),\vec{\underline{S}}^I(t'')]] \quad . \tag{54}$$

This is an exact expression but is still an integral equation. However, we can continue this iteration procedure and each time we do so we represent $\vec{\underline{S}}^I(\tau)$ by $\vec{\underline{S}}^I(t_o)$ in a higher order term in λ. Thus we obtain

$$\vec{\underline{S}}^I(t) = \vec{\underline{S}}^I(t_o) + i\lambda\int_{t_o}^{t} dt'[\underline{\mathcal{H}}_2^I(t'),\vec{\underline{S}}^I(t_o)] - \lambda^2\int_{t_o}^{t} dt'\int_{t_o}^{t} dt''[\underline{\mathcal{H}}_2^I(t'),[\underline{\mathcal{H}}_2^I(t''),\vec{\underline{S}}^I(t_o)]] + O(\lambda^3) \quad . \tag{55}$$

If we neglect all terms of order λ^3 or higher, we have an approximate, non-integral, equation for $\vec{\underline{S}}^I(t)$ since $\vec{\underline{S}}^I(t_o)$ is a constant.

In one of the applications presented in this chapter, $\underline{\mathcal{H}}_2$ will represent the spin-lattice interactions, $\underline{\mathcal{H}}_{SL}$, which do not have an explicit time dependence. However, in Eq. (55) we have introduced a time dependence; if we define the time dependent operator, $\underline{\mathcal{H}}_{SL}^o(t)$, as

$$\underline{\mathcal{H}}_{SL}^o(t) = e^{i\underline{\mathcal{H}}_1 t}\,\underline{\mathcal{H}}_{SL}\,e^{-i\underline{\mathcal{H}}_1 t} \tag{56}$$

then it is readily seen that $\underline{\mathcal{H}}_{SL}^I(t) = \underline{\mathcal{H}}_{SL}^o(-t)$. If we now make use of Eq. (50) to transform $\vec{\underline{S}}^I(t)$ back to $\vec{\underline{S}}(t)$, and if we conveniently allow $t_o = 0$, then

$$\vec{\underline{S}}(t) = \vec{\underline{S}}^{o}(t) + i\lambda\int_{o}^{t} dt'[\underline{\mathcal{H}}^{o}_{SL}(t-t'),\vec{\underline{S}}^{o}(t)] - \lambda^{2}\int_{o}^{t} dt'\int_{o}^{t'} dt''[\underline{\mathcal{H}}^{o}_{SL}(t-t'),[\underline{\mathcal{H}}^{o}_{SL}(t-t''),\vec{\underline{S}}^{o}(t)]] + O(\lambda^{3}) \tag{57}$$

where

$$\underline{S}^{o}(t) = (\exp i\underline{\mathcal{H}}_{1}t)\underline{S}(\exp -i\underline{\mathcal{H}}_{1}t) \;. \tag{58}$$

Since $\underline{\mathcal{H}}_1 = \underline{\mathcal{H}}_S + \underline{\mathcal{H}}_L$, and $\underline{\mathcal{H}}_L$ commutes with both $\vec{\underline{S}}$ and $\underline{\mathcal{H}}_S$,

$$\vec{\underline{S}}^{o}(t) = e^{i\underline{\mathcal{H}}_S t}\vec{\underline{S}}e^{-i\underline{\mathcal{H}}_S t} \;; \tag{58a}$$

i.e. $\vec{\underline{S}}^{o}(t)$ is the simply precessing spin without relaxation.

Note that Eq. (57) is indeed an expansion in the parameter λ, i.e. in $\underline{\mathcal{H}}_{SL}$, that $\mathcal{H}^{o}_{SL}(t)$ and $\vec{\underline{S}}^{o}(t)$ are stationary in the interaction representation, i.e. they are "free-wheeling"[14] and do not include spin-lattice or relaxation effects. We shall turn to this expression in § 8. Note that $\vec{\underline{S}}(t)$ and $\vec{\underline{S}}^{o}(t)$ are rapidly varying quantities but $\vec{\underline{S}}^{I}(t)$ is slowly varying and, therefore, a suitable object to study by perturbation expansions.

Another application of the perturbation expansion involves $\underline{\rho}(t)$ rather than $\underline{S}(t)$; we, therefore, replace $\vec{\underline{S}}$ by $\underline{\rho}$ in Eq. (55) and replace $i(\underline{\mathcal{H}}_1+\underline{\mathcal{H}}_2)$ by $-i(\underline{\mathcal{H}}_1+\underline{\mathcal{H}}_2)$. In this case we let $\underline{\mathcal{H}}_2$ represent the external pulsed or rf field, $-(\mu_o/\hbar)\vec{\underline{S}}\cdot\underline{\underline{g}}\cdot\vec{b}(t)$, and it may be explicitly time dependent. Thus the analogue of Eq. (55) should be

$$\underline{\rho}^{I}(t) = \underline{\rho}^{I}(t_o) - i\lambda\int_{t_o}^{t} dt'[\mathcal{H}^{I}_{2}(t'),\underline{\rho}^{I}(t_o)] + O(\lambda^{2}) \;. \tag{59}$$

Since

$$\underline{\rho}(t) = e^{-i\underline{\mathcal{H}}_1 t}\underline{\rho}^{I}(t)e^{i\underline{\mathcal{H}}_1 t} \;, \tag{60}$$

(notice the sign differences in equations for $\underline{\rho}$ and $\vec{\underline{S}}$),

$$\underline{\rho}(t) = \underline{\rho}^{o}(t) + i\lambda(\mu_o/\hbar)\int_{t_o}^{t} dt'[\vec{\underline{S}}(t'-t)\cdot\underline{\underline{g}}(t'-t)\cdot\vec{b}(t')\underline{\rho}^{o}(t)] + O(\lambda^{2}) \;, \tag{61}$$

where $\underline{\rho}^{o}(t)$, $\vec{\underline{S}}(t)$ and $\underline{\underline{g}}(t)$ are time dependent operators in the absence of the external field $\vec{b}(t)$,

$$\rho^{o}(t) = e^{-i\underline{\mathcal{H}}_1 t}\underline{\rho}^{I}(t_o)e^{i\underline{\mathcal{H}}_1 t} \quad (62a)$$

$$\vec{S}(t) = e^{i\underline{\mathcal{H}}_1 t}\underline{S}e^{-i\underline{\mathcal{H}}_1 t} \quad (62b)$$

$$\underline{\underline{g}}(t) = e^{i\underline{\mathcal{H}}_1 t}\,\underline{\underline{g}}\,e^{-i\underline{\mathcal{H}}_1 t}\ ; \quad (62c)$$

$\bar{b}(t')$ has an externally determined time dependence and is independent of both $\underline{\mathcal{H}}_1$ and $\lambda\underline{\mathcal{H}}_2$. $\underline{\underline{g}}(t)$ is time dependent because of its anisotropy and the tumbling motions of the molecules in a liquid; however, except in very viscous solutions,[15] $\underline{\underline{g}}$ in Eq. (61) can be replaced by the isotropic g-value.

Notice carefully that in the first calculation for $\vec{\underline{S}}(t)$ we let $\underline{\mathcal{H}}_1+\underline{\mathcal{H}}_2$ represent the total hamiltonian in the absence of an external field; thus $\underline{\mathcal{H}}_1 = \underline{\mathcal{H}}_S + \underline{\mathcal{H}}_L$ and $\underline{\mathcal{H}}_2 = \underline{\mathcal{H}}_{SL}$. In the second calculation for $\underline{\rho}(t)$ we let $\underline{\mathcal{H}}_1$ be the total hamiltonian in the absence of an external field and $\underline{\mathcal{H}}_2$ is the interaction with the external field; thus $\underline{\mathcal{H}}_1 = \underline{\mathcal{H}}_S + \underline{\mathcal{H}}_L + \underline{\mathcal{H}}_{SL}$ and $\underline{\mathcal{H}}_2 = -(\mu_o/\hbar)\vec{\underline{S}}\cdot\underline{\underline{g}}\cdot\vec{b}(t)$.

X.6. $\vec{\underline{M}}(t)$ RELATED TO $\langle\vec{\underline{S}}(t)\vec{\underline{S}}\rangle$; PULSE EXPERIMENTS

As mentioned above, in magnetic resonance one measures a component of the magnetization, $\vec{M}(t)$. $\vec{M}(t)$ is usually measured in one of two ways; either by "preparing" the spin system in a given nonequilibrium initial state through the application of an rf pulse, or by perturbing the system continuously through the application of a continuous rf wave. In the first set of experiments the decay of the magnetization is directly observed as a function of time, whereas in the second one the magnetic resonance spectrum is observed and this spectrum, as we shall show, can be associated with the Fourier transform of the time-dependent magnetization. Because of the insight it gives into the entire subject, in this section we shall briefly discuss the pulse experiments even though they have already been thoroughly analyzed in Chapter VI. In later sections we will concentrate exclusively on C.W. experiments.

First we wish to study $\vec{M}(t)$ in an equilibrium system. At equilibrium $\underline{\rho}(t)$ is independent of time and $\underline{\rho}(t) = \underline{\rho}(eq)$. (This is compatible with Eqs. (40) and (49).) Thus, if we examine Eq. (39) for this case we find that

$$M^{(eq)}(t) = g(\mu_o/\hbar)(N/V)\mathrm{Tr}\ \underline{\rho}(eq)\vec{S}\ , \tag{63}$$

i.e. $\vec{M}^{(eq)}(t)$ is time independent. It can readily be shown that $M_x^{(eq)} = M_y^{(eq)} = 0$ whereas $M_z^{(eq)}$ has a finite value, $M^{(eq)}$. In a 90° pulse experiment, a 90° rf pulse at t=0 prepares the system in such a way that $M_x(o) = M(eq)$, $M_y(o) = M_z(o) = 0$, and we can watch the decay of $M_x(t)$ back to its equilibrium value of zero. This time dependence is dependent upon $\underline{S}_x(t)$ and hence upon the various local magnetic interactions and their motional modulations. A 180° rf pulse results in $M_z(o) = -M(eq)$, and $M_x(o) = M_y(o) = 0$; the decay of $M_z(t)$ back to its equilibrium value of +M(eq) can also be observed. Normally, as discussed near Eqs. (1) and (2), the decay of $\vec{M}(t)$ can be expressed by a few exponentials with characteristic spin relaxation times $T_{1a}^{(\alpha)}$ and $T_{2a}^{(\alpha)}$. These experiments can be understood as explained in Chapter VI; the system is "prepared" in a nonequilibrium state at times t<0, at t=0 the external field is removed, and at t>0 the decay of the spin system back to equilibrium is observed. At t=0, the density matrix, $\underline{\rho}(0)$, does not represent equilibrium for the unperturbed system, but we will assume that it is at equilibrium with respect to the applied external pulsing field, $\vec{b}'$. Thus

$$\underline{\rho}(0) = \frac{\exp-\beta[\underline{\mathcal{H}}-g(\mu_o/\hbar)\vec{\underline{S}}\cdot\vec{b}']}{\mathrm{Tr}\ \exp-\beta[\underline{\mathcal{H}}-g(\mu_o/\hbar)\vec{\underline{S}}\cdot\vec{b}']} \tag{64}$$

where $\underline{\mathcal{H}}$ is the total hamiltonian in the absence of the external field, $\vec{\underline{b}}'$. At t=0, the external field $\vec{\underline{b}}'$ is turned off and the prepared system is allowed to decay to equilibrium in the absence of the external field. The time dependent density matrix, $\underline{\rho}(t)$, for the system at times t>0, can readily be obtained from Eq. (61); since the external field $\vec{b}'=0$ for t>0, Eq. (61) yields the result $\underline{\rho}(t)=\underline{\rho}^o(t)$ where

$$\underline{\rho}^o(t) = e^{-i\underline{\mathcal{H}}t}\underline{\rho}(o)e^{i\underline{\mathcal{H}}t} \tag{65}$$

since we have set $t_o=0$ and $\underline{\mathcal{H}}_1=\underline{\mathcal{H}}$ in Eq. (62a).

We now wish to relate $\vec{M}(t)$ to the microscopic autocorrelation function, $<\vec{\underline{S}}(t)\vec{\underline{S}}(0)>$. The relationship between $\vec{M}(t)$ and $<\vec{\underline{S}}\ \vec{\underline{S}}(t)>$ is the fluctuation-dissipation relationship, dissipations referring to $\vec{M}(t)$ and fluctuations to $<\vec{\underline{S}}\ \vec{\underline{S}}(t)>$. Of course, we should really consider $<\underline{\underline{g}}\cdot\vec{S}\ \underline{\underline{g}}(t)\cdot\vec{S}(t)>$, but, as we indicated in the last section, we will assume $\underline{\underline{g}}$ is a scalar, time independent quantity for this calculation. If we combine Eqs. (65) and (39), we obtain

$$\vec{M}(t) = (N/V)(\mu_o/\hbar)g\ \mathrm{Tr}\ \underline{\rho}(0)\vec{\underline{S}}(t) \tag{66}$$

where we have made use of the definition of $\vec{\underline{S}}(t)$ in Eq. (62b) and of the fact that the operator $\exp(-i\underline{\mathcal{H}}t)$ could be cyclically permuted in a trace without altering the value of the trace.

$\vec{M}(t)$ is proportional to $\mathrm{Tr}\ \underline{\rho}(0)\vec{\underline{S}}(t)$ and hence, it involves an average over a non-equilibrium ensemble. It indicates how the spins react to perturbing influences and how they return to their equilibrium distribution when the perturbing influence is removed, i.e. how they return to $<\vec{\underline{S}}(t)> = <\vec{\underline{S}}(0)> = \mathrm{Tr}\ \underline{\rho}(eq)\vec{\underline{S}}$. We will be interested in $\mathrm{Tr}\ \underline{\rho}(0)\vec{\underline{S}}(t)$ only for ensembles that are "not very far" removed from equilibrium, i.e. for small disturbances. Thus we will only consider deviations from equilibrium which are linear in a perturbing influence, a procedure known as linear response theory. See Chapter VI. In the context of this theory, the nonequilibrium behavior of $\mathrm{Tr}\ \underline{\rho}(0)\vec{\underline{S}}(t)$ can be described in terms of equilibrium properties in much the same way as a function f(x) can be described in terms of its properties at $x = x_o$ if only linear variations in $(x - x_o)$ are considered, i.e. $f(x) \approx f(x_o) + (\partial f(x_o)/\partial x)\ (x - x_o)$ where $f(x_o)$ and $\partial f(x_o)/\partial x$ are properties at $x = x_o$. It follows that measurements of $\vec{M}(t)$ can then also yield information concerning dynamic or time dependent processes at equilibrium, processes that exist at equilibrium but are averaged out and can only be detected when the equilibrium balance is disturbed.

In order to obtain an expression for $\vec{M}(t)$, we make use of the formula for $\vec{M}(t)$ in Eq. (66) and the expression for $\underline{\rho}(0)$ in Eq. (64); we then expand $\underline{\rho}(0)$ in powers of $\vec{b}'$ as indicated in Appendix A. In the spirit of linear response theory, we retain only terms linear in $\vec{b}'$:

$$\underline{\rho}(0) = \underline{\rho}(eq)\left\{1 - \int_0^\beta d\lambda e^{\lambda\underline{\mathcal{H}}}(g\mu_o/\hbar)\vec{b}'\cdot[\vec{\underline{S}} - \langle\vec{\underline{S}}\rangle]e^{-\lambda\underline{\mathcal{H}}} + \ldots\ldots\right\} \tag{67}$$

where < > indicates an equilibrium average. We now assume that we are interested in $M_\alpha(t)$, the component of $\vec{M}(t)$ along the α-laboratory axis, and that $\vec{b}'$ is also directed along the α-axis; by combining Eqs. (66), (64) and (67), we obtain

$$\vec{M}_\alpha(t) - \vec{M}_\alpha(eq) = -(g\mu_o/\hbar)^2\int_0^\beta d\lambda\langle e^{\lambda\underline{\mathcal{H}}}[\underline{S}_\alpha - \langle\underline{S}_\alpha\rangle]e^{-\lambda\mathcal{H}}\underline{S}_\alpha(t)\rangle b' \tag{68}$$

where $\vec{M}_\alpha(eq) = \langle(g\mu_o/\hbar)\vec{\underline{S}}_\alpha(0)\rangle$. If we assume high temperatures, we can then set $e^{\pm\lambda\underline{\mathcal{H}}} = \underline{1}$ and

$$M_\alpha(t) - M_\alpha(eq) = -(g\mu_o/\hbar)^2(k_BT)^{-1}\langle[\underline{S}_\alpha - \langle\underline{S}_\alpha\rangle]\underline{S}_\alpha(t)\rangle b' \ . \tag{69}$$

If we set t=0, we obtain an expression for b' which we can then substitute back into Eq. (69):

$$\frac{M_\alpha(t) - M_\alpha(eq)}{M_\alpha(0) - M_\alpha(eq)} = \frac{\langle\underline{S}_\alpha\underline{S}_\alpha(t)\rangle - \langle\underline{S}_\alpha\rangle^2}{\langle\underline{S}_\alpha^2\rangle - \langle\underline{S}_\alpha\rangle^2} \ . \tag{70}$$

And, indeed, we see that the time dependence of $M_\alpha(t)$ is that of $\langle\underline{S}_\alpha\underline{S}_\alpha(t)\rangle$ as stated above.

X.7. LINE WIDTHS: CW EXPERIMENTS

Another means of measuring $\vec{M}_x(t)$ is by a continuous wave rf absorption or dispersion experiment. The power P(t) absorbed if an rf field, $b_x'\cos\omega t$, is applied along the X-axis, is

$$P(t) = b_x'\cos\omega t\,\frac{dM_x(t)}{dt} \ . \tag{71}$$

If b_x' is small, the magnetization is proportional to b_x', (linear response theory), and we can write

$$M_x(t) = [\chi_{xx}'(\omega)\cos\omega t + \chi_{xx}''(\omega)\sin\omega t]b_x'V \tag{72}$$

where χ'_{xx} and χ''_{xx} are the real and imaginary (in-phase and out-of-phase) components of the magnetic susceptibility and V is the sample volume. The susceptibility is independent of b'_x but a function of ω. The average power per cycle, $\overline{P}$, can be obtained by substituting Eq. (72) into Eq. (71) and averaging over one cycle;

$$\overline{P}(\omega) = \frac{\omega}{2}\chi''(\omega)b_x'^2 V \ . \tag{73}$$

$\overline{P}(\omega)$ is the observed spectrum, and the line shape is very nearly given by $\chi''(\omega)$, which in turn is closely related to $M_x(t)$. Note that $\chi'(\omega)$ represents the dispersion of the resonance. If b'_x is large so that linear response theory is not applicable, saturation effects can set in and $\chi(\omega)$ is a function of b'_x.

Now we wish to show that in a continuous wave rf absorption experiment, it is $\langle \underline{S}_x \underline{S}_x(t) \rangle$ that determines the line shape. We will go back to Eq. (39) and investigate $M_x(t)$:

$$M_x(t) = g(\mu_o/\hbar)\mathrm{Tr}\ \underline{\rho}(t)\underline{S}_x \ . \tag{74}$$

We next turn to the perturbation expression for $\underline{\rho}(t)$ given in Eq. (61) and we let $t_o \to -\infty$, at which time $\underline{\rho}^I(-\infty) = \underline{\rho}(-\infty) = \underline{\rho}(eq)$. Since $\underline{\rho}(eq)$ is a function of $\underline{\mathcal{H}}_1$ only, i.e. of the hamiltonian $\underline{\mathcal{H}} = \underline{\mathcal{H}}_S + \underline{\mathcal{H}}_L + \underline{\mathcal{H}}_{SL}$, it can be seen from Eq. (62a) that $\underline{\rho}^o(t) = \underline{\rho}(eq)$, i.e. $\underline{\rho}^o(t)$ is independent of time. The time dependent field, $b'_x(t)$, is the applied rf field, $b'_x \cos\omega t$, where b'_x is small enough for linear response theory to hold. Note that the expansion for $\underline{\rho}(t)$ need not converge but that for $M_x(t)$ it must if the results are to be sensible. (See Chapter VII.) If now the expression for $\underline{\rho}(t)$ in Eq. (61) is inserted into the expression for $M_x(t)$ in Eq. (74), one obtains

$$M_x(t) = ig^2(\mu_o/\hbar)^2 \int_{-\infty}^{t} \mathrm{Tr}[\underline{S}_x(\tau - t), \underline{\rho}(eq)]\underline{S}_x b'_x \cos\omega\tau \tag{75}$$

where we have set $\mathrm{Tr}\ \underline{\rho}(eq)\underline{S}_x = 0$. The commutator can be expanded, the operators in the trace can be permuted cyclically without altering the value of the trace, and the time dependence shifted from one $\underline{S}_x$ to the other by cyclic permutation of the operator, $\exp \pm i\underline{\mathcal{H}}t$, and by noting that $\exp \pm i\underline{\mathcal{H}}t$ commutes with $\underline{\rho}(eq)$; then we write

$$M_x(t) = \int_{-\infty}^{t} d\tau \chi_{xx}(t-\tau) B_x' \cos \omega\tau \tag{76}$$

where the <u>response function</u> $\chi(t-\tau)$ relates a response at time t to all perturbations at times $\tau \leq t$, (Principle of Causality), and

$$\chi_{xx}(t-\tau) = ig^2(\mu_o/\hbar)^2 \langle \underline{S}_x(t-\tau), \underline{S}_x \rangle \, . \tag{77}$$

A change of variable, $t-\tau \rightarrow \tau$, converts Eq. (76) into Eq. (72) where the real and imaginary parts of the susceptibility are

$$\chi_{xx}'(\omega) = \int_0^{\infty} d\tau \chi_{xx}(\tau) \cos \omega\tau \tag{78a}$$

and

$$\chi_{xx}''(\omega) = \int_0^{\infty} d\tau \chi_{xx}(\tau) \sin \omega\tau \, , \tag{78b}$$

respectively. A simple check shows that <u>$\chi'(\omega)$ and $\chi''(\omega)$ are real</u>.

Usually we can make a high temperature approximation. To do this, make use of cyclic permutability of operators in a trace and rewrite Eq. (77) as

$$\chi_{xx}(t) = ig^2(\mu_o/\hbar)^2 \mathrm{Tr}\ \underline{\rho}(eq)[\underline{S}_x(t)\underline{S}_x - \underline{\rho}(eq)^{-1}\underline{S}_x(t)\underline{\rho}(eq)\underline{S}_x] \, . \tag{79}$$

The expansion of $\underline{\rho}(eq)^{-1}\underline{S}_x\underline{\rho}(eq)$ in powers of $(k_BT)^{-1}$ is readily accomplished:

$$\underline{\rho}(eq)^{-1}\underline{S}_x(t)\underline{\rho}(eq) = \underline{S}_x(t) + \beta[\mathcal{H},\underline{S}_x(t)] + \beta^2 \quad + \ldots] \, . \tag{80}$$

Since $\dot{\underline{S}}_x(t) = i[\mathcal{H},\underline{S}_x(t)]$, Eq. (79) can be rewritten as

$$\chi_{xx}(t) = -\beta g^2(\mu_o/\hbar)^2 \langle \dot{\underline{S}}_x(t)\underline{S}_x \rangle \tag{81}$$

and $\chi''(\omega)$ in Eq. (78b) becomes, after a parts integration,

$$\chi_{xx}''(\omega) = \beta g^2(\mu_o/\hbar)^2 \omega \int_0^{\infty} dt \, \langle \underline{S}_x(t)\underline{S}_x \rangle \cos \omega t \, , \tag{82}$$

where we have assumed that $\langle \underline{S}_x(t)\underline{S}_x \rangle \rightarrow 0$ as $t \rightarrow \infty$. Finally if we combine Eq. (82) with Eq. (73), we obtain an expression for the average power absorbed, $\overline{P}(\omega)$, in the high temperature limit:

$$\overline{P}(\omega) = \beta(g\mu_o/\hbar)^2(V/2)\omega^2 b_x'^2 \int_0^{\infty} dt \, \langle \underline{S}_x(t)\underline{S}_x \rangle \cos \omega t \, . \tag{83}$$

Again we see that $\langle \underline{S}_x(t)\underline{S}_x \rangle$ is the fundamental quantity of interest; actually it is the real part of the Fourier transform of the auto-correlation function, $\langle \underline{S}_x(t)\underline{S}_x \rangle$, that is measured. This form is quite characteristic of many transport and relaxation coefficients.

The absorption spectrum is just $\overline{P}(\omega)$; since in most spectra the spectral width, $\Delta\omega$, is narrow compared to the absorption frequency, ω, i.e. $\Delta\omega \ll \omega$, the ω^2 factor in Eq. (83) does not alter the spectral- or line-shapes appreciably. However, we could rewrite Eq. (83) by carrying out two parts integrations and setting $\langle \dot{\underline{S}}_x \underline{S}_x \rangle = \langle \underline{S}_x(\infty)\underline{S}_x \rangle = 0$:

$$\overline{P}(\omega) = \beta(g\mu_o/\hbar)^2(V/2)b_x'^2 \int_o^\infty dt \langle \dot{\underline{S}}_x(t)\dot{\underline{S}}_x \rangle \cos \omega t \; . \qquad (84)$$

Finally, we can readily see that in the very high temperature limit the autocorrelation function, $\langle \underline{S}_x(t)\underline{S}_x \rangle$, is real and an even function of time. In the finite temperature range, i.e. where $\underline{\rho}(eq) \neq \underline{1}$,

$$\langle \underline{S}_x(-t)\underline{S}_x \rangle = \langle \underline{S}_x(t)\underline{S}_x \rangle^* = \langle \underline{S}_x \, \underline{S}_x(t) \rangle \; . \qquad (85)$$

X.8. SPIN AUTOCORRELATION FUNCTION

We now wish to obtain an expression for the autocorrelation function $\langle \underline{S}_\alpha(t)\underline{S}_\alpha(0) \rangle$. In particular, we wish to relate this autocorrelation to an autocorrelation function, $\langle \underline{\mathcal{H}}_{SL}(t)\underline{\mathcal{H}}_{SL} \rangle_o$, for the spin lattice interaction, and we wish to show explicitly how this latter correlation function depends directly upon the dynamic molecular properties of the liquid. We could develop transport equation à la Bloch[16] or Redfield[17] and solve these equations for $\langle \vec{\underline{S}}(t)\vec{\underline{S}}(0) \rangle$; this has been done by Freed and Fraenkel.[18] We shall, however, pursue a method proposed by Kubo and Tomita[1] in which we can solve for $\langle \vec{\underline{S}}(t)\vec{\underline{S}}(0) \rangle$ directly. To do this, we make use of the perturbation expansion in Eq. (57), i.e. we consider the total hamiltonian $\underline{\mathcal{H}} = \underline{\mathcal{H}}_S + \underline{\mathcal{H}}_L + \underline{\mathcal{H}}_{SL}$, and we expand in powers of $\underline{\mathcal{H}}_{SL}$. We then assume that the power series expansion can be summed to an exponential, a procedure justified to some extent by the cumulant theory of Kubo[1,19] and Freed.[20] (See Chapter VIII.)

The expression for $\vec{\underline{S}}(t)$ in Eq. (57) can be used to evaluate the autocorrelation function $\langle \vec{\underline{S}}(t)\vec{\underline{S}}(0) \rangle$:

$$<\vec{\underline{S}}(t)\vec{\underline{S}}(0)> = <\vec{\underline{S}}^{\circ}(t)\vec{\underline{S}}(0)> - i\int_0^t dt'<[\underline{\mathcal{H}}^{\circ}_{SL}(t-t'),\vec{\underline{S}}^{\circ}(t)]\vec{\underline{S}}(0)>$$
$$- \int_0^t dt'\int_0^{t'} dt''<[\underline{\mathcal{H}}^{\circ}_{SL}(t-t'),[\underline{\mathcal{H}}^{\circ}_{SL}(t-t''),\vec{\underline{S}}^{\circ}(t)]]\vec{\underline{S}}(0)> + \ldots\ldots , \tag{86}$$

where $\underline{S}^{\circ}(t)$ is the simply-precessing or free-wheeling spin defined in Eq. (58a), and $\underline{\mathcal{H}}^{\circ}_{SL}(t)$ is defined similarly as

$$\underline{\mathcal{H}}^{\circ}_{SL}(t) = e^{i(\underline{\mathcal{H}}_S+\underline{\mathcal{H}}_L)t}\,\underline{\mathcal{H}}_{SL}\,e^{-i(\underline{\mathcal{H}}_S+\underline{\mathcal{H}}_L)t} . \tag{87}$$

The equilibrium averages in Eq. (86) are defined in Eq. (41) and involve the equilibrium density matrix, $\underline{\rho}(eq)$, defined in Eq. (40). However, we will make the very high temperature approximation and replace $\underline{\rho}(eq)$ by $\underline{\rho}_o$. (See Eqs. 43 and 46.) Thus the averages, $< >$, on the right hand side of Eq. (86) can be replaced by the averages, $< >_o$. (See Eq. 44.) Note that $\underline{\mathcal{H}}_{SL}$ was introduced as a time independent hamiltonian but by shifting to the interaction representation and expanding in powers of $\underline{\mathcal{H}}_{SL}$, we convert it to a time dependent operator.

In order to investigate the result in Eq. (86) more thoroughly we must discuss the spin-lattice terms in somewhat more detail. We will quantize all spin terms along the spatial Z-axis, i.e. along $\vec{B}$. The spin lattice interaction terms can all be written in the form

$$\underline{\mathcal{H}}_{SL} = \sum_q \underline{F}_q \underline{\mathcal{S}}_q \tag{88}$$

where $\underline{F}_q$ is a function of the lattice variables only and $\underline{\mathcal{S}}_q$ is a function of the spin variables only. Typical contributions to $\underline{\mathcal{H}}_{SL}$ are of the form

$$\underline{\mathcal{H}}_{SL} = (-1/3)\Delta g(1-3\cos^2\theta)B\underline{S}_Z + (-1/3)\Delta a(1-3\cos^2\theta)\underline{I}_Z\underline{S}_Z + \ldots \tag{89}$$

where θ is the angle between the molecular and laboratory Z-axes; in this expression, $\underline{\mathcal{S}}_q$ is $\underline{S}_Z$, $\underline{I}_Z\underline{S}_Z$, ... in each successive term, respectively, and $\underline{F}_q$ is $(-1/3)\Delta g(1-3\cos^2\theta)B$, ... If Eq. (87) is combined with Eq. (88), one obtains, because of the commutivity of $\underline{\mathcal{H}}_S$ and $\underline{\mathcal{H}}_L$,

$$\underline{\mathcal{H}}^{\circ}_{SL}(t) = \sum_q \underline{F}^{\circ}_q(t)\underline{\mathcal{S}}^{\circ}_q(t) \tag{90}$$

where

$$\underline{F}^{\circ}_q(t) = e^{i\underline{\mathcal{H}}_L t} \underline{F}_q e^{-i\underline{\mathcal{H}}_L t} \tag{91a}$$

$$\mathcal{S}^{\circ}_q(t) = e^{i\underline{\mathcal{H}}_S t} \mathcal{S}_q e^{-i\underline{\mathcal{H}}_S t} \quad . \tag{91b}$$

Thus $\underline{F}^{\circ}_q(t)$ depends exclusively upon lattice variables and $\mathcal{S}^{\circ}_q(t)$ exclusively upon spin variables. The time dependence of $\underline{F}^{\circ}_q(t)$, i.e. that of the lattice, is assumed to be classical in all molecular liquids except H_2 and perhaps CH_4; on the other hand, the time dependence of the spins, $\mathcal{S}^{\circ}_q(t)$, is strictly quantum mechanical. The superscript (°) indicates that the time evolution of $\mathcal{S}_q$ and $\underline{F}_q$ arise from $\underline{\mathcal{H}}_S$ and $\underline{\mathcal{H}}_L$, respectively, and not from the full hamiltonian.

In an isotropic liquid at sufficiently high temperature, the spin lattice terms average to zero over sufficiently long times; by the ergodic principle this means that the equilibrium ensemble average, $<\ >_o$, vanishes, i.e.

$$<\underline{\mathcal{H}}_{SL}(t)>_o = 0 \ . \tag{92}$$

If $<\underline{\mathcal{H}}_{SL}>_o$ did not vanish, we could define a new spin-lattice hamiltonian, $\underline{\mathcal{H}}_{SL} - <\underline{\mathcal{H}}_{SL}>_o$, and we could add $<\underline{\mathcal{H}}_{SL}>_o$ to the pure spin hamiltonian; in this way the redefined spin-lattice term could be constructed so that its ensemble average vanished. In addition to Eq. (92), we have

$$<\underline{F}^{\circ}_q(t)>_o = 0 \ , \tag{93}$$

where in this case, $<\ >_o$ indicates an ensemble average over lattice variables. It, therefore, follows that the term in Eq. (86) which is linear in $\underline{\mathcal{H}}_{SL}$ vanishes.

Next we can examine the average in the second order term in Eq. (86). If we make use of Eq. (90),

$$<[\underline{\mathcal{H}}^{\circ}_{SL}(t-t'),[\underline{\mathcal{H}}^{\circ}_{SL}(t-t''),\underline{\vec{S}}^{\circ}(t)]]\underline{\vec{S}}(0)>_o$$
$$= \sum_{q,q'} <\underline{F}^{\circ}_q(t-t')\underline{F}^{\circ}_{q'}(t-t'')>_o <[\mathcal{S}^{\circ}_q(t-t'),[\mathcal{S}^{\circ}_{q'}(t-t''),\underline{\vec{S}}^{\circ}(t)]]\underline{\vec{S}}(0)>_o \tag{94}$$

where the averages over the F's just involve the lattice density matrix, $e^{-\beta\underline{\mathcal{H}}_L}/\mathrm{Tr}\ e^{-\beta\underline{\mathcal{H}}_L}$, and the average over the spins involves the density matrix, $e^{-\beta\mathcal{H}_S}/\mathrm{Tr}\ e^{-\beta\mathcal{H}_S}$. [The $\underline{F}$'s commute with each other and with the spins, since we shall treat the F's and the related lattice motion as classical properties.] It can easily be seen, by the invariance of a trace to cyclic permutation, that

$$\langle \underline{F}_q^\circ(t-t')\underline{F}_{q'}^\circ(t-t'')\rangle_o = \langle \underline{F}_q^\circ(t''-t')\underline{F}_{q'}^\circ\rangle_o \ . \tag{95}$$

This correlation function involves only the dynamics of the lattice (liquid) and not of the spins; it is the quantity of prime interest in the study of molecular motions in liquids. Furthermore, we will assume that only those terms for which

$$[\mathcal{H}_S, \mathcal{S}_q^\ell \mathcal{S}_{q'}] = 0 \tag{96}$$

contribute in Eq. (94); these are adiabatic terms. In Appendix B we discuss why the non-adiabatic terms can be neglected. It is easy to show that for the adiabatic terms, the spin correlation function in Eq. (94) becomes $\langle[\mathcal{S}_q^\circ(t''t'), [\mathcal{S}^\circ(0),\vec{\underline{S}}^\circ(t)]\vec{\underline{S}}(0)\rangle$. Finally, we can change the variables of integration in Eq. (86); we introduce $T = t'' + t'$ and

$$\tau = t'' - t' \ . \tag{97}$$

See Appendix B. After all this, Eq. (86) becomes

$$\langle\vec{\underline{S}}(t)\vec{\underline{S}}(0)\rangle = \langle\vec{\underline{S}}^\circ(t)\vec{\underline{S}}(0)\rangle_o - \int_o^t d\tau(t-\tau)\sum_{q,q'}\langle\underline{F}_q\underline{F}_{q'}\rangle f(\tau)\langle[\mathcal{S}_q^\circ(\tau),[\mathcal{S}_{q'},\vec{\underline{S}}^\circ(t)]]\vec{\underline{S}}\rangle_o + \ldots . \tag{98}$$

where the "normalized" lattice correlation function, $f(\tau)$, depends only upon lattice times (molecular motions) and is defined as

$$\langle\underline{F}_q^\circ(\tau)\underline{F}_{q'}(0)\rangle = \langle\underline{F}_q\underline{F}_{q'}\rangle f_{qq'}(\tau) \ , \tag{99}$$

where

$$f(0) = 1 \ . \tag{100}$$

We will assume that

$$\lim_{t\to\infty} f(\tau) \to 0 \tag{101}$$

and that, furthermore, if

$$\tau > \tau_c \ , \tag{102}$$

where τ_c is a characteristic time for the lattice correlation, $f(\tau) = C$ Furthermore, we will assume that τ_c is very short, that all times of interest in spin relaxation are greater than τ_c, i.e.

$$t \gg \tau_c , \tag{103}$$

and, therefore, that $(t-\tau) \to t$ and $\int_0^t \to \int_0^\infty$ in Eq. (98) without appreciable effect on the integral. Thus

$$\langle \vec{\underline{S}}(t)\vec{\underline{S}}(0)\rangle = \langle \vec{\underline{S}}^\circ(t)\vec{\underline{S}}(0)\rangle_o$$
$$- t\sum_{q,q'} \langle \underline{F}_q \underline{F}_{q'}\rangle_o \int_0^\infty d\tau f_{qq'}(\tau)\langle[\mathscr{A}_q^\circ(\tau),[\mathscr{A}_{q'},\vec{\underline{S}}^\circ(t)]]\vec{\underline{S}}\rangle_o + \ldots\ldots \tag{104}$$

In order to complete the expression for $\langle\vec{\underline{S}}(t)\vec{\underline{S}}(0)\rangle$, we note that since $\vec{\underline{S}}^\circ(t)$ is the freely precessing spin defined in Eq. (58a), its matrix elements in the basis $|a\rangle$, where $|a\rangle$ is an eigenfunction of $\underline{\mathcal{H}}_S$, can readily be evaluated:

$$\langle a'|\underline{S}^\circ(t)|a''\rangle = \langle a'|\underline{S}^\circ|a''\rangle e^{i\omega_{a'a''}t} \tag{105}$$

where the Larmor frequency $\omega_{a'a''}$ is

$$\omega_{a'a''} = \langle a'|\underline{\mathcal{H}}_S|a'\rangle - \langle a''|\underline{\mathcal{H}}_S|a''\rangle . \tag{106}$$

We shall see in Appendix C that $|a'\rangle$ and $|a''\rangle$ must both represent the same nuclear spin state, $|I_Z\rangle$, but possibly, different electron spin states, $|S_Z'\rangle$ and $|S_Z''\rangle$, respectively. If there is but one relevant nuclear and one electron spin, $|a\rangle$ represents $|I_Z S_Z\rangle$, but if there are several interacting nuclei the problem is complicated somewhat. (See § 14.) Thus the spin averages in Eq. (104) have the forms

$$\langle\vec{\underline{S}}^\circ(t)\vec{\underline{S}}(0)\rangle_o = \sum_{a',a''} e^{i\omega_{a'a''}t}\langle\vec{\underline{S}}^{a'a''}\vec{\underline{S}}\rangle_o \tag{107a}$$

$$\langle[\mathscr{A}_q^\circ(\tau),[\mathscr{A}_{q'},\vec{\underline{S}}^\circ(t)]]\vec{\underline{S}}\rangle_o = \sum_{a,b} e^{i\omega_{b'b''}\tau} e^{i\omega_{a'a''}t}\langle[\mathscr{A}_q^{b'b''},[\mathscr{A}_{q'},\vec{\underline{S}}^{a'a''}]]\vec{\underline{S}}\rangle_o \tag{107b}$$

where the superscripts, a'a" and b'b", on an operator indicate (a'a") and (b'b") matrix elements, respectively, of the operator. Eq. (106) can then be rewritten as

$$\langle\vec{\underline{S}}(t)\vec{\underline{S}}(0)\rangle = \sum_{a',a''} e^{i\omega_{a'a''}t}\langle\vec{\underline{S}}^{a'a''}\vec{\underline{S}}\rangle_o\left\{1 - t\sum_{b'b''} R_{a'a''b'b''}(\omega_{b'b''}) + \ldots\ldots\right\} \tag{108}$$

where the constant $R_{a'a''b'b''}$ is defined as

$$R_{a'a''b'b''}(\omega_{b'b''}) =$$
$$\sum_{qq'} \langle \underline{F}_q\underline{F}_{q'}\rangle_o \langle[\mathscr{A}_q^{b'b''},[\mathscr{A}_{q'},\vec{\underline{S}}^{a'a''}]]\vec{\underline{S}}\rangle_o \int_0^\infty d\tau f_{qq'}(\tau) e^{i\omega_{b'b''}\tau} / \langle\vec{\underline{S}}^{a'a''}\vec{\underline{S}}\rangle_o . \tag{109}$$

The expansion in Eq. (108) is not necessarily convergent and certainly not rapidly convergent at long times (t) since the next term in the expansion is of order t^2. However, following Kubo,[1] we assume that we can exponentiate, and that

$$<\underline{\vec{S}}(t)\underline{\vec{S}}(0)> = \sum_{a'a''} e^{i(\omega_{a'a''}-\Delta\omega_{a'a''})t} e^{-t/T_{a'a''}} <\underline{\vec{S}}^{a'a''}\underline{\vec{S}}>_o \tag{110}$$

where $T_{a'a''}$ is a spin relaxation time and $\Delta\omega_{a'a''}$ a frequency shift defined by the relation,

$$T^{-1}_{a'a''} + i\Delta\omega_{a'a''} = \sum_b R_{a'a''b'b''}(\omega_{b'b''}) \; . \tag{111}$$

Both $T_{a'a''}$ and $\Delta\omega_{a'a''}$ are real. $\Delta\omega_{a'a''}$ is sometimes called the dynamic nonsecular shift.

The spin-lattice term, $\underline{\mathcal{H}}_{SL}$, as well as the relaxation matrix $R_{a'a''b'b''}(\omega_{b'b''})$, are conveniently divided into three sets of terms; secular, pseudo-secular and non-secular terms. The secular terms in $\underline{\mathcal{H}}_{SL}$ commute with the first order terms in $\underline{\mathcal{H}}_S$,

$$[\underline{\mathcal{H}}_{SL}(\text{sec}), \gamma B\underline{S}_Z + a\underline{I}_Z\underline{S}_Z] = 0 \; . \tag{112}$$

This implies that $\mathcal{A}_q$ is $\underline{S}_Z$, $\underline{I}_Z\underline{S}_Z$ or $\underline{S}_{Z_i}\underline{S}_{Z_j}$; i.e. it is diagonal in the $|\underline{I}_Z\underline{S}_Z>$ representation. The pseudo-secular terms commute with the pure electron spin part of $\underline{\mathcal{H}}_S$,

$$[\underline{\mathcal{H}}_{SL}(\text{psec}), \gamma B\underline{S}_Z] = 0 \; , \tag{113a}$$

but

$$[\underline{\mathcal{H}}_{SL}(\text{psec}), a\, \underline{I}_Z\underline{S}_Z] \neq 0 \; . \tag{113b}$$

This implies that $\mathcal{A}_q$ is $\underline{S}_Z\underline{I}_X$, $\underline{S}_Z\underline{I}_Y$, i.e. it is diagonal in $|S_Z>$ but not in $|I_Z>$. The nonsecular terms do not commute with the pure electron spin part of $\underline{\mathcal{H}}_S$,

$$[\underline{\mathcal{H}}_{SL}(\text{nsec}), \gamma B\underline{S}_Z] \neq 0 \; ; \tag{114}$$

this implies that $\mathcal{A}_q$ contains an $\underline{S}_x$ or an $\underline{S}_y$ and is not diagonal in the $|S_Z>$ representation. The spin part, $\mathcal{A}_q(t)$, of the secular terms are time independent, that of the pseudo secular terms has a slow time dependence with frequency aI_Z, and the nonsecular terms have a rapid time dependence with frequency $\gamma B \gg aI_Z$. (See Appendix C.) These distinctions become important in determining the overall time dependence of $\underline{\mathcal{H}}^{\circ}_{SL}(t)$; if the characteristic frequency of the classical motion of the "lattice" (liquid) is τ_c^{-1}, then τ_c^{-1} is the characteristic frequency of the $f_{qq'}(t)$ factors in Eq. (104). For the secular terms, $\mathcal{S}_q^{\circ}(t)$ has no time dependence and the entire time dependence of $\underline{\mathcal{H}}^{\circ}_{SL}(t)$ comes from $\underline{F}_q^{\circ}(t)$. For the pseudo secular terms, except in very viscous solvents, $\tau_c a \ll 1$ and, again, the predominant time dependence of $\underline{\mathcal{H}}^{\circ}_{SL}(t)$ comes from $\underline{F}_q^{\circ}(t)$. Finally, for the pseudo-secular terms in a typical esr experiment, $\tau_c \gamma B \approx 1$, and the time dependence of $\underline{\mathcal{H}}^{\circ}_{SL}(t)$ comes from both $\underline{F}_q^{\circ}(t)$ and $\mathcal{S}_q^{\circ}(t)$. Some typical numbers for copper acetylacetonate in toluene at room temperature are given below[12]

B	γB	a	τ_c^{-1}
3000 G	6×10^{10} sec^{-1}	2×10^{9} sec^{-1}	10^{11} sec^{-1}

If the hyperfine interaction is large and the applied D.C. field small, so that second order or higher corrections must be included in the calculation of the frequencies, ω_o, corresponding to $\underline{\mathcal{H}}_S$, (see Eq. 10), then the distinctions in the preceding paragraph become less clear since the secular terms do not commute with the entire pure spin Hamiltonian, $\underline{\mathcal{H}}_S$, i.e. they commute with $\gamma B\underline{S}_Z + a\underline{I}_Z\underline{S}_Z$ but not with $\gamma B\underline{S}_Z + a\vec{\underline{I}}\cdot\vec{\underline{S}}$. In this case even the secular spin terms, $\mathcal{S}_q^{\circ}(t)$, have a time dependence.

If $\langle\vec{\underline{S}}(t)\vec{\underline{S}}(0)\rangle$ represents $\langle\underline{S}_Z(t)\underline{S}_Z\rangle$, then $\omega_{a'a''} = 0$, and the $T_{a'a''}$'s are spin lattice relaxation times, $T_{1a'a''}$'s. If $\langle\vec{\underline{S}}(t)\vec{\underline{S}}\rangle$ represents $\langle\underline{S}_x(t)\underline{S}_x(0)\rangle$, then the $T_{a'a''}$'s are $T_{2a'a''}$'s, i.e. transverse relaxation times. One can readily show that for the spin-lattice relaxation times, only non-secular and pseudo-secular spin-lattice terms$(\underline{\mathcal{H}}_{SL})$ contribute, and, consequently, $\omega_{b'b''} \neq 0$. This corresponds to energy transfer between spins and lattice. For transverse relaxation times, secular, pseudo-secular and non-secular spin-lattice terms $(\underline{\mathcal{H}}_{SL})$ contribute; for the secular terms, $\omega_{b'b''}=0$, and no energy transfer takes place, but for the pseudo-secular and non-secular terms $\omega_{b'b''}\neq0$.

In the following sections we shall make use of these expressions and study $\langle \underline{S}_x(t)\underline{S}_x(0)\rangle$ and esr linewidths.

Before leaving this section we wish to make some comments concerning the exponentiation process in Eq. (110). It can readily be shown (Chapters V and VIII) that we can write

$$\langle \vec{\underline{S}}(t)\vec{\underline{S}}(0)\rangle = \sum_{a'a''} e^{i\omega_{a'a''}t-\psi_{a'a''}(t)} \langle \vec{\underline{S}}^{a'a''}\vec{\underline{S}}\rangle_o \tag{115}$$

and that $\psi_{a'a''}(t)$ can be expanded in powers of $\lambda\underline{\mathcal{H}}_{SL}$:

$$\psi_{a'a''}(t) = \sum_{n=1} \lambda^n \varkappa_n(t) \; . \tag{116}$$

On the other hand, Eq. (108) is a perturbation expansion of $\langle \vec{\underline{S}}(t)\vec{\underline{S}}(0)\rangle$:

$$\langle \vec{\underline{S}}(t)\vec{\underline{S}}(0)\rangle = \sum_{a'a''} e^{i\omega_{a'a''}t} \langle \vec{\underline{S}}^{a'a''}\vec{\underline{S}}\rangle_o [1 - \sum_{n=1} (i\lambda)^n r_n^{a'a''}(t)] \; . \tag{117}$$

$r_1=0$ and it can also be shown that $r_3=0$; since the two expansions must be identical, $\varkappa_1=\varkappa_3=0$, and

$$\varkappa_2(t) = r_2(t) \tag{118a}$$

$$\varkappa_4(t) = r_4(t) - \frac{r_2^2}{2}(t) \; . \tag{118b}$$

We note that these results suggest the distinct possibility that $|\varkappa_4| \ll |r_4|$, i.e. that the cumulant expansion in Eq. (116) is much more rapid than the moment expansion in Eq. (117); in Chapters V and VIII, this is seen to be the case provided $1 \gg \tau_c^2 \langle \underline{\mathcal{H}}_{SL}^2\rangle_o$. In particular, note that if $\varkappa_n = 0$ for $n \geq 3$, the factor $\exp-\psi_{aa'}(t)$ in Eq. (115) is a simple exponential, $\exp-\lambda^2\varkappa_2(t)$ where $\varkappa_2(t) = Rt$; whereas the cumulants converge very rapidly, the moment or perturbation expansion does not, i.e. $r_{2n}(t) = (-\lambda^2\varkappa_2(t))^n/n!$. Thus bear in mind that exponentiating the perturbation expansion is not the same as truncating it. More rigorous statements are given in Chapters V and VIII, and it can also be seen in these chapters that if τ_c is no longer short, i.e. $\lambda^2\langle \underline{\mathcal{H}}_{SL}^2\rangle\tau_c^2$ is no longer small with respect to one, then the cumulant expansion must be kept to higher order and, consequently, correlations of order higher than two in $\lambda\underline{\mathcal{H}}_{SL}$ must be included.

X.9. LINE SHAPES IN ABSENCE OF RELAXATION

As an exercise, we will first calculate the line shape in the absence of relaxation for the spin hamiltonian,

$$\underline{\mathcal{H}}_S = (\mu_o g/\hbar)B\underline{S}_z + a\underline{I}_z\underline{S}_z \quad . \tag{119}$$

We will thus have to evaluate the leading term, $<\underline{S}_x^\circ(t)\underline{S}_x>_o$, in the expansion in Eq. (104). To do this, it is convenient to introduce the raising ($\underline{S}_+$, $\underline{I}_+$) and lowering operators ($\underline{S}_-$, $\underline{I}_-$), where

$$\underline{S}_\pm = \underline{S}_x \pm i\underline{S}_y \tag{120a}$$

$$\underline{I}_\pm = \underline{I}_x \pm i\underline{I}_y \quad . \tag{120b}$$

Both $\underline{S}_\pm$ and $\underline{I}_\pm$ are off-diagonal in the $|S_z,I_z>$ representation; (see Appendix C). By means of the rules in Appendix C, it can readily be seen that for a spin S = 1/2 system,

$$<\underline{S}_x^\circ(t)\underline{S}_x(0)> = \frac{1}{4}<\underline{S}_+\underline{S}_- e^{i[\gamma B + a\underline{I}_z]t}> + \frac{1}{4}<\underline{S}_-\underline{S}_+ e^{-i[\gamma B + a\underline{I}_z]t}> \quad . \tag{121}$$

In the very high temperature limit

$$\underline{\rho}(eq) = 1/\text{Tr}\ \underline{1} = \frac{1}{N(2S+1)(2I+1)} \tag{122}$$

where N is the number of paramagnetic particles and S = 1/2. The trace is taken over all electron and nuclear spin states; if Tr_s is the trace over electron spin state

$$\text{Tr}_s\underline{S}_\pm\underline{S}_\mp = \frac{2}{3}(2S+1)(S+1)SN \tag{123}$$

and

$$<\underline{S}_x^\circ(t)\underline{S}_x(0)> = \frac{1}{2}\sum_{I_z}\frac{e^{i[\gamma B + aI_z]t}}{(2I+1)} + \text{complex conj.} \quad . \tag{124}$$

We can readily see that the spectral shape corresponding to Eq. (124) is the real part of the Fourier transform of Eq. (124), i.e. it is a sum of delta functions (sharp lines), each of intensity $(2I+1)^{-1}$ and at frequencies $\gamma B + aI_z$. Harking back to Eq. (107a), we see that $\omega_{a'a''} = \gamma B + aI_z$.

X. 10. LINE WIDTHS AND REORIENTATION: DETAILED DERIVATION

Spin-rotational interactions are discussed in Chapter XI; in this section we discuss anisotropic g-tensor and hyperfine interaction tensors and their contributions to spin relaxation, in particular to esr linewidths. Let us recall that in this case,

$$\underline{\mathcal{H}}_{SL} = (\mu_o/\hbar)\vec{B}\cdot\underline{\underline{G}}\cdot\vec{\underline{S}} + \vec{\underline{I}}\cdot\underline{\underline{A}}\cdot\vec{\underline{S}} \tag{125}$$

where $\underline{\underline{G}}$ and $\underline{\underline{A}}$ are traceless tensors. Recall that $\vec{B}$ is directed along the laboratory Z-axis, that $\vec{\underline{I}}$ and $\vec{\underline{S}}$ are quantized along the same direction, and that $\underline{\underline{A}}$ and $\underline{\underline{G}}$ have identical dependences upon Eulerian angles and hence upon time. Note that whereas we were able, in many cases, to neglect the off-diagonal terms $(\underline{I}_x\underline{S}_x+\underline{I}_y\underline{S}_y)a$ in the isotropic hyperfine term, we cannot neglect the off-diagonal or pseudo-secular and non-secular terms in Eq. (119). This can be understood in terms of second order perturbation theory. For the isotropic terms the diagonal term is of order a and the off-diagonal terms of order $a^2/\gamma B$. For the hyperfine anisotropic terms, the secular terms are diagonal in spin quantum numbers but not in the lattice quantum numbers; thus the anisotropic secular contribution is of order A^2/τ_c^{-1}, where τ_c is the characteristic time for lattice motions, i.e. τ_c^{-1} is the characteristic lattice "resonance energy." The nonsecular anisotropic terms are off-diagonal in both the lattice and spin quantum numbers; thus the anisotropic nonsecular contribution is of order $a^2/(\tau_c^{-1}+\gamma B)$. Typically, τ_c^{-1} and γB are comparable and the secular and nonsecular contributions are also comparable.

Now we wish to obtain specific expression for $<\underline{S}_+(t)\underline{S}_-(0)>$ for the spin-lattice hamiltonian in Eq. (125). These calculations are somewhat involved algebraically and the casual or trusting reader may wish to pass over the remainder of this section and proceed directly to the next one and a discussion of the results.

In order to make the required calculations on $<\underline{S}_+(t)\underline{S}_-(0)>$ we go back to Eqs. (104) through (110). The spin-lattice hamiltonian can be written in the simple form

$$\begin{aligned}\underline{\mathcal{H}}_{SL} &= [(\mu_o/\hbar)BG_{ZZ} + A_{ZZ}\underline{I}_Z]\underline{S}_Z \\ &+ [A_{Z+}\underline{I}_+ + A_{Z-}\underline{I}_-]\underline{S}_Z \\ &+ [(\mu_o/\hbar\; BG_{Z+} + A_{+Z}\underline{I}_Z + A_{++}\underline{I}_+ + A_{+-}\underline{I}_-)\underline{S}_+ \\ &+ (\mu_o/\hbar)BG_{Z-} + A_{-Z}\underline{I}_Z + A_{--}\underline{I}_- + A_{-+}\underline{I}_+)\underline{S}_-]\end{aligned} \tag{126}$$

where the terms in the successive brackets are secular, pseudo-secular and non-secular, respectively. Eq. (126) is deceptively simple because $\underline{\underline{G}}$ and $\underline{\underline{A}}$ have rather involved dependence on lattice variables, i.e. on Eulerian angles. See Appendix D.

First let us consider the secular terms in Eq. (126). In Eq. (104) we can make the identifications, $F_1 = (\mu_o/\hbar)BG_{ZZ}$, $F_2 = A_{ZZ}$, $\mathcal{L}_1^\circ(\tau) = \underline{S}_Z(0)$, $\mathcal{L}_2^\circ(\tau) = \underline{I}_Z\underline{S}_Z$. Thus the spin commutator in Eq. (104) has terms of the form

$$\begin{aligned}<[\underline{I}_Z\underline{S}_Z,[\underline{I}_Z\underline{S}_Z,\underline{S}_+^\circ(t)]\underline{S}_->_o &= <\underline{I}_Z^2\underline{S}_+\underline{S}_-e^{i(\gamma B + a\underline{I}_Z t)}>_o \\ &= \frac{1}{(2I+1)}\,\frac{1}{2}\sum_{I_Z} I_Z^2 e^{i(\gamma B + aI_Z t)}\end{aligned} \tag{127}$$

for the $q = q' = 2$ term. Similar calculations for ($q = q' = 1$) and ($q = 1$, $q' = 2$) and ($q = 2$, $q' = 1$) can be carried out, and the result for all the secular terms is

$$R_{I_Z I_Z I_Z I_Z}(o) = [(\mu_o/\hbar\; B)^2<G_{ZZ}^2>_o + <A_{ZZ}^2>_o I_Z^2 + (2\mu_o/\hbar)BI_Z<G_{ZZ}A_{ZZ}>_o]\,\frac{\tau_{sec}(o)}{(2I+1)} \tag{128}$$

where R is defined in Eqs. (108) and (109), and

$$\tau_{sec}(o) = \int_o^\infty f_{sec}(\tau)\, d\tau \;. \tag{129}$$

Note that G_{ZZ} and A_{ZZ} have the same normalized autocorrelation function, $f_{sec}(\tau)$, and that the normalized cross correlation between them is the same.

Next turn to the pseudo-secular terms in Eq. (126); then the identification with terms in Eq. (104) can be made as $F_3 = A_{Z+}$, $F_4 = A_{Z-}$, $\mathcal{L}_3^\circ(\tau) = \underline{S}_Z\underline{I}_+\exp(ia\underline{S}_Z\tau)$, $\mathcal{L}_4^\circ(\tau) = \underline{S}_Z\underline{I}_-\exp(-ia\underline{S}_Z\tau)$. It can readily be shown that $<F_3^2> = <F_4^2> = 0$. A calculation similar to that in the last paragraph yields

$$R_{I_Z I_Z I_Z, I_Z \pm 1}(\tfrac{1}{2}\, a) = \int_0^\infty d\tau \langle A_{Z\pm} \underline{S}_Z \underline{I}_\pm, [A_{Z\mp}(\tau) \underline{S}_Z \underline{I}_\mp e^{\mp ia\underline{S}_Z \tau}, \underline{S}_+ e^{i(\gamma B + a\underline{I}_Z)t}] \underline{S}_- \rangle_o \;. \tag{130}$$

The integrand of the pseudosecular contribution to R consists of four terms, a typical one having the form

$$-\mathrm{Tr}\left\{ A_{Z\pm} \underline{S}_Z)(\underline{S}_+ e^{i(\gamma B + a\underline{I}_Z)t})(A_{Z\mp}(\tau) \underline{S}_Z \underline{I}_\mp e^{\mp ia\underline{S}_Z \tau}) \underline{S}_- \right\}$$

$$= -\langle A_{Z+}(\tau) A_{Z-} \rangle \mathrm{Tr}\; \underline{I}_\pm \underline{I}_\mp e^{i(\gamma B + a\underline{I}_Z t)} \;\mathrm{Tr}\; \underline{S}_Z \underline{S}_+ \underline{S}_Z e^{\mp ia\underline{S}_Z \tau} \underline{S}_-$$

$$= -\langle S_{Z+}(\tau) A_{Z-} \rangle \sum_{I_Z} [I(I+1) - I_Z \pm I_Z] e^{i(\gamma B + aI_Z)t} (-\tfrac{1}{4}) e^{\pm ia\tau/2} \;.$$

The other terms in the integrand can readily be calculated and the result is

$$\sum_{\pm} R_{I_Z I_Z I_Z, I_Z \pm 1}(a/2) = \langle A_{Z+} A_{Z-} \rangle [I(I+1) - I_Z^2] \tau_{psec}(\tfrac{1}{2}\, a) \tag{131}$$

where

$$\tau_{psec}(\tfrac{1}{2}\, a) = \int_0^\infty f_{psec}(\tau) e^{ia\tau/2} d\tau \;. \tag{132a}$$

Since in most cases the oscillation frequency, (a/2), is slow compared to the decay of $f_{psec}(\tau)$, we can approximate

$$\tau_{psec}(\tfrac{1}{2}\, a) = \tau_{psec}(o) = \int_0^\infty f_{psec}(\tau) d\tau \;. \tag{132b}$$

The nonsecular terms can be treated in similar fashion:

$$\sum_{\pm} R_{I_Z I_Z S_Z, S_Z \pm 1}(\gamma B) = \Big\{ \langle G_{Z+} G_{Z-} \rangle \frac{2}{9} \left(\frac{\mu_o}{\hbar} B\right)^2 + \frac{4}{3} \langle G_{Z+} A_{Z-} \rangle \frac{\mu_o}{\hbar} B I_Z$$
$$+ 2\langle A_{Z+} A_{Z-} \rangle I_Z^2 + 2\langle A_{++} A_{--} \rangle [I(I+1) - I_Z^2]$$
$$+ \frac{1}{8} \langle A_{ZZ}^2 \rangle [\langle I(I+1) - I_Z^2 \rangle] \Big\} \tau_{nsec}(\gamma B) \tag{133}$$

where

$$\tau_{nsec}(\gamma B) = \int_0^\infty f_{nsec}(\tau) e^{i\gamma B\tau} d\tau \;. \tag{134}$$

Actually the frequency γB is $(\gamma B \pm \frac{1}{2}\, a)$ for some of the nonsecular contributions but $a^2 \ll \gamma^2 B^2$ in most cases of interest. Note that since the secular and pseudo-secular terms are diagonal in $|S_Z\rangle$, we have labelled the corresponding R's entirely in terms of I_Z; in the non-secular terms it is sufficient to identify the a'a'' labels on R by $I_Z I_Z$ but it is the change in $|S_Z>$ which dominates the b'b" labels. See Eqs. (108) and (109).

X.11. DISCUSSION OF RESULTS

We assume that the molecular reorientation is isotropic and diffusive, i.e. that there is only one characteristic rotational diffusion constant, τ_c:

$$f_{qq'}(\tau) = \exp -\tau/\tau_c \, . \tag{135}$$

Then the various transforms of $f(\tau)$ that appeared in the last section are

$$\tau_{sec} = \tau_c \tag{136a}$$

$$\tau_{psec} = \tau_c \tag{136b}$$

$$\tau_{nsec}(\gamma B) = \frac{\tau_c}{1+\gamma^2 B^2 \tau_c^2} \, . \tag{136c}$$

If the static lattice correlation functions are evaluated and these values substituted into Eqs. (128), (131) and (133), then Eqs. (110) and (111) can be rewritten as

$$\begin{aligned} T_2^{-1}(I_Z) = \frac{1}{3}\Bigg\{\frac{4}{5}\left(\frac{\Delta g^2}{3} + \delta g^2\right)\left(\frac{\mu_o}{\hbar}\right)^2 B^2 + \frac{1}{10}(\Delta a^2 + 3\delta a^2)I(I+1) \\ + \frac{1}{6}(\Delta a^2 + 3\delta a^2)I_Z^2 \\ + \frac{8}{15}(\Delta a \; \Delta g + 3\delta g \; \delta a)\left(\frac{\mu_o}{\hbar}\right) B \, I_Z\Bigg\}\tau_c \\ + \frac{1}{5}\Bigg\{\left(\frac{\Delta g^2}{3} + \delta g^2\right)\left(\frac{\mu_o}{\hbar}\right)^2 B^2 + \frac{7}{18}(\Delta a^2 + 3\delta a^2)I(I+1) \\ - \frac{1}{18}(\Delta a^2 + 3\delta a^2)I_Z^2 \\ + \frac{2}{3}(\Delta g \; \Delta a + 3\delta g \; \delta a)\left(\frac{\mu_o}{\hbar}\right) B I_Z\Bigg\}\frac{\tau_c}{1+\gamma^2 B^2 \tau_c^2} \, . \end{aligned} \tag{137}$$

Thus each of the hyperfine lines (I_Z) has a width and dynamic shift; the latter is not written out in this chapter.

Now we summarize the assumptions required to obtain Eq. (137).

(1) Non-adiabatic terms were neglected (adiabatic in sense of Eq. (96) and Appendix B).

(2)

$$\tau_c/T \ll 1 ; \qquad (138)$$

this enables us to change limits of integration and set $(t-\tau)$ equal to t in Eq. (98). This is also roughly equal to the condition

$$\langle \mathcal{H}_{SL}^2 \rangle \tau_c^2 \ll 1 \qquad (139)$$

which enables us to relate the expansion in powers of $\mathcal{H}_{SL}$ to a rapidly varying cumulant expansion which in turn permits exponentiation and gives rise to Lorentzian lines. We will call this the motionally narrowed limit or Redfield limit. A different approach is required if these conditions are not satisfied but work on this problem has been carried out, i.e. large molecules in viscous media.[9] See Chapters X and XII. In viscous media, secular shifts in frequency as well as non-Lorentzian line shapes must be considered and correlations in $\mathcal{H}_{SL}$, higher than second order, must be included.

(3) The high temperature approximation was made and it was assumed that the lattice was unaffected by the spins; we will not be concerned with corrections due to finite temperatures but the appropriate calculations can be carried out.

(4) Only one interacting nucleus was assumed. In § 14 we shall extend the results to several interacting nuclei.

(5) Only relaxation due to molecular reorientation was included. Spin-rotations are discussed in Chapter XI, and fluctuations in $\underline{\underline{g}}$-tensors and $\underline{\underline{a}}$-tensors connected with internal molecular motions or solvent bombardment can sometimes be important as can quadrupolar, triplet spin-spin and various spin-orbit relaxations. (See § 3.) The results above are applicable to paramagnetically dilute solutions, i.e. no intermolecular magnetic interactions have been included. A discussion of intermolecular magnetic interactions in concentrated solutions is given in Chapter XVII.

(6) We have assumed isotropic rotational motion, i.e. Eq. (135) holds for all contributions to the line width. The effect of anisotropic rotational motion can be included by noticing that the correlation times are different for rotations about different molecular axes with the result that the various $f_{qq'}(t)$'s are different and

cannot all be described by a single correlation time as in Eq. (135). In Chapter VIII, an extension of these results to molecules where the rotational motion is anisotropic is given.

(7) We have assumed that the rotational motion is diffusive, i.e. that the normalized rotational correlation function has the form $\exp(-t/\tau_c)$; this corresponds to the situation

$$\tau_c \gg \sqrt{\frac{I}{k_B T}} \gg \tau_J \tag{140}$$

where τ_c is the orientational correlation time,

$$\tau_c = \int_0^\infty f(\tau)d\tau \ , \tag{141}$$

$\sqrt{k_B T/I}$ is the classical mean free rotational frequency, τ_J is the angular momentum correlation time,

$$\tau_J = \int_0^\infty \langle J(t)J(0)\rangle d\tau / \langle J^2 \rangle \tag{142}$$

in which J is the molecular angular momentum, and I is the moment of inertia. τ_J is discussed in Chapter XI. If rotational diffusion does not adequately describe molecular reorientation, i.e. if Eq. (140) is not satisfied, then the normalized autocorrelation function, f(t), is often not well-described by a simple exponential, and, consequently, the field-dependence of $\tau_{nsec}(\gamma B)$ as given in Eq. (136c) may no longer be correct. We will have more to say about diffusion below.

(8) We neglected second order isotropic hyperfine contributions, i.e. $a(\underline{I}_x\underline{S}_x + \underline{I}_y\underline{S}_y)$ terms; we have already discussed their effect on the absorption frequency and in § 13 we will discuss their effect on linewidths.

(9) We calculated the autocorrelation function for $\underline{S}_x$ rather than for $\gamma\underline{S}_x$, i.e. we ignored the time dependence of γ in this expression. (Of course, the time dependence of γ was included in $\underline{\mathcal{H}}_{SL}$.) See reference 15.

(10) No cross-correlations between the anisotropic reoreintation terms, $[(\mu_o/\hbar)\vec{\underline{S}}\cdot\underline{\underline{G}}\cdot\vec{B} + \vec{\underline{I}}\cdot\underline{\underline{A}}\cdot\vec{\underline{S}}]$, and other time dependent magnetic interactions have been included, although cross-correlations between $(\mu_o/\hbar)\vec{\underline{S}}\cdot\underline{\underline{G}}\cdot\vec{\underline{B}}$ and $\vec{\underline{I}}\cdot\underline{\underline{A}}\cdot\vec{\underline{S}}$ have, of course, been considered. Cross-

correlations between secular, pseudo-secular and non-secular terms do contribute when correlations higher than second order are retained, i.e. for viscous solutions.[15]

(11) The rf field must be small. (Actually, saturation effects can be included with some modification.)

The results above hold for the motionally narrowed limit or Redfield limit. (See Eq. 139.) The extreme motionally narrowed limit is one for which

$$(\gamma B\tau_c)^2 \ll 1 \; . \tag{143}$$

Note that in Eq. (136c), $\tau_{nsec}(\gamma B) = \tau_c$ in this limit, and it can readily be seen (though we shan't do it here) that

$$T_1 = T_2 \tag{144}$$

in this limit.

X.12. ANALYSIS OF EXPERIMENTS

It is convenient to rewrite Eq. (137) in the form

$$\Delta B_{I_Z} = (\alpha' + \alpha'') + \beta I_Z + \gamma I_Z{}^2 + \delta I_Z{}^3 \tag{145}$$

where ΔB_{I_Z} is the peak to peak width (in Gauss) of the derivative of the esr absorption hyperfine line I_z; the α', β, γ can be obtained by multiplying the coefficients of the various powers of I_Z in Eq. (137) by $(2/\sqrt{3})(\hbar/g\mu_o)$; the coefficient δ arises because of the off-diagonal $a(I_xS_x + I_yS_y)$ terms and will be discussed in § 13; and α'' will be discussed below.

Experimentally we can measure the line widths of each of the hyperfine lines as a function of various physical parameters. For example, in vanadyl acetylacetonate, the vanadyl nucleus has $I = 7/2$ and so $I_Z = \pm 7/2, \pm 5/2, \pm 3/2, \pm 1/2$; the spectrum consists of eight nearly equally spaced lines (not quite equally spaced because of second order $aI_xS_x + aI_yS_y$ terms) and the integrated intensities of each of the eight lines are about equal. The paramagnetic molecule can be diluted in various solvents and if the solutions are sufficiently dilute (10^{-4}M) then intermolecular magnetic interactions can be neglected; this can be checked by noticing that at these low concentrations, ΔB_{I_Z} is not a function of concentration. In each solvent, the parameters $\alpha'+\alpha'',\beta,\gamma,\delta$ can be determined experimentally as a function of temperature, T. The molecular magnetic parameters Δg, δg, Δa, δa can be determined from the dilute crystal or dilute glass spectrum, as explained in § 2, and g and a can be obtained from the liquid spectrum. We will assume that Δg, δg, Δa, δa are relatively unchanged in going from a glass at 77°K to a liquid solution in the same solvent, since we know of no way to determine these parameters directly in liquid solution; we can, however, re-determine the parameters in glasses of each solvent used. Since the variation of the isotropic g and a parameters with solvent phase change and varying temperature is small, since the theory applied under the assumptions above appears to hold quite well, and since we have no ready alternative, we will henceforth assume constancy of Δg, δg, Δa and δa.

The parameter δ is very small and hard to determine accurately, but β and γ can be well measured. Since the magnetic parameters are known, we can express β and γ as functions of τ_c and the applied field, $\vec{B}$. If we assume that the lattice correlation function is exponential, $\exp(-\tau/\tau_c)$, then we can predict the field dependence of β and γ, and we can measure τ_c. Note that besides the dependence on B that enters both β and γ through τ_c(nsec), β is linear in B whereas γ is not. The experimental results fit the predicted field dependence; this leads credence to the theory and analysis, and also seems <u>to confirm the exponential correlation function</u>, at least in some cases. This latter point deserves special comment. Since τ_c for molecules with a radius of about 3A° in various organic solvents at room temperature is typically about 2×10^{-11} sec., $\tau_c \gamma B \approx 1$ for $B \approx 3000$ G. Thus the ESR experiments are relatively sensitive to the shape of the correlation function, (see Eq. 136c) i.e. more sensitive than NMR experiments. However, the experiments up to date are not extensive enough to probe the shape in detail because it is rather difficult to vary the field.

The isotropic rotational diffusion time, τ_c, can be related to molecular parameters by the extended Debye relationship[21,22]

$$\tau_c = \frac{4}{3} \frac{\pi\eta}{k_B T} r^3 \varkappa \qquad (146)$$

where r is the molecular radius and η is the coefficient of viscosity. This expression assumes that the molecule is spherical but it can be generalized to spheroidal molecules. Eq. (146) is useful if $r^3\varkappa$ is independent of temperature; in some situations this is the case[12] whereas in others it is not.[23] The parameter $\varkappa$ is introduced because the radius $r\varkappa^{1/3}$ determined from Eq. (146) is usual[ly] far smaller than the expected molecular radius. (The expected radius, r, may be determined by translational diffusion experiments, covalent atomic radius calculations or molecular model constructions.) It has been found that although $\varkappa$ often does not vary with temperature or pressure for a given solvent, it does vary from solvent to solvent. The value of $\varkappa$ seems to be correlated with molecular size and shape; for a given paramagnetic species $\varkappa$ seems to increase with decreasing solvent molecular radius and perhaps with increasing solvent molecular anisotropy.[21,24] NMR data tends to support these results and to indicate that $\varkappa$ increases with increasing anisotropy of the paramagnetic species.[25] In order to determine $\varkappa$, r must be obtained by some independent method, either by translational diffusion or by molecular models; although $\varkappa$ always seems to be less than one, sometimes as small as 0.01, it should be remembered that it is the relative values of $\varkappa$ rather than the absolute ones which are of importance. Although $\varkappa$ is really the ratio of the correlation time for linear momentum to that for angular momentum (where the correlation times are the zero-frequency transforms of the normalized correlation functions), if the correlation functions for linear and angular momenta have similar time behavior, it can be shown that[22]

$$\varkappa \approx (3/4r^2) \langle \mathcal{T}^2 \rangle / \langle \mathcal{F}^2 \rangle \qquad (147)$$

where $\langle \mathcal{T}^2 \rangle$ and $\langle \mathcal{F}^2 \rangle$ represent the mean square intermolecular torques and forces, respectively. Clearly we would expect $\varkappa \ll 1$ for methane and perhaps $\varkappa \approx 1$ for H_2O. $\mathcal{T}$ can be related to the anisotropic part of the intermolecular potential whereas $\mathcal{F}$ is related to the total intermolecular potential; thus a study of $\varkappa$ yields information concerning the intermolecular potential. It should be noted that $\varkappa$ is really a tensor but one adjustable parameter is more than enough to handle in light of the present state of both experiment and theory. More detailed discussions of $\underline{\varkappa}$ are given in Chapter XI.

Once τ_c is known, α' in Eq. (145) can be calculated, but the experimentally determined quantity, $\alpha' + \alpha''$, is larger than the predicted α'. The residual quantity, α'', is associated with spin-rotational relaxation mechanisms, discussed in Chapter XI. All we need say about α'' here is that

a) it is independent of B as predicted;

b) it appears to vary as T/η whereas α' varies more or less as η/T;

c) it varies in a predictable manner with shift in g-values from the free electron values (e.g. measure for VO^{++} acetylacetonate and predict for Cu^{++} acetylacetonate);

d) it depends on a spin-rotational correlation time which, in a diffusion model, is closely associated with angular momentum correlation times (τ_J) with $\tau_c >> \tau_J$; in this diffusion limit, for spherical molecules,[26]

$$\tau_J \tau_c = \frac{I}{6k_B T} . \tag{148}$$

What more can be said about the results? τ_c is a correlation time for reorientation of a second rank rotational transformation; similar times are obtained from NMR, light scattering and Raman experiments. In magnetic resonance, the magnetic interactions transform as second rank tensors and in light scattering and Raman experiments, it is the molecular polarizabilities which transform as second rank tensors. In dielectric relaxation and infrared experiments, it is the molecular dipole moment which enters, and it transforms as a first rank tensor. In a rotational diffusion limit,

$$\frac{\tau_c^{(1)}}{\tau_c^{(2)}} = 3 \tag{149}$$

where the superscript represents the rank of the tensor involved. In a large jump model

$$\tau_c^{(1)}/\tau_c^{(2)} \approx 1 \ . \tag{150}$$

Experiments seem to indicate that $\tau_c^{(1)}/\tau_c^{(2)} \approx 1.5$ for many small molecules, but this is not well confirmed.[27] It should also be noted that because spins on neighboring molecules are almost completely uncorrelated, the correlation functions giving rise to $\tau_c^{(2)}$ in magnetic resonance experiments are single particle autocorrelation functions, whereas in many other experiments, time-dependent correlations between the orientation of a molecule with that of its neighbors, (pair correlations) must be considered. The single particle correlation functions may have both different static (t=0) and time behavior than do the many particle correlation functions; this effect may account for the apparently different $\tau_c^{(2)}$'s obtained from NMR and light scattering.[28]

Since we have concentrated on the coefficients α and γ, we must still ask whether other relaxation mechanisms could contribute to these coefficients. In particular one might expect fluctuations in the anisotropic g-tensor and hyperfine tensor to contribute terms in I_z and I_z^2. (See Eq. 32.) In relatively rigid molecules, these effects are probably not very important; the fluctuations are usually small and very fast (i.e. small correlation time). However, if internal rotations take place, the effects can be great.[18] We would not expect cross-correlations between fluctuations, reorientations and spin-rotational contributions.

Finally, we might expect the treatment above to break down if the solution becomes too viscous (i.e. if $\mathcal{H}_{SL}^2 \tau_c^2 \gtrsim 1$); the lines are then no longer Lorentzian, and secular, nonsecular and pseudo-secular terms are no longer separable. Anisotropic reorientation will also complicate the results as will a breakdown of rotational diffusion, i.e. if $\tau_c \approx \tau_J$.

It should be emphasized that though the theory appears to be reasonably satisfactory, detailed esr linewidth experiments have been carried out on only a small number of molecules; however, NMR data also seems to confirm the esr results. Nevertheless, all may not be in good shape as indicated by recent experiments.[23]

X.13. SECOND ORDER CORRECTIONS

In § 2 we discussed the effect of the non-diagonal contributions of the isotropic hyperfine interaction, $a(\underline{I}_x\underline{S}_x + \underline{I}_y\underline{S}_y)$, on the absorption frequency, but now we shall discuss its effect upon the linewidth. The effect enters in the following way. The linewidth or inverse spin-relaxation time can, as we have shown, be obtained from time dependent perturbation theory and depends on $\langle m|\underline{\mathcal{H}}_{SL}(t)|m'\rangle\langle m'|\underline{\mathcal{H}}_{SL}|m\rangle$, where $|m\rangle$ is a spin function basis in which the time independent spin hamiltonian, $\underline{\mathcal{H}}_S$, is diagonal. If the $a(\underline{I}_x\underline{S}_x + \underline{I}_y\underline{S}_y)$ term is omitted from $\underline{\mathcal{H}}_S$, then $|m\rangle$ is the simple product function of nuclear, $|I_Z\rangle$, and electronic, $|S_Z\rangle$, spin functions; this is the calculation we have carried out in the preceding sections. However, if the $a(\underline{I}_x\underline{S}_x + \underline{I}_y\underline{S}_y)$ term is included, we must first solve for the correct wave function by time independent perturbation theory, e.g.,

$$|m\rangle = |I_Z S_Z\rangle + \frac{a\langle I_Z S_Z|\underline{I}_x\underline{S}_x + \underline{I}_y\underline{S}_y|I_Z' S_Z'\rangle}{\gamma B}|I_Z' S_Z'\rangle , \qquad (151)$$

and then we must use this function in the time dependent perturbation calculation.

A convenient way to do this is by means of a unitary transformation of the total hamiltonian, $[g(\mu_o/\hbar)\underline{S}_Z B + a\underline{I}_Z\underline{S}_Z] + a(\underline{I}_x\underline{S}_x + \underline{I}_y\underline{S}_y) + \underline{\mathcal{H}}_{SL}(t)$, so that the time independent part is diagonal through second order in a in the $|I_Z S_Z\rangle$ representation; this is a Van Vleck transformation.[12,29] The time dependent operator, $\underline{\mathcal{H}}_{SL}(t)$, is thus transformed to a new time-dependent hamiltonian, $\underline{\mathcal{H}}_{SL}^N(t)$, which we now introduce into the calculations described in previous sections. If this calculation is carried out in a consistent manner to order $(a/\gamma B)$, we find that α', β and γ in Eq. (145) contain, in addition to the terms given in Eq. (137), correction terms of order $(a/\gamma B)$;

furthermore, the δ term in Eq. (145) is now explained and it is of order $(a/\gamma B)$ smaller than β and γ. These calculations are carried out in reference 12.

It is interesting to note that if the calculations are carried out to still higher order in $(a/\gamma B)$ or if quadrupolar interactions are included, terms of the form ϵI_Z^4 must be added to the linewidth expression in Eq. (145).

X.14. SEVERAL INTERACTING NUCLEI

If we have several interacting nuclei, the time independent pure spin hamiltonian becomes

$$\mathcal{H}_{\underline{S}} = (\mu_o g/\hbar)\underline{S}_Z B + \sum_j \underline{I}_Z^{(j)} a^{(j)} \underline{S}_Z \tag{152}$$

where the superscript (j) refers to the jth nucleus within a given paramagnetic molecule and the off-diagonal terms in $\vec{\underline{I}}_j$ have been neglected. The time dependent anisotropic orientation hamiltonian for a given molecule is

$$\mathcal{H}_{\underline{SL}} = (\mu_o/\hbar)\vec{\underline{S}}\cdot\underline{\underline{G}}\cdot\vec{\underline{B}} + \sum_j \vec{\underline{I}}^{(j)}\cdot\underline{\underline{A}}^{(j)}\cdot\vec{\underline{S}} \,. \tag{153}$$

First we will assume that there are no degeneracies, i.e. no two $a^{(j)}$'s are the same. The absorption frequencies become

$$\omega_o(I_Z) = (\mu_o g/\hbar)B + \sum_j a^{(j)} I_Z^{(j)} \,. \tag{154}$$

The line width involves diagonal values of $\mathcal{H}_{\underline{SL}}^2$ which implies that it can contain terms arising from the nuclear spin operator combinations, $\underline{I}_Z^{(j)}$, $\underline{I}_Z^{(j)}\underline{I}_Z^{(k)}$, $\underline{I}_\pm^{(j)}\underline{I}_\mp^{(j)}$, but not $\underline{I}_\pm^{(j)}\underline{I}_\mp^{(k)}$ where $j\neq k$. It, therefore, follows that the line width expression is

$$T_2^{-1} = \alpha' + \alpha'' + \sum_j \beta^{(j)} I_Z^{(j)} + \sum_{j,k} \gamma^{(jk)} I_Z^{(j)} I_Z^{(k)} \,. \tag{155}$$

These parameters differ slightly from those given in previous sections; the parameter, α', is similar to that in Eqs. (145) and (137) except that $(\Delta a)^2 I(I+1)$ and $(\delta a)^2 I(I+1)$ are replaced by $\sum_j (\Delta a^{(j)})^2 I^{(j)} \cdot (I^{(j)}+1)$ and $\sum_j (\delta a^{(j)})^2 I^{(j)} (I^{(j)}+1)$ respectively; the parameter $\beta^{(j)}$ is identical with β except that Δa and δa are replaced by $\Delta a^{(j)}$

and $\delta a^{(j)}$, respectively; the parameter $\gamma^{(jk)}$ is

$$\gamma^{(jk)} = [\Delta a_j \Delta a_k + 3\delta a_j \delta a_k][(\frac{4}{15} - \frac{1}{10}\delta_{jk})\frac{\tau_c}{3} + (\frac{1}{3} - \frac{7}{18}\delta_{jk})\frac{\tau_c/5}{1+\gamma^2 B^2 \tau_c^2}] . \tag{156}$$

Now we will consider a degenerate set of nuclear spins, i.e. all $\langle a^{(j)}\rangle$'s are equal. Here we are careful to write the average Fermi contact constant for each nucleus since the individual instantaneous $a^{(j)}$'s may differ from each other due to "fluctuations"; thus the nuclei are only equivalent on the average--"symmetrically equivalent."[18] Furthermore, the components of $\underline{\underline{A}}^{(j)}$, the traceless anisotropic hyperfine interaction tensor, may be different for the various symmetrically equivalent[30] nuclei; the mean averages, $\langle(\underline{\underline{A}}^{(j)})^2\rangle$ will all be the same but $\langle(A^{(j)})^2\rangle \neq \langle A^{(j)}A^{(k)}\rangle$ if $j \neq k$. This can be seen by looking at the protons on benzene; the dipolar nuclear-electron spin interactions for the various protons can each be diagonalized in a molecular axis system, but three different coordinate systems are needed because the C-H axes points in different directions for the different protons. Thus the $\underline{\underline{A}}^{(j)}$'s for the symmetrically equivalent protons are all different.

The bookkeeping is facilitated by introducing a total nuclear spin, $\vec{\mathcal{I}}$, where

$$\vec{\mathcal{I}} = \sum_j \vec{\underline{I}}^{(j)} . \tag{157}$$

The absorption frequency becomes

$$\omega_o = (\mu_o g/\hbar)B + a\mathcal{I}_Z . \tag{158}$$

As an example, consider two equivalent nuclei with spin I = 1; then $\mathcal{I}_Z = 0, \pm 1, \pm 2$. The $\mathcal{I}_Z = 0$ transition is threefold degenerate, i.e. it corresponds to three transitions at the same frequency, (with $\mathcal{I} = 2, 1, 0$). The $\mathcal{I}_Z = \pm 1$ transitions are each two-fold degenerate (with $\mathcal{I} = 2,1$), and the $\mathcal{I}_Z = \pm 2$ transitions are not degenerate ($\mathcal{I} = 2$). Each of the transitions has equal intensity so if we could ignore relaxation effects, the observed intensities would describe a 1,2,3,2,1 five-line spectrum. If the off-diagonal terms, $\sum_j a(\underline{I}_+^{(j)}\underline{S}_- + \underline{I}_-^{(j)}\underline{S}_+)$, are retained, small shifts cause removal of the degeneracies; this is a rather straightforward effect which we won't pursue here.

We will first look at the secular contribution[31] to the line width of the anisotropic hyperfine interaction tensor, $\sum_j A_{ZZ}^{(j)} \underline{I}_Z^{(j)} \underline{S}_Z$, for the special case given above, i.e. two symmetrically equivalent nuclei, with $A_{ZZ}^{(1)} \neq A_{ZZ}^{(2)}$ and each with spin $I^{(j)} = 1$, and we will concentrate on the two-fold degenerate ($\mathcal{I}_Z = 1$) line. Let us arbitrarily evaluate the matrix elements of $\sum_j A_{ZZ}^{(j)} \underline{I}_Z^{(j)} \underline{S}_Z$ in the $|\bar{\mu}\mathcal{I}_Z>$ nuclear spin representation, i.e. the representation for the total nuclear spin.

$$<\bar{2}\bar{1}|\sum_j A_{ZZ}^{(j)} \underline{I}_Z^{(j)} \underline{S}_Z|\bar{2}\bar{1}> = \frac{1}{2}(A_{ZZ}^{(1)} + A_{ZZ}^{(2)})\underline{S}_Z \tag{159a}$$

$$<\bar{1}\bar{1}|\sum_j A_{ZZ}^{(j)} \underline{I}_Z^{(j)} \underline{S}_Z|\bar{2}\bar{1}> = \frac{1}{2}(A_{ZZ}^{(1)} - A_{ZZ}^{(2)})\underline{S}_Z \ . \tag{159b}$$

(Check this by noting that the $|\bar{\mu}\mathcal{I}_Z>$ functions can be written as a linear sum of $|I_Z^{(1)} I_Z^{(2)}>$ functions, i.e. $|\bar{2}\bar{1}> = \frac{1}{\sqrt{2}}[|10> + |01>]$ and $|\bar{1}\bar{1}> = \frac{1}{\sqrt{2}}[|10> - |10>]$.) A typical term to be evaluated has the form

$$\sum_{j,k} <[A_{ZZ}^{(j)} \underline{I}_Z^{(j)} \underline{S}_Z, [A_{ZZ}^{(k)} \underline{I}_Z^{(k)} \underline{S}_Z, \underline{S}_+(t)]]\underline{S}_->_o$$

$$= <\underline{S}_+\underline{S}_- \sum_j A_{ZZ}^{(j)} \underline{I}_Z^{(j)} \sum_k A_{ZZ}^{(k)} \underline{I}_Z^{(k)} \exp i(\gamma B + a\mathcal{I}_Z)t>_o \ . \tag{160}$$

Since the terms with different time dependent exponentials are separable because of the quite distinct behavior of each of them, we can restrict the average to the $\mathcal{I}_Z = 1$ state; then the exponent is the same for the $|\bar{2}\bar{1}>$ and $|\bar{1}\bar{1}>$ states and we have a two-fold degeneracy. We now pose the question, how can we decide which width goes with which degenerate line? In going from Eq. (108) to (110), we summed all the terms multiplying a given time dependent exponential, $e^{i\omega_{a'a''}t} t\sum_{b'b''} R_{a'a''b'b''}$, and exponentiated to obtain the sum, $e^{i\omega_{a'a''}t} \exp[-t\sum_{b'b''} R_{a'a''b'b''}]$. But in the present, degenerate case, there are two distinct but equal $\omega_{a'a''}$'s, and how do we associate part of the sum as the relaxation of one of the lines and part as the other? The answer lies in the fact that if a trace of an operator is taken in a representation which diagonalized the operator, then the resulting average, can be expressed as a sum of each of the diagonal elements, i.e. there is no mixing between states and all average properties are well represented within each of the states. In this case, $\sum_{a'a''} e^{i\omega_{a'a''}t} e^{tR_{a'a''a'a''}}$ leads to $\sum_{a'a''} e^{i\omega_{a'a''}t} e^{-tR_{a'a''a'a''}}$, even though the $\omega_{a'a''}$'s are degenerate. But the operator averaged in Eqs. (159ab) is not diagonal in the $|\bar{\mathcal{J}}\bar{\mathcal{I}}_Z\rangle$ representation, as illustrated in Eqs. (159ab), and so it must be diagonalized. Since S_+S_- is diagonal, we must first diagonalize $\sum_j A_{ZZ}^{(j)} \underline{I}_Z^{(j)} \sum_k A_{ZZ}^{(k)} \underline{I}_Z^{(k)}$ in a representation that also diagonalizes $\mathcal{I}_Z$. In the example above, where $\mathcal{I}_Z = 1$, the diagonalization involves a 2 x 2 matrix formed from $|\bar{2}\bar{1}\rangle$ and $|\bar{1}\bar{1}\rangle$ states.

It turns out, as is easily verified, that the operator to be averaged above is diagonal in the $|I_Z^{(1)} I_Z^{(2)}\rangle$ representation. Then for $\mathcal{I}_Z = 1$ we have the $|10\rangle$ and $|01\rangle$ states, and

$$\langle 01| \sum_j A_{ZZ}^{(j)} \underline{I}_Z^{(j)} \sum_k A_{ZZ}^{(k)} \underline{I}_Z^{(j)} |01\rangle = [A_{ZZ}^{(2)}]^2 \tag{161a}$$

$$\langle 10| \sum_j A_{ZZ}^{(j)} \underline{I}_Z^{(j)} \sum_k A_{ZZ}^{(k)} \underline{I}_Z^{(j)} |10\rangle = [A_{ZZ}^{(1)}]^2 , \tag{161b}$$

and

$$\langle \underline{S}(t)\underline{S}\rangle = e^{i(\gamma B+a)t} [e^{-\langle (A_{ZZ}^{(1)})^2\rangle_o \tau_c t} + e^{-\langle (A_{ZZ}^{(2)})^2\rangle_o \tau_c t}] . \tag{162}$$

Since $\langle (A_{ZZ}^{(1)})^2\rangle = \langle (A_{ZZ}^{(2)})\rangle^2$, both "degenerate" lines have the same width. However, for the threefold degenerate ($\mathcal{I}_Z = 0$) lines, the operator of interest can readily be evaluated in the $|I_Z^{(1)} I_Z^{(2)}\rangle$ representation (i.e. $|00\rangle$, $|10\rangle$ and $|01\rangle$) which diagonalizes it, and the result is readily seen to be

$$e^{i\gamma Bt} [e^{-0t} + e^{-\langle (A_{ZZ}^{(1)})^2\rangle \tau_c t} + e^{-\langle (A_{ZZ}^{(2)})^2\rangle \tau_c t}] ; \tag{163}$$

in this case two of the degenerate lines have the same width and one is sharper. The spectrum thus consists of a sharp Lorentzian and a superimposed broad one with twice the integrated intensity of the former.

The discussion above can readily be extended to pseudo-secular and non-secular terms. For example, the pseudo-secular interaction contributions to the linewidths in the example above depend upon terms of the form

$$< \sum_j A_{ZZ}^{(j)} \underline{I}_+^{(j)} \underline{S}_Z \sum_k A_{ZZ}^{(k)} \underline{I}_-^{(j)} \underline{S}_Z \underline{S}_+ e^{i(\gamma B + a\mathcal{J}_Z)t} \underline{S}_- > \tag{164a}$$

$$- <(\sum_j A_{ZZ}^{(j)} \underline{I}_+^{(j)} \underline{S}_Z) S_+ e^{i(\gamma B + a\mathcal{J}_Z)t} (\sum_k A_{ZZ}^{(k)} \underline{I}_-^{(k)} \underline{S}_Z) S_- > \tag{164b}$$

etc. If we choose the $|I_Z^{(1)} I_Z^{(2)}>$ representation, which worked so well for the secular terms, we find that the operator enclosed between brackets is not diagonal. It is a simple exercise to show this and we shall not do it here, but note that we only retain terms for which $\exp i(\gamma B + a\mathcal{J}_Z)t \to \exp i(\gamma B + a)t$, i.e. the time-dependent exponential defines the relevant $\mathcal{J}_Z$ manifold ($\mathcal{J}_Z=1$). After the matrix elements are evaluated, the operator in the brackets must be diagonalized in this manifold (3 x 3 matrix) before exponentiating to obtain the spin relaxation time.

Thus, to summarize, if we have several symmetrically equivalent nuclei, we find, as in Eq. (108), that

$$<\underline{S}_+(t)\underline{S}_-(0)> = \sum_{\mathcal{J}_Z} e^{i(\gamma B + a\mathcal{J}_Z)t} <\underline{S}_+\underline{S}_->_o [1 - t \sum_{\mathcal{J}'_Z} R_{\mathcal{J}_Z \mathcal{J}'_Z}(\omega_{\mathcal{J}'_Z}) \ldots] \tag{165}$$

but that $\mathcal{J}_Z$ represents a manifold of N-degenerate states. In turn $R_{\mathcal{J}_Z \mathcal{J}'_Z}$ is the average of an operator ($\underline{\Phi}$) formed from the density matrix, the electron spin operators, and commutators of the relevant spin-lattice hamiltonians, which is itself averaged over all states except the nuclear spin states. We must diagonalize the N x N degenerate matrix, $<\mathcal{J}_Z|\underline{\Phi}|\mathcal{J}_{Z'}>$, i.e. find the appropriate diagonalizing basis functions $|M_N>$ within the $\mathcal{J}_Z$-manifold. The resulting expression for $<\underline{S}_+(t)\underline{S}_->$ is

$$<\underline{S}_+(t)\underline{S}_-> = \sum_M \sum_N e^{i(\gamma B + a\mathcal{J}_Z)t} <\underline{S}_+\underline{S}_->_o \exp[-R_{M_N M_N} t] \ . \tag{166}$$

Freed and Fraenkel[18] and Abragam[10] have treated this problem by means of the Redfield-Bloch equations and, Atkins[31] has developed an approach similar to that above for the secular contributions. The most important effect of this kind occurs when symmetrically equivalent nuclei assume different instantaneous positions corresponding to different instantaneous hyperfine interactions.[18,32]

X.15. BREAKDOWN OF SPIN-HAMILTONIAN; ORBACH PROCESSES

The concept of a spin hamiltonian breaks down if the separation, E_n-E_o, between the ground and relevant excited electronic states, $|o\rangle$ and $|n\rangle$ respectively, is not sufficiently larger than k_BT. This breakdown is similar to that of the Born-Oppenheimer approximation which also occurs when near degeneracies in electronic states exist. A spin hamiltonian cannot be derived for systems with nearly degenerate electronic states because the spin and electronic energy gaps are comparable, and, therefore, the spin effects perturb the electronic states so much that it is meaningless to try to average out the electronic motions independently of the spin motions. Under conditions of near degeneracy, time variations in the crystal or molecular fields of the molecule can induce transitions between the $|o\rangle$ and $|n\rangle$ electronic states; these transitions are coupled to the spin by way of the spin orbit interaction $(\zeta\vec{\underline{\ell}}\cdot\vec{\underline{S}})$, and an electronic transition is therefore accompanied by a spin transition. For the spin transition probability to be appreciable, lattice quanta (phonons) of energy about E_n-E_o must be available to induce the electronic transition, and if $E_n-E_o \gg k_BT$, this cannot be the case because of the small Boltzmann factor controlling the phonon population.

In a combined orbital-spin transition, if the lattice is treated quantum mechanically, the important integral, $\int_0^\infty \langle[\underline{\mathcal{H}}_{SL},[\underline{\mathcal{H}}^\circ_{SL}(\tau),\underline{S}^\circ_+(t)]]\cdot\underline{S}_-\rangle d\tau$, which appears in the relaxation expression in Eq. (109), has the form

$$\int_0^\infty d\tau \; \mathrm{Tr}\; \underline{\rho}_{oo:vv} |\langle ovm_S|\underline{\mathcal{H}}_{SL}|nv'm'_S\rangle|^2 e^{i(E_n-E_o)\tau}\, e^{(\epsilon_{v'}-\epsilon_v)}\, e^{i\gamma B\tau} \langle\tfrac{1}{2}|\underline{S}_+|-\tfrac{1}{2}\rangle\langle-\tfrac{1}{2}|\underline{S}_-|\tfrac{1}{2}\rangle \tag{167}$$

where $|o\rangle$ and $|n\rangle$ refer to electronic states, $|v\rangle$ and $|v'\rangle$ to lattice (vibrations, collisions, rotations) states, and $|m_S\rangle$, $|m'_S\rangle$, $|\tfrac{1}{2}\rangle$, $|-\tfrac{1}{2}\rangle$ to spin states. Clearly only when energy is conserved, i.e. $E_n-E_o+\epsilon_{v'}-\epsilon_v+\gamma B = 0$, is the integral important. If $(E_n-E_o)\geq k_BT$ and

$\gamma B \leq k_B T$, the density matrix, ρ_{oovv}, is

$$\rho_{oovv} \approx \frac{e^{-\beta(E_n+\epsilon_{v'})}}{Tr_{v''} e^{-\beta\epsilon_{v''}}(2S+1)} \tag{168}$$

and high temperature approximations are inappropriate. Nevertheless, Eq. (83) is valid since, as a careful analysis of its derivation will show, only the spin part of ρ need be expanded in powers of $(k_B T)^{-1}$. Eq. (83) is essentially Fermi's Golden rule.

The simplest orbit-spin or "Orbach" interaction can be obtained by letting the spin-lattice hamiltonian, $\mathcal{H}_{SL}$, be equal to $\zeta\vec{\ell}\cdot\vec{S}$, and assuming the combined spin-orbital transition is induced by molecular rotations (rotation-spin-orbit process).[29] Typically one solves for the electronic or orbital wave functions, $|n\rangle$, in a molecular coordinate system whereas $\vec{S}$ is quantized in a laboratory framework. Thus the spin-orbit interaction can take the form, $\zeta\vec{\ell}(\text{mol})\cdot\tilde{\underline{\underline{D}}}\,\vec{S}(\text{lab})$, where $\underline{\underline{D}}$ is the unitary transformation discussed in Eq. (21), which transforms $\vec{S}$ from the laboratory to the molecular coordinate system. $\tilde{\underline{\underline{D}}}$ is a function of the Eulerian angles and hence of time. Thus

$$\mathcal{H}_{SL}(\tau) \rightarrow \zeta\, e^{i(E_o-E_n)\tau} \langle o|\vec{\ell}|n\rangle\cdot\tilde{\underline{\underline{D}}}(\tau)\vec{S} \tag{169}$$

where the time dependence of $\vec{S}$ has been neglected. In non-linear molecules, $\langle o|\vec{\ell}|o\rangle = 0$. (See § 2.) It can readily be seen that the resulting spin relaxation time, T_2, has the form

$$T_2^{-1}(\text{RSO}) \propto \sum_{n\neq 0} \frac{\zeta^2\tau_c^{(1)}|\langle o|\underline{\ell}|n\rangle|^2}{1+(E_o-E_n)^2(\tau_c^{(1)})^2}\, e^{-\beta(E_n-E_o)} . \tag{170}$$

$(E_n-E_o)\tau_c \gg 1$, and it may well be incorrect to assume that the first order orientational autocorrelation function is a simple exponential at such very high frequencies; however, if Eq. (170) is accepted, and only one excited electronic state, $|n\rangle$, is low enough to be considered,

$$T_2^{-1}(\text{RSO}) \propto \frac{\zeta^2}{(E_n-E_o)^2}\,|\langle o|\underline{\ell}|n\rangle|^2\, \frac{e^{-\beta(E_n-E_o)}}{\tau_c^{(1)}} . \tag{171}$$

$\zeta(E_n-E_o)^{-1}$ should be of the order of the g-value shifts (see § 2), which in turn is the order of Δg; therefore,

$$T_2^{-1}(\mathrm{RSO}) \propto (\Delta g)^2 \, e^{-\beta(E_n - E_o)} / \tau_c^{(1)} \, . \tag{172}$$

In hydrocarbon free radicals Δg is usually larger than 10^{-4}, in vanadyl acetylacetonate it is of order 10^{-2} and in $Cu(H_2O)_6^{++}$ about 0.1; at room temperature $\tau_c^{(1)} \approx 10^{-11}$ sec. If $\beta(E_n - E_o) \gg 1$, then clearly T_2^{-1}(RSO) is negligible. In symmetrical radicals, such as $C_6H_6^-$ and $Cu(H_2O)_6^{++}$, with orbital degeneracy, the solvent probably removes this degeneracy but the energy splitting $(E_n - E_o)$ may well be small, i.e. of order k_BT. However, in these cases it appears that $\langle o|\ell|n\rangle$ vanishes if $|o\rangle$ and $|n\rangle$ are the two initially almost degenerate states; thus this mechanism is probably not very important.

Another, and perhaps more interesting Orbach process might be called a second-order statistical[33] or a vibrational spin-orbit (VSO)[34] process. In this process the crystal or molecular field, U, is altered by molecular collisions with the solvent, and the fluctuations in U(t) can be written as $(\partial U/\partial q)q(t)$, where $(\partial U/\partial q)$ is dependent upon the electronic or orbital states, ($|o\rangle$ and $|n\rangle$), and q is dependent only upon lattice motions, e.g. intermolecular distances and orientations. The complete, relevant hamiltonian has the form

$$\mathcal{H} = \mathcal{H}_S + \zeta \vec{\ell}\cdot\vec{S} + \left(\frac{\partial U}{\partial q}\right) q(t) \tag{173}$$

where all terms have been previously defined. A Van Vleck transformation[29] can be carried out to diagonalize the time independent part to second order in $\zeta \vec{\ell}\cdot\vec{S}$, in which case the time dependent part has a second order contribution of the form

$$\mathcal{H}_{SL} = \sum_{n'} \left[\frac{\langle o|\zeta \vec{\ell}\cdot\vec{S}|n'\rangle \langle n'|\frac{\partial U}{\partial q}|n\rangle}{E_o - E_{n'}} + \frac{\langle o|\frac{\partial U}{\partial q}|n'\rangle \langle n'|\zeta \vec{\ell}\cdot\vec{S}|o\rangle}{E_o - E'_n} \right] q \; ; \tag{174}$$

only those time dependent terms which depend upon spin have been retained. If this spin-lattice interaction, rather than the one in Eq. (169), is used in the development of the last paragraph, one obtains a spin relaxation time, T_2, of the form

$$T_2^{-1}(\mathrm{VSO}) \propto \frac{\zeta^2 \left(\frac{\partial U}{\partial q}\right)^2}{(E_o - E_{n'})} \, |\langle o|\ell|n'\rangle|^2 e^{-\beta(E_n - E_o)} \int_o^\infty d\tau \langle q(t)q(0)\rangle e^{i(E_n - E_o)\tau} \, . \tag{175}$$

If we assume the autocorrelation function $\langle q(t)q(0)\rangle$ is a simple exponential, $\langle q^2\rangle \exp(-t/\tau_q)$, then

$$T_2^{-1}(\mathrm{VSO}) \propto \frac{\zeta^2}{(E_o-E'_n)^2}\left(\frac{\partial U}{\partial q}\right)^2 \frac{\langle q^2\rangle |\langle o|\ell|n'\rangle|^2 \, e^{-\beta(E_n-E_o)}}{(E_o-E_n)^2\tau_q} \tag{176}$$

where we have assumed $(E_n-E_o)^2\tau_q^2 \gg 1$. (In crystals a Bose-Einstein factor, $[e^{\beta(E_n-E_o)} - 1]^{-1}$, is used rather than $e^{-\beta(E_n-E_o)}$.) Note the distinction between $|n\rangle$ and $|n'\rangle$ in this process; $|n'\rangle$ need not be low lying and $|n\rangle$ need not enter into the matrix of $\vec{\ell}$, i.e. it need not be related to Δg.

If we let the fractional change in the crystal field due to collisions be ξ,

$$\xi^2 = \left(\frac{\partial U}{\partial q}\right)^2 \langle q^2\rangle/(E_o-E_n)^2, \tag{177}$$

then Eq. (176) becomes

$$T_2^{-1}(\mathrm{VSO}) \propto (\Delta g)^2\, \xi^2 \, \frac{e^{-\beta(E_n-E_o)}}{\tau_q}. \tag{178}$$

Collisional relaxation times, τ_q, may be of order 10^{-13} sec to 5×10^{-13} sec, as determined by light scattering; ξ^2 is difficult to estimate but it might well be of order unity in molecules with degenerate orbital states, such as $C_6H_6^-$ and $Cu(H_2O)_6^{++}$, where (E_n-E_o) varies between 0 and a few k_BT because of solvent interactions. In this case note that

$$\frac{T_2^{-1}(\mathrm{VSO})}{T_2^{-1}(\Delta gB)} \approx \frac{e^{-\beta(E_n-E_o)}}{(\gamma B)^2\tau_q\tau_c^{(2)}} \tag{179}$$

where $T_2^{-1}(\Delta gB)$ is the relaxation time due to anisotropic g-tensor reorientation. At B = 3000 G, if $\tau_q \approx 5\times10^{-13}$ sec and $\tau_c^{(2)} = 10^{-11}$ sec, the VSO mechanism is about 60 $e^{-\beta(E_n-E_o)}$ as effective as the reorientation mechanism; therefore, it might dominate provided $(E_n-E_o) \lesssim 4k_BT$.

Although an effective RSO mechanism must be associated with a large shift in g-value from 2.0023, this is not necessary in the VSO mechanism. Note that neither of these mechanisms are dependent upon the applied field B or upon I_Z. Mechanisms analogous to the VSO process for which the hyperfine interaction, $\langle o|a|n\rangle\, \vec{I}\cdot\vec{S}$, can be substituted for $<o|\zeta\vec{\ell}\cdot\vec{S}|n>$, have been developed; these mechanisms can, however, be grouped with the spin-lattice interaction term, $\mathcal{H}_{FL}$, arising from fluctuations in the hyperfine interaction. (See § 3.)

Other related mechanisms, e.g. the spin-orbit pulse[35] (SOP) and spin-orbit tunneling (SOT) mechanisms[36] have also been studied. These are discussed in more detail in Chapter XVII.

X.16. ACKNOWLEDGEMENT

This work was supported in part by a grant from the National Science Foundation.

APPENDIX A

Expand $\exp(\underline{A}+\underline{B})$ in powers of $\underline{B}$, where $\underline{A}$ and $\underline{B}$ are non-commuting operators. Consider an operator function, $\underline{G}(\lambda)$:

$$\underline{G}(\lambda) = \exp(-\lambda\underline{A})\exp(\lambda\underline{A}+\lambda\underline{B}) .$$

Differentiate with respect to λ:

$$\frac{d\underline{G}(\lambda)}{d\lambda} = \exp(-\lambda\underline{A})\cdot\underline{B}\exp(\lambda\underline{A}+\lambda\underline{B}) .$$

Integrate from $\lambda=0$ to $\lambda=1$:

$$\underline{G}(1) - \underline{G}(0) = \int_0^1 d\lambda\, e^{-\lambda\underline{A}}\underline{B}\, e^{\lambda(\underline{A}+\underline{B})} .$$

Note that $\underline{G}(0) = 1$. Premultiply by $\exp \underline{A}$

$$e^{(\underline{A}+\underline{B})} = e^{\underline{A}} + e^{\underline{A}}\int_0^1 d\lambda\, e^{-\lambda\underline{A}}\underline{B}\, e^{\lambda(\underline{A}+\underline{B})} .$$

If we interate once:

$$e^{(\underline{A}+\underline{B})} = e^{\underline{A}}[1 + \int_0^1 d\lambda\, e^{-\lambda\underline{A}}\underline{B}\, e^{\lambda\underline{A}}$$

$$+ \int_0^1 d\lambda\, e^{-\lambda\underline{A}}\underline{B}e^{\lambda\underline{A}}\int_0^\lambda d\lambda'\, e^{-\lambda'\underline{A}}\underline{B}\, e^{\lambda'(\underline{A}+\underline{B})}] .$$

We can continue this interaction process and we can generate, in this way, an expansion in $\underline{B}$.

APPENDIX B

Consider the integrations in Eq. (86). To do this, rotate the time coordinates by 45°, i.e. introduce two new variables

$$\tau' = \frac{t'+t''}{\sqrt{2}} ; \qquad \tau'' = \frac{t''-t'}{\sqrt{2}} .$$

See Figs. 1 and 2. The equivalent integration in terms of the new coordinates has τ' ranging from τ'' to $(-\tau'' + \sqrt{2}\, t)$ and τ'' ranging from 0 to $t/\sqrt{2}$. Thus

$$\int_0^t dt' \int_0^{t'} dt''\, h(t',t'') = \int_0^{t/\sqrt{2}} d\tau'' \int_{\tau''}^{-\tau''+\sqrt{2}\,t} d\tau'\, h\left(\frac{\tau'-\tau''}{\sqrt{2}} , \frac{\tau'+\tau''}{\sqrt{2}}\right) .$$

We have inserted the factor of $\sqrt{2}$ in the transformations above to keep the picture of a 45° rotation in mind; now we shall, however, change scale to the variables

$$\tau = \sqrt{2}\, \tau'' = t''-t'$$

$$T = \sqrt{2}\, \tau' = t''+t' .$$

Then

$$\int_0^t dt \int_0^{t'} dt''\, h(t',t'') = \frac{1}{2} \int_0^t d\tau \int_\tau^{-\tau+2t} dT\, h\left(\frac{T-\tau}{2} , \frac{T+\tau}{2}\right).$$

If $h = h(t''-t')$,

$$\int_0^t dt \int_0^{t'} dt''\, h(t''-t') = \int_0^t d\tau (t-\tau) h(\tau) .$$

Now note that in Eq. (86), for those terms that have $h = h(\tau,T)$, i.e. the non-adiabatic terms mentioned near Eq. (96), the T dependence is exponential ($e^{i\omega T}$), and the integrations above lead to terms of order $(\omega t)^{-1}$ smaller than those for which $h = h(\tau)$. Since $T \approx T_2$ and ω is a Larmor frequency, $(\omega t)^{-1}$ is usually very small. Furthermore, the static averages $\langle \underline{F}_q \underline{F}_{q'} \rangle$ and $\langle \mathcal{J}_q \mathcal{J}_{q'} \rangle$ usually vanish for the non-adiabatic terms.

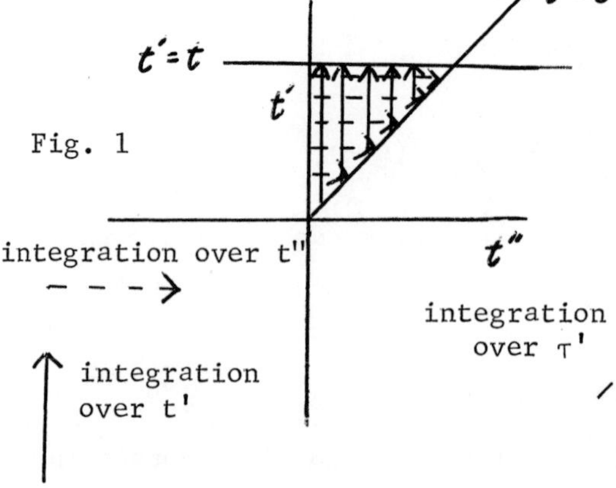

Fig. 1

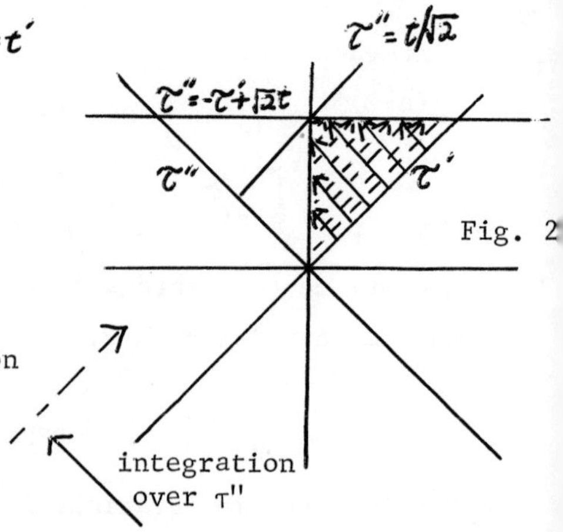

Fig. 2

APPENDIX C

The nonvanishing spin matrix elements for $S = \frac{1}{2}$ and $I = \frac{1}{2}$ are

$$<S_Z I_Z|\underline{S}_Z|S_Z I_Z> = S_Z$$

$$<S_Z I_Z|\underline{I}_Z|S_Z I_Z> = I_Z$$

$$<S_Z \pm 1, I_Z|\underline{S}_\pm|S_Z I_Z> = 1$$

$$<S_Z, I_Z \pm 1|\underline{I}_\pm|S_Z I_Z> = 1 \; .$$

learly $\underline{S}_+\underline{S}_+$ and $\underline{S}_-\underline{S}_-$ are not diagonal but $\underline{S}_+\underline{S}_-$ and $\underline{S}_-\underline{S}_+$ are. The ommutation rules are

$$[\underline{S}_+,\underline{S}_Z] = -\underline{S}_+$$

$$[\underline{S}_-,\underline{S}_Z] = +\underline{S}_-$$

$$[\underline{S}_+,\underline{S}_-] = 2\underline{S}_Z \; .$$

urthermore, $\underline{S}_\pm\underline{S}_\mp$ is diagonal and

$$\underline{S}_\pm\underline{S}_\mp = S(S+1) - S_Z^2 \pm S_Z \; .$$

he same rules follow for $\vec{\underline{I}}$. Finally, using (29) and (58a)

$$\underline{S}_\pm^\circ(t) = e^{\pm i[\gamma B + a\underline{I}_Z]t}$$

$$\underline{I}_\pm^\circ(t) = e^{\pm ia\underline{S}_Z t} \; .$$

APPENDIX D

n Eq. (126)[12]

$$A_{ZZ} = \frac{1}{3}\,\Delta a(3\cos^2\theta - 1) - \delta a \sin^2\theta \cos 2\phi$$

$$A_{+-} = A_{-+} = -\frac{1}{4}\,A_{ZZ}$$

$$A_{Z\pm} = [\frac{1}{4}\,\Delta a \sin 2\theta + \frac{1}{2}\,\delta a \sin\theta(-\sin 2\phi \pm i\cos\theta\cos 2\phi)]e^{\mp i\zeta}$$

$$_{\pm\pm} = \left\{-\frac{1}{4}\,\Delta a \sin^2\theta + \frac{1}{4}\,\delta a[(1+\cos^2\theta)\cos 2\phi \mp 2i\cos\theta\sin 2\phi]\right\}e^{\mp 2i\zeta}$$

$$G_{ZZ} = [\frac{1}{3}\,\Delta g\,(3\cos^2\theta - 1) - \delta g \sin^2\theta \cos 2\phi]$$

$$G_{Z\pm} = [\pm\frac{1}{4}\,i\,\Delta g \sin^2\theta + \frac{1}{2}\,\delta g \sin\theta(-\sin 2\phi \pm i\cos\theta\cos 2\phi]$$

here θ, ϕ and ζ are Eulerian angles between molecular and laboratory oordinates, and we have assumed that $\underline{\underline{A}}$ and $\underline{\underline{G}}$ have been simultaneously iagonalized. (Note that in reference 12, Eq. (20) is incorrect.)

REFERENCES

1. R. Kubo and K. Tomita, J. Phys. Soc. Japan 9, 888 (1959). R. Kubo, J. Phys. Soc. Japan 12, 570 (1957).

2. $a \propto |\psi(o)|^2$ where $\psi(o)$ is the wave function of the unpaired electron evaluated at the interacting nucleus. Thus if ψ is a molecular orbital, "a" will depend upon the s-character of the constituent atomic orbital on the relevant atom.

3. R. H. Sands, Phys. Rev. 99, 1722 (1955). E. L. Cochran, F. J. Adrian and V. A. Bowers, J. Chem. Phys. 34, 1161 (1961).

4. R. Neiman and D. Kivelson, J. Chem. Phys. 35, 162 (1961).

5. M. E. Rose, Elementary Theory of Angular Momentum, (John Wiley and Sons, Inc., New York, 1957).

6. R. D. Rataiczak and M. T. Jones, J. Chem. Phys. (To be published

7. J. Norris and S. Weissman, J. Phys. Chem. 73, 3119 (1969).

8. J. Sinclair, Thesis, UCLA (1966).

9. A. Abragam and B. Bleaney, Electron Paramagnetic Resonance of Tr sition Ions, Oxford, Clarendon Press, 1970.

10. A. Abragam, The Principles of Nuclear Magnetism, (Oxford Univ. Press, London, 1961).

11. C. Schlicter, Principles of Magnetic Resonance (Harper & Row, Ne York. 1963).

12. R. Wilson and D. Kivelson, J. Chem. Phys. 44, 154 (1966).

13. A. Messiah, Quantum Mechanics, North Holland Publ. Co., Amsterda (1965).

14. P. W. Atkins, "Theories of Electron Spin Relaxation in Solution, Advances in Molecular Relaxation Processes, 1971.

15. H. Sillescu and D. Kivelson, J. Chem. Phys. 48, 3493 (1968). H. Sillescu, J. Chem. Phys. 54, 2110 (1971). J. Freed, J. Chem. Phys.

16. F. Bloch, Phys. Rev., 70, 460 (1946). F. Bloch, Phys. Rev., 102, 104 (1956).

17. A. G. Redfield, I.B.M. J. Res. & Development, 1, 19 (1957). A. G. Redfield, Adv. Mag. Res., 1, 1 (1965).

18. J. Freed and G. Fraenkel, J. Chem. Phys. 39, 326 (1963).

19. R. Kubo, J. Phys. Soc. Japan, 17, 1100 (1967).

20. J. Freed, J. Chem. Phys. 49, 376 (1968).

21. R. E. D. McClung and D. Kivelson, J. Chem. Phys. 49, 3380 (1968)

22. D. Kivelson, M. Kivelson and I. Oppenheim, J. Chem. Phys. 52, 1810 (1970).

23. R. Huang, To be included in Thesis, UCLA, 1971; and others.

24. J. Hwang, To be included in Thesis, UCLA, 1971.

25. T. E. Bull and J. Jonas, J. Chem. Phys. 52, 4553 (1970).
T. E. Bull, J. S. Barthel and J. Jonas, J. Chem. Phys. 54, 3663 (1971).
H. J. Parkhurst, Y. Lee and J. Jonas, J. Chem. Phys. (Submitted).

26. P. S. Hubbard, Phys. Rev. 131, 1155 (1963).

27. A. Ben-Reuven and N. D. Gershon, J. Chem. Phys. 51, 893 (1969).

28. G. I. A. Stegeman, Ph.D. Thesis, Univ. of Toronto (1969).

29. D. Kivelson, J. Chem. Phys. 45, 1324 (1966).
P. W. Atkins, Molec. Phys. 13, 37 (1967).

30. J. Schreurs and D. Kivelson, J. Chem. Phys. 36, 117 (1962).

31. P. W. Atkins, Molec. Phys. 21, 97 (1971).

32. A. Carrington and H. C. Lonquet-Higgins, Mol. Phys. 5, 447 (1962).

33. J. P. Lloyd and G. E. Pake, Phys. Rev. 94, 579 (1954).

34. D. Kivelson and G. Collins, Proc. Item. Conf. Magnetic Resonance, Jerusalem 1961, 496 (1962).

35. H. M. McConnell and A. D. McLaughlan, J. Chem. Phys. 34, 1 (1961).
H. M. McConnell, J. Chem. Phys. 34, 13 (1961).

36. J. H. Freed and R. G. Kooser, J. Chem. Phys. 49, 4715 (1968).
R. G. Kooser, W. V. Volland and J. H. Freed, J. Chem. Phys. 50, 5243 (1969).
M. R. Das, S. B. Wagner and J. H. Freed, J. Chem. Phys. 52, 5404 (1970).

PIN - ROTATION INTERACTION

P.W. Atkins

Physical Chemistry Laboratory, University of Oxford

XI.1 INTRODUCTION

The line width arising from the rotational modulation of nisotropic interaction tensors has the general form[1-3] ntroduced in Chapter X (see X.10.1):

$$\Delta = (2\hbar/g\mu_B\sqrt{3})T_2^{-1} = \alpha' + \beta m_I + \gamma m_I^2 + \delta m_I^3 + \varepsilon m_I^4 \qquad (1)$$

here α' arises from the modulation of the g and hyperfine nisotropies. A careful test of this formula[3] indicated that α' oes not always account for the m_I-independent contribution to the ine width and that there is in some systems a residual line width, α'', which is the deviation of α' from the experimental value, α_{ex}. he viscosity and temperature dependence of β, γ, δ, and ε was ound to be quite adequately accounted for by the rotational ebye correlation time (and all the coefficients α', β, ... ε aried as η/T) but at low values of η/T it was found that α_{ex} aried in a peculiar manner: when α' was extracted the residual ine width varied as T/η. The variation with η/T of α_{ex}, α', and α'' for vanadyl acetylacetonate in toluene is illustrated in Fig.1[3]. his kind of behaviour ($\alpha'' \propto T/\eta$) points to the importance of a

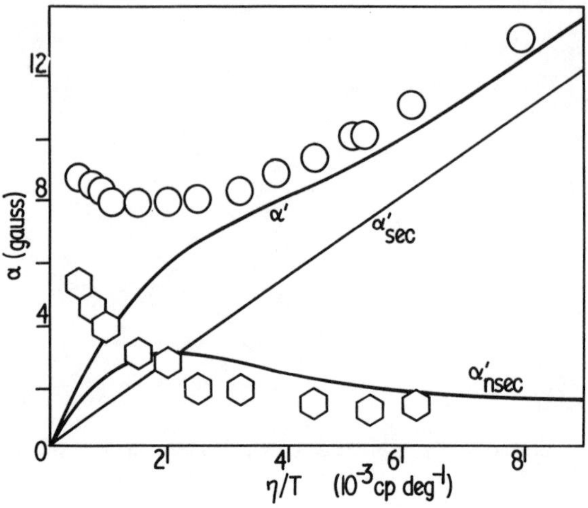

Figure 1. α as a function of η/T for vanadyl acetylacetonate in toluene[3]. α' is the calculated value of α and the circles are the experimental values α_{ex}. The residual line width α'' is the difference between the experimental and the calculated values and is represented by the hexagons.

spin-rotation interaction as the relaxation modulation because in n.m.r. such a T/η dependence has been widely observed[4-26]. It is plausible that the line-width should be proportional to T/η because a perturbation of the type $(C/\hbar)\mathbf{J}\cdot\mathbf{S}$ will lead to a spectral density of the form $(C/\hbar)^2\langle J^2\rangle\tau_J$, where τ_J is an angular momentum correlation time; the average value of J^2 is proportional to T, and the correlation time τ_J might be expected to be proportional to $1/\eta$ because it depends inversely on the strength of the intermolecular forces, and so the overall dependence is T/η.

In this chapter we shall begin by deducing an expression for α'' in a simple manner, and then proceed to a more complete derivation, which in turn will generate a number of problems which we shall attempt to examine. The most important of these is the value of τ_J, and we shall describe one of the approaches which has had some, but by no means complete, success: the connection between

τ_J and τ_K (the reorientation correlation time for a spherical tensor of rank K) and the factorisability of a bivariate correlation function remain the major problems of spin-rotation theory.

XI.2 BASIC THEORY OF THE INTERACTION

It is convenient to work with a spherical tensor form of operators[27-29], and so for angular momenta we take the components

$$j_o^{(1)} = j_z \quad j_{\pm 1}^{(1)} = \mp (1/\sqrt{2})(j_x \pm ij_y). \tag{2}$$

This implies that the magnetic moment operator is

$$m_x = -(1/\sqrt{2}) \sum_{\alpha=\pm 1} \alpha m_\alpha = -(\gamma/\sqrt{2}) \sum_{\alpha=\pm 1} \alpha S_\alpha^{(1)} \tag{3}$$

and that the magnetic moment correlation function, which the preceding chapters have shown to be at the root of a line width calculation, is

$$G(t) = \langle\langle\{m_x(t),m_x\}\rangle\rangle = G^{(0)}(t) + G^{(2)}(t) + \ldots \tag{4}$$

with

$$G^{(0)}(t) = -\tfrac{1}{2}N \sum_{\alpha=\pm 1} \langle\{m_\alpha,m_{-\alpha}\}\rangle \exp\{i\alpha\omega_Z t\} \tag{5}$$

$$G^{(2)}(t) = -\tfrac{1}{2}N \sum_{\alpha=\pm 1} \sum_\eta \int_o^t d\tau(t-\tau)\langle\{H_{I,\eta}^L(0)^\times H_{I,-\eta}^L(\tau)^\times m_\alpha,m_{-\alpha}\}\rangle$$

$$\times \exp\{i\alpha\omega_Z t + i\eta\omega_Z\tau\}, \tag{6}$$

where N is the number of spins in the sample, ω_Z is the Zeeman frequency, <<•••>> denotes the presence of H_I in both the density matrix and the motion, <•••> implies the absence of H_I, and the relaxation perturbation H_I has been written

$$H_I = \sum_\eta H_{I,\eta} = \sum_\eta (-1)^\eta w_{-\eta}^{(1)} S_\eta^{(1)} \tag{7}$$

($w^{(1)}$ being a first rank spherical tensor). In (6) {a,b} is the symmetrised product of a and b, and the superoperator operates as far as the comma. The interaction evolves in time under the influence of the lattice hamiltonian H_L according to (I.1.11):

$$H_I^L(t) = \exp\{(it/\hbar)H_L^{\times}\}H_I \ . \tag{8}$$

Eqn. (4) can be written as[30]

$$G(t) = -\tfrac{1}{2}N \sum_{\alpha} <\{m_\alpha, m_{-\alpha}\}>\exp\{i\alpha\omega_Z t + \psi_\alpha(t)\} \tag{9}$$

if

$$\exp\{\psi_\alpha(t)\} = 1 + (i/\hbar)^2 \sum_{\eta} \int_o^t d\tau(t-\tau)\sigma^2_{\eta\alpha}(\tau)\exp\{i\eta\omega_Z\tau\} \tag{10}$$

and

$$\sigma^2_{\eta\alpha}(\tau) = <\{H^L_{I,\eta}(0)^{\times}H^L_{I,-\eta}(\tau)^{\times}m_\alpha, m_{-\alpha}\}>/<\{m_\alpha, m_{-\alpha}\}>. \tag{11}$$

If the second term on the right of (10) is small $\psi_\alpha(t)$ may be set equal to $\exp\{\psi_\alpha(t)\} - 1$; alternatively the right of (10) may be considered to be the start of a rapidly converging cumulant series[30] (see Chapters V and VIII).

The central problem is the evaluation of $\sigma(\tau)$, and this is simplified by using the properties of the spherical tensor operators[31]. In particular it may be noted that the commutation relations for the components of $j^{(1)}$ may be written as[29]

$$j^{(1)\times}_{\eta_2} j^{(1)}_{\eta_1} = (\eta_2{:}\eta_1)\hbar j^{(1)}_{\eta_2+\eta_1} \tag{12}$$

where

$$(\eta_2{:}\eta_1) = \frac{-(\eta_2-\eta_1)}{(|\eta_2|+|\eta_1|)} \tag{13}$$

From this it follows that

$$j^{(1)\times}_{\eta_3} j^{(1)\times}_{\eta_2} j^{(1)}_{\eta_1} = (\eta_3{:}\eta_2+\eta_1)(\eta_2{:}\eta_1)\hbar^2 j^{(1)}_{\eta_3+\eta_2+\eta_1} \tag{14}$$

and in general

$$\prod_{i=2}^{n} \{j_{\eta_i}^{(1)\times}\} j_{\eta_1}^{(1)} = \{\prod_i (\eta_i : \sum_{k=1}^{i-1} \eta_k)\} \hbar^{n-1} \, j^{(1)}_{\{\sum_{i=1}^{n} \eta_i\}} \tag{15}$$

The advantage of this notation is that it is obvious that $s_{\eta}^{(1)\times} s_{-\eta}^{(1)\times}$ is a superoperator which satisfies

$$s_{\eta}^{(1)\times} s_{-\eta}^{(1)\times} s_{\alpha}^{(1)} = \hbar^2 f(\eta,\alpha) s_{\alpha}^{(1)} , \tag{16}$$

the eigenvalue $\hbar^2 f(\eta,\alpha)$ being deduced from (12) and (13) to be

$$f(\eta,\alpha) \equiv (\eta:\alpha-\eta)(-\eta:\alpha) = \frac{(\alpha-2\eta)(\alpha+\eta)}{(|\alpha-\eta|+|\eta|)(|\alpha|+|\eta|)}$$

$$f(0,0) \equiv 0 . \tag{17}$$

Then $\sigma^2(\tau)$ acquires a simple form:

$$\sigma^2_{\eta\alpha}(\tau) = \langle\{w^L_{-\eta}(0) s_{\eta}^{(1)\times} w^L_{\eta}(\tau) s_{-\eta}^{(1)\times} m_{\alpha}, m_{-\alpha}\}\rangle / \langle\{m_{\alpha}, m_{-\alpha}\}\rangle$$

$$= \langle w^L_{-\eta}(0) w^L_{\eta}(\tau)\rangle \langle\{s_{\eta}^{(1)\times} s_{-\eta}^{(1)\times} m_{\alpha}, m_{-\alpha}\}\rangle / \langle\{m_{\alpha}, m_{-\alpha}\}\rangle$$

$$= \langle w^L_{-\eta}(0) w^L_{\eta}(\tau)\rangle \hbar^2 f(\eta,\alpha) \tag{18}$$

The second equality is valid for high lattice energies, and the third follows from the use of (16). The magnetisation correlation function may now be deduced from

$$\psi_{\alpha}(t) \sim - \sum_{\eta} \int_0^t d\tau (t - \tau) g_{\eta}(\tau) f(\eta,\alpha) \exp\{i\eta\omega_Z\tau\}$$

$$\sim - t \sum_{\eta} f(\eta,\alpha) \int_0^{\infty} d\tau \, g_{\eta}(\tau) \exp\{i\eta\omega_Z\tau\} \tag{19}$$

where $g_{\eta}(\tau)$ is the correlation function for the motion of the relaxation interaction induced by the lattice:

$$g_\eta(\tau) = \langle w^L_{-\eta}(0) w^L_\eta(\tau)\rangle \ . \qquad (20)$$

Direct examination of (17) shows that

$$f(-\eta, -\alpha) = f(\eta,\alpha) \qquad (21)$$

and the properties of correlation functions enables one to deduce that

$$g^*_\eta(\tau) = g_{-\eta}(-\tau); \quad g_\eta(\tau) = g_{-\eta}(-\tau) \text{ at high temperatures;} \qquad (22)$$

consequently[29,30]

$$\text{re } \psi_\alpha(t) = -\tfrac{1}{2}t \sum_\eta f(\eta,\alpha) j_\eta(\eta\omega_Z) \equiv -t\phi'_\alpha \qquad (23)$$

$$\text{im } \psi_\alpha(t) = \left(\frac{1}{2\pi}\right) t \sum_\eta f(\eta,\alpha) \mathcal{P}\!\!\int_{-\infty}^{\infty} d\omega' \left\{\frac{j_\eta(\omega')}{\omega' - \eta\omega_Z}\right\} \equiv -t\phi''_\alpha \qquad (24)$$

where

$$j_\eta(\omega) = \int_{-\infty}^{\infty} d\tau g_\eta(\tau) \exp\{i\omega\tau\} \ . \qquad (25)$$

The magnetization correlation function G(t) may be written

$$G(t) = -\tfrac{1}{2}N \sum_\alpha \langle\{m_\alpha, m_{-\alpha}\}\rangle \exp\{i[\alpha\omega_Z + i\phi'_\alpha - \phi''_\alpha]t\} \ , \qquad (26)$$

and so ϕ'_α may be identified as the line width and ϕ''_α as the shift. We note that $\phi'_\alpha = \phi'_{-\alpha}$ and $\phi''_\alpha = -\phi''_{-\alpha}$ for high temperatures. Fourier transformation of (26) yields the line width parameter as

$$T_2^{-1} = \tfrac{1}{2}\, j_o(0) - \tfrac{1}{2}\, j_1(\omega_Z) \qquad (27)$$

$$= T^{-1}_{2\ \text{sec}} + T^{-1}_{2\ \text{nsec}} = T^{-1}_{2\ \text{sec}} + \tfrac{1}{2}T_1^{-1} \qquad (28)$$

and so

$$T_1^{-1} = -\, j_1(\omega_Z) \ . \qquad (29)$$

The calculation is now specialised[32] to the case of a simple form of the spin-rotation interaction in which the coupling constant is scalar:

$$H_I = \hbar^{-1}\, J\cdot C\cdot S = \hbar^{-1} C \sum_{\eta} (-1)^{\eta} J^{(1)}_{-\eta} S^{(1)}_{\eta} \;. \tag{30}$$

Clearly the tensor $w^{(1)}_{-\eta}$ of (7) is $(C/\hbar)J^{(1)}_{-\eta}$. The line width is determined by the correlation function (20) which becomes

$$g_{\eta}(\tau) = (C/\hbar)^2 \langle J^{(1)L}_{\eta}(\tau) J^{(1)L}_{-\eta}(0)\rangle. \tag{31}$$

The angular momentum correlation function may be estimated by constructing and solving the rotational analogue of the Langevin equation[13]. The linear Langevin equation for the velocity of a particle of mass m under the influence of a stochastic force A(t) is

$$m\dot{u}(t) = -\zeta u(t) + mA(t) \tag{32}$$

where ζ is the frictional constant. Hubbard[13] wrote the analogue of (32) as

$$I\dot{\omega}(t) = -\zeta\omega(t) + IA(t) \tag{33}$$

and applied the standard method of solving this equation[33] to deduce that

$$g_{\eta}(\tau) = (-1)^{\eta}\{C^2 IkT/\hbar^2\}\exp\{-\tau/\tau_J\} \tag{34}$$

$$\tau_J = I/\zeta = I/8\pi r_e^3\eta \tag{35}$$

where r_e is the hydrodynamic radius of the molecule and η the viscosity of the medium. The spectral densities are therefore

$$j_{\eta}(\eta\omega_Z) = (-1)^{\eta}\{2C^2 IkT/\hbar^2\}\{\tau_J/(1+\eta^2\omega_Z^2\tau_J^2)\} \tag{36}$$

and the line width parameter and relaxation time are

$$T_2^{-1} = \{C^2 IkT/\hbar^2\}\tau_J \left\{1 + \frac{1}{1+\omega_Z^2\tau_J^2}\right\} \tag{37}$$

$$T_1^{-1} = \{2C^2 IkT/\hbar^2\}\left\{\frac{\tau_J}{1+\omega_Z^2\tau_J^2}\right\} \quad . \tag{38}$$

In the limit $\omega_Z^2\tau_J^2 << 1$ one finds

$$T_2^{-1} = T_1^{-1} = \{2C^2 IkT/\hbar^2\}\tau_J$$

$$= \{C^2 I^2/4\pi r_e^3\hbar^2\}(kT/\eta) \tag{39}$$

and the anticipated and required T/η dependence is explicit.

For most radicals the value of the spin-rotation coupling constant is unknown, but it may be related to the deviation of g from the free spin value[32,34]. This relation arises by finding the value of δg and C from second order perturbation theory in two isomorphous calculations. The hamiltonian for a rotating doublet molecule in a magnetic field B is

$$H = (\zeta/\hbar)L\cdot S + (\mu_B/\hbar)g_e B\cdot S + (\mu_B/\hbar)B\cdot L + (1/\hbar)N\cdot A\cdot N \tag{40}$$

where A is the rotational tensor for the supposedly rigid molecule which is rotating with angular momentum N. On writing J = N + L and omitting L^2 terms one may rewrite (40) as

$$H = (\zeta/\hbar)L\cdot S + (1/\hbar)J\cdot A\cdot J - (2/\hbar)J\cdot A\cdot L + (\mu_B/\hbar)B\cdot L + (\mu_B/\hbar)g_e B\cdot S \tag{41}$$

Bilinear combinations of B•L with L•S and J•A•L with L•S in second-order perturbation theory (see X.1) will yield an effective hamiltonian within the electronic ground state manifold of the form

$$H_{eff} = (1/\hbar)J\cdot A\cdot J + (\mu_B/\hbar)B\cdot g\cdot S + (1/\hbar)J\cdot C\cdot S \quad . \tag{42}$$

The second-order bilinear energies arising from (41) are

$$E^{(2)}(B,S) = 2\sum_n \langle 0|\left\{\frac{(\mu_B/\hbar)B\cdot L|n\rangle\langle n|(\zeta/\hbar)L\cdot S}{(E_o - E_n)}\right\}|0\rangle \qquad (43)$$

$$E^{(2)}(J,S) = -4\sum_n \langle 0|\left\{\frac{(1/\hbar)J\cdot A\cdot L|n\rangle\langle n|(\zeta/\hbar)L\cdot S}{(E_o - E_n)}\right\}|0\rangle \qquad (44)$$

and so

$$g = g_e 1 - (2/\hbar)\sum_n \left\{\frac{\langle 0|L|n\rangle\langle n|\zeta L|0\rangle}{(E_n - E_o)}\right\} = g_e 1 + \delta g \qquad (45)$$

$$C = (4/\hbar)A\cdot\sum_n \left\{\frac{\langle 0|L|n\rangle\langle n|\zeta L|0\rangle}{(E_n - E_o)}\right\} \qquad (46)$$

Obviously δg and C are related by

$$C = -\ 2A\cdot\delta g\ , \qquad (47)$$

and for an isotropic molecule this has the simple form

$$IC = -\ \hbar\delta g = -\ \hbar(g - g_e). \qquad (48)$$

The calculation in the last paragraph enables (39) to be written in the simple form[32]

$$T_1^{-1} = T_2^{-1} = \{\delta g^2/4\pi r_e^3\}(kT/\eta),\ \omega_Z^2\tau_J^2 \ll 1 \qquad (49)$$

and it should be noted that in this limit both T_2 and T_1 are independentof the moment of inertia of the molecules. The residual line width itself in the same limit is

$$\alpha'' = (2\hbar/g\mu_B\sqrt{3})T_2^{-1} = \{\hbar\delta g^2/2\pi g\mu_b r_e^3\sqrt{3}\}(kT/\eta). \qquad (50)$$

A test of this theory has been provided by studies of a number of paramagnetic species[3,32,35-38]. For example, in the case of vanadyl acetylacetonate[3] the calculated value of α'' (taking into account the anisotropy of the molecule) is 2.9×10^{-5} (T/η) gauss and the measured value has been found to be 3.2×10^{-5} (T/η) gauss. Even better agreement has been found in the case of copper acetylacetonate[35] where the calculated and measured values of α'' were both equal to $2.95 \times 10^{-4}(T/\eta)$ gauss. Deviations from the predictions of the theory have been observed in a number of cases, especially where the interaction between the solute and solvent is weak: a particular species which has been studied closely is chlorine dioxide[36,39,40] in, for example, hydrocarbon solvents. But it is not surprising that there are deviations in some cases, for the theory outlined above is obviously a simplification. In particular the relaxation modulation in the example was the change of the angular momentum itself (either its amplitude or its direction) because the interaction tensor was supposed to be scalar: in an anisotropic molecule C will be anisotropic and so the reorientation of the molecule (as well as the change in the angular momentum) will modulate the spin-rotation interaction. Furthermore the use of the Langevin equation and the macroscopic viscosity is contentious and the role of the intermolecular torques (as distinct from the forces) has not been sufficiently analysed. These points are investigated in the following sections: first we set up the full problem and then consider the dynamical problem involved. We shall see that a number of problems are revealed in this way, and some (indeed all) remain unsolved.

XI.3. THE FULL RELAXATION PROBLEM

In the full relaxation problem we consider the line width that arises from a perturbation of the form

$$H_I = (1/\hbar)\mathbf{J}\cdot\mathbf{C}\cdot\mathbf{S} \qquad (51)$$

and observe that C may be divided into an isotropic (scalar) component $C^{(0)}$ and a traceless second rank tensor $C^{(2)}$ which is zero if C is isotropic:

$$J\cdot C\cdot S = C^{(0)} J\cdot S + C^{(2)} : JS$$

$$= C^{(0)} J\cdot S + \sum_{\mu} (-1)^{\mu} C^{(2)}_{-\mu} (SJ)^{(2)}_{\mu} \qquad (52)$$

where $(SJ)^{(2)}$ is the second rank combination of S and J. The second term may be expressed as

$$\sum_{\mu} (-1)^{\mu} C^{(2)}_{-\mu} (SJ)^{(2)}_{\mu} = \sqrt{5} \sum_{\mu,\eta} \begin{pmatrix} 1 & 1 & 2 \\ \eta & \mu-\eta & -\mu \end{pmatrix} S^{(1)}_{\eta} J^{(1)}_{\mu-\eta} C^{(2)}_{-\mu} . \qquad (53)$$

In order to use the basic theory of the preceding section we need to write each component of (52) in the form (7) : the scalar component was treated in the preceding section, and so all that is necessary to do is to determine the quantities $c^{(2)}$ by a re-coupling so that

$$(1/\hbar) \sum_{\eta} (-1)^{\eta} (c^{(2)} J)^{(1)}_{-\eta} S^{(1)}_{\eta} = (1/\hbar) \sum_{\mu} (-1)^{\mu} C^{(2)}_{-\mu} (SJ)^{(2)}_{\mu} . \qquad (54)$$

The left hand side of this equation may be written

$$(1/\hbar) \sum_{\eta} (-1)^{\eta} (c^{(2)} J)^{(1)}_{-\eta} S^{(1)}_{\eta} = (-\sqrt{3}/\hbar) \sum_{\mu,\eta} \begin{pmatrix} 1 & 1 & 2 \\ \eta & \mu-\eta & -\mu \end{pmatrix} S^{(1)}_{\eta} J^{(1)}_{\mu-\eta} c^{(2)}_{-\mu} \qquad (55)$$

and comparison with (53) shows that

$$c^{(2)}_{\mu} = - (5/3)^{\frac{1}{2}} C^{(2)}_{\mu} . \qquad (56)$$

It is convenient to defer the expression of C in terms of its spherical components $C^{(2)}_{\mu}$ until later.

The perturbation component $w^{(1)}$ in (7) has now been shown to be given by

$$w^{(1)}_{-\eta} = (1/\hbar)\{C^{(0)}J^{(1)}_{-\eta} + (c^{(2)}J)^{(1)}_{-\eta}\} \tag{57}$$

and so the correlation function we require is

$$\begin{aligned} g_\eta(\tau) = (1/\hbar^2)\{&C^{(0)2}\langle J^{(1)L}_{\eta}(\tau)J^{(1)L}_{-\eta}(0)\rangle \\ &+ C^{(0)}[\langle J^{(1)L}_{\eta}(\tau)(c^{(2)}J)^{(1)}_{-\eta}\rangle + \langle (c^{(2)}J)^{(1)L}_{\eta}(\tau)J^{(1)}_{-\eta}\rangle] \\ &+ \langle (c^{(2)}J)^{(1)L}_{\eta}(\tau)(c^{(2)}J)^{(1)L}_{-\eta}(0)\rangle\} \end{aligned} \tag{58}$$

There now appear to be two methods of proceeding from this point: the first is to uncouple $c^{(2)}$ and $J^{(1)}$ and then refer $c^{(2)}$ to the (rotating) molecular axes (in which we know its components), and the second is to refer the contraction $(c^{(2)}J)^{(1)}$ to the molecular axes and then to effect the uncoupling. In the approximation we shall use these procedures lead to different results and it is worth considering them in order to see one of the difficulties that approximations can introduce.

Method 1. The uncoupling procedure yields

$$\langle J^{(1)L}_{\eta}(\tau)(c^{(2)}J)^{(1)}_{-\eta}\rangle = -\sqrt{3}\sum_{\mu}(-1)\begin{pmatrix}1 & 1 & 2\\ \eta & \mu-\eta & -\mu\end{pmatrix}\langle J^{(1)L}_{\eta}(\tau)c^{(2)}_{-\mu}J^{(1)}_{\mu-\eta}\rangle \tag{59}$$

and if $c^{(2)}$ is expressed in terms of its components $c^{(2)}$ referred to the molecular axes (which are related to the laboratory axes by the eulerian angles Ω) through

$$c^{(2)}_{-\mu} = \sum_{\xi}\tilde{c}^{(2)}\mathcal{D}^{(2)}_{\xi,-\mu}(\Omega) \tag{60}$$

the right hand side of (59) will contain a component for which we may make an approximation which is true at $\tau = 0$ and also presumably true for large τ when any short time correlations built up initially will have disappeared; that is

$$\tilde{c}_{\xi}^{(2)} \langle J_{\eta}^{(1)L}(\tau) J_{\mu-\eta}^{(1)} \mathscr{D}_{\xi,-\mu}^{(2)}(\Omega)\rangle \sim \tilde{c}_{\xi}^{(2)} \langle J_{\eta}^{(1)L}(\tau) J_{\mu-\eta}^{(1)}\rangle\langle \mathscr{D}_{\xi,-\mu}^{(2)}(\Omega)\rangle = 0. \tag{61}$$

This suggests that it is a good approximation to consider only the first and last terms in (58) and henceforth we neglect the rest.

The uncoupling of the last term proceeds as follows:

$$\begin{aligned}
&\langle (c^{(2)}J)_{\eta}^{(1)L}(\tau)(c^{(2)}J)_{-\eta}^{(1)L}(0)\rangle \\
&= 3\sum_{\mu\mu'} \begin{pmatrix} 2 & 1 & 1 \\ -\mu & \mu+\eta & -\eta \end{pmatrix}\begin{pmatrix} 2 & 1 & 1 \\ -\mu' & \mu'-\eta & \eta \end{pmatrix} \langle (c_{-\mu}^{(2)} J_{\mu+\eta}^{(1)})^{L}(\tau) \\
&\qquad \times (c_{-\mu'}^{(2)} J_{\mu'-\eta}^{(1)})^{L}(0)\rangle \\
&= 3\sum_{\mu\mu'}\sum_{\xi\xi'} \tilde{c}_{\xi}^{(2)}\tilde{c}_{\xi'}^{(2)} \begin{pmatrix} 2 & 1 & 1 \\ -\mu & \mu+\eta & -\eta \end{pmatrix}\begin{pmatrix} 2 & 1 & 1 \\ -\mu' & \mu'-\eta & \eta \end{pmatrix} \\
&\qquad \times \langle [\mathscr{D}_{\xi,-\mu}^{(2)}(\Omega) J_{\mu+\eta}^{(1)}]^{L}(\tau)[\mathscr{D}_{\xi',-\mu'}^{(2)}(\Omega) J_{\mu'-\eta}^{(1)}]^{L}(0)\rangle
\end{aligned} \tag{62}$$

and, as we anticipated, we are led to a bivariate correlation function in which the joint dynamical variables are the orientation and the angular momentum of the molecules. If the correlation time for the angular momentum differs greatly from that for the reorientation an approximation that has been made is[13]

$$\begin{aligned}
&\langle [\mathscr{D}_{\xi,-\mu}^{(2)}(\Omega) J_{\mu+\eta}^{(1)}]^{L}(\tau)[\mathscr{D}_{\xi',-\mu'}^{(2)}(\Omega) J_{\mu'-\eta}^{(1)}]^{L}(0)\rangle \\
&\qquad \sim \langle [\mathscr{D}_{\xi,-\mu}^{(2)}(\Omega)]^{L}(\tau)[\mathscr{D}_{\xi',-\mu'}^{(2)}(\Omega)]^{L}(0)\rangle\langle J_{\mu+\eta}^{(1)L}(\tau) J_{\mu'-\eta}^{(1)L}(0)\rangle .
\end{aligned} \tag{63}$$

(Alternatively one may use the extended diffusion model of Gordon[41] and McClung[42] to avoid the factorisation : we do not do so here.)

Next one supposes that the rotational motion is diffusional in character so that one may employ the standard result[43-46]

$$<[\mathcal{D}^{(K)}_{\xi,-\mu}(\Omega)]^L(\tau)[\mathcal{D}^{(K')}_{\xi',-\mu'}(\Omega)]^L(0)>$$

$$= (-1)^{\xi+\mu}<[\mathcal{D}^{(K)}_{-\xi,\mu}(\Omega)^*]^L(\tau)[\mathcal{D}^{(K')}_{\xi',-\mu'}(\Omega)]^L(0)>$$

$$= \{1/(2K+1)\}(-1)^{\xi+\mu}\delta_{K,K'}\delta_{\xi',-\xi}\delta_{\mu',-\mu}\exp\{-\tau/\tau_K\} \qquad (64)$$

with

$$\tau_K = 1/DK(K+1) = 8\pi r_e^3\eta/kTK(K+1) \qquad (65)$$

and K = 2. Likewise, for the angular momentum correlation function one may use the appropriate modification of (34):

$$<J^{(1)L}_{\mu+\eta}(\tau)J^{(1)L}_{\mu'-\eta}(0)> = \delta_{\mu',-\mu}IkT(-1)^{\mu+\eta}\exp\{-\tau/\tau_J\} \qquad (66)$$

(τ_J being given by (35)). The correlation function in (62) then becomes

$$<(c^{(2)}J)^{(1)L}_{\eta}(\tau)(c^{(2)}J)^{(1)L}_{-\eta}(0)>$$

$$= (3/5)IkT\ \exp\{-\tau/\tau_J-\tau/\tau_2\}\sum_{\mu\xi}(-1)^{\xi+\eta}\tilde{c}^{(2)}_{\xi}\tilde{c}^{(2)}_{-\xi}\begin{pmatrix} 2 & 1 & 1 \\ -\mu & \mu+\eta & -\eta \end{pmatrix}^2$$

$$= (1/5)IkT\ \exp\{-\tau/\tau_{J2}\}(-1)^{\eta}\sum_{\xi}(-1)^{\xi}\tilde{c}^{(2)}_{\xi}\tilde{c}^{(2)}_{-\xi} \qquad (67)$$

where the third equality has drawn on the orthonormality of the 3j-symbols

$$\sum_{m_1}\begin{pmatrix} j_1 & j_2 & j_3 \\ m_1 & m_3-m_1 & m_3 \end{pmatrix}\begin{pmatrix} j_1 & j_2 & j_3' \\ m_1 & m'_3-m_1 & m_3' \end{pmatrix} = (2j_3+1)^{-1}$$

$$\times\ \delta_{j_3j_3'}\delta_{m_3m_3'} \qquad (68)$$

The final component of the last line of (67) will be recognized as a spherical tensor scalar product

$$\sum_{\xi} (-1)^{\xi} \tilde{c}_{\xi}^{(2)} \tilde{c}_{-\xi}^{(2)} = (\tilde{c}^{(2)} \cdot \tilde{c}^{(2)}) = (5/3)(\tilde{C}^{(2)} \cdot \tilde{C}^{(2)}) \qquad (69)$$

and so

$$\langle (c^{(2)}J)_{\eta}^{(1)L}(\tau)(c^{(2)}J)_{-\eta}^{(1)L}(0)\rangle = (1/3)(-1)^{\eta} IkT(\tilde{C}^{(2)} \cdot C^{(2)}) \times \exp\{-\tau/\tau_{J2}\} \qquad (70)$$

where

$$\tau_{J2}^{-1} = \tau_{J}^{-1} + \tau_{2}^{-1} . \qquad (71)$$

Method 2. As the first step in this method one refers the contraction $(c^{(2)}J)^{(1)}$ to the molecular frame and then decouples:

$$\langle (c^{(2)}J)_{\eta}^{(1)L}(\tau)(c^{(2)}J)_{-\eta}^{(1)L}(0)\rangle = \sum_{\mu\mu'} \langle [(\tilde{c}^{(2)}\tilde{J})_{\mu}^{(1)} \mathcal{D}_{\eta\mu}^{(1)}(\Omega)]^{L}(\tau) [(\tilde{c}^{(2)}\tilde{J})_{\mu'}^{(1)} \mathcal{D}_{-\eta\mu'}^{(1)}(\Omega)]^{L}(0)\rangle$$

$$= 3 \sum_{\mu\mu'} \sum_{\xi\xi'} \tilde{c}_{-\xi}^{(2)} \tilde{c}_{-\xi'}^{(2)} \begin{pmatrix} 2 & 1 & 1 \\ -\xi & \xi+\mu & -\xi \end{pmatrix} \begin{pmatrix} 2 & 1 & 1 \\ -\xi' & \xi'+\mu' & -\mu \end{pmatrix}$$

$$\langle [\mathcal{D}_{\eta,\mu}^{(1)}(\Omega)\tilde{J}_{\xi+\mu}^{(1)}]^{L}(\tau) [\mathcal{D}_{-\eta,\mu'}^{(1)}(\Omega)\tilde{J}_{\xi'+\mu'}^{(1)}]^{L}(0)\rangle . \qquad (72)$$

In order to deal with the bivariate correlation function one again assumes that the angular momentum and orientation are uncorrelated and writes

$$\langle [\mathcal{D}_{\eta\mu}^{(1)}(\Omega)\tilde{J}_{\xi+\mu}^{(1)}]^{L}(\tau) [\mathcal{D}_{-\eta,\mu'}^{(1)}(\Omega)\tilde{J}_{\xi'+\mu'}^{(1)}]^{L}(0)\rangle$$

$$\sim \langle [\mathcal{D}_{\eta\mu}^{(1)}(\Omega)]^{L}(\tau) [\mathcal{D}_{-\eta,\mu'}^{(1)}(\Omega)]^{L}(0)\rangle \langle \tilde{J}_{\xi+\mu}^{(1)}(\tau)\tilde{J}_{\xi'+\mu'}^{(1)L}(0)\rangle . \qquad (73)$$

The calculation now proceeds as in Method 1 : one employs (64) for the reorientation correlation function (but with K = 1) and for the angular momentum function one may, if desired, generalise (66) to allow for anisotropy in the damping by writing

$$\langle \tilde{J}^{(1)L}_{\xi+\mu}(\tau)\tilde{J}^{(1)L}_{\xi'+\mu'}(0)\rangle = \delta_{\xi'+\mu',-\xi-\mu}(-1)^{\xi+\mu}I_{\xi+\mu}kT\exp\{-\tau/\tau_{J_{\xi+\mu}}\} . \tag{74}$$

We shall disregard this complication and use (66) as it is written. In this way one obtains

$$\langle (c^{(2)}J)^{(1)L}_{\eta}(\tau)(c^{(2)}J)^{(1)L}_{-\eta}(0)\rangle = (1/3)(-1)^{\eta}IkT(\tilde{C}^{(2)}\cdot\tilde{C}^{(2)})\exp\{-\tau/\tau_{J1}\} \tag{75}$$

where

$$\tau^{-1}_{J1} = \tau^{-1}_{J} + \tau^{-1}_{1} \tag{76}$$

$$\tau_1 = 4\pi r_e^3\eta/kT . \tag{77}$$

The difference between the two methods is the appearance of τ_2 in the first but τ_1 in the second : when $\tau_J \ll \tau_K$ so that $\tau^{-1}_{JK} = \tau^{-1}_{J}+\tau^{-1}_{K} \sim \tau^{-1}_{J}$ the choice of approach is clearly irrelevant, but when τ_J is comparable to or exceeds τ_K the choice is important for then the angular momentum correlation does not dominate the reorientation. One way of deciding which is the correct version is to consider the validity of the factorisation of the bivariate correlation function as represented by (63) or (73). In the former one must suppose that the angular momentum components in the laboratory frame are uncorrelated with the orientation; in the latter one supposes that the angular momentum components in the molecular (rotating) frame are uncorrelated with the orientation. I am inclined to think that the better approximation is to suppose that the reorientation is more completely decoupled from the angular momentum components in the molecular frame than from the components in the laboratory frame, because whereas a change of orientation of the molecule implies a change of the projection of J on a laboratory fixed frame and a change in the projection does not imply a re-orientation, when the projections refer to the molecular frame neither implication applies; but the problem needs closer investigation and it would be helpful if the factorisations in (63) and (73) were more fully examined for example by McClung's method[42] or in 2-dimensions. From now on we shall suppose that $\tau_J << \tau_K$ (angular momenta are quenched very rapidly) and take $\tau_{JK} \sim \tau_J$ and thereby avoid the factorisation ambiguity.

The correlation function of (58) has therefore been found to have the approximate form

$$g_\eta(\tau) \sim (-1)^\eta(IkT/h^2)\{(C^{(0)}\cdot C^{(0)}) + \tfrac{1}{3}(\tilde{C}^{(2)}\cdot\tilde{C}^{(2)}\}\exp\{-\tau/\tau_J\} \tag{78}$$

and the only remaining problem is the determination of the tensors $C^{(0)}$ and $\tilde{C}^{(2)}$. If the principal axes of C are x, y and z then its scalar component is

$$C^{(0)} = \tfrac{1}{3}\,\mathrm{tr}\,\tilde{C} = \tfrac{1}{3}(C_{xx} + C_{yy} + C_{zz}) \tag{79}$$

and the components of $\tilde{C}^{(2)}$ are[29]

$$\begin{aligned}\tilde{C}^{(2)}_2 &= (1/\sqrt{6})\{2C_{zz} - C_{xx} - C_{yy}\}\\ \tilde{C}^{(2)}_{\pm 1} &= 0\\ \tilde{C}^{(2)}_{\pm 2} &= (1/2)\{C_{xx} - C_{yy}\}\end{aligned} \tag{80}$$

It is a convenient simplification to assume that the molecule is axially symmetric with $C_{zz} = C_{||}$ and $C_{xx} = C_{yy} = C_{\perp}$: then the only non-zero component of $\tilde{C}^{(2)}$ is

$$\tilde{C}_0^{(2)} = (2/3)^{\frac{1}{2}}(C_{||} - C_{\perp}) \tag{81}$$

and so

$$(C^{(0)}\cdot C^{(0)}) + \tfrac{1}{3}(\tilde{C}^{(2)}\cdot\tilde{C}^{(2)}) = \tfrac{1}{3}(C_{||}^2 + 2C_{\perp}^2) . \tag{82}$$

Consequently

$$g_{\eta}(\tau) = \tfrac{1}{3}(-1)^{\eta}(IkT/\hbar^2)\{C_{||}^2 + 2C_{\perp}^2\}\exp\{-\tau/\tau_J\} \tag{83}$$

and the line width will be determined by

$$T_2^{-1} = \{(C_{||}^2 + 2C_{\perp}^2)IkT/3\hbar^2\}\tau_J\left\{1 + \frac{1}{1+\omega_Z^2\tau_J^2}\right\} \tag{84}$$

as the required generalisation of (37) : T_1^{-1} is given by the analogous modification of (38).

Although this section has examined the geometrical structure of the problem it has used Hubbard's Langevin approach to the dynamical problem, and this is not wholly satisfactory. Therefore we now turn to the dynamical problem.

XI.4 THE DYNAMICAL PROBLEM

One aspect of the dynamics of the system that has been omitted so far is the precessional motion of the radical if it is an asymmetric top : when the torques exerted by the environment are weak the evolution of the angular momentum components in time may be determined by the intrinsic asymmetry of the molecule[36,45,46]. This may be illustrated by writing the 'lattice' hamiltonian (which determines the evolution of J) as a sum of two parts:

$$H_L = H_{L,ex} + H_R \quad . \tag{85}$$

$H_{L,ex}$ is the hamiltonian for the molecules other than the radical of interest (and their interaction with the radical), and H_R is the rotational hamiltonian of the radical which we take to be asymmetric with principal moments of inertia I_{xx}, I_{yy}, I_{zz} :

$$H_R = \sum_q (1/2\hbar I_{qq})\tilde{J}_q^2 . \tag{86}$$

$H_{L,ex}$ we write as the sum of a translational part H_T and an interaction part which represents the influence of the solvent on the solute radical:

$$H_{L,ex} = H_T + H_{int} \quad . \tag{87}$$

H_{int} and H_T do not commute for the translational motion of the solvent modulates its interaction with the solute.

The equation of motion for the angular momentum in the molecular frame is

$$\begin{aligned} \dot{\tilde{J}}_q &= (i/\hbar)H_L^{\times}\tilde{J}_q \\ &= (i/\hbar)H_{int}^{\times}\tilde{J}_q + (i/\hbar)H_R^{\times}\tilde{J}_q \end{aligned} \tag{88}$$

The first term is just the torque, T_q, exerted on the solute by the solvent, and direct calculation enables second term to be calculated (one must observe that (88) implies the use of commutation relations for momenta referred to the rotating frame; these differ in their sign from the relations referred to the fixed frame[49]). Therefore one obtains

$$\tilde{J}_q = T_q + (i/2\hbar^2) \sum_{q'} (1/I_{q'q'})\tilde{J}^{2\times}_{q'}\tilde{J}_q$$

$$= T_q + \{I_{qq}/I_{q'q'}I_{q''q''}kT\}^{\frac{1}{2}} \Delta_q\{\tilde{J}_{q'},\tilde{J}_{q''}\} \tag{89}$$

with

$$\Delta_q = (I_{q'q'} - I_{q''q''}) \left\{ \frac{kT}{I_{xx}I_{yy}I_{zz}} \right\}^{\frac{1}{2}} . \tag{90}$$

This result[36,48] displays clearly the importance of precessional terms when the torques are small and the molecular asymmetry large.

One approach to the calculation of the correlation function for such a system is to use the Baker-Hausdorff formula (see eqn. VII.2.14) in the following way[36]. We wish to calculate the correlation function $<\tilde{J}^L_q(t)\tilde{J}_{q'}>$ where

$$\tilde{J}^L_q(t) = \exp\{(it/\hbar)[H_{L,ex} + H_R]^\times\}\tilde{J}_q . \tag{91}$$

If it is assumed that precessional effects are weak but not negligible one may write

$$<\tilde{J}^L_q(t)\tilde{J}_{q'}> = <\tilde{J}^{L,ex}_q(t)\tilde{J}_{q'}>$$

$$+ (i/\hbar)\int_0^t dt_1 <\{H^{L,ex}_R(t_1)^\times \tilde{J}^{L,ex}_q(t)\}\tilde{J}_{q'}>$$

$$+ (i/\hbar)^2\int_0^t dt_1\int_0^{t_1} dt_2 <\{H^{L,ex}_R(t_2)^\times H^{L,ex}_R(t_1)^\times \tilde{J}^{L,ex}_q(t)\}\tilde{J}_{q'}> \tag{92}$$

where

$$H^{L,ex}_R(t) = \exp\{(it/\hbar)H^\times_{L,ex}\}H_R \tag{93}$$

and

$$\tilde{J}_q^{L,ex}(t) = \exp\{(it/\hbar)H^{\times}_{L,ex}\}\tilde{J}_q \ . \tag{94}$$

The term linear in H_R vanishes under the ensemble average, and the first term is the angular momentum correlation function due to intermolecular torques. The remaining term contains the effects of precession and is difficult to evaluate. One may, however, make the assumption[36] that the correlation time for $H_R^{L,ex}(t)$ is much less than that for $\tilde{J}_q$ so that

$$\begin{aligned}
&\langle\{H_R^{L,ex}(t_2)^{\times}H_R^{L,ex}(t_1)^{\times}\tilde{J}_q^{L,ex}(t)\}\tilde{J}_{q'}\rangle \\
&= -\langle\{H_R^{L,ex}(t_1)^{\times}\tilde{J}_q^{L,ex}(t)\}\{H_R^{L,ex}(t_2)^{\times}\tilde{J}_{q'}\}\rangle \\
&= -\langle\{H_R^{L,ex}(t_1-t_2)^{\times}\tilde{J}_q^{L,ex}(t-t_2)\}\{H_R^{L,ex}(0)^{\times}\tilde{J}_{q'}^{L,ex}(-t_2)\}\rangle \\
&\sim -\langle\{H_R^{L,ex}(t_1-t_2)^{\times}\tilde{J}_q^{L,ex}(t_1-t_2)\}\{H_R^{L,ex}(0)^{\times}\tilde{J}_{q'}^{L,ex}(0)\}\rangle \\
&= -(\hbar/i)^2\langle\dot{\tilde{J}}_q^{L,ex}(t_1-t_2)\dot{\tilde{J}}_{q'}^{L,ex}(0)\rangle
\end{aligned} \tag{95}$$

Consequently

$$\langle\tilde{J}_q^L(t)\tilde{J}_{q'}\rangle \sim \langle\tilde{J}_q^{L,ex}(t)\tilde{J}_{q'}\rangle - \int_0^t d\tau(t-\tau)\langle\dot{\tilde{J}}_q^{L,ex}(\tau)\dot{\tilde{J}}_{q'}^{L,ex}\rangle \tag{96}$$

and from (89)

$$\langle\dot{\tilde{J}}_q^{L,ex}(\tau)\dot{\tilde{J}}_q^{L,ex}\rangle = \left\{\frac{I_{qq}\Delta_q^2}{I_{q'q'}I_{q''q''}kT}\right\}\langle\{\tilde{J}_{q'},\tilde{J}_{q''}\}^{L,ex}(\tau)\{\tilde{J}_{q'},\tilde{J}_{q''}\}^{L,ex}(0)\rangle. \tag{97}$$

The first term in (96) can be obtained by a further use of an analogue of (92)

$$\langle\tilde{J}_q^{L,ex}(t)\tilde{J}_q\rangle = \langle\tilde{J}_q^T(t)\tilde{J}_q\rangle$$

$$+ (i/\hbar)^2\int_o^t dt_1 \int_o^t dt_2 \langle\{H_{int}^T(t_2)^\times H_{int}^T(t_1)^\times \tilde{J}_q^T(t)\}\tilde{J}_q\rangle \quad (98)$$

where

$$H_{int}^T(t) = \exp\{(it/\hbar)H_T^\times\}H_{int} \quad (99)$$

and

$$\tilde{J}_q^T(t) = \exp\{(it/\hbar)H_T^\times\}\tilde{J}_q = \tilde{J}_q \quad (100)$$

because H_T and $\tilde{J}_q$ commute. It is now possible to cast (98) into the form

$$\langle\tilde{J}_q^{L,ex}(t)\tilde{J}_q\rangle = \langle\tilde{J}_q\tilde{J}_q\rangle - (i/\hbar)^2\int_o^t d\tau(t-\tau)\langle\{H_{int}^T(\tau)^\times\tilde{J}_q\}\{H_{int}^{(T)}(0)^\times\tilde{J}_q\}\rangle$$

$$= \langle\tilde{J}_q\tilde{J}_q\rangle - \int_o^t d\tau(t-\tau)\langle T_q^T(\tau)T_q^T(0)\rangle \ . \quad (101)$$

Now (97) and (101) may be combined to give the total correlation function for the angular momentum as

$$\langle\tilde{J}_q^L(t)\tilde{J}_q\rangle = \langle\tilde{J}_q\tilde{J}_q\rangle\Bigg\{1 - \int_o^t d\tau(t-\tau)\left[\frac{\langle T_q^T(\tau)T_q^T(0)\rangle}{\langle\tilde{J}_q^2\rangle}\right]$$

$$- \int_o^t d\tau(t-\tau)\left[\frac{I_{qq}\Delta_q^2}{I_{q'q'}I_{q''q''}kT}\right]\left[\frac{\langle\{\tilde{J}_{q'},\tilde{J}_{q''}\}^{L,ex}(\tau)\{\tilde{J}_{q'},\tilde{J}_{q''}\}\rangle^{L,ex}}{\langle\tilde{J}_q^2\rangle}\right]\Bigg\}$$

$$\sim \langle\tilde{J}_q\tilde{J}_q\rangle\left\{1 - t\int_o^\infty d\tau\left[\frac{\langle T_q^T(\ )T_q\rangle}{\langle\tilde{J}_q^2\rangle}\right]\right.$$

$$- t \int_0^\infty d\tau \Delta_q^2 \left[\frac{\langle \{\tilde{J}_{q'},\tilde{J}_{q''}\}^{L,ex}(\tau)\{\tilde{J}_{q'},\tilde{J}_{q''}\}\rangle}{\langle \tilde{J}_{q'}^2\rangle\langle \tilde{J}_{q''}^2\rangle} \right]\Bigg\}$$

$$\sim \langle \tilde{J}_q \tilde{J}_q \rangle \exp\{-t/\tau_{J_q}\} \; , \tag{102}$$

with[36]

$$\tau_{J_q}^{-1} = \int_0^\infty d\tau \{\langle T_q^T(\tau) T_q \rangle / \langle \tilde{J}_q^2 \rangle\}$$
$$+ \Delta_q^2 \int_0^\infty d\tau \{\langle \{\tilde{J}_{q'},\tilde{J}_{q''}\}^{L,ex}(\tau)\{\tilde{J}_{q'},\tilde{J}_{q''}\}\rangle / \langle \tilde{J}_{q'}^2\rangle\langle \tilde{J}_{q''}^2\rangle\} . \tag{103}$$

Next one assumes that $\tilde{J}_{q'}^{L,ex}$ and $\tilde{J}_{q''}^{L,ex}$ are uncorrelated so that

$$\int_0^\infty d\tau \left\{ \frac{\langle \{\tilde{J}_{q'},\tilde{J}_{q''}\}^{L,ex}(\tau)\{\tilde{J}_{q'},\tilde{J}_{q''}\}\rangle}{\langle \tilde{J}_{q'}^2\rangle\langle \tilde{J}_{q''}^2\rangle} \right\}$$

$$\int_0^\infty d\tau \left\{ \frac{\langle \tilde{J}_{q'}^{L,ex}(\tau)\tilde{J}_{q'}\rangle}{\langle \tilde{J}_{q'}^2\rangle} \right\} \left\{ \frac{\langle \tilde{J}_{q''}^{L,ex}(\tau)\tilde{J}_{q''}\rangle}{\langle \tilde{J}_{q''}^2\rangle} \right\}$$

$$\sim \int_0^\infty d\tau \, \exp\{-\tau(\tau_{J_{q'}}^{-1} + \tau_{J_{q''}}^{-1})\} = \tau_{J_{q'}} \tau_{J_{q''}} / (\tau_{J_{q'}} + \tau_{J_{q''}}) . \tag{104}$$

Thus the result for the angular momentum correlation function that this implies is[36]

$$\tau_{J_q}^{-1} = \int_0^\infty d\tau \langle T_q^T(\tau) T_q \rangle / \langle \tilde{J}_q^2 \rangle + \Delta_q^2 \tau_{J_{q'}} \tau_{J_{q''}} / (\tau_{J_{q'}} + \tau_{J_{q''}}) \; . \tag{105}$$

Kivelson _et al._[36,48] have introduced the parameter κ_{qq} by the relation

$$\kappa_{qq} = (3/4r_e^2) \int_0^\infty d\tau \langle T_q^T(\tau) T_q \rangle / \int_0^\infty d\tau \langle F_q^T(\tau) F_q \rangle \tag{106}$$

(where F_q is the component of the force on the solute radical), and it requires only a simple calculation to show that the correlation time for linear momentum is

$$\tau_c^{-1} = \int_0^\infty d\tau \langle F_q^T(t) F_q \rangle / \langle P_q^2 \rangle \tag{107}$$

Combining (105) - (107) one obtains[36]

$$\tau_{J_q}^{-1} = \tfrac{4}{3} r_e^2 \kappa_{qq} \tau_{P_q}^{-1} \{\langle P_q^2 \rangle / \langle \tilde{J}_q^2 \rangle\} + \Delta_q^2 \tau_{J_{q'}} \tau_{J_{q''}} / (\tau_{J_{q'}} + \tau_{J_{q''}})$$

$$= \tfrac{4}{3} r_e^2 \kappa_{qq} \tau_c^{-1} (m/I_{qq}) + \Delta_q^2 \tau_{J_{q'}} \tau_{J_{q''}} / (\tau_{J_{q'}} + \tau_{J_{q''}})$$

$$= (kT/D)(4r_e^2/3I_{qq})\kappa_{qq} + \Delta_q^2 \tau_{J_{q'}} \tau_{J_{q''}} / (\tau_{J_{q'}} + \tau_{J_{q''}}) \tag{108}$$

because

$$D = (kT/m)\tau_c . \tag{109}$$

When $\kappa_{qq} = 1$ and precessional effects are absent ($\Delta_q = 0$) the correlation time is the same as that given by Hubbard's theory[7], and a more complete calculation[48] has demonstrated that $0<\kappa<1$ for a reasonable model of the intermolecular potential. Experimental analysis of the residual line width of chlorine dioxide in various solvents[36] gives values of κ in the range 0.0178 to 0.372, but it is unfortunate that κ still remains an empirical parameter. Precessional effects may be important, especially when κ is small (so that the intermolecular torques are weak), and the second term in (108) accounts for the deviation from linearity in the dependence of α'' on

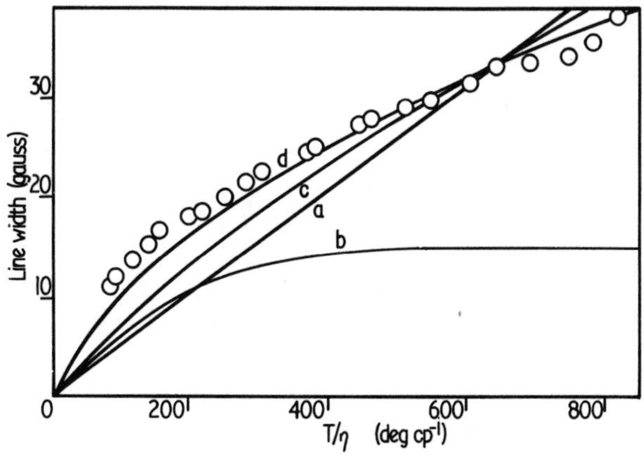

Figure 2. The variation of the spin rotational line width for ClO_2 in n-pentane[36]: a) no reorientation or precession contributions; b) no reorientation contribution; c) no precession contribution; d) precession and orientation effects included.

T/η : a typical experimental analysis is illustrated in Fig. 2[36].

A more detailed approach to the problem of the calculation of the angular momentum correlation functions is that of Kivelson *et al.*[48] which is based on the use of projection operators[50-52]. To proceed with the calculation we first normalise the angular momentum correlation function and write

$$\gamma_q(\tau) = \langle \tilde{J}_q^L(\tau)\tilde{J}_q^L(0)\rangle / \langle \tilde{J}_q^2\rangle \ . \tag{110}$$

If one writes

$$\phi_q(\tau) = \langle \dot{\tilde{J}}_q^L(\tau)\dot{\tilde{J}}_q^L(0)\rangle / \langle \tilde{J}_q^2\rangle \tag{111}$$

$$\ddot{\gamma}_q(\tau) = -\,\phi_q(\tau) \ . \qquad (\gamma_q(0) = 1). \tag{112}$$

Laplace transforms we write $\tilde{\gamma}_q(s)$ and $\tilde{\phi}_q(s)$; therefore the Laplace transform of (112) yields

$$s^2\tilde{\gamma}_q(s) - s = -\tilde{\phi}(s) \tag{113}$$

or[50]

$$s\tilde{\gamma}_q(s) - 1 = -\tilde{K}(s)\tilde{\gamma}(s) \tag{114}$$

where

$$\tilde{K}(s) = \{1 - (1/s)\tilde{\phi}(s)\}^{-1}\tilde{\phi}(s) . \tag{115}$$

Rearrangement of (114) yields

$$\gamma_q(\tau) = \mathcal{L}^{-1}[\{s + \tilde{K}(s)\}^{-1}] \tag{116}$$

and so if $K(\tau)$ is known $\gamma_q(\tau)$ may be calculated. The correlation time τ_{J_q} is given by

$$\tau_{J_q} = \int_0^\infty d\tau\gamma_q(\tau) = \tilde{\gamma}_q(0) = \tilde{K}(0)^{-1} \tag{117}$$

and so τ_{J_q} is known if $K(\tau)$ is known.

The method used by Berne et al.[52] and Kivelson et al.[40], is to introduce the projection operators (see Chapter IV)

$$P = \tilde{J}_q \mathrm{tr}\rho\tilde{J}_q = \tilde{J}_q\rangle\langle\tilde{J}_q \qquad Q = 1-\tilde{J}_q\mathrm{tr}\rho\tilde{J}_q = 1-\tilde{J}_q\rangle\langle\tilde{J}_q \tag{118}$$

(we have modified their notation[28] and for simplicity of notation we have assumed that $\langle\tilde{J}_q{}^2\rangle = 1$) and to observe that

$$\tilde{\phi}_q(s) = \mathcal{L}\langle\dot{\tilde{J}}_q \exp\{(i\tau/\hbar)H_L^\times\}\dot{\tilde{J}}_q\rangle$$

$$= \langle\dot{\tilde{J}}_q\{s - (i/\hbar)H_L^\times\}^{-1}\dot{\tilde{J}}_q\rangle. \tag{119}$$

Using the identity

$$a^{-1} = b^{-1} + a^{-1}(b - a)b^{-1} \tag{120}$$

with $a = s - (i/\hbar)H_L^\times$ and $b = s - (i/\hbar)Q\cdot H_L^\times$ the preceding equation becomes

$$\tilde{\phi}_q(s) = \langle\dot{\tilde{J}}_q\{s-(i/\hbar)Q\cdot H_L^\times\}^{-1}\dot{\tilde{J}}_q\rangle + \langle\dot{\tilde{J}}_q\{s-(i/\hbar)H_L^\times\}^{-1}(i/\hbar)PH_L^\times \{s-(i/\hbar)QH_L^\times\}^{-1}\dot{\tilde{J}}_q\rangle$$

$$= \langle\dot{\tilde{J}}_q\{s-(i/\hbar)QH_L^\times\}^{-1}\dot{\tilde{J}}_q\rangle + \langle\dot{\tilde{J}}_q\{s-(i/\hbar)H_L^\times\}^{-1}\tilde{J}_Q\rangle\langle\tilde{J}_q(i/\hbar)H_L^\times \{s-(i/\hbar)QH_L^\times\}^{-1}\dot{\tilde{J}}_q\rangle$$

$$= \langle\dot{\tilde{J}}_q\{s-(i/\hbar)QH_L^\times\}^{-1}\dot{\tilde{J}}_q\rangle\{1 - (1/s)\langle\tilde{J}_q\{s - (i/\hbar)H_L^\times\}^{-1}\dot{\tilde{J}}_q\rangle\}$$

$$= \langle\dot{\tilde{J}}_q\{s-(i/\hbar)QH_L^\times\}^{-1}\dot{\tilde{J}}_q\rangle\{1 - (1/s)\tilde{\phi}_q(s)\}. \tag{121}$$

Comparing this with (115) one finds that

$$\tilde{K}(s) = \langle\dot{\tilde{J}}_q\{s-(i/\hbar)QH_L^\times\}^{-1}\dot{\tilde{J}}_q\rangle \tag{122}$$

and so

$$\tau_{J_q}^{-1} = \langle\dot{\tilde{J}}_q\{s-(i/\hbar)QH_L^\times\}^{-1}\dot{\tilde{J}}_q\rangle/\langle\tilde{J}_q^2\rangle\Big|_{s=o}$$

$$\equiv\{\dot{\tilde{J}}_q|\dot{\tilde{J}}_q\}/\langle\tilde{J}_q^2\rangle . \tag{123}$$

One may now proceed rapidly to the principal intermediate result by using (89): this yields[48]

$$\tau_{J_q}^{-1} = (kTI_{qq})^{-1}\{T_q|T_q\}$$

$$+ (1/kT)^{\frac{3}{2}}(I_{xx}I_{yy}I_{zz})^{-\frac{1}{2}}\Delta_q[\{\{\tilde{J}_{q'},\tilde{J}_{q''}\}|T_q\} + \{T_q|\{\tilde{J}_{q'},\tilde{J}_{q''}\}]$$

$$+ (I_{q'q'}I_{q''q''}k^2T^2)^{-\frac{1}{2}}\Delta_q^2\{\{\tilde{J}_{q'},\tilde{J}_{q''}\}|\{\tilde{J}_{q'},\tilde{J}_{q''}\}\}. \tag{124}$$

The correlation time for linear momentum is

$$\tau_c^{-1} = \{\dot{P}_q|\dot{P}_q\}/\langle P_q^2\rangle = (mkT)^{-1}\{F_q|F_q\} \tag{125}$$

and so by writing

$$\kappa_q = (3/4r_e^2)\{T_q|T_q\}/\{F_q|F_q\} \tag{126}$$

$$\tau_{q'q''} = \{\{\tilde{J}_{q'},\tilde{J}_{q''}\}|\{\tilde{J}_{q'},\tilde{J}_{q''}\}\}/I_{q'q'}I_{q''q''}k^2T^2 \tag{127}$$

$$\lambda_q = \left\{\frac{\{\{\tilde{J}_{q'},\tilde{J}_{q''}\}|T_q\} + \{T_q|\{\tilde{J}_{q'},\tilde{J}_{q''}\}\}}{2\Delta_q\tau_{q'q''}(kT)^{3/2}(I_{xx}I_{yy}I_{zz})^{\frac{1}{2}}}\right\} \tag{128}$$

one obtains[48]

$$\tau_{J_q}^{-1} = (8\pi\eta r_e^3\kappa_q/I_{qq}) + \Delta_q^2\tau_{q'q}\lambda_{q''} + \Delta_q^2\tau_{q'\,q''}. \tag{129}$$

Kivelson _et al_.[48] proceed to show that

$$\tau_{q'q''}^{-1} \sim \tau_{J_{q'}}^{-1} + \tau_{J_{q''}}^{-1} \tag{130}$$

$$4\kappa_q r_e^2/3 \sim \langle T_q^2\rangle/\langle F_q^2\rangle \tag{131}$$

$$\lambda_q \sim 0 \tag{132}$$

and so finally they obtain

$$\tau_{J_q}^{-1} = (8\pi\eta r_e^3\kappa_q/I_{qq}) + \Delta_q^2\tau_{J_{q'}}\tau_{J_{q''}}/(\tau_{J_{q'}} + \tau_{J_{q''}}) \tag{133}$$

which is the same as (108).

Although this approach (which is developed in greater detail in the original paper[48]) has had reasonable success, recent work[53] has demonstrated that it is not always applicable.

In the Brown, Gutowsky, Shimomura model of spin rotation interactions in n.m.r.[54] it is supposed that the interaction is zero except during its flight into a new orientation when it rises suddenly to some random value at which it remains constant during the duration Δ of the flight. For a scalar interaction the strength of the perturbation is proportional to J_q and so we consider the angular momentum to be the modulated quantity. Following Brown *et al.*[54], when $\tau > \Delta$

$$\langle J_q(0)J_q(\tau)\rangle = 0 \tag{134}$$

and when $\tau < \Delta$

$$\begin{aligned}\langle J_q(0)J_q(\tau)\rangle &= \langle J_q^2\rangle \times \{\text{probability that } J_q(0) \neq 0\} \\ &\quad \times \{\text{probability that } J_q(\tau) \neq 0 \text{ assuming } J_q(0) \neq 0\} \\ &= \langle J_q^2\rangle P_1 P_2 \end{aligned} \tag{135}$$

where P_1 is the probability that a flight began between $t - \Delta$ and t and P_2 is the conditional probability that the beginning of the flight assumed to have occurred between t - Δ and t actually occurred between t - Δ + τ and t (so that the flight is still in progress at t + τ) :

$$P_1P_2 = \{(\Delta/\tau_J')\exp(-\Delta/\tau_J')\}\{(\Delta-|\tau|)/\Delta\} \tag{136}$$

where τ_J' is the average interval between starts of flights. If a flight is brief (in the sense $\Delta << \tau_J'$)

$$\langle J_q(0)J_q(\tau)\rangle \sim \langle J_q^2\rangle\{(\Delta-|\tau|)/\tau_J\} . \tag{137}$$

The spectral density is therefore

$$j(\omega) = \int_{-\infty}^{\infty} d\tau\langle J_q^2\rangle\{(\Delta - |\tau|)/\tau_J'\}\exp\{i\omega\tau\}$$

$$= 2\langle J_q^2\rangle \int_0^{\Delta} d\tau\left\{\frac{\Delta - \tau}{\tau_J'}\right\} \cos\omega\tau$$

$$= 2\langle J_q^2\rangle\{(1 - \cos\omega\Delta)/\omega^2\tau_J'\} . \tag{138}$$

In the limit $\omega\tau_J' << 1$, which implies $\omega\Delta << 1$, this yields

$$j(\omega) \sim \langle J_q^2\rangle\{\Delta^2/\tau_J'\} . \tag{139}$$

(It is noted[54] that if $\Delta = \tau_J'$ the model is identical to the description of spin-rotation interactions in gases, Δ being the time between events which change the molecule from one rotational state to another.) The number of events that affect the angular momentum might decrease with the viscosity, and so it is plausible to suppose that $\tau_J' \propto \eta$; in this way $j(\omega)$, and hence T_2^{-1}, is proportional, in this model, to T/η, as required. The frequency of events that enable a flight to occur may be profoundly affected by the structure of the liquid, particularly the polarity of the molecules, and in some cases it might be necessary to suppose that some kind of defect (for example a hole) must be formed at the site of the radical before a flight can occur : this problem of conditional events has been treated in several places[55-58].

APPENDIX

The properties of spherical tensors are fully discussed in several places[27,28], but some properties are collected below for convenience. Useful properties of the 3j-symbols[57] and the Wigner rotation matrices are also included.

1. 3j-symbols

Symmetry:
$$\begin{pmatrix} j_1 & j_2 & j_3 \\ m_1 & m_2 & m_3 \end{pmatrix} = \begin{pmatrix} j_1 & j_3 & j_2 \\ m_1 & m_3 & m_2 \end{pmatrix} (-1)^{j_1+j_2+j_3}, \text{ etc.}$$

$$= \begin{pmatrix} j_1 & j_2 & j_3 \\ -m_1 & -m_2 & -m_3 \end{pmatrix} (-1)^{j_1+j_2+j_3}$$

$$= 0 \text{ unless } \begin{cases} \Delta(j_1 j_2 j_3) \\ m_1+m_2+m_3 = 0 \end{cases} .$$

Orthogonality:
$$\sum_{j_3 m_3} (2j_3+1) \begin{pmatrix} j_1 & j_2 & j_3 \\ m_1 & m_2 & m_3 \end{pmatrix} \begin{pmatrix} j_1 & j_2 & j_3 \\ m_1' & m_2' & m_3 \end{pmatrix} = \delta_{m_1 m_1'} \delta_{m_2 m_2'}$$

$$\sum_{m_1 m_2} \begin{pmatrix} j_1 & j_2 & j_3 \\ m_1 & m_2 & m_3 \end{pmatrix} \begin{pmatrix} j_1 & j_2 & j_3' \\ m_1 & m_2 & m_3' \end{pmatrix} = (2j+1)^{-1} \delta_{j_3 j_3'} \delta_{m_3 m_3'} .$$

Numerical values : see reference 57.

2. Spherical tensors

Definition
$$J_z^{\times} T_q^{(K)} = q T_q^{(K)}$$

$$J_{\pm}^{\times} T_q^{(K)} = \{K(K+1) - q(q\pm 1)\}^{\frac{1}{2}} T_{q\pm 1}^{(K)} .$$

Scalar product:
$$(T^{(K)} \cdot T^{(K)}) = \sum_q (-1)^q T_q^{(K)} T_{-q}^{(K)} .$$

Tensor product:

$$T_q^{(K)} = (2K+1)^{\frac{1}{2}} \sum_{q_1 q_2} (-1)^{-K_1+K_2-q} \begin{pmatrix} K_1 & K_2 & K \\ q_1 & q_2 & -q \end{pmatrix} T_{q_1}^{(K_1)} T_{q_2}^{(K_2)} .$$

Transformation properties:

$$T_q^{(K)} = \sum_{q'} T_{q'}^{(K)} \mathscr{D}_{q'q}^{(K)}(\Omega) .$$

3. Rotation matrices

Definition:
$$\mathscr{D}_{q'q}^{(K)}(\Omega) = \langle Kq' | \mathscr{D}(\Omega) | Kq \rangle$$

$$\mathscr{D}(\Omega) = \exp\{-i\alpha J_z\} \exp\{-i\beta J_y\} \exp\{-i\gamma J_z\}$$

Symmetry:
$$\mathscr{D}_{q'q}^{(K)}(\alpha,\beta,\gamma)^* = (-1)^{q'-q} \mathscr{D}_{-q',-q}^{(K)}(\alpha,\beta,\gamma) = \mathscr{D}_{qq'}^{(K)}(-\gamma,-\beta,-\alpha)$$

Orthogonality:
$$\frac{1}{8\pi^2} \int d\Omega \mathscr{D}_{q_1' q_1}^{(K_1)}(\Omega)^* \mathscr{D}_{q_2' q_2}^{(K_2)}(\Omega) = (2K_1+1)^{-1} \delta_{K_1 K_2} \delta_{q_1' q_2'} \delta_{q_1 q_2}$$

$$\frac{1}{8\pi^2}\int d\Omega\, \mathscr{D}^{(K_1)}_{q_1'q_1}(\Omega)\, \mathscr{D}^{(K_2)}_{q_2'q_2}(\Omega)\, \mathscr{D}^{(K_3)}_{q_3'q_3}(\Omega) = \begin{pmatrix} K_1 & K_2 & K_3 \\ q_1' & q_2' & q_3' \end{pmatrix}\begin{pmatrix} K_1 & K_2 & K_3 \\ q_1 & q_2 & q_3 \end{pmatrix}$$

Products:

$$\mathscr{D}^{(K_1)}_{q_1'q_1}(\Omega)\, \mathscr{D}^{(K_2)}_{q_2'q_2}(\Omega) = \sum_{Kq'q} (2K+1)\mathscr{D}^{(K)}_{q'q}(\Omega) \begin{pmatrix} K_1 & K_2 & K \\ q_1 & q_2 & q \end{pmatrix}\begin{pmatrix} K_1 & K_2 & K \\ q_1' & q_2' & q \end{pmatrix}$$

REFERENCES

1) D. Kivelson, J. Chem. Phys. 27, 1087 (1957).
2) D. Kivelson, J. Chem. Phys. 33, 1094 (1960).
3) R. Wilson and D. Kivelson, J. Chem. Phys. 44, 154 (1966).
4) H.S. Gutowsky, I.J. Lawrenson and K. Shimomura, Phys. Rev. Letts. 6, 349 (1961).
5) C.S. Johnson, J.S. Waugh and J.N. Pinkerton, J. Chem. Phys. 35, 2020 (1961).
6) C.S. Johnson and J.S. Waugh, J. Chem. Phys. 35, 2020 (1961).
7) M. Bloom and M. Lipsicas, Canad. J. Phys. 39, 881 (1961).
8) J.G. Powles and D.J. Neale, Proc. Phys. Soc. 77, 739 (1961); 78, 377 (1961); 85, 87 (1965).
9) M. Bloom and H.S. Sandhu, Canad. J. Phys. 40, 289 (1962).
10) J.G. Powles and D.K. Green, Phys. Letts. 3, 134 (1962).
11) J.S. Blicharski and K. Krynicki, Acta Phys. Polon. 22, 409 (1962).
12) G.W. Flynn and J.D. Baldeschwieler, J. Chem. Phys. 38, 226 (1962).
13) P.S. Hubbard, Phys. Rev. 131, 1155 (1963).
14) J.H. Rugheimer and P.S. Hubbard, J. Chem. Phys. 39, 552 (1963).
15) K. Krynicki and J.G. Powles, Phys. Letts. 4, 260 (1963).
16) K.F. Kuhlmann and J.D. Baldeschwieler, J. Chem. Phys. 43, 572 (1965).
17) P.S. Hubbard, J. Chem. Phys. 42, 3546 (1965).
18) G.A. deWit and M. Bloom, Canad. J. Phys. 43, 986 (1965).
19) R.H. Faulk and M. Eisner, J. Chem. Phys. 44, 2926 (1966).
20) A.S. Dubin and S.I. Chan, J. Chem. Phys. 46, 4533 (1967).
21) S.I. Chan, J. Chem. Phys. 47, 1191 (1967).
22) M.K. Ahn and C.S. Johnson, J. Chem. Phys. 50, 641 (1969).
23) J. Jonas and T.M. DiGennaro, J. Chem. Phys. 50, 2392 (1969).
24) T.E. Bull and J. Jonas, J. Chem. Phys. 52, 1978 (1970).
25) T.E. Bull, J.S. Barthel and J. Jonas, J. Chem. Phys. 54, 3663 (1971).

26) D.E. O'Reilly, E.M. Peterson, D.L. Hogenboom and C.E. Scheie, J. Chem. Phys. 54, 4194 (1971).
27) A.R. Edmonds, Angular momentum in quantum mechanics, Princeton University Press (1957).
28) B.R. Judd, Operator techniques in atomic spectroscopy, McGraw-Hill, New York (1963).
29) P.W. Atkins, Adv. in Molec. Relaxation Processes, in press.
30) R. Kubo and K. Tomita, J. Phys. Soc. Japan 9, 888 (1954).
31) P.W. Atkins and J.N.L. Connor, Mol. Phys. 13, 201 (1967).
32) P.W. Atkins and D. Kivelson, J. Chem. Phys. 44, 169 (1966).
33) S. Chandrasekhar, Rev. Mod. Phys. 15, 1 (1943).
34) R.F. Curl, Mol. Phys. 9, 585 (1965).
35) R. Wilson and D. Kivelson, J. Chem. Phys. 44, 4445 (1966).
36) R.E.D. McClung and D. Kivelson, J. Chem. Phys. 49, 3380 (1968).
37) G. Nyberg, Mol. Phys. 12, 69 (1967); 17, 87 (1968).
38) L. Burlamacchi, Mol. Phys. 16, 369 (1969).
39) P.W. Atkins, A. Horsfield and M.C.R. Symons, J. Chem. Soc. 5220 (1964).
40) J.Q. Adams, J. Chem. Phys. 45, 4167 (1966).
41) R.G. Gordon, J. Chem. Phys. 44, 1830 (1966).
42) R.E.D. McClung, J. Chem. Phys. 51, 3842 (1969).
43) A. Abragam, The principles of nuclear magnetism, Clarendon Press, Oxford (1961).
44) P. Debye, Polar molecules, The Chemical Catalog Co. Inc., New York 1929.
45) N. Bloembergen, E.M. Purcell and R.V. Pound, Phys. Rev. 73, 679 (1948).
46) W.H. Furry, Phys. Rev. 107, 7 (1957).
47) R.E.D. McClung, Ph.D. Thesis, U.C.L.A. (1967).
48) D. Kivelson, M.G. Kivelson and I. Oppenheim, J. Chem. Phys. 52, 1810 (1970).
49) J.H. Van Vleck, Rev. Mod. Phys. 23, 213 (1951).
50) R. Zwanzig, Lectures in theoretical physics (Vol. III), Interscience, New York (1961).
51) H. Mori, Prog. Theoret. Phys. 34, 399 (1965).
52) B.J. Berne, J.P. Boon and S.A. Rice, J. Chem. Phys. 45, 1086 (1966).
53) R. Poupko, H. Gilboa, B.L. Silver and A. Loewenstein, in press.
54) R.J.C. Brown, H.S. Gutowsky and K. Shimomura, J. Chem. Phys. 38, 76 (1963).
55) P.W. Atkins, Mol. Phys. 17, 321 (1969).
56) P.W. Atkins, A. Loewenstein and Y. Margalit, Mol. Phys. 17, 329 (1969).
57) J.E. Anderson, J. Chem. Phys. 47, 4879 (1967).
58) S.H. Glarum and J.H. Marshall, J. Chem. Phys. 46, 55 (1967).
59) M. Rotenberg, R. Bivins, N. Metropolis, J.K. Wooten, The 3j- and 6j-symbols, Technology Press, M.I.T. (1959).

ELECTRON SPIN RELAXATION IN ^6S STATE IONS

G.R. Luckhurst

Department of Chemistry, The University

Southampton SO5 9NH, England

XII.1 INTRODUCTION

The rotational diffusion of paramagnetic species in dilute solution is often responsible for the relaxation of the electron spin and hence broadening of the lines in the electron resonance spectrum.[1] Of course, spin relaxation can only occur if the magnetic interactions are anisotropic for then, as the molecule rotates, transitions are induced between the spin levels and the energy of these levels are modulated. In the case of doublet state species such as vanadyl acetylacetonate the important anisotropic interactions for relaxation are the electron Zeeman coupling and the electron-nuclear hyperfine interaction.[2] When the components of the g tensor deviate from the free-spin value there may also be significant contributions to the linewidth resulting from spin-rotation relaxation[3] as we saw in Chapter XI. The theory of spin relaxation in doublet states is now reasonably well understood and has provided a novel technique for probing the nature of molecular dynamics in fluids.

When the paramagnetic species contains two or more unpaired electrons another mode of spin relaxation is possible and usually dominant. Even in the absence of a magnetic field the degeneracy of a particular spin multiplet is partially removed by a combination of spin-spin and spin-orbit coupling forces. This zero-field splitting is anisotropic and, when coupled to the molecular rotational diffusion, dominates the spin relaxation processes for transition metal ions in solution.[4] Although the resulting linewidths can be extremely large, their magnitude has usually[5,6] been estimated within the framework of Redfield's relaxation theory[7] which was discussed in Chapter VIII; we also adopt this technique in order to calculate the linewidths. In the next section vector coupling techniques are used to modify the expressions for the elements of the relaxation matrix, and so reduce the labour involved in calculating this matrix.[8] The application of the theory to species with five unpaired electrons in ^6S states is considered in section 3. These calculations are of practical importance since manganese (II) is being employed as a dynamic probe in a wide variety of systems ranging from molten salts[9] to aqueous solutions of nucleotides.[10]

Of course when an ion possesses cubic symmetry the zero-field splitting vanishes and the quartic terms provide the first non-zero spin-spin interaction.[11] The spin relaxation produced by modulation of the quartic terms via rotational diffusion is also discussed, although this mode of relaxation turns out to be of little importance in most systems. In fact, for species such as $Mn^{2+}(H_2O)_6$ the dominant relaxation is almost certainly caused by fluctuations in the zero-field splitting which are induced by collisional distortion of the complex.[12] A simple dynamic model for this process is introduced in section 5 in order to calculate the appropriate correlation functions. In the last two sections we shall be concerned with extending the relaxation theory when the nuclear hyperfine interaction is large.

XII.2. ADAPTION OF REDFIELD'S THEORY

We begin by considering a paramagnetic species with n unpaired electrons and ignore, for the moment, all hyperfine interactions. The scalar spin hamiltonian is then

$$\mathcal{H}^{(o)} = g\beta B S_z \ , \qquad \text{(XII.1.)}$$

where S_z is the z-component of the total electron spin operator. The spin states $|S,m_s\rangle$ are eigenfunctions of $\mathcal{H}^{(o)}$ with eigenvalues

$$E_{|S,m_s\rangle} = g\beta B m_s , \qquad \text{(XII.2.)}$$

where S is the total spin angular momentum and m_s is the component in the z-direction. When the oscillating microwave field is orthogonal to the static magnetic field B only the transitions

$$|m_s\rangle \longleftrightarrow |m_s \pm 1\rangle$$

are allowed and their intensities are proportional to $\{S(S+1)-m_s(m_s \pm 1)\}$. The solution electron resonance spectrum is therefore centred on the resonant field

$$B_r = \hbar\omega_o/g\beta \ , \qquad \text{(XII.3.)}$$

where ω_o is the microwave frequency. The use of species with high spin multiplicities as dynamic probes does not look encouraging since all of the information is contained in a single line. However, as we shall see, the widths of the component lines may differ.

The dynamic perturbation results from fluctuations in the anisotropic electron-electron interaction, and we shall write this perturbation, using irreducible tensor notation,[13] as

$$\mathcal{H}'(t) = \sum_{L;p} (-1)^p F^{(L,p)} T^{(L,-p)} . \qquad \text{(XII.4.)}$$

Here $T^{(L,p)}$ is the pth component of the Lth rank combination of the total electron spin operators and $F^{(L,p)}$ depends only on the spatial variables. The zero-field splitting hamiltonian[4] is obtained when the summation is restricted to terms with L equal to two and the spin hamiltonian with L = 4 represents the quartic terms.[11] Since $\mathcal{H}'(t)$ is written in a spatial coordinate system the spin operators $T^{(L,p)}$ are time independent and the time dependence is restricted to the interaction tensors $F^{(L,p)}$. We shall assume, for the moment, that the fluctuations in $F^{(L,p)}$ originate in the rotational diffusion of the molecule for then this time dependence can be restricted to a Wigner rotation matrix:

$$\mathcal{H}'(t) = \sum_{L;p,q} (-1)^p F'^{(L,q)} \mathcal{D}^{(L)}_{q,p}(t) T^{(L,-p)} , \qquad (\text{XII.5.})$$

where the prime denotes the value in a molecular coordinate system.

The linewidths may be calculated from Redfield's relaxation matrix provided certain conditions concerning the magnitude of the perturbation and the rate of fluctuations are satisfied.[7] The elements of this relaxation matrix are related to the spectral densities J by

$$R_{\varkappa\varkappa',\lambda\lambda'} = 2J_{\varkappa\lambda,\varkappa'\lambda'}(\varkappa-\lambda) - \sum_{\gamma} \delta_{\varkappa'\lambda'} J_{\gamma\lambda,\gamma\varkappa}(\gamma-\lambda) - \sum_{\gamma} \delta_{\varkappa\lambda} J_{\gamma\lambda',\gamma\varkappa'}(\gamma-\lambda') , \qquad (\text{XII.6.})$$

where $\delta_{\varkappa\lambda}$ vanishes unless the eigenstates $|\varkappa\rangle$ and $|\lambda\rangle$ are identical (c.f. equation VIII.35.). The spectral densities depend on the matrix elements of the spin operators as well as on the Fourier transforms of the autocorrelation functions of the spatial variables:

$$J_{\varkappa\lambda,\varkappa'\lambda'}(\omega) = \sum_{\substack{L,L';p,p',\\ q,q'}} (-1)^{p+p'} \langle\varkappa|T^{(L,-p)}|\lambda\rangle\langle\varkappa'|T^{(L',-p')}|\lambda'\rangle^* \, F'^{(L,q)} F'^{(L',q')*} \tfrac{1}{2}\int_{-\infty}^{\infty} \overline{\mathcal{D}^{(L)}_{q,p}(0)\mathcal{D}^{(L')*}_{q',p'}(t)} e^{-i\omega t} dt. \qquad (\text{XII.7.})$$

The ensemble average of the autocorrelation function of the rotation matrix may be written as

$$\overline{\mathcal{D}^{(L)}_{q,p}(0)\mathcal{D}^{(L')*}_{q',p'}(t)} = \int \mathcal{D}^{(L)}_{q,p}(\Omega_o)P(\Omega_o)d\Omega_o \int \mathcal{D}^{(L')^*}_{q',p'}(\Omega)P(\Omega_o|\Omega,t)d\Omega, \quad (XII.8.)$$

where Ω_o is the orientation at zero time and Ω that at time t. For a macroscopically isotropic system the angular distribution function is

$$P(\Omega) = 1/8\pi^2, \quad (XII.9.)$$

and for isotropic rotational diffusion the conditional probability is[14]

$$P(\Omega_o|\Omega,t) = \sum_{L,p,q} \frac{2L+1}{8\pi^2} \mathcal{D}^{(L)^*}_{p,q}(\Omega_o)\, \mathcal{D}^{(L)}_{p,q}(\Omega)\exp(-|t|/\tau_L). \quad (XII.10.)$$

The desired ensemble average is then

$$\overline{\mathcal{D}^{(L)}_{q,p}(0)\mathcal{D}^{(L')*}_{q',p'}(t)} = \delta_{LL'}\delta_{pp'}\delta_{qq'}\exp(-|t|/\tau_L), \quad (XII.11)$$

where τ_L is the rotational correlation time appropriate for an Lth rank interaction. In Debye's theory [14] of rotational diffusion these correlation times are related to the diffusion coefficient D by[15]

$$\tau_L = 1/DL(L+1)\ . \quad (XII.12.)$$

The spectral density may now be written in the simplified form:

$$J_{\varkappa\lambda,\varkappa'\lambda'}(\omega) = \sum_{L;p,q} \langle\varkappa|T^{(L,-p)}|\lambda\rangle\langle\varkappa'|T^{(L,-p)}|\lambda'\rangle^* F'^{(L,q)}F'^{(L,q)*}(2L+1)^{-1}j(\omega), \quad (XII.13.)$$

$$\text{where } j(\omega) = \tau_L/(1+\omega^2\tau_L^2)\ . \quad (XII.14.)$$

In general the evaluation of the matrix elements involved in the spectral densities can be both tedious and laborious. However, the effort can be reduced in problems such as this where the eigenstates are of the form $|S,m_s\rangle$ by use of the Wigner-Eckart theorem.[13] This states that the matrix elements $\langle S',m_s'|T^{(L,p)}|S,m_s\rangle$ are related to a reduced matrix element $\langle S'||T^{(L)}||S\rangle$ and the appropriate Clebsch-Gordan coefficient by

$$\langle S',m_s'|T^{(L,p)}|S,m_s\rangle = C(SLS';m_s,-m_s')\langle S'\|T^{(L)}\|S\rangle. \qquad \text{(XII.15.)}$$

Substitution of this relationship into equation (13) gives the spectral density as[8]

$$J_{\varkappa\lambda,\varkappa'\lambda'}(\omega) = \sum_{L;q} C(SLS;\lambda,\varkappa-\lambda)C(SLS;\lambda',\varkappa'-\lambda')|\langle S\|T^{(L)}\|S\rangle|^2$$

$$\times F'^{(L,q)}F'^{(L,q)*}(2L+1)^{-1}j(\omega), \qquad \text{(XII.16.)}$$

which can now be employed to calculate the relaxation matrix. We shall proceed directly to illustrate the use of these equations to describe spin relaxation in ^6S state ions.

XII.3. RELAXATION VIA ROTATIONAL MODULATION OF THE ZERO-FIELD SPLITTING

The spin hamiltonian for the zero-field splitting is usually written in a molecule fixed coordinate system as

$$\mathcal{H} = D(S_a^2 - \tfrac{1}{3}S^2) + E(S_b^2 - S_c^2), \qquad \text{(XII.17.)}$$

where a, b, c are the principal axes. However, for rotational relaxation problems it is more convenient to use the expressions obtained from equations (4) and (5) with L = 2. In this notation the spatial operators $F'^{(2,p)}$ are related to D and E by

$$F'^{(2,0)} = (2/3)^{\frac{1}{2}}D,$$

$$F'^{(2,\pm1)} = 0, \qquad \text{(XII.18.)}$$

and

$$F'^{(2,\pm2)} = E.$$

The irreducible components of the spin operators are

$$T^{(2,0)} = (1/6)^{\frac{1}{2}}\{3S_z^2 - S(S+1)\},$$

$$T^{(2,\pm 1)} = \mp (1/2)(S_{\pm}S_z + S_zS_{\pm}) \,, \qquad \text{(XII.19.)}$$

and $$T^{(2,\pm 2)} = (1/2)S_{\pm}^2 \,.$$

For the sextet state problem the reduced matrix element $\langle 5/2\|T^{(2)}\|5/2\rangle$ is readily calculated from equation (15) by using the known matrix element of the operator $T^{(2,2)}$ which is just $\frac{1}{2}S_+^2$ and so

$$\langle 5/2\|T^{(2)}\|5/2\rangle = \langle 5/2|T^{(2,2)}|\tfrac{1}{2}\rangle / C(5/2\ 2\ 5/2;\tfrac{1}{2},2)$$

$$= (140/3)^{\frac{1}{2}}. \qquad \text{(XII.20.)}$$

The Clebsch-Gordan coefficient used in this calculation was evaluated, like the others, from tables for $C(5/2\ 2\ 5/2;m_1,m_2)$ given by Condon and Shortly.[16] The total relaxation matrix calculated from equations (6) and (16) with L = 2 is straightforward and yields:[5,8]

$$\begin{array}{c} \\ \langle 2| \\ \langle 1| \\ \langle 0| \\ \langle -1| \\ \langle -2| \end{array} \begin{array}{c} \begin{array}{ccccc} |2\rangle & |1\rangle & |0\rangle & |-1\rangle & |-2\rangle \end{array} \\ \left[\begin{array}{ccccc} A & D & E & 0 & 0 \\ D & B & 0 & F & 0 \\ E & 0 & C & 0 & E \\ 0 & F & 0 & B & D \\ 0 & 0 & E & D & A \end{array}\right] \end{array}, \qquad \text{(XII.21.)}$$

where the basis functions $|2\rangle$, $|1\rangle$ etc. represent the five allowed transitions $|5/2\rangle \leftrightarrow |3/2\rangle \leftrightarrow |\tfrac{1}{2}\rangle$ ..... The matrix elements are given, in units of (D:D)/5, by[17]

$$A = -(24J_o + 48J_1 + 28J_2) \,,$$

$$B = -(6J_o + 36J_1 + 46J_2) \,,$$

$$C = -(16J_1 + 56J_2) \,, \qquad \text{(XII.22.)}$$

$$D = 8\sqrt{10}\, J_1,$$

$$E = 12\sqrt{5}\, J_2,$$

and $$F = 36 J_2,$$

where (D:D) denotes the inner product of the zero-field splitting tensor $\sum_q F'^{(2,q)} F'^{(2,q)*}$. The argument ω of the function $j(\omega)$ is always a multiple of the electron resonance frequency ω_o and so J_n is defined by

$$J_n = j(n\omega_o)$$
$$= \tau_2/(1 + n^2\omega_o^2\tau_2^2). \qquad \text{(XII.23.)}$$

As we saw in Chapter VIII the line shape, which is a sum of Lorentzians,is calculated from the eigenvalues and eigenvectors of the relaxation matrix. The eigenvalues when multiplied by minus one give the widths of the component lines and their relative intensities are obtained from the eigenvectors and the matrix elements of S_+.[15] The relaxation matrix given in equation (21) cannot be diagonalised analytically although there are two limiting solutions. The first situation occurs when the molecular reorientation is fast in the sense that $\omega_o\tau_2 \ll 1$ for then

$$J_o = J_1 = J_2 \,, \qquad \text{(XII.24.)}$$

and there is a single linewidth $32(D{:}D)\tau_2/5$ corresponding to the eigenvector

$$(35)^{-\frac{1}{2}}\{3|0\rangle + \sqrt{8}(|1\rangle + |-1\rangle) + \sqrt{5}(|2\rangle + |-2\rangle)\}. \qquad \text{(XII.25.)}$$

The other limit is when $\omega_o\tau_2 > 1$ for then the off-diagonal elements vanish and the linewidths are

$$T_2^{-1} = 24(D:D)\tau_2/5, \qquad \text{(XII.26.)}$$

for the $|\pm 5/2\rangle \leftrightarrow |\pm 3/2\rangle$ transitions,

$$T_2^{-1} = 6(D:D)\tau_2/5\ , \qquad \text{(XII.27.)}$$

for the $|\pm 3/2\rangle \leftrightarrow |\pm\frac{1}{2}\rangle$ transition and the $|\frac{1}{2}\rangle \leftrightarrow |-\frac{1}{2}\rangle$ transition is not broadened in this approximation. For all other rotational correlation times the linewidths and the intensities must be obtained by numerical diagonalisation of the relaxation matrix. The results[18] of such calculations are shown in Figure 1, and these demonstrate that there is no 1:1 correspondence between the linewidths and the degenerate $|m_s\rangle \leftrightarrow |m_s \pm 1\rangle$ transitions as has previously been supposed.[19]

This is not, of course, the final stage in the analysis for in order to compare experiment and theory the total line shape should be reconstructed using the theoretical widths and intensities given in Figure 1. Such comparisons are not without difficulty and a simpler approach would be welcome. One possibility is to define an average linewidth and to compare this with some experimental measure for the width of the line. This procedure will only be possible if the total line shape does not deviate too seriously from a Lorentzian shape. The technique would therefore be quite inappropriate when the widths of the component lines are widely different as in systems exhibiting an alternating linewidth effect.[1] One such average has been introduced by McLachlan who weights the component widths according to their intensities.[5] This definition is potentially important since the average width $\langle T_2^{-1}\rangle$ may be evaluated without diagonalising the relaxation matrix:

$$\langle T_2^{-1}\rangle = \frac{\sum \langle \varkappa'|S_-|\varkappa\rangle R_{\varkappa\varkappa',\lambda\lambda'}\langle\lambda|S_+|\lambda'\rangle}{\sum \langle\varkappa'|S_-|\varkappa\rangle\langle\lambda|S_+|\lambda'\rangle}\ , \qquad \text{(XII.28.)}$$

where the summations are over the allowed transitions. For this particular problem the average width is found to be:

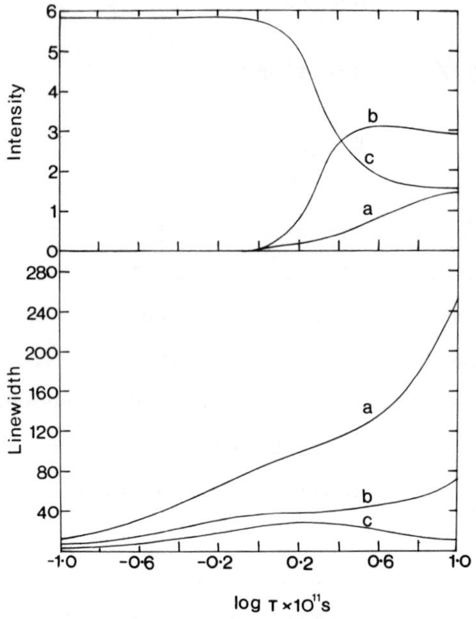

Figure XII.1. The component linewidths and intensities for the spectrum of a ^6S state ion as a function of the correlation time.

$$\langle T_2^{-1} \rangle = 16(D:D)(3J_o + 5J_2)/25. \qquad \text{(XII.29.)}$$

In principle this theoretical average is related to the observed absorption line shape L(B-Br) by

$$\langle T_2^{-1} \rangle = \lim_{(B-Br)\to\infty} \{\pi(B-Br)^2 L(B-Br)\}. \qquad \text{(XII.30.)}$$

Since such a quantity is virtually impossible to determine experimentally it is tempting to associate $2\langle T_2^{-1}\rangle/\sqrt{3}$ with the spacing between the extema of the derivative curve. Unfortunately, this assumption appears to be poor, at least for ^6S states, where the width of the theoretical line shape has been compared with that calculated from equation (29).[20] The concept of an average linewidth is apparently not valid in these systems, and so the experimental results can only be interpreted by simulating the total line shape.

The theory which we have discussed is appropriate when the symmetry of the complex is less than cubic. However, for complexes, such as $Mn^{2+}(H_2O)_6$, the cubic symmetry forces the zero-field splitting to vanish and we must look elsewhere for the relaxation process. Provided the time dependence is produced entirely by rotational diffusion of the molecule the higher order quartic terms in the spin-spin interaction might be responsible for spin relaxation.[6,8] We consider the extent of line broadening produced by these terms in the next section.

XII.4. RELAXATION VIA THE QUARTIC TERMS

In an undistorted octahedral complex the ligands generate the cubic potential:

$$Y_{4,0} + (5/14)^{\frac{1}{2}}(Y_{4,4} + Y_{4,-4}) \quad , \qquad \text{(XII.31.)}$$

where $Y_{4,m}$ is a fourth order spherical harmonic.[11] The first non-zero spin-spin coupling between the electrons is the so-called quartic term which is usually represented by the spin hamiltonian[21]

$$\mathcal{H} = \frac{a}{6}\{S_a^4 + S_b^4 + S_c^4 - \frac{1}{5}S(S+1)(3S^2+3S-1)\}, \qquad \text{(XII.32.)}$$

in a molecular coordinate system where a is now the fine structure parameter. The effect of the quartic interaction in zero field is to split the six degenerate spin levels of the ^6S state into a set of two and a set of four separated by 3a.

The relaxation matrix for rotational modulation of the quartic terms is calculated from equations (6) and (16) as in the preceeding section, but with L = 4. The necessary Clebsch-Gordan coefficients $C(5/2\ 4\ 5/2;\ m_1,m_2)$ can be calculated from the expressions given by Saito and Morita.[22] The only other quantity required is the reduced matrix element $\langle 5/2\|T^{(4)}\|5/2\rangle$ and this is obtained by constructing the irreducible spin operator $T^{(4,4)}$ using vector coupling techniques:[13]

$$T^{(4,\ 4)} = \sum_{m_1,m_2} T^{(2,m_1)}T^{(2,m_2)}C(224;m_1,m_2) \ ,$$

$$= \frac{1}{4}\ S_+^4 \quad . \qquad \text{(XII.33.)}$$

The reduced matrix element is then

$$\langle 5/2 \| T^{(4)} \| 5/2 \rangle = \langle 5/2 | T^{(4,4)} | -3/2 \rangle / C(5/2\ 4\ 5/2; -3/2, 4),$$

$$= (540)^{\frac{1}{2}}. \qquad \text{(XII.34.)}$$

The remaining calculation of the relaxation matrix is now straightforward and gives[8]:

$$\begin{array}{c|ccccc|} & |2\rangle & |1\rangle & |0\rangle & |-1\rangle & |-2\rangle \\ \langle 2| & A & D & E & O & F \\ \langle 1| & D & B & O & G & O \\ \langle 0| & E & O & C & O & E \\ \langle -1| & O & G & O & B & D \\ \langle -2| & F & O & E & D & A \end{array}\ , \qquad \text{(XII.35.)}$$

where

$$A = -(16J_o + 18J_1 + 14J_2 + 14J_3 + 28J_4),$$

$$B = -(25J_o + 24J_1 + 19J_2 + 14J_3 + 14J_4),$$

$$C = -(20J_1 + 28J_2 + 28J_3)\ ,$$

$$D = -4\sqrt{10}\ J_1\ , \qquad \text{(XII.36.)}$$

$$E = -6\sqrt{5}\ J_2,$$

$$F = 28J_4,$$

$$G = 10J_2\ .$$

Here the matrix elements are in units of $a^2/21$: and this is related to the inner product $\sum_q F'^{(4,q)} F'^{(4,q)*}$ which is $a^2/30$. In addition the J_n are defined in terms of the rotational correlation time τ_4 and not τ_2. As before there is no general analytic expression for the line shape but in the region of extreme narrowing there is just a single line whose width is equal to $32a^2J_o/7$.[8]

This theoretical estimate of the contribution to the linewidth is sufficient for our purpose since in the relevant experimental situations the rotational correlation time is indeed small. Two ^{6}S state ions are of practical importance and the first of these is manganese (II). In aqueous solution the linewidth of the octahedral aquo complex of manganese is about 20 gauss. Now the fine structure parameter for manganese[23] is only about $1.5 \times 10^8 s.^{-1}$ and consequently, since the rotational correlation time is approximately $3 \times 10^{-11} s.$ in water at room temperature, the linewidth should be 0.1 gauss, which is clearly too small to account for the observed width. This is not necessarily the case for iron (III) complexes where a can be as large as $2.5 \times 10^9 s.^{-1}$ and so for the same rotational correlation time the resulting linewidth is 24 gauss. Although this is negligible for the hexa aquo complex of iron where the linewidth is about 1000 gauss it may well be important for the hexafluoride where the linewidth is sufficiently small to resolve fluorine hyperfine structure.[20]

XII.5. COLLISIONAL FLUCTUATIONS OF THE ZERO-FIELD SPLITTING

Since the zero-field splitting of an undistorted octahedral complex of iron (III) or manganese (II) must vanish and because the linewidth caused by rotational modulation of the quartic term is small compared with the observed width an additional relaxation process must be involved. One suggestion is that the instantaneous symmetry of the complex is less than cubic and so at any instant the zero-field splitting is not zero.[12] Since the distortion is not expected to be permanent the geometry of the complex will fluctuate in time presumably as a consequence of molecular collisions. The resulting modulation of the components of the zero-field splitting tensor as well as the reorientation of the complex could constitute a powerful spin relaxation process. We shall derive the auto-correlation function $\psi(t)$ for these processes by assuming a rather simple dynamical model.[18] The starting point is the dynamic spin hamiltonian in equation (5) in which the rotation matrix $\mathcal{D}^{(L)}_{q,p}(t)$

is now taken to connect the space fixed coordinate system and one set in the molecule. Thus $\mathcal{D}^{(L)}_{q,p}(t)$ contains the time dependence of $\mathcal{H}'(t)$ resulting from rotational diffusion. However, the components $F'^{(L,q)}$ also fluctuate because the principal components, as well as the principal axes, of the zero-field splitting tensor are changing. In order to simplify the calculation we assume that only the orientation of the principal axes, with respect to a given molecular coordinate system, changes as a result of the molecular collisions. The time dependence of $F'^{(L,q)}(t)$ can therefore be removed by transforming from this molecular coordinate system to one containing the principal axes for the zero-field splitting. The dynamic spin hamiltonian is then

$$\mathcal{H}'(t) = \sum_{L;p,q,r} (-1)^p F''^{(L,r)} \mathcal{D}^{(L)}_{r,q}(t) \mathcal{D}^{(L)}_{q,p}(t) T^{(L,-p)} , \qquad \text{(XII.37.)}$$

where the principal components $F''^{(L,r)}$ are time independent and so the autocorrelation function $\psi(t)$ is

$$\psi(t) = \sum_{q,q'} \overline{\mathcal{D}^{(L)}_{r,q}(0) \mathcal{D}^{(L')*}_{r',q'}(t) \mathcal{D}^{(L)}_{q,p}(0) \mathcal{D}^{(L')*}_{q',p'}(t)} . \qquad \text{(XII.38.)}$$

If we boldly assert that the molecular rotations and collisions are uncoupled then the two correlation functions may be averaged separately. The autocorrelation function for isotropic rotational diffusion was encountered in section 2 and if the result given by equation (11) is used in equation (38) we find

$$\psi(t) = \sum_{q} \overline{\mathcal{D}^{(L)}_{r,q}(0) \mathcal{D}^{(L)*}_{r',q}(t)}\ (2L+1)^{-1} \delta_{LL'} \delta_{pp'} \exp(-|t|/\tau_L) . \qquad \text{(XII.39.)}$$

The problem now is to determine the collisional correlation function $\psi_{coll}(t)$ and to do this we assume that the complex can distort in n equivalent ways. For example, when n = 3 each form would correspond to an axial distortion along one of the three C_4 axes of the octahedron. Similar behaviour is exhibited by octahedral copper (II) complexes although here the fluctuations result from a dynamic Jahn-Teller effect.[24] The correlation function may be written as

$$\psi_{coll.}(t) = \sum_{q;i,j} \mathcal{D}^{(L)}_{r,q}(\Omega_i)P(\Omega_i)\mathcal{D}^{(L)*}_{r',q}(\Omega_j)P(\Omega_i|\Omega_j,t) , \tag{XII.40.}$$

where the subscripts i and j denote the n possible orientations of the principal axes with respect to a given molecular coordinate system. Since the n distortions are equivalent the probability of a given distortion is just

$$P(\Omega_i) = 1/n . \tag{XII.41.}$$

The conditional probability $P(\Omega_i|\Omega_j,t)$ is obtained by solving the familiar kinetic equations for the n-site problem:

$$\frac{dP(\Omega_\alpha)}{dt} = -\frac{1}{\tau_c} P(\Omega_\alpha) + \frac{1}{(n-1)\tau_c} \sum_{\beta\neq\alpha} P(\Omega_\beta) \tag{XII.42.}$$

where τ_c is the lifetime of a distortion. The solution of these equations gives[18]

$$P(\Omega_i|\Omega_j,t) = \frac{1}{n}\{1 + (n\delta_{ij}-1)\exp[-n|t|/(n-1)\tau_c]\}. \tag{XII.43.}$$

Upon substitution of this result and equation (41) into the expression for the collisional correlation function we find

$$\psi_{coll.}(t) = \frac{1}{n^2} \sum_{q;i,j} \mathcal{D}^{(L)}_{r,q}(\Omega_i)\mathcal{D}^{(L)*}_{r',q}(\Omega_j)\{1+(n\delta_{ij}-1)\exp(-n|t|/(n-1)\tau_c)\}, \tag{XII.44.}$$

but because

$$\sum_j \mathcal{D}^{(L)}_{r,q}(\Omega_j) = 0 , \tag{XII.45.}$$

for a uniform distribution of equivalent orientations:

$$\psi_{coll}(t) = \frac{1}{n} \sum_{q,i} \mathcal{D}^{(L)}_{r,q}(\Omega_i)\mathcal{D}^{(L)*}_{r',q}(\Omega_i) \exp(-n|t|/(n-1)\tau_c) \,. \tag{XII.46.}$$

Finally we obtain the total correlation function

$$\psi(t) = \delta_{LL'}\delta_{pp'}\delta_{rr'}(2L+1)^{-1}\exp\{-(\frac{1}{\tau_L} + \frac{n}{n-1}\frac{1}{\tau_c})|t|\}, \tag{XII.47.}$$

because the sum over q can be evaluated with the orthonormality relationship:[13]

$$\sum_{q} \mathcal{D}^{(L)}_{r,q}(\Omega_i)\mathcal{D}^{(L)*}_{r',q}(\Omega_i) = \delta_{rr'} \,. \tag{XII.48.}$$

This result is equivalent to that for rotational modulation but with the rotational correlation time replaced by the composite quantity τ where[18]

$$\frac{1}{\tau} = \frac{1}{\tau_L} + \frac{n}{n-1}\,\frac{1}{\tau_c} \,. \tag{XII.49.}$$

Since this is the only effect of collisional fluctuations the general form of the relaxation matrix and hence of the linewidths is identical to that derived in sections 3 and 4.

In order to test the theory when L = 2 we need to know either the correlation time τ or the components of the zero-field splitting tensor. As neither of these quantities is available experimentally at least two measurements of the linewidth are required. The best way of obtaining these parameters is to determine the widths at different microwave frequencies because varying ω_o should change the parameters J_1 and J_2 while keeping (D:D) and τ constant. Thus, on going from X-band where ω_o is 0.58×10^{11} s.$^{-1}$ to Q-band where ω_o is 2.2×10^{11} s.$^{-1}$ the product $\omega_o\tau$ will increase and according to the results in Figure 1 this should cause the linewidth to decrease.

This frequency dependence is indeed observed for $Fe^{3+}(H_2O)_6$ in aqueous solution; at X-band the separation between the extrema is 1100 gauss which decreases to 650 gauss at Q-band.[18] These two linewidths may be interpreted quantitatively in the following way. Firstly, the theoretical line shapes are simulated for a range of correlation times for both X- and Q-band frequencies using the theoretical widths and intensities of the component lines. The widths of the simulated lines are then measured, although they are in arbitary units of (D:D). However, this unknown inner product may be eliminated by taking the ratio of the theoretical widths at X- and Q-band for a given correlation time. This theoretical ratio depends only on τ, and so by comparison with the experimental ratio for these widths the correlation time can be determined. It is then a simple task to obtain the inner product (D:D) from the observed linewidth at either X- or Q-band.[20] For this particular iron (III) complex

$$(D:D)^{\frac{1}{2}} = 3.81 \times 10^{9} s.^{-1} \text{ and } \tau = 4.6 \times 10^{-11} s.$$

If the rotational correlation time τ_2 is taken to be 3×10^{-11}s. then τ_c must lie in the range $5\text{-}8 \times 10^{-12}$s. since n/n-1 only varies from 1.5 to 1 as n goes from three to infinity. A correlation time of $5\text{-}8 \times 10^{-12}$s. is too small to be attributed to rotational diffusion, but is the correct order of magnitude for molecular collisions. These results, taken together with studies of nuclear spin relaxation for the water molecules in the system seem to confirm that collisional fluctuations in the zero-field splitting dominate the electron spin relaxation processes for ^{6}S state ions with cubic symmetry.[18]

In principle it should be possible to employ a similar procedure to analyse the widths of the six hyperfine lines observed for solutions of manganese (II). The first attempt[18] at such an analysis has not been encouraging, possibly because the relaxation theory was not modified to include the effect of the hyperfine interaction.[8] In the last two sections we shall consider some of the complications caused by the large manganese hyperfine interaction.

XII.6. MODIFICATIONS DEMANDED BY A HYPERFINE INTERACTION

In the presence of a single hyperfine interaction the scalar spin hamiltonian for a ^6S state becomes

$$\mathcal{H}^{(o)} = g\beta B S_z + a\underset{\sim}{I}.\underset{\sim}{S} \quad , \qquad \text{(XII.50.)}$$

and if the hyperfine coupling is large, as in complexes of manganese (II), then the non-secular terms must be retained. Their retention results in the mixing of the spin states and this has two distinct effects. The first is simply to make the spacings between adjacent lines unequal as in the spectra of doublet state species. The second is of greater importance for it corresponds to the removal of the degeneracy of the five electron spin transitions for a given nuclear quantum number, m_I.[25] The magnitude of these two effects is best gauged by using perturbation theory to calculate the resonant field for the transition

$$|m_s, m_I\rangle \longleftrightarrow |m_s+1, m_I\rangle .$$

The result, correct to second order in a/ω_o is

$$B_r = B_o - a\frac{\hbar}{\beta} m_I - \frac{a^2\hbar}{2\omega_o\beta}\{I(I+1) - m_I^2 + m_I(2m_s+1)\} , \qquad \text{(XII.51.)}$$

where $$B_o = \hbar\omega_o/g\beta. \qquad \text{(XII.52.)}$$

The most striking feature of this result is that the removal of the degeneracy depends on the nuclear quantum number and so will vary amongst the various hyperfine lines. Experimentally the splitting of the degeneracy is small compared with the linewidth and so the separate transitions have not been resolved. Instead the hyperfine lines are inhomogeneously broadened to an extent which depends on

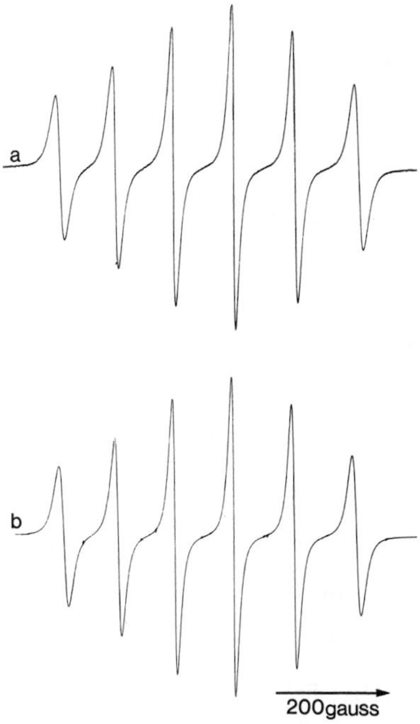

Figure XII.2. The experimental (a) and theoretical (b) X-band spectra of manganese (II) ions in aqueous solution at 363 K.

$|m_I|$. The spectrum of $Mn^{2+}(H_2O)_6$ at 363K shown in Figure 2 clearly demonstrates the expected linewidth dependence on the nuclear quantum number. Although the linewidth variation is reminiscent of the asymmetric broadening exhibited by doublet state species[1] it is important to realise that this variation for manganese (II) spectra is not associated with any relaxation process. The lifting of the degeneracy means that calculation of the linewidths using Redfield's theory presents certain problems. On the one hand the transitions are not degenerate and so the linewidths cannot be equated with minus one times the eigenvalues of the relaxation matrix given by equation (21). On the other hand the removal of the degeneracy is insufficient, compared with the linewidth, to use just the diagonal relaxation matrix elements to calculate the linewidth. In view of these difficulties it is necessary to turn to a more general formulation of the line shape problem.[26]

For a low power, slow passage electron resonance experiment the line shape function $L(\omega)$ is given by[27]

$$L(\omega) \propto \mathrm{Re}\{\underset{\sim}{S}- . \underset{\sim}{M}_o^{-1} . \underset{\sim}{\sigma}\}, \qquad \text{(XII.53.)}$$

where $\underset{\sim}{S}-$ is a vector composed of the matrix elements $\langle\varkappa| S- |\varkappa'\rangle$. The other vector $\underset{\sim}{\sigma}$ is defined by

$$\sigma_{\varkappa\varkappa'} = \langle\varkappa| S- |\varkappa'\rangle (\rho_{\varkappa\varkappa} - \rho_{\varkappa'\varkappa'}) , \qquad \text{(XII.54.)}$$

and the elements of both vectors are only evaluated for eigenstates involved in the allowed transitions. The quantity involving the density matrix is proportional to the population difference for the eigenstates $|\varkappa'\rangle$, which for a non-saturated sample is just a constant. The matrix $\underset{\sim}{M}_o$ is related to the elements of the relaxation matrix and the resonant frequencies $\omega_{\varkappa\varkappa'}$ by

$$M_{o;\varkappa\varkappa',\lambda\lambda'} = i(\omega_{\varkappa\varkappa'} - \omega)\delta_{\varkappa\lambda}\delta_{\varkappa'\lambda'} + R_{\varkappa\varkappa',\lambda\lambda'} , \qquad \text{(XII.55.)}$$

provided Redfield's theory is applicable. The line shape could be obtained by inverting a complex non-Hermitian matrix at each point ω in the spectrum. This method of calculating the spectrum is time consuming and a faster procedure is possible.[27,28] The matrix $\underset{\sim}{M}_o$ can be written as

$$\underset{\sim}{M}_o = \underset{\sim}{C} - i\omega\underset{\sim}{E} , \qquad \text{(XII.56.)}$$

where $\underset{\sim}{E}$ is the unit vector and $\underset{\sim}{C}$ is independent of the frequency. Since the transformation $\underset{\sim}{U}$ which diagonalises $\underset{\sim}{C}$ also diagonalises $\underset{\sim}{M}_o$

$$\begin{aligned} \underset{\sim}{U}^{-1} . \underset{\sim}{M}_o . \underset{\sim}{U} &= \underset{\sim}{U}^{-1} . \underset{\sim}{C} . \underset{\sim}{U} - i\omega\underset{\sim}{E} \\ &= \underset{\sim}{\Lambda} - i\omega\underset{\sim}{E} . \qquad \text{(XII.57.)} \end{aligned}$$

Consequently the inverse $\underset{\sim}{M}_o^{-1}$ is readily calculated from

$$\underset{\sim}{M}_o^{-1} = \underset{\sim}{U}(\underset{\sim}{\Lambda} - i\omega E)^{-1}\underset{\sim}{U}^{-1} , \qquad \text{(XII.58.)}$$

where $\underset{\sim}{\Lambda}$ and $\underset{\sim}{U}$ need only be calculated once for each simulation. The factorisation of the frequency dependence from $\underset{\sim}{M}_o$ has a further advantage since it enables us to calculate the first derivative line shape which is usually measured in electron resonance experiments:

$$L'(\omega) \propto \mathrm{Im}\{\underset{\sim}{S}\text{-}\underset{\sim}{U}(\underset{\sim}{\Lambda} - i\omega\underset{\sim}{E})^{-2}\underset{\sim}{U}^T\underset{\sim}{\sigma}\} . \qquad \text{(XII.59.)}$$

For $Mn^{2+}(H_2O)_6$ the dominant relaxation process is almost certainly modulation of the zero-field splitting caused for fluctuations in the geometry of the complex and so the relaxation matrix is given by equation (21). The resonant frequencies may be inferred from the resonant fields determining by diagonalising the hamiltonian matrix for the magnetic field at each hyperfine line. In this way there is a linear relationship between magnetic field and microwave frequency. The theoretical manganese spectrum shown in Figure 2 was calculated with the parameters[29]

$$(D:D)^{\frac{1}{2}} = 4.3 \times 10^9 s.^{-1} \text{ and } \tau = 1.18 \times 10^{-12} s.$$

The agreement between experiment and theory is extremely good; in fact the two spectra are virtually superimposable. The quantitative results of the calculation are encouraging for the correlation time is comparable to that obtained from a straightforward analysis of the electron resonance spectrum of $Fe^{3+}(H_2O)_6$ and close to the value expected for molecular collisions. In addition the magnitude of the inner product (D:D) is similar to that estimated from an analysis of proton spin relaxation times in the same system.[18] It should now be possible to realise the full potential of manganese (II) as a spin probe for the investigation of the precise details of the molecular dynamics in a wide range of systems.

XII.7. SYMMETRIC LINEWIDTH VARIATIONS IN THE SPECTRA OF MANGANESE(II) IONS

In concluding our survey of the various relaxation effects encountered in the study of ^6S states we should consider one consequence of the large hyperfine interaction which we chose to ignore in the previous section. The retention of the non-secular hyperfine terms means that the product spin funtions $|S,m_s;m_I\rangle$ are not eigenfunctions of the scalar spin hamiltonian. We can arrange for these states to be eigenfunctions of a transformed spin hamiltonian:[30]

$$\tilde{\mathcal{H}}^{(o)} = e^{i\epsilon F}\mathcal{H}^{(o)}e^{-i\epsilon F} \quad . \qquad \text{(XII.60.)}$$

The operator F is then chosen to yield the correct eigenvalues of $\mathcal{H}^{(o)}$ so that $\tilde{\mathcal{H}}^{(o)}$ is diagonal within the basis $|S,m_s;m_I\rangle$ through order ϵ^3 and for the ^6S state problem[31]

$$\epsilon F = \frac{ia}{2\omega_o}(S_-I_+ - S_+I_-). \qquad \text{(XII.61.)}$$

The relaxation matrix can then be calculated within the same simple basis, but with a transformed dynamic perturbation $\tilde{\mathcal{H}}'(t)$ given by the series[2]

$$\tilde{\mathcal{H}}'(t) = \mathcal{H}'(t) + I\epsilon[F,\mathcal{H}'(t)] - \tfrac{1}{2}\epsilon^2\Big[[\mathcal{H}'(t),F],F\Big] + .. \qquad \text{(XII.62)}$$

In general the higher order terms in $\tilde{\mathcal{H}}'(t)$ introduce additional small corrections to the zeroth order elements of the relaxation matrix without producing any qualitative changes in the linewidth. This is not the case when the correlation time is large for then the zeroth order terms J_1 and J_2 may well be small in comparison with the higher order secular terms in ϵJ_o and $\epsilon^2 J_o$. Fortunately it is not necessary to calculate corrections to the entire relaxation matrix for when $\omega_o\tau > 1$ the differences in the diagonal elements are large compared with the off-diagonal elements which may therefore be ignored. In addition, the widths of the transitions $|\pm 5/2\rangle \leftrightarrow |\pm 3/2\rangle$ and $|\pm 3/2\rangle \leftrightarrow |\pm\tfrac{1}{2}\rangle$ are so great that they are not observed and this accounts for the decrease in the height of the electron resonance signal with decreasing temperature.[32] In the limit just the $|\frac{1}{2}\rangle \leftrightarrow |-\frac{1}{2}\rangle$ transition is observed and so we need only calculate the corrections to the single element $R_{-\frac{1}{2}\frac{1}{2},-\frac{1}{2}\frac{1}{2}}$. Of course any higher order corrections in J_1 and J_2 will be negligible in comparison with the zeroth order matrix element $-(D:D)(16J_1+16J_2)/5$ and so only secular electron spin operators will be retained in $\tilde{\mathcal{H}}'(t)$.

The only term of this form in the first commutator is obtained from the $T^{(2,\pm1)}$ components of $\mathcal{H}'(t)$ and takes the form:

$$\pm \frac{a}{\omega_o}\left(\frac{3S_z^2-S^2}{2}\right)I_{\pm} \quad . \qquad \text{(XII.63.)}$$

The first order corrections will therefore depend on products of the matrix elements for this operator, and those of $\mathcal{H}'(t)$ but subject to the condition imposed by equation (11) that only the same components of the operators are used to evaluate the matrix elements. Since the operator in equation (63) implicity contains $T^{(2,\pm1)}$ components the only term in $\mathcal{H}'(t)$ to give a non-zero contribution is also $T^{(2,\pm1)}$. However, the operator $T^{(2,\pm1)}$ produces just non-secular terms and so there are no secular first order corrections to the linewidth. There is, of course, a second order correction from the operator in equation (63) and this makes the contribution

$$-64\left(\frac{a}{\omega_o}\right)^2\frac{(D:D)}{5}\{I(I+1) - m_I^2\}J_o \, , \qquad \text{(XII.64.)}$$

to $R_{-\frac{1}{2}\frac{1}{2},-\frac{1}{2}\frac{1}{2}}$. There could also be a second order correction from the double commutator if there was a non-vanishing cross-term with $\mathcal{H}'(t)$ which involved the secular operator $T^{(2,0)}$. We need only evaluate the double commutator for the $T^{(2,0)}$ term in $\mathcal{H}'(t)$ and this gives the operator:

$$(2/3)^{\frac{1}{2}}(a/\omega_o)^2\{(3S_z^2-S^2)(I^2-I_z^2) - I_zS_z(2S^2-2S_z^2-1)\}.$$

$$\text{(XII.65.)}$$

However, this does not make any contribution to the relaxation matrix element $R_{-\frac{1}{2}\frac{1}{2},-\frac{1}{2}\frac{1}{2}}$ and so the width of the $|\frac{1}{2}\rangle \leftrightarrow |-\frac{1}{2}\rangle$ transition is

$$T_2^{-1}(m_I) = \frac{(D:D)}{5} \{16J_1 + 56J_2 + 64 \left(\frac{a}{\omega_o}\right)^2 [I(I+1)-m_I^2]J_o\}. \tag{XII.66.}$$

This result predicts that, in contrast to manganese spectra measured at high temperature, those at low temperature should exhibit an unusual linewidth variation. The quadratic dependence on the nuclear quantum number m_I preserves the symmetric appearance of the spectrum and the negative coefficient will make the lines at the ends of the spectrum sharper than those in the centre. In addition, the symmetric linewidth variation will become more pronounced at lower temperatures since J_1 and J_2 should both decrease whereas J_o increases. These predictions are completely verified[31] by the experimental spectra of manganous perchlorate dissolved in methanol which were measured at X-band and are shown in Figure 3. At 296 K there is a slight asymmetric linewidth effect caused by the inhomogeneous broadening discussed in section 6. As the temperature is lowered the quadratic dependence of the linewidths on m_I is observed and at 181 K the line heights approach the limiting values of 11.6 : 1.7 : 1:1 : 1.7 : 11.6. At first sight the linewidth dependence on the nuclear quantum number might be expected to decrease on passing from X- to Q-band since the coefficient $(a/\omega_o)^2$ will decrease. However, the spectral densities J_1 and J_2 are also frequency dependent and decrease with increasing frequency. As a consequence the symmetric linewidth variation will not necessarily decrease in importance at Q-band since the total linewidth will also be reduced at this frequency. Detailed calculations suggest that the observed frequency dependence of the symmetric linewidth variation is in agreement with the theory.[33]

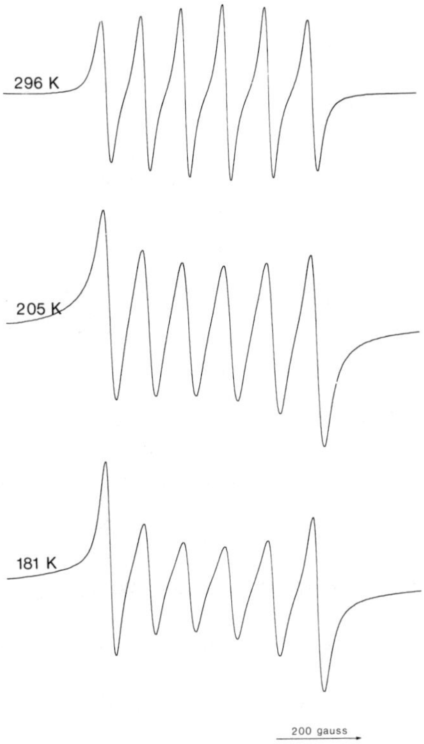

Figure XII.3. The temperature dependence of the electron resonance spectrum of manganese perchlorate dissolved in methanol.

REFERENCES

1. See, for example, Chapter X as well as A. Hudson and G.R. Luckhurs Chem. Rev., 69, 191 (1969).

2. R. Wilson and D. Kivelson, J. Chem. Phys., 44, 154 (1966).

3. P.W. Atkins and D. Kivelson, J. Chem. Phys., 44, 169 (1966).

4. S.I. Weissman, J. Chem. Phys., 29, 1189 (1958); W. Moffitt and M. Gouterman, ibid, 30, 1107 (1959); B.R. McGarvey, J. Phys. Chem. 61, 1232 (1957).

5. A.D. McLachlan, Proc. Roy. Soc. A, 280, 271 (1964).

6. A. Carrington and G.R. Luckhurst, Mol. Phys., 8, 125 (1964).

7. A.G. Redfield, Adv. Mag. Res., 1, 1 (1965).

8. A. Hudson and G.R. Luckhurst, Mol. Phys., 16, 395 (1969).

9. L. Yarmus, M. Kukk, and B.R. Sundheim, J. Chem. Phys., 40, 33 (196

10. See, for example, A.S. Mildvan and M. Cohn, Advan. Enzymol., 33, 1 (1970).

11. B. Bleaney and K.W.H. Stevens, Rep. Prog. Phys., 16, 108 (1953).

12. N. Bloembergen and L.O. Morgan, J. Chem. Phys., 34, 842 (1961).

13. M.E. Rose, Elementary Theory of Angular Momentum (John Wiley and Sons, New York, 1957).

14. P. Debye, Polar Molecules (Dover Publications, New York, 1929).

15. A. Abragam, The Principles of Nuclear Magnetism (Oxford: Clarendon Press, 1961) p.299.

16. E.V. Condon and G.H. Shortley, The Theory of Atomic Spectra (Cambridge University Press, 1935).

17. The expression for B in ref.5 is incorrect, as is that for D in refs. 8 and 31.

18. M. Rubinstein, A. Baram and Z. Luz, Mol. Phys., 20, 67 (1971).

19. B.B. Garrett and L.O. Morgan, J. Chem. Phys., 44, 890 (1966); C.C. Hinckley and L.O. Morgan, ibid, 44, 898 (1966).

20. H. Levanon, G. Stein and Z. Luz, J. Chem. Phys., 53, 876 (1970).

21. G.F. Coster and H. Statz, Phys. Rev., 113, 445 (1959).

22. R. Saito and M. Morita, Prog. Theor. Phys., Osaka, 13, 540 (1955).

23. W. Low, Paramagnetic Resonance in Solids (Academic Press, New York, 1960).

24. A. Hudson, Mol. Phys., 10, 575 (1966).

25. F.K. Hurd, M. Sachs and W.D. Hershberger, Phys. Rev., 93, 373, (1954).

26. G.K. Fraenkel, J. Chem. Phys., 42, 4275 (1965); R.M. Lynden-Bell, Prog. Nuc. Mag. Res., 2, 163 (1967).

27. G. Binsch, Mol. Phys., 15, 469 (1968)

28. R.G. Gordon and R.P. McGinnis, J.Chem. Phys., 49, 2455 (1968).

29. G.R. Luckhurst and G.F. Pedulli, Mol.Phys., in submission.

30. E.C. Kemble, The Fundamental Principles of Quantum Mechanics (Dover Publications, New York, 1937).

31. G.R. Luckhurst and G.F. Pedulli, Chem. Phys. Lett., 7, 49 (1970).

32. L. Burlamacchi and E. Tiezzi, Chem. Phys. Lett., 4, 173 (1969).

33. L. Burlamacchi, private communication.

MAGNETIC RESONANCE LINE SHAPES IN SLOWLY TUMBLING MOLECULES

Roy G. Gordon and Thomas Messenger

Department of Chemistry, Harvard University

Cambridge, Massachusetts 02138

XIII.1. INTRODUCTION

The line shapes of some ESR and NMR spectra are affected by motion of the molecules. Naturally, for motion to have any effect, the resonant frequencies of a single molecule must depend in some way on the motion. For instance, the ESR transitions of a diradical depend on the orientation of the molecule relative to the applied field, and the orientation is changed during the thermally agitated tumbling of the molecule in solution.[1] Any orientation dependent resonance frequency is likewise affected by this rotational diffusion process. Another example is the ESR spectrum of radicals with nuclear hyperfine splittings, such as nitroxide radicals.[2]

There are two limiting cases in which the magnetic resonance line shapes are well understood. One is the limit of essentially fixed molecular orientation, as in a crystal or a very viscous solution or glass. For a single crystal in some specific orientation, one sees only a single sharp spectrum. Many single crystal spectra have been taken,[3] particularly to measure the dependence of the spectra on orientations. If instead of a crystal we consider a powder or glass, the spectrum can be reconstructed by adding together single crystal spectra at all orientations.[4] The other limit which is well understood is the very rapid rotation typical of liquid solutions of small molecules at ordinary viscosities. Line shape theories appropriate for this case have been developed by Freed and Fraenkel,[5] and Kivelson,[6] among others.

The intermediate case typically occurs in very viscous solutions of small molecules, and ordinary solutions of proteins and other large biological molecules. Theories for such intermediate tumbling rates have been developed[1,7-10] using an adiabatic assumption, similar to that used earlier by Anderson.[11] Unfortunately, as we will demonstrate below, this adiabatic assumption often breaks down seriously, in the case of hyperfine splittings in ESR spectra. A formally complete treatment of the intermediate case has been presented by Sillescu and Kivelson,[12] but a perturbation assumption in their approach limits their results to the case of small deviations from the rapid tumbling limit.

Here we present a more complete treatment of spectra in the intermediate range of tumbling rates, using neither adiabatic nor perturbation approximations.[13]

We use the conventional description of molecular tumbling, with the rotational diffusion equation.[14] Some of the mathematical methods developed in the course of this study may also be applied to determining resonance line shapes for other, more elaborate, descriptions of molecular motion.

Since the theory relates diffusion rate to line shape, the rotational correlation time or rotational diffusion rate can be determined from the spectra. Thus, the theory can be used to provide important information[15] about the rotational motion of molecules, over the complete range of behavior, from slow reorientation to rapid tumbling.

XIII.2. EXPRESSIONS FOR THE LINE SHAPES

The absorption shape $I(\omega)$ in an unsaturated resonance spectrum is given by[16]

$$I(\omega) = \frac{1}{\pi} \operatorname{Re} \int_0^\infty dt e^{-i\omega t} C(t) \tag{2.1}$$

where $C(t)$, the correlation function, is itself given by

$$C(t) = \langle \mu_x(0) \mu_x(t) \rangle_0 . \tag{2.2}$$

$\mu(t)$ is the magnetic transition dipole and $\langle\rangle_0$ indicates an average over the initial equilibrium ensemble. The motion of the x component of the magnetic dipole moment of the spins, is given by the Heisenberg equation of motion

$$i\hbar(\dot{\mu}_x)_{if} = \langle i | [\mu_x, \underline{H}] | f \rangle . \tag{2.3}$$

where $(\mu_x)_{if}$ is the familiar shorthand for $\langle i|\mu_x|f\rangle$ and $\underline{H}$ is the

Hamiltonian. This system of equations can be thought of as the equation of motion of a single vector $\underset{\sim}{\mu}$ whose components are the μ_{if} above. (We hereafter suppress the subscript x.) The states $|i\rangle$ and $|f\rangle$ are here thought of as those of the total spin rotation and translation of the system. (Below, Eq. (2.6) and following, the quantum description will be restricted to the spin states alone.)

The Liouville operator of the Hamiltonian is defined by

$$L_{mn,m'n'} = \hbar^{-1}[\underline{H}_{mm'}\delta_{nn'} - \underline{H}_{n'n}\delta_{mm'}] \tag{2.4}$$

As discussed in Chapter I, the resonance frequencies are the eigenvalues of this Liouville operator. With this notation, we may rewrite the equation of motion as

$$\dot{\underset{\sim}{\mu}} = i\,\underset{\sim}{L}\,\underset{\sim}{\mu}.$$

i.e.,

$$\dot{\mu}_{(if)} = \frac{i}{\hbar}\sum_{(i'f')}(\underline{H}_{ii'}\delta_{ff'} - \underline{H}_{f'f}\delta_{ii'})\mu_{i'f'}$$

$$= \frac{i}{\hbar}\{\sum_{i'}\underline{H}_{ii'}\mu_{i'f} - \sum_{f'}\underline{H}_{f'f}\mu_{if'}\}$$

$$= \frac{i}{\hbar}\{\sum_{i'}\langle i|\underline{H}|i'\rangle\langle i'|\mu|f\rangle - \sum_{f'}\langle i|\mu|f'\rangle\langle f'|\underline{H}|f\rangle\}$$

$$= \frac{i}{\hbar}\langle i|[\underline{H},\mu]|f\rangle = \frac{-i}{\hbar}\langle i|[\mu,\underline{H}]|f\rangle \tag{2.5}$$

Thus,

$$i\hbar\langle i|\dot{\mu}|f\rangle = \langle i|[\mu,\underline{H}]|f\rangle, \text{ as it should.}$$

This fully quantum-mechanical description is of course too difficult to apply to resonance spectra in liquids. To simplify the description sufficiently to make calculations possible, we make four commonly used approximations. First, the <u>Born-Oppenheimer approximation</u> allows us to separate the "high-frequency" electronic motion from that of the nuclei, with the electrons simply providing the "glue" bonding the molecules together. In fact, we shall even consider the vibrations of the nuclei as high-frequency coordinates, and thus only the rigid rotation and transation of the molecules are separated for consideration.

The second important assumption is to describe these separated molecular orientations by <u>classical mechanics</u>, so that we may speak of the exact positions and velocities of the nuclei, as functions of time. The quantum-mechanical description is retained for the electron and nuclear spins.

Next, we assume that the time-dependence of the molecular rotation takes place without regard to which states the electron and nuclear spins may be in. This is a reasonable assumption, since the typical energies of molecular rotation and translation are about kT ($\approx 200\ cm^{-1}$ at room temperature), whereas the total spin energies rarely exceed 1 cm^{-1}. Thus, even if the spin system underwent a transition from its highest to its lowest state, this little energy added to the rotation and translation of the molecules would hardly have any noticeable effect. Thus, the problem of the molecular motion is separated from that of the spins, while of course the spin motion still depends strongly on the molecular motion.

Finally, the fourth of the general approximations is the use of stochastic, Markovian models for the molecular motion, specifically the rotational motion. The stochastic model which we will analyze in detail is that of rotational diffusion, but the techniques for solution are applicable to many other Markovian models, large-angle rotational diffusion, and gas-phase collisions models.

By these assumptions, the magnetic spin dipole matrix element at time t factors into a product

$$\mu_{if}(\Omega,t)P(\Omega,t) \tag{2.6}$$

in which $\mu_{if}(\Omega,t)$ is the spin matrix element for a molecule oriented at angles Ω, and $P(\Omega,t)$ is the probability that a molecule has orientation Ω at time t. According to the third assumption above, this orientational probability is independent of the spins, and may be calculated from a spin-independent equation. For the case of rotational diffusion of a spherical molecule, the orientational probability is assumed to satisfy the rotational diffusion equation

$$\frac{\partial P}{\partial t}(\Omega,t) = D\nabla_\Omega^2 P(\Omega,t) \tag{2.7}$$

where ∇_Ω^2 is the angular Laplacian, and D is the rotational diffusion constant.

The total time dependence of the dipole in Eq. (2.6) thus arises from two processes: (1) time dependence of the spins for molecules at a given orientation, described by Eqs. (2.3) or (2.4), with the spin Hamiltonian $\underline{H}(\Omega)$, and (2) the time dependence of the molecular orientation, as given by the rotational diffusion equation. Because the dipole contribution, Eq. (2.6), is a product of factors, its time derivative is simply the sum of these effects:[17]

$$\frac{\partial}{\partial t}(\mu_{if}(\Omega,t)P(\Omega,t)) = (i/\hbar)[H(\Omega),\mu_x]_{if}P(\Omega,t)$$

$$+ \mu_{if}D\nabla_\Omega^2 P(\Omega,t) \tag{2.8}$$

or, in the matrix notation of Eq. (2.5)

$$\frac{\partial}{\partial t}(\underset{\sim}{\mu}P(\Omega,t)) = i\underset{\sim}{L}(\Omega)\cdot\underset{\sim}{\mu}P(\Omega,t) + \underset{\sim}{\mu}D\nabla_\Omega^2 P(\Omega,t) \tag{2.9}$$

The solution to these equations is to be averaged over initial angles Ω_0 and over initial spin states, as in Eq. (2.2), and then Fourier transformed, according to Eq. (2.1), to obtain the spectrum. It is more convenient, however, in discussing the solutions, to interchange the order of averaging and Fourier transformation. Thus, we first define the one-sided Fourier transform of the solution

$$\underset{\sim}{f}(\Omega,\omega) \equiv i\int_0^\infty e^{-i\omega t}\underset{\sim}{\mu}(\Omega,t)P(\Omega,t)dt \tag{2.10}$$

In terms of this, the spectral intensity is the equilibrium average over initial orientations and spin states,

$$I(\omega) = \frac{1}{\pi}\,\mathrm{Im}\langle\underset{\sim}{\mu}(0)\cdot\underset{\sim}{f}(\Omega,\omega)\rangle_0 \tag{2.11}$$

Applying the Fourier transformation to the differential equation (2.9), and integrating by parts to remove the time derivative, gives

$$[\omega\underset{\sim}{1}-(\underset{\sim}{L}(\Omega)-i\underset{\sim}{1}D\nabla_\Omega^2)]\cdot\underset{\sim}{f}(\Omega,\omega) = \underset{\sim}{\mu}(0)P(\Omega,0) \tag{2.12}$$

In this equation, $P(\Omega,0)$ is the initial distribution of orientations at the initial time t=0. Because of the angle-dependent spin energies, there will be a small preference for some molecular orientations but, according to our third assumption, this should be a negligible effect. Thus, we may safely assume all molecular orientations are equally likely, and take $P(\Omega,0)$ as a constant, say unity.

The following sections are devoted to analysis and solution of Eq. (2.12).

XIII.3. REDUCTION TO ALGEBRAIC EQUATIONS

There are two commonly used approaches to the approximate solution of linear differential equations: finite-difference methods, and expansions of the solution in a complete set of known functions. Both of these approaches reduce the problem to that of solving coupled linear equations. In this section we shall consider examples of both of these methods of reduction. Section IV is then devoted to developing the most efficient method for solving the algebraic equations which result from either approach.

To simplify the presentation, we shall first apply both methods to a simple case. We suppose that the spin Hamiltonian is axially symmetric about a molecular symmetry axis, and that the dipole matrix elements are independent of orientation about the laboratory magnetic field direction. Then there is only one significant angle remaining, the angle θ between the molecular axis and laboratory field. Thus, the diffusion term in Eq. (2.13) becomes

$$D\nabla_{\Omega}^2 f(\Omega) = D[(1-x^2)\frac{\partial^2}{\partial x^2} - 2x\frac{\partial}{\partial x}]f(x) \qquad (3.1)$$

in which $x = \cos\theta$, and we have suppressed writing the other arguments θ_0 and ω in f.

Now the finite-difference approximation can be made for these derivatives. In the simplest approach, we take

$$\frac{\partial f(x)}{\partial x} \simeq \frac{1}{2\Delta}[f(x+\Delta)-f(x-\Delta)] \qquad (3.2)$$

and

$$\frac{\partial^2 f}{\partial x^2} \simeq \frac{1}{\Delta^2}[f(x+\Delta)-2f(x)+f(x-\Delta)] \qquad (3.3)$$

in which Δ is a small increment in the cosθ variable describing molecular orientation. Thus, the solution is known only at a grid of points equally spaced in cosθ. By making the spacing Δ smaller, we obtain a more accurate representation of the derivatives, as well as the knowledge of the solution at more orientations.

The finite-difference approximation to the diffusion term thus becomes

$$D\nabla_{\Omega}^2 f(\Omega) \simeq D\Delta^{-2}[(1-x_\ell^2-x_\ell\Delta)f(x_{\ell+1}) \qquad (3.4)$$
$$-2(1-x_\ell^2)f(x_\ell) + (1-x_\ell^2+x_\ell\Delta)f(x_{\ell-1})]$$

This approximation scheme is thus easily visualized as jumps taking place between neighboring points on the grid of orientations.[1] By taking grid points equally spaced in $x = \cos\theta$, the areas in the bands between successive latitudes θ_ℓ = constant are equal, so we may for convenience set them equal to unity, on the right-hand sides of the difference equations.

The finite-difference approximation to Eq. (2.12), for this simple axial case, is thus

$$[\omega \underset{\sim}{1}_s - \underset{\sim}{L}(\theta_\ell) - 2iD\Delta^{-2}(1-x_\ell^2)\underset{\sim}{1}_s]\underset{\sim}{f}(x_\ell) + iD\Delta^{-2}(1-x_\ell^2-x_\ell\Delta)\underset{\sim}{f}(x_{\ell+1})$$
$$+ iD\Delta^{-2}(1-x_\ell^2+x_\ell\Delta)\underset{\sim}{f}(x_{\ell-1}) = \underset{\sim}{\mu}(\theta_\ell) \tag{3.5}$$

in which $\underset{\sim}{1}_s$ is a unit matrix acting on the space of the spin dipole matrix elements. The frequency (or Liouville) matrix $\underset{\sim}{L}(\theta_\ell)$ is just the matrix $\underset{\sim}{L}(\Omega)$ in Eq. (2.4) evaluated at the orientation of the grid point θ_ℓ. It is thus easy to calculate $\underset{\sim}{L}(\theta_\ell)$ from any given spin Hamiltonian, whatever its angle-dependence.

If there are N grid points, these linear equations hold for $\ell = 2,3,\ldots,N-1$. The first and last equation are simpler, since we do not allow diffusion "past" the end grid point. For $\ell=1$, we have

$$[\omega \underset{\sim}{1}_s - \underset{\sim}{L}(\theta_1) - iD\Delta^{-2}(1-x_1^2-x_1\Delta)\underset{\sim}{1}_s]\underset{\sim}{f}(x_1) + iD\Delta^{-2}(1-x_1^2-x_1\Delta)\underset{\sim}{f}(x_2) = \underset{\sim}{\mu}(\theta_1) \tag{3.6}$$

and for $\ell=N$

$$[\omega \underset{\sim}{1}_s - \underset{\sim}{L}(\theta_N) - iD\Delta^{-2}(1-x_N^2+x_N\Delta)\underset{\sim}{1}_s]\underset{\sim}{f}(x_N) + iD\Delta^{-2}(1-x_N^2-x_N\Delta)\underset{\sim}{f}(x_{N-1}) = \underset{\sim}{\mu}(\theta_N) \tag{3.7}$$

Since Eqs. (3.5) to (3.7) are in matrix form with respect to the spin dipole matrix elements, the number of unknowns $\underset{\sim}{f}$ and the number of coupled equations is the product of the number of grid points, N, <u>times</u> the number of dipole matrix elements of interest. To have a concise and convenient notation for discussing these equations, we write them in matrix form also with respect to the angular grid, in the form

$$[\omega \underline{\underline{1}} - (\underline{\underline{L}} + i\underline{\underline{\Pi}})]\underline{\underline{f}} = \underline{\underline{\mu}} \tag{3.8}$$

$\underline{\underline{f}}$ and $\underline{\underline{\mu}}$ are now vectors in this direct product space of orientations and spin matrix elements. For example, if we had two spin matrix elements and three orientations, $\underline{\underline{f}}$ would have six components: $f_1(\theta_1)$, $f_2(\theta_1)$, $f_1(\theta_2)$, $f_2(\theta_2)$, $f_1(\theta_3)$, and $f_2(\theta_3)$. $\underline{\underline{\mu}}$ is the direct product vector of the spin dipole matrix elements at t=0, and the equilibrium orientational populations (all equal, in this case). The Liouville matrix $\underline{\underline{L}}$ is diagonal in the orientations.

$$\underline{\underline{L}} = \begin{pmatrix} \underset{\sim}{L}(\theta_1) & 0 & 0 & \cdot & \cdot \\ 0 & \underset{\sim}{L}(\theta_2) & 0 & \cdot & \cdot \\ 0 & 0 & \underset{\sim}{L}(\theta_3) & & \\ \cdot & \cdot & & & \\ \cdot & \cdot & & & \end{pmatrix} \tag{3.9}$$

where each of these elements is a matrix in the spin dipole space. Conversely, the diffusion matrix $\underline{\underline{\Pi}}$ couples the orientations but is diagonal in the spin dipoles. Specifically, we have, from Eqs. (3.5) to (3.8),

$$\underline{\underline{\Pi}} = \begin{pmatrix} \underline{\Pi}_{11} & \underline{\Pi}_{12} & 0 & 0 & \cdot & \cdot & \cdot \\ \underline{\Pi}_{21} & \underline{\Pi}_{22} & \underline{\Pi}_{23} & 0 & \cdot & \cdot & \cdot \\ 0 & \underline{\Pi}_{32} & \underline{\Pi}_{33} & \underline{\Pi}_{34} & \cdot & \cdot & \cdot \\ 0 & 0 & \underline{\Pi}_{43} & \underline{\Pi}_{44} & & & \\ \cdot & \cdot & \cdot & \cdot & & & \\ \cdot & \cdot & \cdot & \cdot & & & \end{pmatrix} \tag{3.10}$$

in which the diagonal blocks of elements are

$$\underline{\Pi}_{\ell\ell} = 2D\Delta^{-2}(1-x_\ell^2)\underset{\sim}{1}_s, \quad \ell = 2,3,\ldots N-1 \tag{3.11}$$

and

$$\underline{\Pi}_{11} = D\Delta^{-2}(1-x_1^2-x_1\Delta)\underset{\sim}{1}_s \tag{3.12}$$

$$\underline{\Pi}_{NN} = D\Delta^{-2}(1-x_N^2+x_N\Delta)\underset{\sim}{1}_s. \tag{3.13}$$

The off-diagonal blocks are

$$\underline{\Pi}_{\ell,\ell+1} = -D\Delta^{-2}(1-x_\ell^2-x_\ell\Delta)\underset{\sim}{1}_s \tag{3.14}$$

and

$$\underline{\Pi}_{\ell+1,\ell} = \underline{\Pi}_{\ell,\ell+1}. \tag{3.15}$$

From this last equation, we see that the diffusion matrix in the finite difference approximation is symmetric. This symmetry is of considerable help in the numerical algorithms developed in Sect. IV. The symmetry is due to choosing the grid angles in such a way that they correspond to equal populations at equilibrium.

After the linear equations (3.8) have been solved, as discussed in Sect. IV, the absorption spectrum is formed from the solutions $\underline{\underline{f}}$ according to Eq. (2.11), from the dot product

$$I(\omega) = \pi^{-1}\mathrm{Im}\underline{\underline{\mu}}^{T}\cdot\underline{\underline{f}} \tag{3.16}$$

The equilibrium average over all orientations is carried out in the process of summing over all grid points. Similarly, the implied sum over all spin dipole moments forms the required spin average.

More accurate finite-difference approximations are available for the derivatives,[18] but these involve jumps between non-nearest neighbor grid points. This means that diffusion matrix elements like $\Pi_{\ell,\ell+2}$ are also non-zero, and this increases the work involved in numerical solutions (see Sect. IV). In most cases, unless very high accuracy is required, it is more efficient simply to use approximations (3.2) and (3.3), along with a sufficiently fine grid.

In the simple axially symmetric case considered above, diffusion is followed in a single angular variable, θ. When there is less symmetry in either the diffusion or spin parts of the problem, the diffusion must be followed in two or three angular variables. Similar finite-difference formulas are available for spaces of two or more variables,[18] so the extension to these cases is straightforward. However, many more grid points are required to obtain similar accuracy, in more than one dimension. Also, no matter how the grid points are numbered, diffusion jumps take place between non-consecutive grid points. Thus, more diffusion matrix elements are non-zero than in the one-dimensional case. Both the larger number of grid points and the non-nearest neighbor jumps considerably increase the numerical work over effectively one-dimensional cases.

In the grid method, it is quite easy to set up other diffusion models than the Brownian (small jump) case considered above. For example, in a strong collision (random orientation) model, all the off-diagonal Π_{ij} values are equal. Alternatively, one may consider a distribution of jump lengths, by making a band of Π_{ij}, $|i-j|<n$ non-zero. Also, a restricted angular region of diffusion may be selected by elimination of the grid points outside some allowed range.

Next we consider the reduction of Eq. (2.12) to coupled linear equations, by expansion of the solution in a series of known functions.

This approach, which has been widely used in quantum-mechanical problems, has been used by several authors[8-10,12,13] in the study of line shapes in spin resonance. The angle-dependence of the solutions is written as a linear combination

$$\underset{\sim}{f}(\Omega) = \Sigma\, \underset{\sim}{f}_m \phi_m(\Omega) \tag{3.17}$$

of a fixed set of functions $\phi_m(\Omega)$, with coefficients $\underset{\sim}{f}_m$ which are vectors in the "line space" of the dipole matrix elements. For our later convenience we suppose the known basis functions ϕ_m to be orthogonal to each other, and normalized:

$$\int \phi_m^*(\Omega)\phi_n(\Omega)d\Omega = \delta_{mn} \tag{3.18}$$

Other restrictions and specifications for the basis functions will be noted below, as needed.

To find equations for determining the unknown coefficients $\underset{\sim}{f}_m$, we substitute the trial function (3.17) into Eq. (2.12), multiply by ϕ_ℓ^* and integrate over all angles: This standard procedure yields the coupled linear equations

$$\sum_m (\omega \underset{\sim}{1}_s - \underset{\sim}{L}_{\ell m} - i\underset{\sim}{\Pi}_{\ell m})\underset{\sim}{f}_m = \int \phi_\ell^*(\Omega)\underset{\sim}{\mu} d\Omega \tag{3.19}$$

In this, the frequency matrix $\underset{\sim}{L}_{\ell m}$ is given by

$$\underset{\sim}{L}_{\ell m} = \int \phi_\ell^* \underset{\sim}{L}(\Omega)\phi_m d\Omega \tag{3.20}$$

and the diffusion matrix $\underset{\sim}{\Pi}_{\ell m}$ is defined by

$$\underset{\sim}{\Pi}_{\ell m} = -\underset{\sim}{1}_s D \int \phi_\ell^* \nabla^2 \phi_m d\Omega \tag{3.21}$$

To simplify the right-hand sides of Eqs. (3.19), we assume that the functions ϕ_m contain one (and only one) member ϕ_0, which satisfies

$$\nabla_\Omega^2 \phi_0 = 0 \tag{3.22}$$

This function, which is time-independent according to the rotational diffusion equation (2.17), represents the equilibrium distribution over orientations (ordinarily, ϕ_o = constant). In most cases of interest, the strength of a resonance line does not depend significantly on molecular orientation, so that the right-hand side of Eq. (3.19) is proportional to

$$\underset{\sim}{\mu}\delta_{\ell 0} \tag{3.23}$$

by the use of the orthogonality relation (3.18).

If the other ϕ_m's (m>0) are also chosen to be eigenfunctions of the rotational diffusion operator ∇_Ω^2, then $\underset{\sim}{\Pi}_{\ell m}$ becomes a diagonal matrix. These eigenfunctions in all applications made so far, this choice has indeed been made.[8-10,12,13] We might note, however, that it is not really necessary for the method. There are probably some cases, such as fully anisotropic rotational diffusion, in which such a choice would probably not be convenient. Rather, one might still employ the basis functions of a symmetric molecule, and use the non-diagonal diffusion matrix which results.

To give a specific illustration of such an expansion in orthogonal functions, we consider again the case discussed above: an axially symmetric spin Hamiltonian, and a dipole matrix independent of orientation. The eigenfunctions of ∇_Ω^2 in this case are Legendre polynomials, which satisfy the eigenvalue equation

$$\begin{aligned}\nabla_\ell^2\, \hat{P}_n(x) &\equiv [(1-x^2)\frac{\partial^2}{\partial x^2} - 2x\frac{\partial}{\partial x}]\hat{P}_n(x) \\ &\equiv n(n+1)\hat{P}_n(x)\end{aligned} \tag{3.24}$$

Normalization according to Eq. (3.18) requires

$$\hat{P}_n(1) \equiv (\frac{2n+1}{2})^{1/2}, \tag{3.25}$$

which differs from the conventional normalization.[19] The importance of this choice of normalization will be seen below.

The diffusion matrix for this choice of basis, is diagonal, as expected when using eigenfunctions of ∇_Ω^2:

$$\underset{\sim}{\Pi}_{\ell m} = -\delta_{\ell m}\underset{\sim}{1}_s Dm(m+1) \tag{3.26}$$

The frequency matrix, however, couples different orientation indices. To find the frequency matrix in this approach requires taking a specific form for the angle-dependence of the spin Hamiltonian. The simplest cases to treat are those in which this angle-dependence is directly expressible in terms of one or a few of the angle-dependent eigenfunctions A simple model of this sort, which has been widely studied[1-8,10,13] is that of a single line whose frequency depends on orientation,

$$L(\Omega) = \alpha\hat{P}_2(\cos\theta) \tag{3.27}$$

A slightly anisotropic g tensor has a spectrum of this form. Also, a pair of spins 1/2 with dipole-dipole coupling, in the high-field, adiabatic limit, has a spectrum consisting of a superposition of two such spectra with opposite signs of the coefficients α. The frequency matrix for this P_2 spectrum couples only even-order Legendre polynomials to the equilibrium distribution P_0. Thus, we may drop from the start the odd-order Legendre polynomials, since their coefficients would vanish. The trial solution (3.17) thus becomes

$$f(\Omega) = \sum_{n=0}^{\infty} f_n \hat{P}_{2n}(\cos\theta) \tag{3.28}$$

The relevant diffusion matrix elements in (3.21) then become

$$\Pi_{n,n'} = -\delta_{n,n'} D2n(2n+1) \tag{3.29}$$

Since there is only a single line, each of the f_n's is only a one-component vector, a single number.

The frequency matrix for this P_2 spectrum is found from Eq. (3.29) and (3.20), which in this case is

$$L_{\ell m} = \alpha \int_{-1}^{1} \hat{P}_{2\ell}(x)\hat{P}_2(x)\hat{P}_{2m}(x)dx \tag{3.30}$$

The integrals of three Legendre polynomials are well known,[20] and give the following non-vanishing frequency matrix elements:

$$L_{m,m} = \frac{\alpha 10^{1/2}\ m(2m+1)}{2(4m+3)(4m-1)} \tag{3.31}$$

$$L_{m,m+1} = \frac{3\alpha(5/2)^{1/2}(2m+1)(m+1)}{(4m+5)^{1/2}(4m+3)(4m+1)^{1/2}} \tag{3.32}$$

$$L_{m+1,m} = L_{m,m+1} \tag{3.33}$$

The symmetry of the frequency matrix, expressed in this last equation, is most helpful in the numerical solutions discussed in the following section. The symmetry of the frequency matrix results from the use of the normalization given by Eq. (3.18). Other applications, using different normalizations, have given asymmetric matrices.[8-10,13]

To see the structure of these linear equations, we define a direct product space of the orientation function indices and the spectral lines (spin dipole matrix elements). Then Eq. (3.19) takes the form

$$[\omega \underline{\underline{1}} - (\underline{\underline{L}}+i\underline{\underline{\Pi}})]\underline{\underline{f}} = \underline{\underline{\mu}} \qquad (3.34)$$

in which $\underline{\underline{1}}$ is a unit matrix in the direct product space, and the dipole matrix $\underline{\underline{\mu}}$ has spin components $\tilde{\mu}$ for orientation index 0, and vanishes for all other orientation indices [cf. Eq. (3.23)].

Thus the same form of linear equations results from the finite difference (3.8) and orthogonal function (3.34) methods. However, the structure of the coefficient matrices is quite different, in the two cases. In the finite difference approach, the frequency matrix $\underline{\underline{L}}$ is diagonal in the orientation indices (but not in spin), while the diffusion matrix $\underline{\underline{\Pi}}$ is non-diagonal in orientation, but diagonal in spin. In the orthogonal function expansion, just the reverse is true: the diffusion matrix $\underline{\underline{\Pi}}$ is entirely diagonal, while the frequency matrix $\underline{\underline{L}}$ is non-diagonal in both spin and orientation.

Finally, we should also compare the final forms of the spectral intensity in the two methods. For the orthogonal expansion method, the integration of the solution over initial orientation angles, indicated in Eq. (2.11), picks out only the coefficient of the equilibrium distribution over angles, ϕ_0. Thus,

$$I(\omega) = \pi^{-1}\mathrm{Im}(\underline{\mu}^T \cdot \underline{f}_0) \qquad (3.35)$$

But this is equivalent to writing

$$I(\omega) = \pi^{-1}\mathrm{Im}(\underline{\underline{\mu}}^T \cdot \underline{\underline{f}}) \qquad (3.36)$$

in the direct product space, since the higher orientational components are removed by the vanishing of the corresponding components of $\underline{\underline{\mu}}$. Thus, the formula for the spectral intensity also has the same form, in the finite difference (3.16) and orthogonal function (3.36) cases. Again, we should emphasize that the magnitudes of the individual components in these equations bear no relation in the two approaches. Only the final summed intensities should be the same.

Thus, we see that the form of the linear equations is the same, whether we use a finite-difference approximation or an orthogonal function expansion. The advantage of the finite-difference approach is its generality and ease of application: Any spin Hamiltonian may be used, just by specifying values of its matrix elements for molecules at various orientations. Likewise, any model for the motion may be used, just by giving the appropriate transition probabilities between all pairs of configurations.

The orthogonal function expansion must be tailored more specifically to each application. However, if the spin Hamiltonian may be written simply in terms of one or a few of the chosen orthogonal functions, then the relevant frequency matrix elements are usually easy to evaluate and use.

XIII.4. COMPUTATIONAL ALGORITHMS

The linear equations (3.34) from which we may calculate the resonance line shapes (3.36) can of course be solved in many different ways. Numerical methods are usually required, because of the large number of equations. However, most numerical methods for the solution of linear equations do not take advantage of the special properties of these equations. These special properties are (1) the solutions are needed at <u>many values of</u> ω, the frequency, to plot out the spectrum. In most methods, this means a whole new calculation at each value of ω. (2) The coefficient matrices are <u>symmetric</u>. (3) The coefficient matrices are <u>band-form</u>; i.e., they vanish for coefficients removed by more than a certain fixed number of elements from the diagonal elements. In this section we develop a method of solution which takes advantage of these special properties of the solutions. Its use results in a dramatic reduction of both computer time and storage required for these calculations of line shapes.

A. Reduction to an Eigenvalue Problem

We first want to exhibit the frequency-dependence of the solutions. Multiplying the equations (3.34) on the left by a nonsingular matrix $\underline{\underline{R}}^{-1}$, gives

$$\underline{\underline{R}}^{-1}[\omega\underline{\underline{1}}-(\underline{\underline{L}}+i\underline{\underline{\Pi}})]\underline{\underline{f}} = \underline{\underline{R}}^{-1}\underline{\underline{\mu}} \tag{4.1}$$

Inserting the unit matrix $\underline{\underline{R}}\,\underline{\underline{R}}^{-1}$ between the square bracket matrix and the solution vector $\underline{\underline{f}}$ gives

$$[\omega\underline{\underline{1}}-\underline{\underline{R}}^{-1}(\underline{\underline{L}}+i\underline{\underline{\Pi}})\underline{\underline{R}}](\underline{\underline{R}}^{-1}\underline{\underline{f}}) = \underline{\underline{R}}^{-1}\underline{\underline{\mu}} \tag{4.2}$$

Now we suppose that a suitable matrix $\underline{\underline{R}}$ can be found to diagonalize the matrix $\underline{\underline{L}}+i\underline{\underline{\Pi}}$:

$$\underline{\underline{R}}^{-1}(\underline{\underline{L}}+i\underline{\underline{\Pi}})\underline{\underline{R}} = \underline{\underline{\Lambda}} \tag{4.3}$$

where $\underline{\underline{\Lambda}}$ is a <u>diagonal</u> matrix. Algorithms for the construction of $\underline{\underline{R}}$

are given in the following sections. Then the solution to the modified linear equations (4.2) is trivially accomplished by inverting the diagonal elements:

$$\underline{\underline{R}}^{-1}\underline{\underline{f}} = (\omega\underline{\underline{1}}-\underline{\underline{\Lambda}})^{-1}(\underline{\underline{R}}^{-1}\underline{\underline{\mu}}) \tag{4.4}$$

Then the spectral intensity I(ω) is easily formed from Eq. (3.36):

$$\begin{aligned} I(\omega) &= \pi^{-1}\mathrm{Im}(\underline{\underline{\mu}}^{T}\cdot\underline{\underline{f}}) \\ &= \pi^{-1}\mathrm{Im}(\underline{\underline{\mu}}^{T}\cdot\underline{\underline{R}})\cdot(\underline{\underline{R}}^{-1}\cdot\underline{\underline{f}}) \\ &= \pi^{-1}\mathrm{Im}(\underline{\underline{\mu}}^{T}\cdot\underline{\underline{R}})\cdot(\omega\underline{\underline{1}}-\underline{\underline{\Lambda}})^{-1}\cdot(\underline{\underline{R}}^{-1}\underline{\underline{\mu}}) \end{aligned} \tag{4.5}$$

The advantage of formula (4.5) is easily seen when written in component form:

$$I(\omega) = \pi^{-1}\mathrm{Im}\sum_{\ell}(\underline{\underline{R}}^{T}\underline{\underline{\mu}})_{\ell}(\underline{\underline{R}}^{-1}\underline{\underline{\mu}})_{\ell}/(\omega-\Lambda_{\ell\ell}) \tag{4.6}$$

A single summation over the eigenvalues produces the spectrum at any frequency. The complicated part of the calculation is the eigenvalue problem (4.3), but this is independent of the frequency, and need only be done once for any spectrum.[21] Another advantage of this form is that it is simply differentiated analytically with respect to frequency ω, to obtain the derivative spectrum usually observed experimentally.

The form the spectrum (4.6) is a sum of complex Lorentzian functions. Because of the complex number $(\underline{\underline{R}}^{T}\underline{\underline{\mu}})_{\ell}(\underline{\underline{R}}^{-1}\underline{\underline{\mu}})_{\ell}$ in the numerator, both absorption and dispersion-like components occur. Of course, the overall absorption sum is positive. The real parts of $\Lambda_{\ell\ell}$ give the main frequencies of the contribution, while the imaginary part gives the characteristic width of that contribution.

If all the complex eigenvalues $\Lambda_{\ell\ell}$ are distinct, then the required non-singular matrix $\underline{\underline{R}}$ certainly exists. The algorithms below are designed to find it. If some of the eigenvalues are degenerate, then it is possible, in principle, that such a diagonalization does <u>not</u> exist. (For <u>Hermitian</u> matrices, a diagonalization always exists, even in the presence of degeneracy, but $\underline{\underline{L}}+i\underline{\underline{\Pi}}$ is <u>not</u> Hermitian.) In practice, the eigenvalues are always perturbed or uncertain by small numerical roundoff errors, so mathematically exact degeneracy cannot be distinguished from near-degeneracy. The only symptoms of such problems are occasional large matrix elements in $\underline{\underline{R}}$ or $\underline{\underline{R}}^{-1}$. Using double precision (16-digit) computation, such occasional large transformation elements have caused no difficulty.

In the case of most interest, complex symmetric matrices $\underline{\underline{L}}+i\underline{\underline{\Pi}}$, the transformation $\underline{\underline{R}}$ is complex orthogonal; i.e., $\underline{\underline{R}}^{-1}=\underline{\underline{R}}^{T}$. In this case, the inverse transformation need not be computed, and the formula (4.6) simplifies to

$$I(\omega) = \pi^{-1}\mathrm{Im}\sum_{\ell} (\underline{\underline{R}}^{T}\underline{\underline{\mu}})_{\ell}^{2}/(\omega-\Lambda_{\ell\ell}) \tag{4.7}$$

We now turn to the actual construction of the diagonalizing transformation $\underline{\underline{R}}$ in Eq. (4.3). The strategy consists of two main steps. First, a transformation $\underline{\underline{R}}_1$ is found which reduces to zero all off-diagonal elements except those immediately adjacent to the diagonal. Such a matrix, with only three non-zero elements in any row, is called a tridiagonal matrix. Then a second transformation $\underline{\underline{R}}_2$ is applied to the tridiagonal matrix, and is designed to eliminate the remaining off-diagonal elements bordering the diagonal one. The desired transformation $\underline{\underline{R}}$ is then the product $\underline{\underline{R}}_1\underline{\underline{R}}_2$.

Different strategies are required for these two stages of transformation. There are many possible methods for each of these stages. One procedure has been successfully used.[21] However, here we should like to maintain the symmetric and bandform character of the matrix $\underline{\underline{L}}+i\underline{\underline{\Pi}}$ at each stage of transformation. These restrictions minimize both the storage required and the number of arithmetic operations. Also, in many cases of interest, there are many zero elements scattered within the bandwidth considered. Thus, we would also like a method which, as much as possible, takes advantage of elements which are already zero. These requirements are met by the algorithm discussed below. All others of which the authors are aware either fail to maintain symmetry, or increase the bandwidth at intermediate stages, or cannot take advantage of existing zeros, or combine several of these disadvantages.

B. Reduction of Band-Width to Tridiagonal

First, we consider a transformation developed by Rutishauser[22] to symmetrically tridiagonalize a symmetric bandform matrix by Jacobi rotations. A Jacobi rotation in the plane of i and j is the orthogonal transformation

$$J(i,j,\phi) = \begin{pmatrix} ..\cos\phi...\sin\phi.. \\ .-\sin\phi...\cos\phi.. \end{pmatrix} \begin{matrix} i \\ j \end{matrix} \tag{4.8}$$

All of the Jacobi rotations used here have j = i+1; thus the right-hand transformation matrix is

$$
\underset{\sim}{J} = \begin{pmatrix} \ldots 1 & 0 & 0 & 0 & 0 & 0\ldots \\ \ldots 0 & 1 & 0 & 0 & 0 & 0\ldots \\ \ldots 0 & 0 & \cos\phi & \sin\phi & 0 & 0\ldots \\ \ldots 0 & 0 & -\sin\phi & \cos\phi & 0 & 0\ldots \\ \ldots 0 & 0 & 0 & 0 & 1 & 0\ldots \\ \ldots 0 & 0 & 0 & 0 & 0 & 1\ldots \\ \vdots & \vdots & \vdots & \vdots & \vdots & \vdots \end{pmatrix} \begin{matrix} \\ \\ i \\ i+1 \\ \\ \\ \\ \end{matrix} \qquad (4.9)
$$

Let $\underset{\sim}{A}$ be the bandform matrix, of dimension N, being trigiagonalized. Let $A_{1,B+1}$ be the rightmost nonzero element in the first row; i.e., B is the half bandwidth excluding the diagonal. If B=3, then the matrix $\underset{\sim}{A}$ is

$$
\begin{matrix} A_{11} & A_{12} & A_{13} & A_{14} & 0 & 0 & 0 & 0 & \ldots \\ & A_{22} & A_{23} & A_{24} & A_{25} & 0 & 0 & 0 & \ldots \\ & & A_{33} & A_{34} & A_{35} & A_{36} & 0 & 0 & \ldots \\ & & & A_{44} & A_{45} & A_{46} & A_{47} & 0 & \ldots \\ & & & & A_{55} & A_{56} & A_{57} & A_{58} & \ldots \end{matrix} \qquad (4.10)
$$

where we suppress the elements below the diagonal, because the matrix is symmetric and stays symmetric after each complete transformation. Thus, $\underset{\sim}{AJ}$ is

$$
\begin{matrix} A_{11} & A_{12} & A_{13}{*} & (A_{13}\sin\phi + A_{14}\cos\phi) & 0 & 0 & 0 & 0 \\ & A_{22} & A_{23}^{*} & A_{24}^{*} & A_{25} & 0 & 0 & 0 \\ & & A_{33}{*} & A_{34}{*} & A_{35} & A_{36} & 0 & 0 \\ & & & A_{44}{*} & A_{45} & A_{46} & A_{47} & 0 \\ & & & & A_{55} & A_{56} & A_{57} & A_{58} \end{matrix} \qquad (4.11)
$$

where * means that the element of $\underset{\sim}{AJ}$ is different from the corresponding element of $\underset{\sim}{A}$ but is not given explicitly. The point of writing A_{14}^{*} out explicitly as $A_{13}\sin\phi + A_{14}\cos\phi$ will become apparent.

Upon multiplying on the left by $\underset{\sim}{J}^{-1}$, we have

$$
\underset{\sim}{A}' = \left\{
\begin{array}{llllllll}
A_{11} A_{12} A_{13}*(A_{13}\sin\phi+A_{14}\cos\phi) & 0 & 0 & 0 & 0 \\
A_{22} A_{23}*A_{24}* & & A_{25} & 0 & 0 & 0 \\
A_{33}*A_{34}* & & A_{35}* & A_{36}*(\sin\phi A_{47}) & 0 \\
A_{44}* & & A_{45}* & A_{46}* & A_{47}* & 0 \\
& & A_{55} & A_{56} & A_{57} & A_{58}
\end{array}
\right\} \quad (4.12)
$$

where * again indicates change from $\underset{\sim}{A}$. Let $\cos\phi/\sin\phi = -A_{13}/A_{14}= -R$. The $1-\sin^2\phi = R^2\sin^2\phi$, or $\sin\phi = (1+R^2)^{-1/2}$. Therefore, $\cos\phi = -R\sin\phi = -R(1+R^2)^{-1/2}$. Notice that this choice of ϕ makes $A_{14}^* = 0$. Thus, we see that we can annihilate the first element of the outermost superdiagonal by rotation in the plane 3,4. But now we have a new non-zero element $A_{37}^* = -A_{47}\sin\phi$. This blight can be exterminated by rotation in the plane 6,7, where R is now $-A_{36}^*/A_{46}^*$, but only at the cost of a new pest at (6,10). Now we can wipe out (6,10) by rotation in (9,10), but we get (9,13), etc. Eventually, however, we come to the right-hand edge of the matrix, and the new element is formed "over the edge," i.e., we annihilate without creating. We show this process diagrammatically in Fig. 1.

So we have pushed A_{14} over the edge. Now we push (2,5) over the edge, beginning rotations in the plane 4,5 with R chosen analogously. After the uppermost superdiagonal has been annihilated, we destroy the next uppermost one, etc. If, at any stage, an element to be eliminated already vanishes, the corresponding transformations may be omitted.

```
The general expression of the algorithm is
B' = B
loop over k from 1 to 2 less than the original bandwidth B
    loop over i from 1 to N - current bandwidth, B'
    rotate in the plane i+B'-1, i+B' with
        R = -A_{i,i+B'-1}/A_{i,i+B'}
        loop over j from i+B'-1 to N-B'-1 by steps of B'
        rotate in the plane B'+j, B'+j+1 with
            R = -A_{j,B'+j}/A_{j,B'+j+1}
        end of j loop
    end of i loop
B' = B' - 1
end of k loop
```

Notice that the relative positions of created and annihilated elements depend on the _current_ bandwidth.

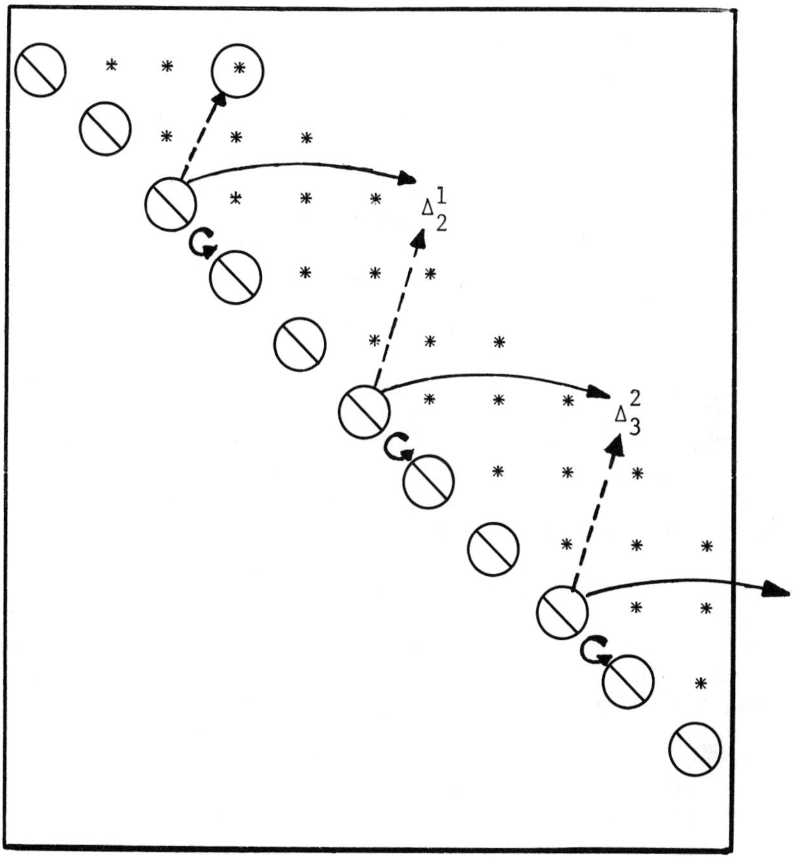

Fig. 1: Strategy for reduction of matrix band width by Jacobi rotations. ⊛ is the first element to be eliminated. * are the non-zero matrix elements, and ⊘ are the diagonal elements. Jacobi rotations in the i, i+1 plane are represented by G between the corresponding diagonal elements. Δ^a_{a+1} is the element created by rotation a and destroyed by the next.

According to Eq. (4.7), the transformation R^T must also be applied to μ. This is most efficiently done inside the j loop, in order to avoid storing the whole transformation R.

This transformation to tridiagonal form takes a fixed number of operations to find R_1. The number of multiplications is approximately given by B(2N−B)4N, where N is the order of the matrix, and B is the number of non-zero columns to be eliminated, one one side of the diagonal. (The other side, being symmetric, is not used explicitly.) If zero elements are found within these columns, then some of these operations are omitted, so this estimate gives an upper limit to the number of operations required. The matrix storage required is about N(B+3) complex elements.

C. Diagonalization of a Complex Symmetric Tridiagonal Matrix

We next turn to the diagonalization of a tridiagonal matrix. The matrix $L+i\Pi$ for the simple P_2 spectrum discussed in Sect. 3 is already tridiagonal, using either finite difference or Legendre polynomial expansions. For most other cases $L+i\Pi$ must first be "squeezed" (reduced in bandwidth) by the algorithm of the previous section, to make it tridiagonal.

We now use a modification of the QR algorithm of Francis,[23] to diagonalize this complex symmetric tridiagonal matrix. For the following discussion, let A_1 be a general N×N matrix to be diagonalized. Then A_1 has the factorization $A_1 = Q_1R_1$ where Q_1 is complex orthogonal and R_1 is uppertriangular. Then $A_2 \equiv R_1Q_1 = Q_1^{-1}A_1Q_1$ is similar to A_1. Similarly, $A_3 = Q_2^{-1}A_2Q_2 = Q_2^{-1}Q_1^{-1}A_1Q_1Q_2$, etc. It can be shown that this series of matrices similar to A_1 generally tends to an upper triangular matrix.[23,24] The diagonal elements of this matrix are, of course, the eigenvalues of A_1, since the characteristic polynomial of the triangular matrix with diagonal elements μ_i is $\prod_{i=1}^{N}(\mu_i-\lambda)$.

We use the Jacobi rotations to effect the decomposition $A_k=Q_kR_k$. Actually, we consider the problem $Q_k^{-1}A_k = R_k$. That is, we pick our Jacobi rotations (whose product is Q_k^{-1}) such that they triangularize A_k. For a general matrix we triangularize row r+1, by rotating successively in the planes i, r+1 where i=1,2,...r. For each rotation we pick ϕ such that $a_{r+1,i}$ becomes zero.

The above algorithm for a general matrix is almost hopelessly slow and wasteful of space as it stands. An upper Hessenberg matrix has the form $a_{ij} = 0$ for $i<j-1$. For such a matrix, however, we need only <u>one</u> rotation to triangularize each row. Since a general matrix may be reduced to upper Hessenberg form,[24] it is possible to diagonalize a general matrix with the Hessenberg QR. The Hessenberg form permits another economy. Suppose that after a QR step the resulting matrix has the form

$$
\begin{array}{llllllll}
* & \text{———} & & & & & & \\
* & * & \text{———} & & & & & \\
0 & * & * & \text{———} & & & & \\
0 & 0 & * & * & \text{———} & & & \\
0 & 0 & 0 & * & * & \text{———} & & \\
0 & 0 & 0 & 0 & n & * & \text{———} & \\
0 & 0 & 0 & 0 & 0 & * & * & \text{———} \\
0 & 0 & 0 & 0 & 0 & 0 & * & * \;\text{———}
\end{array}
$$

where n indicates a "negligibly small" element. Our matrix has thus assumed block Hessenberg form, and each block may be diagonalized separately. This means fewer rotations for each decomposition. Of course, if n occurs in the last row, the diagonal element of that row is an eigenvalue. The succeeding decompositions need not tri-diagonalize this row. That is, later operations take place only on the first N-1 rows and N-1 columns of the matrix. If n occurs in the second to last row, we can easily diagonalize the two by two matrix so isolated. Then we can omit further operations on the last two rows and columns.

Unfortunately, the rate of convergence of this algorithm is only linear. However, applications of the algorithm to $\mathbf{A}_1 - s\mathbf{I}$ where s is a "shift" close to an eigenvalue greatly accelerates convergence.[23,24] We estimate s by solving for the eigenvalues of the two by two submatrix at the bottom right-hand corner and using the root of smaller modulus.

Returning to our complex symmetric tridiagonal matrix, we note that this matrix is already upper Hessenberg. (It is lower Hessenberg, too, of course.) Additionally, we can take advantage of the symmetry by using the transformation of Ortega and Kaiser.[25] Let

$$
\mathbf{L} + i\mathbf{\Pi} = \mathbf{A} = \left\{ \begin{array}{l} a_1 b_2 \\ b_2 a_2 b_3 \\ \quad b_3 a_3 b_4 \\ \quad \cdots\cdots \; b_n a_n \end{array} \right\} \tag{4.13}
$$

$$
\text{and } \mathbf{A} = \mathbf{R}\mathbf{Q} = \mathbf{Q}^{-1}\mathbf{A}\,\mathbf{Q} = \left\{ \begin{array}{l} \bar{a}_1 \bar{b}_2 \\ \bar{b}_2 \bar{a}_2 \bar{b}_3 \\ \quad \cdots\cdots \\ \qquad \bar{b}_n \bar{a}_n \end{array} \right\} \tag{4.14}
$$

$$\text{and } \underset{\sim}{R} = \begin{Bmatrix} r_1 q_1 t_1 & & & \\ & r_2 q_2 t_2 & & \\ & & \cdot\cdot\cdot\cdot\cdot & \\ & & \cdot\cdot\cdot & \\ & & \cdot\cdot & \\ & & & r_n \end{Bmatrix} \tag{4.15}$$

(The similarity transformation does preserve symmetry, and bandwidth, and each row of R has only three elements.) We left out shifts for simplicity in the exposition, but they are used in the calculation. The exposition above shows that A can be reduced to $\underset{\sim}{R}$ by N-1 rotations in the planes (i,i+1) where i=1 to N-1. Let the sine and cosine of the i-th rotation be c_i and s_i. Then,

$$s_i = b_{i+1}/(p_i^2+b_{i+1}^2)^{1/2}, \quad c_j = p_j/(p_j^2+b_{j+1}^2)^{1/2} \tag{4.16}$$

where $p_1 = a_1$, $p_2 = c_1a_2-s_1b_2$, and $p_i = c_{i-1}a_i-s_{i-1}c_{i-2}b_i$ for i>2, are the rotations which remove the b's in order. That is, for example,

$$\begin{bmatrix} \cos\phi_1 & \sin\phi_1 & & \\ -\sin\phi_1 & \cos\phi_1 & & \\ & & 1 & \\ & & & 1 \end{bmatrix} \begin{bmatrix} a_1 & b_2 & & \\ b_2 & a_2 & b_3 & \\ 0 & b_3 & a_3 & b_4 \\ & & & \end{bmatrix}$$

$$= \begin{bmatrix} (a_1\cos\phi_1 + b_2\sin\phi_1) & (b_2\cos\phi_1 + a_2\sin\phi_1) & 0 \\ (-a_1\sin\phi_1 + b_2\cos\phi_1) & (-b_2\sin\phi_1 + a_2\cos\phi_1) & b_3\cos\phi_1 \\ 0 & b_3 & a_3 \end{bmatrix} \tag{4.17}$$

and

$$-a_1\sin\phi_1 + b_2\cos\phi_1 = -a_1b_2/(a_1^2+b_2^2)^{1/2}+b_2a_1/(a_1^2+b_2^2)^{1/2} = 0 \tag{4.18}$$

Also,

$$r_i = c_i p_i + s_i b_{i+1} \qquad i = 1,\dots N-1$$
$$r_N = p_N$$
$$q_1 = c_1 b_2 + s_1 a_2 \qquad (4.19)$$
$$q_i = c_i c_{i-1} b_{i+1} + s_i a_{i+1} \qquad i = 2,\dots N-1$$
$$t_i = s_i b_{i+2} \qquad i = 1,\dots N-2$$

and, upon forming $\underset{\sim}{R}\underset{\sim}{Q}$, we get

$$\bar{a}_i = c_1 r_1 + s_1 q_1$$
$$\bar{a}_i = c_{i-1} c_i r_i + s_i q_i \qquad i = 2,\dots N-1 \qquad (4.20)$$
$$\bar{a}_N = c_{N-1} r_N$$
$$\bar{b}_{i+1} = s_i r_{i+1} \qquad i = 1,\dots N-1$$

Let

$$\gamma_1 = p_1, \quad \gamma_i = c_{i-1} p_i \qquad i = 2,\dots,N \qquad (4.21)$$

Then we may eliminate q and r from the expressions (20) by using (21) and (19)

$$\bar{a}_i = (1+s_i^2)\gamma_i + s_i^2 a_{i+1}, \qquad a_N = \gamma_N$$
$$\bar{b}_{i+1} = s_i (p_{i+1}^2 + b_{i+2}^2)^{1/2} \qquad i = 1,\dots,N-2; \ \bar{b}_N = s_{N-1} p_N$$
$$\gamma_1 = p_1 = a_1, \ \gamma_i = a_i - s_{i-1}^2 (a_i + \gamma_{i-1}) \qquad (i = 2,\dots N)$$
$$p_1 = a_1, \ p_2 = c_1 a_2 - s_1 b_2, \ p_i = c_{i-1} a_i - s_{i-1} c_{i-2} b_i \quad (i=3,\dots,N)$$

Let $\delta_i = s_{i-1} c_{i-2}$, then $p_i = c_{i-1} a_i - \delta_i b_i$. Notice that only one δ, γ, p, c, or s is needed at any time. Thus we need not make these variables into arrays, since we can store the new value on the old. Also the $\bar{a}$ and $\bar{b}$ may be overlaid on the a and b. In the algorithm below, we will use the arrow $\leftarrow$ to indicate storage as in

$$a_i \leftarrow \bar{a}_i = \text{expression}$$

which means that $\bar{a}_i$ as evaluated from the expressions is overlaid on a_i. We will also use square brackets around expressions using subscripted δ, γ, p, c, s to make clear which $\delta_i, \gamma_i, p_i, c_i, s_i$ are currently stored in δ, γ, p, c, s. The algorithm as programmed is given without brackets.

Decide size of negligible element on basis of norm times machine accuracy.

Repeat this loop until all eigenvalues are found.

solve lower right-hand 2×2 matrix and use smaller root as shift
subtract current shift from diagonal elements
cumulate total shift - i.e., add current shift to total shift
set some initial values

$$\delta = 0 \qquad (\text{i.e., } \delta_1 = 0)$$

$$u = 0 \qquad (\text{i.e., } u_o = 0)$$

$$c = 1 \qquad (\text{i.e., } c_o = 1)$$

$$b_{N+1} = 0$$

$$a_{N+1} = 0$$

loop over i from current lower matrix index N'' to current upper index N'

$$\gamma \leftarrow \gamma_1 = [a_i - u_{i-1}] = a_i - u$$

$$p \leftarrow p_i = [c_{i-1} a_i - \delta_i b_i] = c a_i - \delta b_i$$

if i = N' go to end of i loop

$$z = [(p_i^2 + b_{i+1}^2)^{1/2}] = (p^2 + b_{i+1}^2)^{1/2}$$

if i > N'', then $\bar{b}_i = [s_{i-1} z] = sz$

$$\zeta = 1/|z|^2$$

$$s \leftarrow s_i = z\zeta b_{i+1} = b_{i+1}/z$$

$$\delta \leftarrow \delta_{i+1} = [c_{i-1} s_i] = cs$$

$$u \leftarrow u_i = s_i^2(\gamma_i + a_{i+1})$$

$$a_i \leftarrow \bar{a}_i = \gamma_i + u_i$$

end of loop on i
for i = N' we must finish up here

$$b_{N'} \leftarrow \bar{b}_{N'} = [s_{N'-1}p_i]$$

$$a'_N \leftarrow \bar{a}_{N'} = [\gamma_{N'}] = \gamma$$

We now check along the b's for a negligible element. If none is found, we iterate again (repeat loop over i). If there is one, then we deal with the block tridiagonal matrix as already discussed, altering N" and N' as required. The shift is added back to each diagonal element as it is isolated as a eigenvalue.

Again recall that the vectors $\underline{\underline{\mu}}$ must be transformed by $\underline{\underline{R}}^T\underline{\underline{\mu}}$ whenever the matrix is obtained.

These iterations are continued until all of the off-diagonal elements of the tridiagonal matrix are reduced below a preset tolerance, usually a norm of the matrix times the relative accuracy of a floating point number. The number of iterations required depends somewhat on the matrix being diagonalized. In practice, around three iterations per eigenvalue are required, and the computational effort is about

$$8N^2 \text{ (\# iterations)}[1+R]$$

complex multiplications, where N is the order of the matrix, and R is the ratio of computational effort required for one complex square root, divided by 16 complex multiplications.

Comparing this estimate with that given above for the reduction of band width to tridiagonal, we see that for narrow banded matrices (half band width less than about six) the diagonalization of the tridiagonal matrix requires more calculations. Conversely, for matrices with a broader band of non-zero elements, the reduction to tridiagonal form is the longest part of the calculation. This is the case for the nitroxide line shape calculations discussed in Sect. VI, for which the half band width is 13.

The matrix storage requirements for diagonalizing a tridiagonal matrix and transforming the $\underline{\underline{\mu}}$ vector, are only about 3N complex elements. This is always less than that estimated above for the reduction of band width to tridiagonal.

XIII.5. ESTIMATING RATES OF CONVERGENCE

First we consider the line shapes for the adiabatic model of dipolar broadening[1,7-10,13] discussed in Sect. 3. Calculations were made using both the finite-difference and orthogonal function expansions. For rapidly tumbling molecules there is complete agreement between the results, even when only a few terms are kept in either approach. For slower tumbling rates (where the rotational diffusion constant D is smaller than the width of the spectrum, α, in Eq. (3.31)) the convergence of both approximations is slower. The differences between the two results are most pronounced near the outer edges of the spectrum. The finite difference results tend to oscillate above and below the limiting results, while the polynomial expansion exhibits an "overshoot," similar to the Gibbs phenomenon in summing Fourier series. Figure 2 gives some typical curves. It is difficult to give a single quantitative measure for the rate of convergence of the curves. However, qualitatively, the polynomial function expansion seems to approach the limit somewhat faster and more smoothly.

It is desirable to have some more quantitative estimates for the rates of convergence of the two expansions. First we consider the finite difference approach. If we have too few grid points for convergence, spurious grid "lines" remain in the spectrum. As the tumbling rate increases, the rate of "jumping" between these lines increases, and the line structure blends together. A rough condition for this motional averaging to occur, is that the rate of jumping must exceed the splitting between lines[11]. In the present case, this rate of jumping is given by minus the off-diagonal $\underline{\underline{\Pi}}$ matrix element in Eq. (3.14). The minimum jumping rate occurs for the first and last grid points (near $\cos\theta = \pm 1$), and it has the value

$$-\Pi_{12} = -\Pi_{N,N-1} = \frac{1}{4} D \tag{5.1}$$

It also happens that the frequency spacing between neighboring grid points is also the largest near $\cos\theta = \pm 1$, having the value

$$\Delta\omega = 4\omega_f/N \tag{5.2}$$

where ω_f is the full width of the spectrum ($\omega_f = \alpha(3/2)(5/2)^{1/2}$, where α is defined in Eq. (3.27)). Since for convergence to a smooth spectrum requires that the jumping rate exceed the frequency spacing, we see that the frequency region corresponding to $\cos\theta = \pm 1$ should be the last to converge, which agrees with the numerical results discussed above. The condition for convergence of this last part of the spectrum is thus

$$\frac{1}{4} D \gtrsim 4\omega_f/N \tag{5.3}$$

or

$$N \gtrsim 16(\omega_f/D) \tag{5.4}$$

This convergence condition (5.4) also agrees with the numerical results, in showing that more points are required at slower tumbling rates (smaller D). We should also emphasize that most of the spectrum has converged long before this condition is met. Only the points near $\cos\theta = \pm 1$ converge this slowly.

This convergence discussion suggests that a non-uniform grid of points, with closer spacings near $\cos\theta = \pm 1$, would give a finite difference approximation which is more rapidly and uniformly convergent. Initially, non-uniform grids were rejected because they lead to non-symmetric $\underline{\underline{\Pi}}$ matrices, which cannot be treated directly by the methods of Sect. IV. However, it would be easy to symmetrize the equations with one additional similarity transform by the diagonal matrix $\underline{\underline{p}}^{1/2}$, where $\underline{\underline{p}}$ has as its diagonal elements the equilibrium populations of the grid points, and then the methods of Sect. IV could again be used. We expect that non-uniform grids can achieve substantially improved convergence over the uniform one used here. For example, a grid uniformly spaced in angles θ, rather than $\cos\theta$, would have much finer frequency spacings near $\cos\theta = \pm 1$, varying as N^{-2} rather than N^{-1} as in Eq. (5.2). Then the convergence criterion would be

$$N \gtrsim \text{constant} \times (\omega_f/D)^{1/2} \tag{5.5}$$

which is a substantial improvement over Eq. (5.4) at small tumbling rates.

Next we turn to the rate of convergence for the polynomial expansion [Eq. (3.17)] for the P_2 spectrum. The higher order coefficients become negligible when the coupling $L_{m,m+1}$ in Eq. (3.32) no longer is effective in coupling the linear equations. As m becomes large, L approaches the constant value

$$L_{m,m+1} \rightarrow (3/8)(5/2)^{1/2}\alpha = \frac{1}{4}\,\omega_f \tag{5.6}$$

Thus the uncoupling does not come about because the coupling element approaches zero, but rather because the diagonal elements of $\underline{\underline{\Pi}}$ become large [cf. Eq. (3.28)]

$$\begin{aligned} \Pi_{mm} &= -Dm(m+1) \\ &\approx -Dm^2 \end{aligned} \tag{5.7}$$

at large m. Thus the solution components are strongly coupled together up to a value of the index, m_o, such that diagonal and off-diagonal elements are of similar size:

$$\frac{1}{4}\,\omega_f \approx Dm_o^2 \tag{5.8}$$

The solution components f_n in Eq. (3.28) then decrease for $m>m_o$. To see how soon they become small, we write the largest terms in a typical equation for these coefficients, for large m, as

$$\frac{1}{4}\omega_f f_{m-1} - iDm^2 f_m + \frac{1}{4}\omega_f f_{m+1} = 0 \tag{5.9}$$

For large m, the last term is negligible, since f_{m+1} is smaller than f_{m-1}. Rearranging first two terms in Eq. (5.9) gives

$$\frac{|f_m|}{|f_{m-1}|} \simeq \frac{\omega_f}{4Dm^2} = \left(\frac{m_o}{m}\right)^2 \tag{5.10}$$

Iterating this reduction factor up to the N-th term gives

$$\frac{|f_N|}{|f_{m_o}|} \simeq \left(\frac{m_o}{m_o+1}\right)^2 \left(\frac{m_o}{m_o+2}\right)^2 \cdots \left(\frac{m_o}{N}\right)^2 = \left|\frac{m_o^{N-m_o}\, m_o!}{N!}\right|^2 \tag{5.11}$$

Approximating the factorials with Stirling's formula, this reduction factor becomes

$$\frac{|f_N|}{|f_{m_o}|} \simeq \left(\frac{em_o}{N}\right)^{2N} e^{-2m_o} \tag{5.12}$$

Now if we take N large enough so that

$$N \gtrsim em_o \tag{5.13}$$

then the factor in parentheses in Eq. (5.12) is smaller than unity and the exponential factor is also small, if m_o is reasonably large. Thus the final component of the solution f_N is small, and we may safely set to zero the higher components of the solution, obtaining a finite system of linear equations. Combining Eqs. (5.10) and (5.13), we obtain

$$N \gtrsim e(\omega_f/4D)^{1/2} \tag{5.14}$$

as the number of terms required for convergence of the polynomial approximation. This is the <u>same</u> order of magnitude as estimated for a finite difference approximation based on equal angular steps, Eq. (5.5), and a considerable <u>improvement</u>, at slow tumbling rates, over the finite difference grid with equal increments in cosθ, Eq. (5.4). It is also interesting to note that Fixman[8] obtained for a convergence condition for the number of terms

$$N \gtrsim \text{constant} \times (\omega_f/D)^{1/2} + 2 \tag{5.15}$$

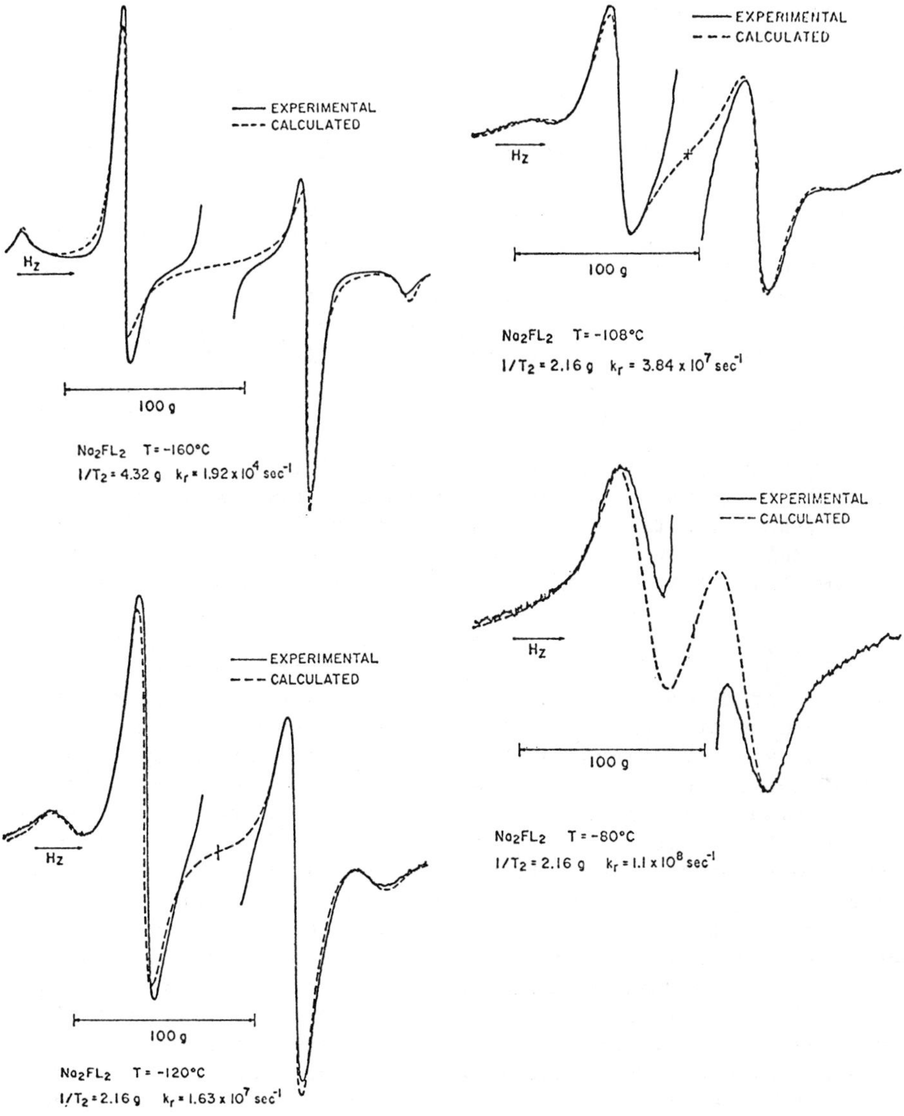

Fig. 2: Comparisons of experimental and theoretical line shapes for triplet spectra (adiabatic P_2 model), from ref. 1.

from a numerical study of the convergence of the polynomial expansion. At small tumbling rates, this agrees well with our theoretical estimate, Eq. (5.14). At large tumbling rates, the added 2 in Eq. (5.15) guarantees that at least two terms are always kept, so that the usual perturbation results are obtained in the limit of fast rotational diffusion.

XIII.6. APPLICATIONS AND COMPARISON WITH EXPERIMENTS

Some experimental ESR spectra corresponding to dipole-dipole interactions in a tumbling triplet molecule have been obtained recently.[1] Figure 2 shows that the main features do indeed correspond to the P_2 model discussed in Sect. III and V. From comparisons of this sort, one may estimate the rotational diffusion constants for these molecules. The main experimental problem is the large central peak, which represents molecules with spin 1/2, which do not show any strong intramolecular dipole-dipole splitting. However, the outer unobscured portions of the triplet spectrum are sensitive to the motion, and are sufficient to estimate the rate of tumbling.

Another example, of great current interest, is that of nitroxide radicals,[2,15] in which the main structure is due to the hyperfine coupling with the nitrogen nucleus. This coupling is nearly axial,[2,15] and for simplicity we assume exact axial symmetry. We also assume axial symmetry for the $\underset{\approx}{g}$ tensor, with the same axis as the nitrogen hyperfine tensor. This axial assumption about the $\underset{\approx}{g}$ tensor is not too accurate, but significant effects of the g tensor x-y anisotropy might only be expected in certain portions of the spectra, mainly near the central component, which has the smallest hyperfine broadening. Because of these assumptions of axial symmetry, the orthogonal expansion functions can be taken in this case to be normalized spherical harmonics $\hat{Y}_{\ell m}(\theta,\phi)$. The necessary matrix elements required for the frequency matrix in Eq. (3.20) are well known.[20] To reduce the number of equations we include only the dipole matrix elements for the upward electron spin transition. This means that we neglect certain "non-secular" contributions to line widths, but these are only important at very rapid tumbling rates, which can cause the electron spin to flip. These rapid tumbling cases can be treated readily by the usual perturbation theory,[5,6] and will not concern us here. We do include all spin matrix elements of the I=1 spin of the ^{14}N nucleus. Thus all the "pseudo-secular" contributions to the line shape are included. These terms, which are not included in previous adiabatic treatments,[7-10] have important effects on the line shape in the slow and intermediate rates of tumbling which concern us here. Thus nine spin dipole matrix elements are included, including the three hyperfine lines allowed at high field (corresponding to no change in the ^{14}N spin quantum number), and the six lines forbidden at high field. The "pseudo-secular" off-diagonal matrix elements of $\underset{\sim}{L}$ mix together all of these nine lines.

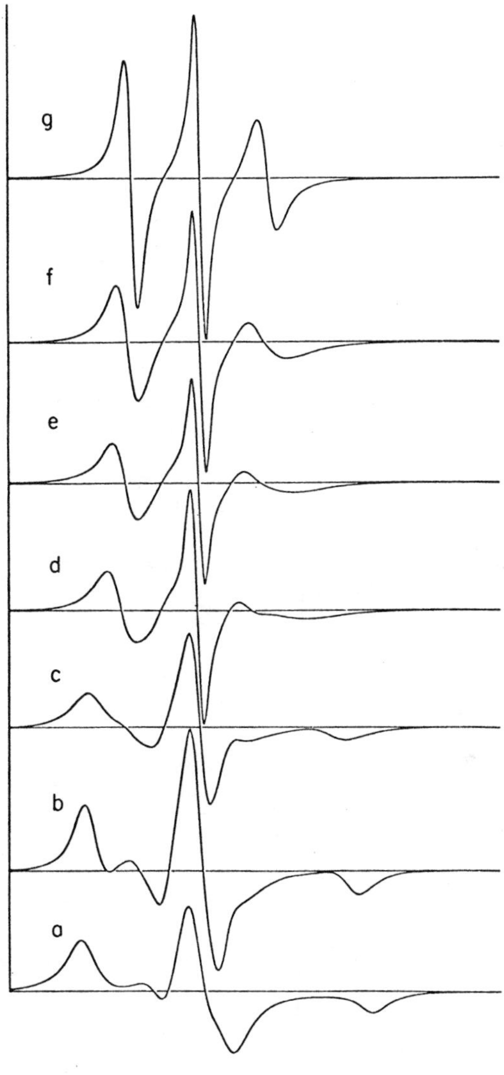

Fig. 3: Calculated nitroxide radical spectra. The assumed constants in the spin Hamiltonian are $g_x = 2.0075$, $g_y = 2.0075$, $g_z = 2.0027$, $A_x = 6.35$ gauss, $A_y = 6.35$ gauss, $A_z = 32.0$ gauss. The assumed rotational correlation times (τ, seconds) and intrinsic width (Δ, gauss). (a) $\tau = 2.1\times10^{-7}$ sec, $\Delta = 4$ gauss; (b) $\tau = 2.6\times10^{-8}$ sec, $\Delta = 2$ gauss; (c) $\tau = 1.1\times10^{-8}$, $\Delta = 2$; (d) $\tau = 3.5\times10^{-9}$, $\Delta = 1$; (e) $\tau=2.6\times10^{-9}$, $\Delta = 1$; (f) $\tau = 2.1\times10^{-9}$, $\Delta = 1$; (g) $\tau = 1.1\times10^{-9}$, $\Delta = 1$.

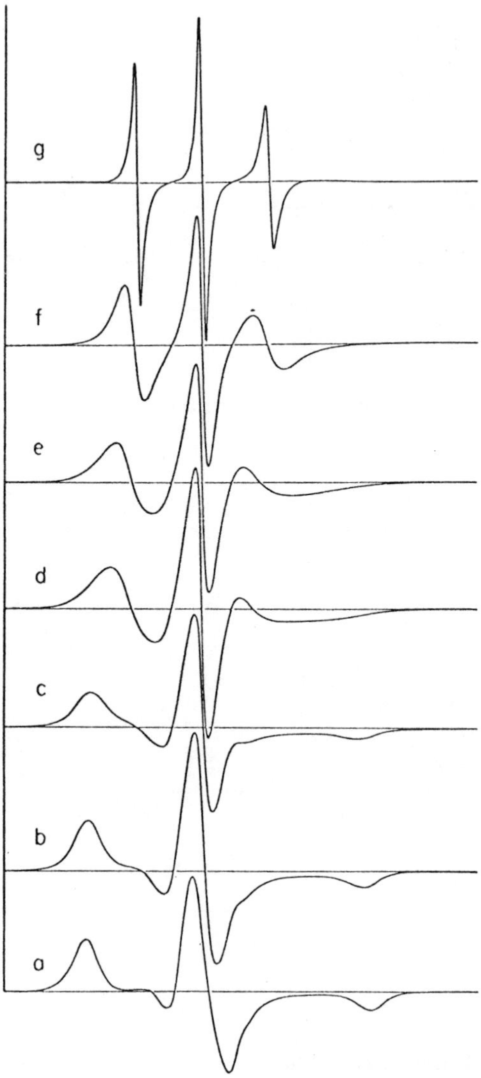

Fig. 4: Experimental di-tertiary butyl nitroxide derivative spectra in glycerol, as a function of temperature, after O.H. Griffith (private communication). Temperatures, in °C: (a) -49. (b) -26. (c) -17.5 (d) -3.5 (e) +0.3 (f) 9.4 (g) 26.5.

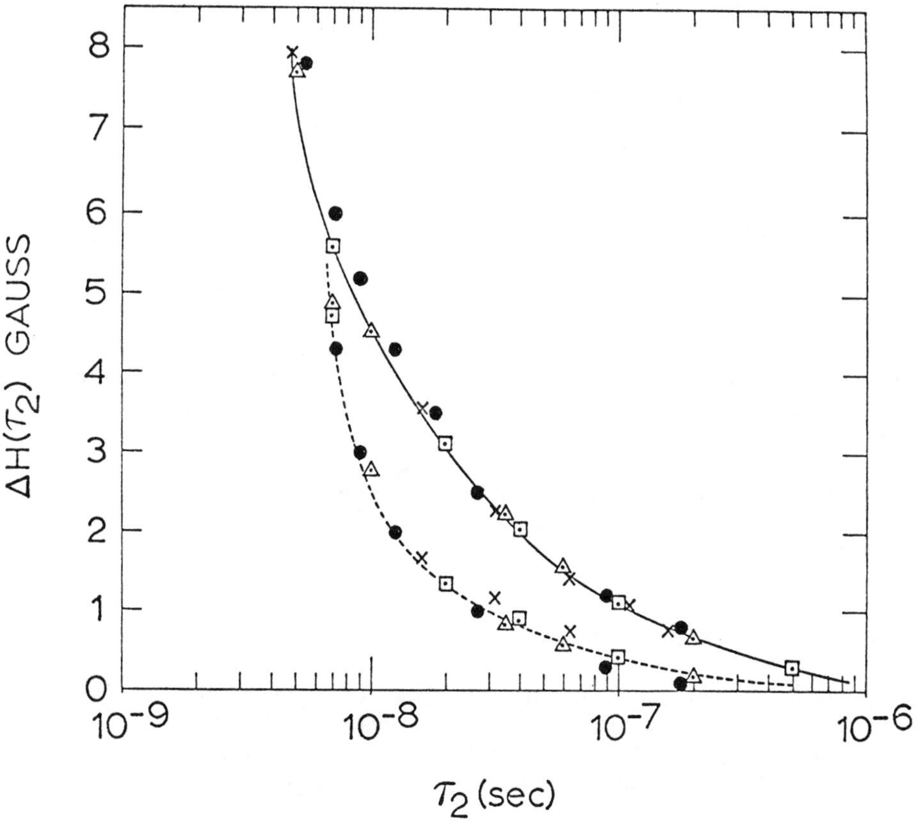

Fig. 5:

Inward shifts (ΔH, in gauss) of the high field (—) and low field (---) extrema of calculated nitroxide derivative spectra as a function of the second-order rotational correlation time $\tau_2=(6D)^{-1}$; points X, present results for $A_{\parallel}=30.8$G., $A_{\perp}=5.8$G, $\Delta g=.0053$ and intrinsic broadening from $T_2=3.3\times10^{-8}$sec; points • from S. A. Goldman, G. V. Bruno, C. F. Polnaszek, and J. H. Freed, J. Chem. Phys. (to be published), with $A_{\parallel}=32.0$G, $A_{\perp}=6.0$G, $\Delta g = 0.0048$, and $T_2=2.2\times10^{-6}$ sec; points Δ from an approximate calculation by R. C. McCalley, E. J. Shimshick, and H. M. McConnell (to be published), for $A_{\parallel}$ = 30.8G, $A_{\perp}=5.8$G, $\Delta g = 0.0053$, and $T_2 = 2.4\times10^{-8}$sec, and points □ for $A_z = 34.3$G, $A_x=6.8$G, $A_y=6.2$G, $g_z=2.0024$, $g_x=2.0090$, $g_y=2.0060$, and $T_2 = 2.4\times10^{-8}$ sec.; also from McCalley <u>et al</u>. are calculated lines (—— and ---) for $A_{\parallel}=34.3$G, $A_{\perp}=6.5$G, $\Delta g=0.0051$ and $T_2=2.4\times10^{-8}$ sec.

Some typical results are plotted in Fig. 3, for parameters that correspond reasonably to the experimental curves in Fig. 4. In general, the agreement between theory and experiment is excellent, and all the main features of the experimental curves are well reproduced by the theory. Rotational diffusion constants, or rotational relaxation times,[14] are well determined, especially at the intermediate rates, where the form of the spectrum is changing rapidly.

In the region of slow tumbling, Shimshick and McConnell[26] have suggested a convenient parameter to characterize the changes in the spectrum, as the decrease in distance between the outermost peaks in the derivative of the spectrum, from the distance they would be in a rigid glass or solid, without tumbling. This quantity is convenient to measure from experimental spectra, and it correlates monotonically with tumbling rate, as shown in Fig. 5, which is calculated from the theoretical results. This "calibration curve" allows one to estimate easily rotational diffusion rates in the region of slow tumbling. For intermediate tumbling rates, Fig. 3 allows the rates to be estimated. For more rapid tumbling rates, the linewidths may be interpreted by the usual perturbation theory[5,6] which is valid in this limit.

Together, these methods permit interpretation of ESR spectra of tumbling molecules over a wide range of relaxation times, from about 10^{-6} sec to less than 10^{-11} sec.

XIII.7. ACKNOWLEDGEMENTS

The authors appreciate the stimulus and continuing interest of Professor Harden M. McConnell. The numerical work was aided by suggestions and programs from Walter Neilsen. Support for this work from the National Science Foundation is gratefully acknowledged.

REFERENCES

1. J. R. Norris and S. I. Weissman, J. Phys. Chem. 73, 3119 (1969).

2. L. Stryer and O. H. Griffith, Proc. Natl. Acad. Sci. (U.S.) 54, 1785 (1965).

3. For some typical single-crystal spectra, see L. J. Libertini and O. H. Griffith, J. Chem. Phys. 53, 1359 (1970) and references cited therein.

4. R. Lefebvre and J. Maruani, J. Chem. Phys. 42, 1480 (1965).

5. J. H. Freed and G. K. Fraenkel, J. Chem. Phys. 39, 326 (1963).

6. D. Kivelson, J. Chem. Phys. 33, 1094 (1960).

7. M. S. Itzkowitz, J. Chem. Phys. 46, 3048 (1967).

8. M. Fixman, J. Chem. Phys. 48, 223 (1968).

9. N. N. Korst and A. V. Lazarev, Mol. Phys. 17, 481 (1969).

10. I. V. Alexandrov, A. N. Ivanova, N. N. Korst, A. V. Lazarev, A. I. Prinkhozhenko, and V. B. Stryukov, Mol. Phys. 18, 681 (1970).

11. P. W. Anderson, J. Phys. Soc. (Japan) 9, 316 (1954).

12. H. Sillescu and D. Kivelson, J. Chem. Phys. 48, 3493 (1968).

13. Closely related results have been obtained very recently by J. H. Freed, G. V. Bruno, and C. Polnaszek (to be published).

14. P. Debye, Polar Molecules (Dover Publications, Inc., N. Y., 1929); A. Abragam, The Principles of Magnetic Resonance (Oxford University Press, London, 1961), Ch. VIII.

15. For recent reviews of experimental applications, mainly in biochemical problems, see O. H. Griffith and A. S. Waggoner, Accounts of Chemical Research 2, 17 (1969); H. M. McConnell and B. G. McFarland, Quarterly Reviews of Biophysics 3, 91 (1970).

16. See, for example, R. Kubo, in Fluctuation, Relaxation and Resonance in Magnetic Systems, D. ter Haar, ed. (Oliver & Boyd, Edinburgh, 1962), p. 23.

17. Similar equations have been discussed by R. Kubo, in "Stochastic Processes in Chemical Physics," Advances in Chemical Physics, Vol. XVI, K. E. Shuler, ed. (John Wiley and Sons, New York, 1969), p. 101.

18. J. Albrecht, Z. Angew. Math. Mech. 42, 397-402 (1962); L. V. Kantorovich and V. I. Krylov, Approximate Methods in Higher Analysis (Interscience, New York, 1958).

19. Handbook of Mathematical Functions, Applied Math Series No. 55 (National Bureau of Standards, U.S.), p. 774.

20. A. R. Edmonds, Angular Momentum in Quantum Mechanics (Princeton University Press, Princeton, N. J., 1960), p. 63.

21. R. G. Gordon and R. P. McGinnis, J. Chem. Phys. 49, 2455 (1968).

22. H. Rutishauser, Proc. Amer. Math. Soc. Symposium in Applied Mathematics 15, 219 (1963).

23. J. G. F. Francis, Computer J. <u>4</u>, 265, 332 (1962).

24. J. H. Wilkinson, <u>The</u> <u>Algebraic</u> <u>Eigenvalue</u> <u>Problem</u> (Oxford University Press, Oxford, 1965).

25. J. M. Ortega and H. F. Kaiser, Computer J. <u>6</u>, 99 (1963).

26. E. J. Shimshick and H. M. McConnell, Biochem. Biophys. Research Comm. (to be published).

Appendix A

DIAGONALIZATION PROGRAMS

In this appendix we give FORTRAN IV programs which diagonalize complex symmetric band-form matrices, using che algorithms presented in Sect. IV.   These versions have been run on IBM-360 series computers.

```
      SUBROUTINE CSQZ(A,AMP,IDIM,JDIM,N,M,SQTOL)            CSQZ    1
      IMPLICIT REAL*8(A-H,O-Z)                              CSQZ    2
C                                                           CSQZ    3
C         SUBROUTINE CSQZ TRANSFORMS, BY A SERIES OF COMPLEX JACOBI  CSQZ    4
C         ROTATIONS, AN N BY N COMPLEX SYMMETRIC BAND MATRIX, A,     CSQZ    5
C         OF BAND WIDTH 2M-1 INTO A COMPLEX SYMMETRIC TRIDIAGONAL    CSQZ    6
C         MATRIX, T = TRANSPOSE(R)*A*R                      CSQZ    7
C                                                           CSQZ    8
C         ENTERING THE SUBROUTINE,A(I,1) CONTAINS THE ITH DIAGONAL   CSQZ    9
C         ELEMENT AND A(I,J+1) CONTAINS THE JTH ELEMENT FROM THE DIAGONCSQZ 10
C         IN THE ITH ROW.                                   CSQZ   11
C         N IS THE LENGTH OF THE COLUMNS OF A, AND M IS THE HALF BAND CSQZ 12
C         WIDTH PLUS ONE.                                   CSQZ   13
C                                                           CSQZ   14
C         LEAVING THE SUBROUTINE A(I,1) CONTAINS THE DIAGONAL ELEMENTS CSQZ 15
C         AND A(I,J) CONTAIN THE NEW OFF-DIAGONAL ELEMENTS ORDERED AS  CSQZ 16
C         BEFORE                                            CSQZ   17
C                                                           CSQZ   18
C         ON ENTERING THE VECTOR TO BE ROTATED IS IN AMP.   AT EXIT    CSQZ 19
C         AMP CONTAINS THE ROTATED VECTOR                   CSQZ   20
C                                                           CSQZ   21
      COMPLEX*16 A(IDIM,JDIM), AMP(IDIM),                   CSQZ   22
     1 B, G, S, SX, C, CX, CS, U, V, TEMP                   CSQZ   23
      DIMENSION RG(2)                                       CSQZ   24
      EQUIVALENCE (G,RG(1))                                 CSQZ   25
      SQTOL=0.                                              CSQZ   26
      GRASS=0.                                              CSQZ   27
      DO 101 LJM=1,N                                        CSQZ   28
      GRASS=GRASS+1.                                        CSQZ   29
  101 SQTOL=CDABS(A(LJM,1))+SQTOL                           CSQZ   30
      DO 102 KID=2,M                                        CSQZ   31
      LOW=N+1-KID                                           CSQZ   32
      DO 102 LJM=1,LOW                                      CSQZ   33
      GRASS=GRASS+2.                                        CSQZ   34
  102 SQTOL=2.*CDABS(A(LJM,KID))+SQTOL                      CSQZ   35
      SQTOL=1.E-15*SQTOL/DSQRT(GRASS)                       CSQZ   36
      MX=M-2                                                CSQZ   37
```

```
C              EACH PASS THROUGH THIS LOOP WILL DESTROY A SET OF OFF-    CSQZ  38
C              DIAGONAL ELEMENTS A(I,NQ)                                 CSQZ  39
        DO 20 L=1,MX                                                     CSQZ  40
        NR=M-L                                                           CSQZ  41
        NQ=NR+1                                                          CSQZ  42
        NX=N-NR                                                          CSQZ  43
C          THIS STATEMENT SAVES COMPUTATION WHEN THE BAND IS LARGER THANCSQZ  44
C          THE MATRIX                                                    CSQZ  45
        IF(NX) 20,20,5                                                   CSQZ  46
C              EACH PASS THROUGH THIS LOOP DESTROYS ONE OFF-DIAGONAL ELEMENTCSQZ  47
C              A(K,NQ)                                                   CSQZ  48
      5 DO 15 K=1,NX                                                     CSQZ  49
C          THE FIRST PASS THROUGH THIS LOOP DESTROYS AN OFF-DIAGONAL     CSQZ  50
C          ELEMENT A(K,NQ) AND CREATES ANOTHER ELEMENT G BY THE JACOBI   CSQZ  51
C          ROTATION OF COLUMNS AND ROWS I AND I+1  SUBSEQUENT PASSES     CSQZ  52
C          DESTROY THE ELEMENT G AND CREATE ANOTHER UNTIL THE LAST PASS  CSQZ  53
C          WHICH DESTROYS WITHOUT CREATING                              CSQZ  54
        DO 25 J=K,NX,NR                                                  CSQZ  55
        IF(J-K) 40,50,40                                                 CSQZ  56
     50 IF(CDABS(A(J,NQ))-SQTOL) 15,15,60                                CSQZ  57
     60 B= -A(J,NR)/A(J,NQ)                                              CSQZ  58
        GO TO 80                                                         CSQZ  59
     40 IF(RG(1)**2+RG(2)**2-SQTOL**2) 15,15,70                          CSQZ  60
     70 B= -A(J-1,NQ)/G                                                  CSQZ  61
C          THESE FIVE STATEMENTS COMPUTE THE CORRECT SINE S AND COS C    CSQZ  62
     80 SX=1.0/(1.0+B**2)                                                CSQZ  63
        S=CDSQRT(SX)                                                     CSQZ  64
        C=B*S                                                            CSQZ  65
        CX=C**2                                                          CSQZ  66
        CS=C*S                                                           CSQZ  67
        I=J+NR-1                                                         CSQZ  68
C          THESE THREE STATEMENTS ROTATE AMP                            CSQZ  69
        TEMP=AMP(I)                                                      CSQZ  70
        AMP(I)=C*TEMP-S*AMP(I+1)                                         CSQZ  71
        AMP(I+1)=C*AMP(I+1)+S*TEMP                                       CSQZ  72
C          THESE FIVE STATEMENTS ROTATE THE CROSS ELEMENTS              CSQZ  73
        U=A(I,1)                                                         CSQZ  74
        V=A(I+1,1)                                                       CSQZ  75
        A(I,1)=CX*U-2.0*CS*A(I,2)+SX*V                                   CSQZ  76
        A(I+1,1)=SX*U+2.0*CS*A(I,2)+CX*V                                 CSQZ  77
        A(I,2)=CS*(U-V)+(CX-SX)*A(I,2)                                   CSQZ  78
     65 IZ=I-1                                                           CSQZ  79
C              THIS LOOP TRANSFORMS THE APPROPRIATE COLUMNS              CSQZ  80
        DO 90 IX=J,IZ                                                    CSQZ  81
        IK=I-IX+1                                                        CSQZ  82
        U=A(IX,IK)                                                       CSQZ  83
        V=A(IX,IK+1)                                                     CSQZ  84
        A(IX,IK)=C*U-S*V                                                 CSQZ  85
     90 A(IX,IK+1)=S*U+C*V                                               CSQZ  86
     75 IF(J-K) 95,85,95                                                 CSQZ  87
     95 A(J-1,NQ)=C*A(J-1,NQ)-S*G                                        CSQZ  88
     85 IR=MINO(NR,N-I)                                                  CSQZ  89
        IF(IR-2) 55,45,45                                                CSQZ  90
C              THIS LOOP TRANSFORMS THE APPROPRIATE ROWS                CSQZ  91
     45 DO 10 IX=2,IR                                                    CSQZ  92
        U=A(I,IX+1)                                                      CSQZ  93
        A(I,IX+1)=C*U-S*A(I+1,IX)                                        CSQZ  94
     10 A(I+1,IX)=S*U+C*A(I+1,IX)                                        CSQZ  95
     55 IF(I+NR-N) 30,25,25                                              CSQZ  96
     30 G= -S*A(I+1,NQ)                                                  CSQZ  97
        A(I+1,NQ)   =C*A(I+1,NQ)                                         CSQZ  98
     25 CONTINUE                                                         CSQZ  99
C          END OF THE LOOP OVER J    ONE OFF-DIAGONAL ELEMENT DESTROYED  CSQZ 100
     15 CONTINUE                                                         CSQZ 101
C          END OF THE LOOP OVER K    ONE SET OF OFF-DIAGONAL ELEMENTS    CSQZ 102
C          DESTROYED                                                    CSQZ 103
     20 CONTINUE                                                         CSQZ 104
C          END OF THE  LOOP OVER L   MX SETS OF OFF DIAGONAL ELEMENTS    CSQZ 105
C          DESTROYED                                                    CSQZ 106
        RETURN                                                          CSQZ 107
        END                                                            CSQZ 108
```

```
      SUBROUTINE CQRT(A,B,AMP,IDIM,N,M,TOL,EIGVAL)                    CQRT   1
      IMPLICIT REAL*8(A-H,O-Z)                                        CQRT   2
CCQRT                                                                 CQRT   3
C                                                                     CQRT   4
C        CQRT IS INTENDED TO PRODUCE THE EIGENVALUES OF A COMPLEX     CQRT   5
C     SYMMETRIC (NOT HERMITIAN) TRIDIAGONAL MATRIX BY ITERATIVE       CQRT   6
C     QR TRANSFORMS AS DESCRIBED IN WILKINSON, 'THE ALGEBRAIC         CQRT   7
C     EIGENVALUE PROBLEM', P.565                                      CQRT   8
C                                                                     CQRT   9
C        ENTERING THE SUBROUTINE THE ARRAY A OF DIMENSION N CONTAINS  CQRT  10
C     THE DIAGONAL ELEMENTS, B THE OFF-DIAGONAL ELEMENTS.             CQRT  11
C     THE VECTOR TO BE ROTATED IS AMP.  N IS THE DIMENSION            CQRT  12
C     OF THE MATRIX AND THE LENGTH OF AMP.                            CQRT  13
C     ITERATIONS ARE CONTINUED UNTIL ALL OFF-DIAGONAL ELEMENTS        CQRT  14
C     HAVE BEEN REDUCED TO LESS THAN 'TOL' IN MAGNITUDE.              CQRT  15
C     M IS THE MAXIMUM NUMBER OF ITERATIONS ALLOWED PER EIGENVALUE    CQRT  16
C     M=20 HAS ALWAYS BEEN MORE THAN SUFFICIENT                       CQRT  17
C     THE NUMBER OF ITERATIONS REQUIRED MAY BE REDUCED IF ACCURATE    CQRT  18
C     ESTIMATES OF THE EIGENVALUES ARE ALREADY KNOWN.  THESE          CQRT  19
C     INITIAL GUESSES SHOULD BE PLACED IN THE ARRAY 'EIGVAL'.         CQRT  20
C     IF NO GUESSES ARE FURNISHED, EIGVAL(1) SHOULD BE SET EQUAL      CQRT  21
C     TO A COMPLEX ZERO, (0.D0,0.D0), IN THE CALLING PROGRAM,         CQRT  22
C     AND CQRT WILL GENERATE ITS OWN GUESSES.                         CQRT  23
C                                                                     CQRT  24
C        LEAVING THE SUBROUTINE, A CONTAINS THE EIGENVALUES (B CON-   CQRT  25
C     TAINS THE TRANSFORMED OFF-DIAG ELEMENTS, AND AMP                CQRT  26
C     CONTAINS THE ROTATED VECTOR                                     CQRT  27
C                                                                     CQRT  28
C        THE MATRICES A,B,AMP ARE TREATED AS REAL ARRAYS             CQRT  29
C     OF DIMENSION N*2 INSIDE THE SUBROUTINE.  OUTSIDE THEY          CQRT  30
C     THEY ARE TREATED AS COMPLEX ARRAYS OF DIMENSION N.             CQRT  31
      INTEGER HELL                                                    CQRT  32
      DIMENSION A(1), B(1), AMP(1), SM(2), EIGVAL(1)                  CQRT  33
      COMPLEX*16 SC                                                   CQRT  34
C        THE EQUIVALENCING OF THE ARRAY SM TO THE COMPLEX NUMBER SC   CQRT  35
C     PERMITS THE USE OF THE COMPLEX SQUARE ROOT ROUTINE CDSQRT       CQRT  36
      EQUIVALENCE (SC,SM(1))                                          CQRT  37
C        NX WILL BE THE UPPER LIMIT FOR THE ITERATION , NI WILL BE    CQRT  38
C     THE LOWER LIMIT AND BOTH MAY VARY BETWEEN ITERATIONS            CQRT  39
      NX=2*N                                                          CQRT  40
      NI=2                                                            CQRT  41
C        THE TOLERANCE -- THAT IS, THE SIZE TO WHICH OFF-DIAGONAL     CQRT  42
C     ELEMENTS ARE REDUCED -- IS SET IN CQRT ON THE BASIS OF THE      CQRT  43
C     NORM OF THE TRIDIAGONAL MATRIX                                  CQRT  44
      TOL=0.                                                          CQRT  45
      GRASS=N                                                         CQRT  46
      DO 200  KID=1,NX,2                                              CQRT  47
  200 TOL=DSQRT(A(KID)**2+A(KID+1)**2)+TOL+                           CQRT  48
     1 2.*DSQRT(B(KID)**2+B(KID+1)**2)                                CQRT  49
      TOL=1.E-15*TOL/DSQRT(3.*GRASS)                                  CQRT  50
C        SHR AND SHI WILL CONTAIN THE TOTAL SHIFT                     CQRT  51
      SHR=0.0                                                         CQRT  52
      SHI=0.0                                                         CQRT  53
C        K WILL COUNT THE NUMBER OF ITERATIONS PER EIGENVALUE         CQRT  54
      K=0                                                             CQRT  55
C        IF A SET OF 'EIGVALS' IS AVAILABLE FOR USE AS INITIAL SHIFTS CQRT  56
C     -- I.E. IF THE FIRST 'EIGVAL' IS NOT (0.,0.) -- THEN WE USE     CQRT  57
C     THE GUESSES IN 'EIGVAL' AS SHIFTS BY SETTING 'HELL' EQUAL 101   CQRT  58
      ASSIGN 100 TO HELL                                              CQRT  59
      IF(DABS(EIGVAL(1))+DABS(EIGVAL(2)) .NE. 0.)  ASSIGN 101 TO HELL CQRT  60
      GO TO HELL, (100,101)                                           CQRT  61
C        EACH NEW ITERATION BEGINS AT THIS STATEMENT OR AT STATEMENT  CQRT  62
C     101 -- WHICHEVER 'HELL' IS                                      CQRT  63
  100 K=K+1                                                           CQRT  64
C        THESE SEVEN STATEMENTS SOLVE THE TWO BY TWO MATRIX IN THE    CQRT  65
C     LOWER LEFT CORNER AND USE ONE ROOT AS THE SHIFT                 CQRT  66
      ATR=A(NX-1) + A(NX-3)                                           CQRT  67
      ATI=A(NX) + A(NX-2)                                             CQRT  68
      SM(1)=ATR**2-ATI**2-4.0*(A(NX-1)*A(NX-3)-A(NX)*A(NX-2)-         CQRT  69
     1     B(NX-3)**2+B(NX-2)**2)                                     CQRT  70
      SM(2)= 2.0*ATR*ATI-4.0*(A(NX-1)*A(NX-2)+A(NX)*A(NX-3)-          CQRT  71
     1     2.0*B(NX-3)*B(NX-2))                                       CQRT  72
      SC=CDSQRT(SC)                                                   CQRT  73
      STR=(ATR-SM(1))*0.5                                             CQRT  74
      STI=(ATI-SM(2))*0.5                                             CQRT  75
```

```
         STIR=(ATR+SM(1))*.5                                    CQRT  76
         STTI=(ATI+SM(2))*.5                                    CQRT  77
         TEMP=CDABS(DCMPLX(A(NX-1)-STR,A(NX)-STI))              CQRT  78
         TEMPO=CDABS(DCMPLX(A(NX-1)-STTR,A(NX)-STTI))           CQRT  79
         IF(TEMPO .GT. TEMP)   GO TO 102                        CQRT  80
         STR=STTR                                               CQRT  81
         STI=STTI                                               CQRT  82
         GO TO 102                                              CQRT  83
   101   K=K+1                                                  CQRT  84
         IF(K .GE. M)  GO TO 999                                CQRT  85
         STR=EIGVAL(NX-1)-SHR                                   CQRT  86
         STI=EIGVAL(NX)-SHI                                     CQRT  87
   102   IF(K .LE. M)  GO TO 104                                CQRT  88
   999   WRITE(6,103) M                                         CQRT  89
   103   FORMAT(26H CQRT HAS NOT CONVERGED IN,I3,11H ITERATIONS)CQRT  90
         CALL EXIT                                              CQRT  91
C             THESE TWO STATEMENTS INCREASE THE TOTAL SHIFT BY THE CQRT 92
C             TEMPORARY SHIFT                                   CQRT  93
   104   SHR=SHR+STR                                            CQRT  94
         SHI=SHI+STI                                            CQRT  95
C             THIS LOOP SUBTRACTS THE TEMPORARY SHIFT FROM THE DIAGONAL CQRT 96
C             ELEMENTS                                          CQRT  97
         DO 20 I=2,NX,2                                         CQRT  98
         A(I-1)=A(I-1)-STR                                      CQRT  99
   20    A(I)=A(I)-STI                                          CQRT 100
C             THESE FOUR STATEMENTS SUPPLY INITIAL VALUES FOR THE ITERATION CQRT 101
         CR=1.0                                                 CQRT 102
         CI=0.0                                                 CQRT 103
         CSR=0.                                                 CQRT 104
         CSI=0.                                                 CQRT 105
         UR=0.0                                                 CQRT 106
         UI=0.0                                                 CQRT 107
C             THIS LOOP COMPLETES ONE ITERATION, THAT IS ONE QR TRANSFORM CQRT 108
C             I IS DOUBLE THE I IN THE ALGORITHM                CQRT 109
         DO 10 I=NI,NX,2                                        CQRT 110
C             G IS THE GAMMA IN WILKINSON                       CQRT 111
         GR=A(I-1)-UR                                           CQRT 112
         GI=A(I)-UI                                             CQRT 113
C             Q IS THE P IN WILKINSON                           CQRT 114
         QR=CR*A(I-1)-CI*A(I)-CSR*B(I-3)+CSI*B(I-2)             CQRT 115
         QI=CI*A(I-1)+CR*A(I)-CSI*B(I-3)-CSR*B(I-2)             CQRT 116
C             THIS BRANCH AVOIDS UNNECESSARY COMPUTATION AT THE END CQRT 117
   15    IF(I.EQ.NX) GO TO 10                                   CQRT 118
         SM(1)=QR**2-QI**2+B(I-1)**2-B(I)**2                    CQRT 119
         SM(2)=2.0*(QR*QI+B(I-1)*B(I))                          CQRT 120
         SC=CDSQRT(SC)                                          CQRT 121
         RR=SM(1)                                               CQRT 122
         RI=SM(2)                                               CQRT 123
         IF(I.EQ.NI) GO TO 18                                   CQRT 124
C             THESE TWO STATEMENTS ROTATE AN OFF-DIAG ELEMENT   CQRT 125
         B(I-3)=SR*RR-SI*RI                                     CQRT 126
         B(I-2)=SR*RI+SI*RR                                     CQRT 127
   18    RS=1.0/(RR**2+RI**2)                                   CQRT 128
C             THESE TWO STATEMENTS COMPUTE THE NEW SINE FOR THE  JACOBI CQRT 129
C             ROTATION                                          CQRT 130
         SR=(RR*B(I-1)+B(I)*RI)*RS                              CQRT 131
         SI=(RR*B(I)-B(I-1)*RI)*RS                              CQRT 132
C             THESE TWO STORE THE PRODUCT OF THE NEW SINE AND THE OLD CQRT 133
C             COSINE                                            CQRT 134
         CSR=CR*SR-CI*SI                                        CQRT 135
         CSI=CR*SI+CI*SR                                        CQRT 136
C             THESE TWO COMPUTE THE NEW COSINE                  CQRT 137
         CR=(RR*QR+RI*QI)*RS                                    CQRT 138
         CI=(RR*QI-RI*QR)*RS                                    CQRT 139
C             THESE SIX COMPUTE A NEW U                         CQRT 140
         TR=GR+A(I+1)                                           CQRT 141
         TI=GI+A(I+2)                                           CQRT 142
         SXR=SR**2-SI**2                                        CQRT 143
         SXI=2.0*SR*SI                                          CQRT 144
         UR=SXR*TR-SXI*TI                                       CQRT 145
         UI=SXR*TI+SXI*TR                                       CQRT 146
C             THESE TWO ROTATE THE DIAG. ELEMENT                CQRT 147
         A(I-1)=GR+UR                                           CQRT 148
         A(I)=GI+UI                                             CQRT 149
C             THESE NINE ROTATE AMP                             CQRT 150
```

```
                L=I                                                          CQRT 151
                TAR=AMP(L-1)                                                 CQRT 152
                TAI=AMP(L)                                                   CQRT 153
                TBR=AMP(L+1)                                                 CQRT 154
                TBI=AMP(L+2)                                                 CQRT 155
                AMP(L-1)=TAR*CR-TAI*CI+TBR*SR-TBI*SI                         CQRT 156
                AMP(L)=TAI*CR+TAR*CI+TBI*SR+TBR*SI                           CQRT 157
                AMP(L+1)=TBR*CR-TBI*CI-TAR*SR+TAI*SI                         CQRT 158
                AMP(L+2)=TBI*CR+TBR*CI-TAI*SR-TAR*SI                         CQRT 159
             10 CONTINUE                                                     CQRT 160
C                    THIS ENDS ONE ITERATION                                 CQRT 161
C                    THESE TWO COMPUTE THE LAST OFF-DIAG ELEMENT             CQRT 162
                B(NX-3)=SR*QR-SI*QI                                          CQRT 163
                B(NX-2)=SR*QI+SI*QR                                          CQRT 164
C                    THESE TWO COMPUTE THE LAST DIAG ELEMENT                 CQRT 165
                A(NX-1)=GR                                                   CQRT 166
                A(NX)=GI                                                     CQRT 167
C                    AT THIS POINT WE BEGIN CHECKING UPWARD THROUGH THE OFF-DIAG  CQRT 168
C                    ELEMENTS TO FIND THOSE LESS THAN TOL                    CQRT 169
             85 IT=NX                                                        CQRT 170
C                    THESE THREE STATEMENTS CONSTITUTE AN EFFECTIVE BACKWARD  CQRT 171
C                    LOOP                                                    CQRT 172
             30 IT=IT-2                                                      CQRT 173
C                    THIS LOOP IS LEFT WHEN AN ELEMENT LESS THAN TOL IS FOUND  CQRT 174
                IF(DABS(B(IT-1))+DABS(B(IT)).LE.TOL) GO TO 40               CQRT 175
                IF(IT-NI) 100,100,30                                         CQRT 176
C                    IF NO OFF-DIAG ELEMENTS LESS THAN TOL ARE FOUND WE PERFORM  CQRT 177
C                    ANOTHER ITERATION                                      CQRT 178
C                    THIS CONDITIONAL BRANCHES ACCORDING TO WHETHER THE MATRIX  CQRT 179
C                    ISOLATED BY THE SMALL OFF-DIAG ELEMENT IS OF DIMENSION  CQRT 180
C                    ONE,  TWO,  OR GREATER THAN TWO                        CQRT 181
             40 IF(NX-IT-4) 50,60,70                                        CQRT 182
C                    THESE TWO STATEMENTS EXTRACT THE EIGENVALUE OF A ONE BY ONE  CQRT 183
C                    MATRIX, ADDING BACK THE SHIFT                          CQRT 184
             50 A(NX-1)=A(NX-1)+SHR                                         CQRT 185
                A(NX)=A(NX)+SHI                                             CQRT 186
C                    THIS DECREASES THE SIZE OF THE PORTION OF THE MATRIX AFFECT  CQRT 187
C                    ED BY LATER ITERATIONS                                 CQRT 188
                NX=NX-2                                                     CQRT 189
C                    THIS RESETS THE ITERATION COUNTER                      CQRT 190
                K=0                                                        CQRT 191
                GO TO 80                                                   CQRT 192
C                    THIS SECTION EXTRACTS THE EIGENVALUES FROM A TWO BY TWO  CQRT 193
C                    MATRIX SECTION WHICH HAS BECOME ISOLATED FROM THE REST.  CQRT 194
C                    IT ALSO PERFORMS THE CORRESPONDING ROTATIONS ON AMP.   CQRT 195
C                    THE FIRST TWENTY-NINE STATEMENTS COMPUTE THE PROPER VALUES  CQRT 196
C                    FOR THE SINE AND COSINE                                CQRT 197
             60 ALR=-B(NX-3)                                                CQRT 198
                ALI=-B(NX-2)                                                CQRT 199
                AMR=0.5*(A(NX-3)-A(NX-1))                                   CQRT 200
                AMI=0.5*(A(NX-2)-A(NX))                                     CQRT 201
                SM(1)=ALR**2-ALI**2+AMR**2-AMI**2                           CQRT 202
                SM(2)=2.0*(ALR*ALI+AMR*AMI)                                 CQRT 203
                SC=CDSQRT(SC)                                               CQRT 204
                ANR=SM(1)                                                   CQRT 205
                ANI=SM(2)                                                   CQRT 206
                TAR=AMR+ANR                                                 CQRT 207
                TAI=AMI+ANI                                                 CQRT 208
                TBR=ANR-AMR                                                 CQRT 209
                TBI=ANI-AMI                                                 CQRT 210
                SIG=1.0                                                     CQRT 211
C                    THIS BRANCH CHOOSES THE ROOT OF THE IMPLICITLY SOLVED  CQRT 212
C                    QUADRATIC SO THAT THE COSINE HAS THE LARGER ABSOLUTE VALUE  CQRT 213
                IF(TAR**2+TAI**2.GT.TBR**2+TBI**2) GO TO 65                 CQRT 214
                TAR=TBR                                                     CQRT 215
                TAI=TBI                                                     CQRT 216
                SIG=-1.0                                                    CQRT 217
C                    THESE SIX COMPUTE THE COSINE                           CQRT 218
             65 TBR=0.5/(ANR**2+ANI**2)                                     CQRT 219
                SM(1)=(ANR*TAR+ANI*TAI)*TBR                                 CQRT 220
                SM(2)=(ANR*TAI-ANI*TAR)*TBR                                 CQRT 221
                SC=CDSQRT(SC)                                               CQRT 222
                CR=SM(1)                                                    CQRT 223
                CI=SM(2)                                                    CQRT 224
C                    THESE FIVE COMPUTE THE SINE                            CQRT 225
```

```
      TAR=ANR*CR-ANI*CI                                               CQRT 226
      TAI=ANR*CI+ANI*CR                                               CQRT 227
      TBR=0.5/(TAR**2+TAI**2)                                         CQRT 228
      SR=SIG*(TAR*ALR+TAI*ALI)*TBR                                    CQRT 229
      SI=SIG*(TAR*ALI-ALR*TAI)*TBR                                    CQRT 230
C         THESE SIXTEEN STORE DATA NEEDED IN THE ROTATION OF THE      CQRT 231
C         MATRIX ELEMENTS                                             CQRT 232
      TAR=A(NX-3)                                                     CQRT 233
      TAI=A(NX-2)                                                     CQRT 234
      TBR=A(NX-1)                                                     CQRT 235
      TBI=A(NX)                                                       CQRT 236
      TCR=B(NX-3)                                                     CQRT 237
      TCI=B(NX-2)                                                     CQRT 238
      CXR=CR**2-CI**2                                                 CQRT 239
      CXI=2.0*CR*CI                                                   CQRT 240
      SXR=SR**2-SI**2                                                 CQRT 241
      SXI=2.0*SR*SI                                                   CQRT 242
      CSR=CR*SR-CI*SI                                                 CQRT 243
      CSI=CR*SI+CI*SR                                                 CQRT 244
      TR=CXR-SXR                                                      CQRT 245
      TI=CXI-SXI                                                      CQRT 246
      UR=2.0*(TCR*CSR-TCI*CSI)                                        CQRT 247
      UI=2.0*(TCI*CSR+TCR*CSI)                                        CQRT 248
C         THESE FOUR ROTATE THE DIAG ELEMENTS                         CQRT 249
      A(NX-3)=TAR*CXR-TAI*CXI+TBR*SXR-TBI*SXI-UR+SHR                  CQRT 250
      A(NX-2)=TAR*CXI+TAI*CXR+TBR*SXI+TBI*SXR-UI+SHI                  CQRT 251
      A(NX-1)=TAR*SXR-TAI*SXI+TBR*CXR-TBI*CXI+UR+SHR                  CQRT 252
      A(NX)=TAR*SXI+TAI*SXR+TBR*CXI+TBI*CXR+UI+SHI                    CQRT 253
C         THESE TWO ROTATE THE OFF-DIAG ELEMENTS                      CQRT 254
      B(NX-3)=2.0*(AMR*CSR-AMI*CSI)+TCR*TR-TCI*TI                     CQRT 255
      B(NX-2)=2.0*(AMR*CSI+AMI*CSR)+TCR*TI+TCI*TR                     CQRT 256
      I=NX-2                                                          CQRT 257
C         THESE NINE ROTATE AMP                                       CQRT 258
C         NOTICE THAT THE SENSE OF ROTATION IS OPPOSITE TO THAT       CQRT 259
C         OF THE FIRST ROTATION                                       CQRT 260
      L=I                                                             CQRT 261
      TAR=AMP(L-1)                                                    CQRT 262
      TAI=AMP(L)                                                      CQRT 263
      TBR=AMP(L+1)                                                    CQRT 264
      TBI=AMP(L+2)                                                    CQRT 265
      AMP(L-1)=TAR*CR-TAI*CI-TBR*SR+TBI*SI                            CQRT 266
      AMP(L)=TAI*CR+TAR*CI-TBI*SR-TBR*SI                              CQRT 267
      AMP(L+1)=TBR*CR-TBI*CI+TAR*SR-TAI*SI                            CQRT 268
      AMP(L+2)=TBI*CR+TBR*CI+TAI*SR+TAR*SI                            CQRT 269
C         THIS RESETS THE ITERATION COUNT                             CQRT 270
      K=0                                                             CQRT 271
C         THIS DECREASES THE SIZE OF THE PORTION OF THE MATRIX        CQRT 272
C         AFFECTED BY THE LATER ITERATIONS                            CQRT 273
      NX=NX-4                                                         CQRT 274
      GO TO 80                                                        CQRT 275
C         THIS STATEMENT IS REACHED WHEN THE PORTION OF THE MATRIX    CQRT 276
C         ISOLATED IS GREATER THAN TWO BY TWO.  IT CHANGES THE LOWER  CQRT 277
C         LIMIT OF THE ITERATION SO THAT ONLY THIS PORTION WILL BE    CQRT 278
C         AFFECTED BY SUBSEQUENT ROTATIONS UNTIL ALL ITS EIGENVALUES  CQRT 279
C         ARE FOUND                                                   CQRT 280
   70 NI=IT+2                                                         CQRT 281
C         THIS STATEMENT TRANSFERS TO THE BEGINNING OF ANOTHER        CQRT 282
C         ITERATION                                                   CQRT 283
      GO TO HELL, (100,101)                                           CQRT 284
C         THIS STATEMENT IS REACHED AFTER EITHER ONE OR TWO EIGENVALUES CQRT 285
C         HAVE JUST BEEN FOUND. IT TRANSFERS IF ALL THE EIGENVALUES   CQRT 286
C         IN THIS PORTION OF THE MATRIX HAVE BEEN FOUND               CQRT 287
   80 IF(NX.LT.NI) GO TO 90                                           CQRT 288
C         THIS BRANCH TRANSFERS ACCORDING TO WHETHER ONE , TWO, OR    CQRT 289
C         MORE EIGENVALUES REMAIN TO BE FOUND IN THIS PORTION         CQRT 290
   95 IF(NX-NI-2) 50,60,85                                            CQRT 291
C         THIS CONDITIONAL IS REACHED WHEN ALL THE EIGENVALUES IN     CQRT 292
C         THIS PORTION OF THE MATRIX HAVE BEEN FOUND.  IT RETURNS IF  CQRT 293
C         THIS IS THE LAST PORTION OF THE MATRIX                      CQRT 294
   90 IF(NI.EQ.2) RETURN                                              CQRT 295
C         THIS ENLARGES THE PORTION OF THE MATRIX BEING TREATED TO    CQRT 296
C         INCLUDE THE BEGINNING OF THE MATRIX                         CQRT 297
      NI=2                                                            CQRT 298
      GO TO 95                                                        CQRT 299
      END                                                             CQRT 300
```

SYMMETRY AND THE SLOWLY TUMBLING SPIN SYSTEM

R.M. Lynden-Bell

School of Molecular Sciences, University of Sussex

XIIIA.1. APPLICATION OF GROUP THEORY

The use of group theory can simplify the computation of the spectra of slowly tumbling spin systems. The spectrum is obtained by computing the function

$$I(\omega) = \pi^{-1}\mathrm{Im}\{\underline{\mu}^T \cdot [\omega \underline{\underline{1}} - (\underline{\underline{L}} + i\underline{\underline{\Pi}})]^{-1} \cdot \underline{\mu}\} \qquad (1)$$

(equations (XIII.3.34) and (XIII.3.35)).

When the symmetry group of ($\underline{\underline{L}} + i\underline{\underline{\Pi}}$) has been determined, the eigenvalues and eigenvectors for each irreducible representation of the group can be found separately by using symmetry adapted basis functions, and if $\underline{\mu}$ transforms according to one of the irreducible representations only the eigenvectors of $\underline{\underline{L}} \neq i\underline{\underline{\Pi}}$ corresponding to that irreducible representation are needed.

The nitroxide radical provides an example of the use of symmetry. As the electron has an anisotropic g tensor and is anisotropically coupled to the nitrogen nucleus (I = 1) the complete Liouville operator $\hat{\hat{L}}_{tot}(=H^{\times})$ is

$$\hat{\hat{L}}_{tot} = \omega_o \hat{\hat{S}}_o + \omega_N \hat{\hat{I}}_o + \sum_{m,\mu} (-1)^m g_\mu D^{(2)}_{-m\mu} \hat{\hat{S}}_m + \frac{1}{2} a \sum_m (-1)^m (\hat{\hat{I}}_m \hat{\hat{T}}_{-m} + \hat{\hat{J}}_m \hat{\hat{S}}_{-m})$$

$$+ \frac{1}{2} \sum_{m,\mu,n} (5)^{\frac{1}{2}} A_\mu D^{(2)}_{-m\mu} (\hat{\hat{I}}_n \hat{\hat{T}}_{m-n} + \hat{\hat{J}}_n \hat{\hat{S}}_{m-n}) \begin{pmatrix} 2 & 1 & 1 \\ -m & n & m-n \end{pmatrix} \qquad (2)$$

where $D^{(2)}_{m\mu}$ are Wigner rotation matrix elements, functions of the Euler angles (η,θ,ϕ) relating the molecular coordinate system to the laboratory coordinate system, and g_μ and A_μ are the components of the tensors representing the anisotropic part of the g and nuclear coupling tensors in molecule fixed axes. $\hat{\hat{S}}_m$ and $\hat{\hat{T}}_m$ are the spherical components of the electron spin superoperators formed from the commutator and the anticommutator of the spin operators.

$$\hat{\hat{S}}_m \hat{Q} = [\hat{S}_m,\hat{Q}]_- = \hat{S}_m^{\times}\hat{Q} \qquad (3)$$

$$\hat{\hat{T}}_m \hat{Q} = [\hat{S}_m,\hat{Q}]_+, \qquad (4)$$

$\hat{\hat{I}}$ and $\hat{\hat{J}}$ are the corresponding nuclear spin superoperators.

This Liouville operator may be simplified in the high field situation by dropping all terms which do not commute with $\hat{\hat{S}}_o$, and by dropping the term $\omega_N \hat{\hat{I}}_o$ which is negligible in this case. As $\underline{\mu}$ in equation 1 is an eigenoperator of $\hat{\hat{S}}_o$ with eigenvalue 1 and of $\hat{\hat{T}}_o$ with eigenvalue zero, we may define an effective Liouville operator $\hat{\hat{L}}$ in the space of nuclear spin operators and molecular orientation:

$$\hat{\hat{L}} = \omega_o + \sum_\mu 2\pi\beta\, g_\mu D^{(2)}_{o\mu} + \frac{1}{2} a \hat{\hat{J}}_o + \frac{1}{2}\sum (5)^{\frac{1}{2}} A_\mu D^{(2)}_{m\mu} \hat{\hat{J}}_m \begin{pmatrix} 2 & 1 & 1 \\ -m & o & m \end{pmatrix} \qquad (5)$$

Symmetry operations of $\hat{\hat{L}}$ and $\underline{\underline{\pi}}$ include:

1) $\hat{\hat{L}}$ is invariant under simultaneous rotations of the molecule and the nuclear spin coordinate system about the field direction.

2) $\hat{\hat{L}}$ is invariant under the operation $U\hat{\hat{R}}$, where U is rotation of the molecule by π about the laboratory axis, $Uf(\eta,\theta,\phi) = f(-\eta,\pi-\theta,\pi+\phi)$ and $\hat{\hat{R}}$ is the nuclear spin operator reversal defined by:

$$\hat{\hat{R}}|m\rangle\langle n| = (-1)^{m-n}|n\rangle\langle m| \qquad (6)$$

where $|m\rangle$, $|n\rangle$ are eigenkets of $\hat{I}_z$ with eigenvalues m,n.

If the fluid is macroscopically isotropic (not an aligned liquid crystal or a single crystal) $\underline{\underline{\Pi}}$ is invariant under operations 1) and 2). In addition, provided that g, A and the diffusion tensor have the same principal axes, there are three further symmetry operations which leave $\hat{L}$ and $\underline{\underline{\Pi}}$ invariant:

3) V_x,V_y and V_z, rotations of the molecule by π about the principal axes x,y and z.

$$V_z f(\eta,\theta,\phi) = f(\eta,\theta,\phi+\pi) \tag{7}$$

$$V_x f(\eta,\theta,\phi) = f(\eta+\pi,\pi-\theta,-\phi) \tag{8}$$

$$V_y f(\eta,\theta,\phi) = f(\eta+\pi,\pi-\theta,\pi-\phi) \tag{9}$$

We do not need to investigate the structure of the group generated by these symmetry operations in any detail as $\underline{\mu}$ is unaltered by them and so belongs to the totally symmetric irreducible representation which is one dimensional. Symmetric basis functions can be constructed for either of the representations discussed by Gordon and Messenger in Chapter XIII. For example, a representation based on the averaging of spin operators over the set of angular dependent functions can be constructing by considering the action of U,V_x and V_y on functions:

$$u(L,\lambda,K) = [(2L+1)/8\pi^2]^{\frac{1}{2}} D^{(L)}_{\lambda K}(\eta,\theta,\phi); \tag{10}$$

which form a complete normalised set over all orientations of a rigid body relative to a laboratory coordinate system. Their transformation properties under the operations $\frac{\partial}{\partial\phi}$,$V_z$ and V_x are:

$$\frac{\partial}{\partial\phi} u(L,\lambda,K) = \lambda u(L,\lambda,K) \tag{11}$$

$$U\, u(L,\lambda,K) = (-1)^L u(L,-\lambda,K) \tag{12}$$

$$V_z u(L,\lambda,K) = (-1)^K u(L,\lambda,K) \tag{13}$$

$$V_x u(L,\lambda,K) = (-1)^L u(L,\lambda,-K) \tag{14}$$

The nine nuclear spin operators $\hat{b}(s,\sigma)$,

$$\hat{b}(s,\sigma) = (2S+1)^{\frac{1}{2}}\sum|m\rangle\langle n|(-1)^{I+n-\sigma}\begin{pmatrix} I & I & S \\ m & -n & \sigma \end{pmatrix}, \qquad (15)$$

which are irreducible tensor operators transforming according to $D^{(s)}_{\sigma}$, form a complete set over nuclear Liouville space and transform under $\hat{\hat{R}}$ to:

$$\hat{\hat{R}}\,\hat{b}(S,\sigma) = \hat{b}(S,-\sigma) \qquad (16)$$

Combining these properties we see that the set of operators

$$(LS\sigma K) = \frac{1}{2}\{[u(L\sigma K)+(-1)^{L}u(L\sigma\text{-}K)]\hat{b}(S\text{-}\sigma)+[u(L\text{-}\sigma\text{-}K) + (-1)^{L}u(L\text{-}\sigma K)]\hat{b}(S\sigma)\} \qquad (17)$$

are totally symmetric under the operations of the group. These also have the advantage that $\underline{\underline{\Pi}}$ is diagonal in the basis unless the diffusion has less than axial symmetry as discussed by Freed, equation XIV.21a. The values of the numbers L,S,σ and K have considerable restrictions. They are all non-negative integers, $S\leq 2$; $\sigma\leq$(S and L); $K\leq L$ and is even, and if L is odd K and σ cannot be zero. These imply that for given values of L and K, there are only 6 possible operators if L is even and 3 if L is odd. If the spin Hamiltonian is axially symmetric, then only L even and K = 0 operators need be considered.

ESR LINESHAPES AND SATURATION IN THE SLOW MOTIONAL REGION—THE STOCHASTIC LIOUVILLE APPROACH

Jack H. Freed

Department of Chemistry, Cornell University

Ithaca, New York 14850

XIV.1. GENERAL APPROACH[1]

Let us define an orientation-dependent spin-density matrix, $\sigma(\Omega, t)$ by:

$$\sigma(\Omega, t) = e^{-(i\mathcal{H}^{\times}+\Gamma_{\Omega})t}\,\sigma(0) \tag{1}$$

such that $\langle P_0|\sigma(\Omega, t)|G_0\rangle = \sigma(t)$ as given by eq. VIII-119. Then it obeys the stochastic Liouville equation of motion:[1,2,3]

$$\frac{\partial}{\partial t}\sigma(\Omega, t) = -[i\mathcal{H}(\Omega)^{\times}+\Gamma_{\Omega}]\sigma(\Omega, t). \tag{2}$$

We again consider the steady-state spectrum in the presence of a single rotating rf-field. One has for the λth (multiple) hyperfine line at "orientation" specified by Ω (cf. eq. XVIII-16):

$$P_{\lambda}(\Omega) = 2\mathfrak{N}\hbar\omega\sum_j d_{\lambda_j} Z_{\lambda_j}^{(1)''}(\Omega) \tag{3}$$

where P_{λ} is the power absorbed, $\mathfrak{N}$ is the concentration of electron spins, d_{λ_j} is the "transition moment" for the λ_jth component line and is usually given by $d_{\lambda_j} = \frac{1}{2}|\gamma_e|B_1$ (see below), and $Z_{\lambda_j}^{(1)''}$ is defined by the series of equations:

$$(\sigma - \sigma_0)_{\lambda_j} \equiv \chi_{\lambda_j} \tag{4}$$

$$\chi_{\lambda_j} = \sum_{n=-\infty}^{\infty} e^{in\omega t} Z_{\lambda_j}^{(n)} \tag{5}$$

and

$$Z_{\lambda_j}^{(n)} = Z_{\lambda_j}^{(n)\prime} + iZ_{\lambda_j}^{(n)\prime\prime} \tag{6}$$

In eq. 4, $\sigma_0(\Omega)$ is the equilibrium spin-density matrix whose Ω dependence is such that $\Gamma_\Omega \sigma_0 = 0$. In eq. 5 the steady-state solution, χ_{λ_j} has been expanded in a Fourier series with time independent coefficients $Z_{\lambda_j}^{(n)}$. Thus $\sigma_0(\Omega) = \sigma_0^{(0)}$. Eq. 3 displays the fact that it is the n = 1 harmonic which is directly observed. (We will also use $\chi_a^{(0)}$ as the 0th harmonic for the diagonal element corresponding to state a.)

In the above notation σ_{λ_j} means the matrix element of σ:

$$\sigma_{\lambda_j} \equiv \langle \lambda_j - |\sigma| \lambda_j + \rangle \tag{7}$$

where λ_j- and λ_j+ are the two levels between which the λ_jth transition occurs, and a "raising convention" is implied. Some aspects of this notation are discussed in Ch. XVIII.

The total absorption is then obtained as an equilibrium average of eq. 3 over all Ω. Thus we introduce averages such as:

$$\bar{Z}_{\lambda_j}^{(n)} \equiv \langle P_o(\Omega) | Z_{\lambda_j}^{(n)}(\Omega) | P_o(\Omega) \rangle \equiv \int d\Omega\, Z_{\lambda_j}^{(n)}(\Omega) P_o(\Omega) \tag{8}$$

so that

$$P_\lambda = 2\mathfrak{N}\hbar\omega \sum_j d_{\lambda_j} \bar{Z}_{\lambda_j}^{(1)\prime\prime}, \tag{9}$$

where we have taken d_{λ_j} essentially independent of orientation. We now separate $\mathcal{H}$ into three components:

$$\mathcal{H} = \mathcal{H}_0 + \mathcal{H}_1(\Omega) + \epsilon(t) \tag{10}$$

where in the high-field approximation:

$$\hbar\mathcal{H}_0 = \bar{g}_s B_e B_o S_z - \hbar \sum_i \gamma_i I_{z_i} B_o - \hbar\gamma_e \sum_i \bar{a}_i S_z I_{z_i} \tag{11}$$

yields the zero-order energy levels and transition frequencies,

$$\mathcal{H}_1(\Omega) = \sum_{L,m,\mu,i} F'^{(L,m)}_{\mu,i}(\Omega) A'^{(L,-m)}_{\mu,i} \tag{12}$$

is the perturbation depending on orientation angles Ω, and $\epsilon(t)$ is given by:

$$\hbar\epsilon_1(t) = \tfrac{1}{2}\hbar\gamma_e B_1\,[S_+\exp(-i\omega t) + S_-\exp(i\omega t)] \tag{13}$$

and is the interaction with the radiation field.

When one takes the $\langle\lambda_{j-}|\ \ |\lambda_{j+}\rangle$ matrix elements of eq. 2, and utilizes eqs. 4-6, the steady-state equation for $Z^{(1)}_{\lambda_j}$ is found to be in the high temperature approximation (cf. Ch. XVIII):

$$\Delta\omega_\lambda Z^{(1)}_{\lambda_j} + [H_1(\Omega)^\times Z^{(1)}(\Omega)]_{\lambda_j} - i[\Gamma_\Omega Z^{(1)}(\Omega)]_{\lambda_j} + d_{\lambda_j}\left(\chi^{(0)}_{\lambda_{j+}} - \chi^{(0)}_{\lambda_{j-}}\right) = q\omega_\lambda d_{\lambda_j} \tag{14}$$

The equation for $\bar{Z}^{(1)}_{\lambda_j}$ then becomes:

$$\Delta\omega_\lambda \bar{Z}^{(1)}_{\lambda_j} + \int d\Omega\,[\mathcal{H}_1(\Omega)^\times Z^{(1)}(\Omega)]_{\lambda_j} P_0(\Omega) - i\int d\Omega\,[\Gamma_\Omega Z^{(1)}(\Omega)P_0(\Omega)]_{\lambda_j} + d\left(\bar{\chi}^{(0)}_{\lambda_j+} - \bar{\chi}^{(0)}_{\lambda_j-}\right) = q\,\omega_\lambda d_{\lambda_j} \tag{15}$$

In order to perform the integrations over Ω in eq. 15, we expand the matrix element $Z(\Omega)^{(n)}_{\lambda_j}$ in a complete set of orthogonal eigenfunctions of Γ_Ω, call them $G_m(\Omega)$, with eigenvalues E_m:

$$Z(\Omega,\omega)^{(n)}_{\lambda_j} = \sum_m [C^{(n)}_m(\omega)]_{\lambda_j} G_m(\Omega), \tag{16}$$

or in operator notation:

$$Z(\Omega)^{(n)} = \sum_m C^{(n)}_m(\omega) G_m(\Omega) \tag{16a}$$

where $C_m^{(n)}$ is still an operator in spin space, and is a function of ω, but is independent of Ω. Then eq. 15 becomes:

$$\Delta\omega_\lambda \bar{Z}^{(1)}_{\lambda_j} + \sum_m \int d\Omega P_0(\Omega) G_m(\Omega)\,[\mathcal{H}_1(\Omega)^\times C^{(1)}_m]_{\lambda_j} + d_\lambda\left(\bar{\chi}^{(0)}_{\lambda_{j+}} - \bar{\chi}^{(0)}_{\lambda_{j-}}\right) = q\omega_\lambda d_\lambda \tag{17}$$

In obtaining eq. 17 we have assumed that $P_0(\Omega)$ gives an isotropic distribution of orientations.

Note also that:

$$\bar{Z}^{(n)}_{\lambda_j} = [C^{(n)}_0]_{\lambda_j} \tag{18}$$

from the definitions of eqs. 8 and 16. Thus the absorption (eq. 9) depends only on the $[C_o^{(1)}]_{\lambda_j}$ for all allowed transitions λ_j.

When we premultiply eq. 14 by $G_m^*(\Omega)$ and integrate over Ω, we obtain for $[C_m^{(1)}]$ and isotropic orientations:

$$N_{m'}(\Delta\omega_\lambda - iE_{m'})[C_{m'}^{(1)}]_{\lambda_j} + \sum_m \int d\Omega G_{m'}^*(\Omega)G_m(\Omega)[\mathcal{H}_1(\Omega)^\times C_m^{(1)}]_{\lambda_j} +$$

$$N_{m'}d_{\lambda_j}([C_{m'}^{(0)}]_{\lambda_{j+}} - [C_{m'}^{(0)}]_{\lambda_{j-}}) = q\omega_\lambda d_\lambda \delta(m', 0)N_{m'}. \quad (19)$$

Here $N_{m'}$ is a normalizing factor:

$$N_m = \int d\Omega G_m^*(\Omega)G_m(\Omega). \quad (20)$$

Thus the coupling to the Markovian relaxation process of Γ_Ω comes about only if the perturbation $\mathcal{H}_1(\Omega)$ can couple $[C_o^{(1)}]_{\lambda_j}$ to some coefficient $C_m^{(1)}$ where $m \neq 0$.

The above approach, hence eq. 19, is valid for any Markovian or diffusive process. Eq. 19 will yield coupled algebraic equations for the coefficients $[C_m^{(n)}]_{\lambda_j}$, and one attempts to solve for the $[C_o^{(1)}]_{\lambda_j}$ utilizing only a finite number of such coefficients. The convergence depends partially on the ratio $|\mathcal{H}_1(\Omega)|/E_{m'}$. The larger the value of this ratio the more terms $[C_m^{(1)}]_{\lambda_j}$ are needed. The results obtained by second order relaxation theory (cf. Ch. VIII) are recovered when only one order beyond $[C_o^{(1)}]_{\lambda_j}$ is included.

When we apply the method to rotational modulation, then Ω refers to the values of the Euler angles for a tumbling molecular axis with respect to a fixed laboratory axis system. Thus we have for isotropic rotational diffusion (cf. eq. VIII.86):

$$-\Gamma_\Omega \rightarrow R\nabla_\Omega^2 \quad (21)$$

where ∇_Ω^2 is the rotational diffusion operator and R is the diffusion coefficient. Although the method is fully applicable to problems involving completely anisotropic rotational diffusion, we shall assume axially symmetric diffusion for simplicity. Then the complete set of eigenfunctions of Γ_Ω for eq. 21 are the Wigner rotation matrices:

$$E_m = E_{L,K,M} = R_1 L(L+1) + (R_3 - R_1)K^2 \quad (21a)$$

(cf. Sect. VIII.5.B). We now express eq. 12 as (cf. eq. VIII.101):

$$\mathcal{H}_1(t) = \sum_{L,m,m',\mu} \mathcal{D}^{L}_{-m,m'}(\Omega) F'^{(L,m)}_{\mu} A^{(L,m')}_{\mu} \tag{22}$$

where both the $F'^{(L,m)}_{\mu}$ and the $A^{(L,m')}_{\mu}$ are irreducible tensor components of rank L and component m and m'. The F' in eq. 22 are expressed in molecule-fixed co-ordinates, while A is a spin operator quantized in space-fixed axes. The $\mathcal{D}^{L}_{-m,m'}(\Omega)$ terms include the transformation from space-fixed to molecule-fixed axes. It follows from the orthogonality relation[4] of the $\mathcal{D}^{L}_{KM}$'s that:

$$N^{L}_{KM} = \frac{8\pi^2}{2L+1} \tag{23}$$

and

$$P_o(\Omega) = \frac{1}{8\pi^2} = \frac{1}{8\pi^2}\mathcal{D}^{0}_{0,0}(\Omega). \tag{24}$$

The evaluation of the integral on the lhs of eq. 19 is obtained utilizing:

$$\int d\Omega \mathcal{D}^{L_1}_{m_1 m_1'}(\Omega) \mathcal{D}^{L_2}_{m_2 m_2'}(\Omega) \mathcal{D}^{L_3}_{m_3 m_3'}(\Omega) = 8\pi^2 \begin{pmatrix} L_1 & L_2 & L_3 \\ m_1 & m_2 & m_3 \end{pmatrix} \begin{pmatrix} L_1 & L_2 & L_3 \\ m_1' & m_2' & m_3' \end{pmatrix} \tag{25}$$

where the terms in parenthesis in eq. 25 are the 3j symbols;[4] also

$$\mathcal{D}^{L^*}_{m,m'} = (-)^{m-m'} \mathcal{D}^{L}_{-m,-m'} . \tag{26}$$

XIV.2. FREE RADICALS OF S = ½—NO SATURATION[1]

The case of no saturation is achieved by setting $\chi_{\lambda_j+} = \chi_{\lambda_j-} = 0$, so the last term on the lhs of eq. 19 is zero.

A. Axially Symmetric Secular G-Tensor

A particularly simple example of the above formulation is for a one-line ESR spectrum broadened mainly by the secular anisotropic g-tensor term, for which $g_x = g_y = g_\perp$ and $g_z = g_\parallel$. (We assume the z axis of the diffusion tensor is coincident.) When $\omega_o^2\tau_R^2 \gg 1$ the non-secular term will make a negligible contribution compared to the secular term.

For this case eq. 22 is:

$$\mathcal{H}_1(t) = \mathcal{D}^{2}_{0,0}(\Omega)\tfrac{2}{3}\hbar^{-1}\beta_e B_0(g_\parallel - g_\perp)S_z \equiv \mathcal{D}^{2}_{0,0}(\Omega)\mathcal{F}S_z . \tag{27}$$

We have

$$-[S_z^x C_m^{(1)}]_{\lambda_j} = [C_m^{(1)}]_{\lambda_j} \tag{28}$$

for a doublet state. When eqs. 27 and 28 are substituted into eq. 19 (for no saturation), and eqs. 23-26 are utilized, then one obtains:

$$[(\omega-\omega_o)-iR_1 L(L+1)][C^L_{o,o}(\omega)]_{\lambda_j} - (2L+1)\mathcal{F}\sum_{L'}\begin{pmatrix}L2L'\\000\end{pmatrix}^2 [C^{L'}_{o,o}(\omega)]_{\lambda_j}$$

$$= q\omega_\lambda d_{\lambda_{\lambda_j}}\delta(L,0). \tag{29}$$

Now from eq. 9 the absorption is proportional to:

$$\overline{Z}^{(1)''}_{\lambda_j} = \frac{1}{\pi}\mathrm{Im}[C^{0}_{0,0}{}^{(1)}]_{\lambda_j}. \tag{30}$$

Equation 29 defines an infinite set of coupled algebraic equations for the complex coefficients $C^L_{0,0}(\omega)$ (where we have dropped the superscript (1) for simplicity). The triangle property of the 3j symbols means, however, that the Lth equation is coupled only to the $L\pm2^{th}$ equations, so only even L values appear. Approximations to the complete solution may be obtained by terminating the coupled equations at some finite limit by letting $C^L_{0,0} = 0$ for $L>n$ where, $r = \frac{n}{2}+1$ gives the order of the equations.

One finds that as $|\mathcal{F}/R|$ increases to unity (the region where relaxation theory applies) the main effect on the line shape is a broadening. In the region $|\mathcal{F}/R|$ of 1 to 10 the resonance peak shifts downfield and the shape becomes markedly asymmetric. Between 10 and 100 a new high field peak appears, and at 1,000 the solid-like spectrum is sharpening up (one is observing a decrease of "motional broadening") as it is approaching the powder spectrum. (Jensen in Ch. III Fig. 2 obtains similar results.)

B. Asymmetric Secular G-Tensor

If now we let $g_x \neq g_y$, then

$$\mathcal{H}_1(\Omega) = \mathcal{F}_o\mathcal{D}^2_{o,o}S_z + \mathcal{F}_2 S_z[\mathcal{D}^2_{-2,o} + \mathcal{D}^2_{2,o}] \tag{31}$$

where

$$\mathcal{F}_o = \tfrac{2}{3}[g_z-\tfrac{1}{2}(g_x+g_y)]\hbar^{-1}\beta_e B_o \tag{32a}$$

and

$$\mathcal{F}_2 = \frac{1}{\sqrt{6}} (g_x - g_y)\hbar^{-1}\beta_e B_0 \tag{32b}$$

The relevant coupled eqs. are given by:

$$(2L+1)^{-1}[(\omega-\omega_0)-iT_2^{-1}-iE_{L,K})]\bar{C}^L_{K,0} - \mathcal{F}_0 \sum_{L'} \begin{pmatrix} L & 2 & L' \\ K & 0 & -K \end{pmatrix}\begin{pmatrix} L & 2 & L' \\ 0 & 0 & 0 \end{pmatrix}\bar{C}^{L'}_{K,0}$$

$$-\mathcal{F}_2 \sum_{L'} \begin{pmatrix} L & 2 & L' \\ 0 & 0 & 0 \end{pmatrix}\left[\begin{pmatrix} L & 2 & L' \\ K & -2 & -(K-2) \end{pmatrix}\bar{C}^{L'}_{K-2,0} + \begin{pmatrix} L & 2 & L' \\ K & 2 & -(K+2) \end{pmatrix}\bar{C}^{L'}_{K+2,0}\right] = q\omega_\lambda d_{\lambda_j}\delta_{L,0}\delta_{K,0} \tag{33}$$

where K is positive, $E_{L,K}$ is given by eq. 21a, and eq. 30 as well as the triangle rule requiring $L' = L\pm2$ or L still holds. Also one must have $K \leq L$, etc. The order of the equations, when one terminates for $L>n$ is now $1+n(\frac{n}{4}+1)$.

The progress of the line shapes as $|\mathcal{F}_0/R|$ increases, is similar to case 1, but in the region of >10 three peaks now appear. Typical results are shown in Fig. 1.

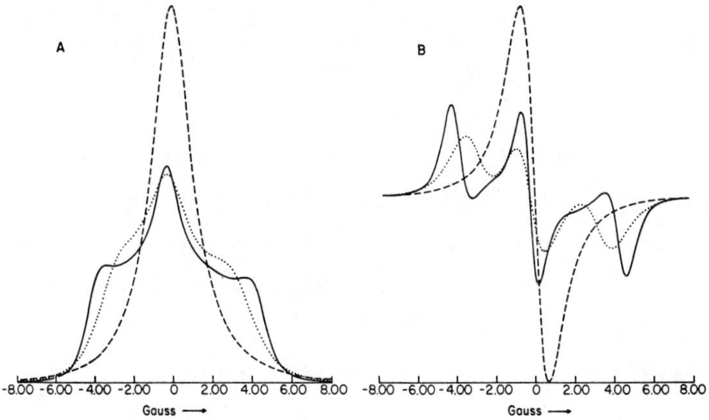

Fig. 1. Lineshapes for an asymmetric g-tensor as a function of $|\mathcal{F}|/R$. A) Absorption Lineshapes; B) First derivative. The different $\mathcal{F}/R$ values are _ _ _5, ...25, _______100. [By permission from Ref. 1.]

C. G-Tensor plus END Tensor Including Pseudosecular Terms

The retention of the pseudo-secular terms (i.e. terms in $I_{\pm}S_z$) results in $[\mathcal{H}_1, \mathcal{H}_0] \neq 0$, unlike the previous cases.

i) <u>One Nuclear Spin of $I = \frac{1}{2}$</u>. This is the simplest case for illustrating the method. The labelling of the energy levels and

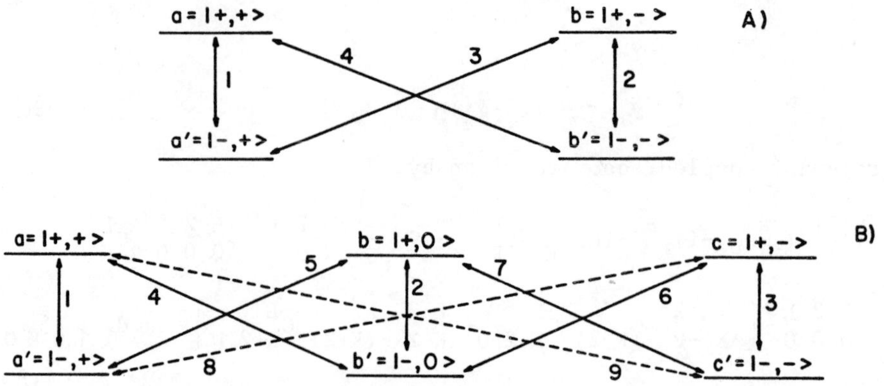

Fig. 2. Energy levels and transitions for A) $S = \frac{1}{2}$, $I = \frac{1}{2}$; B) $S = \frac{1}{2}$, $I = 1$. The notation is $|M_S, M_I\rangle$.

relevant transitions are given in Fig. 2a. The resonance frequencies for the two allowed transitions (1 and 2) and the two forbidden transitions (3 and 4) are:

$$\begin{aligned} \omega_1 &= \omega_{aa'} = \omega_e - \tfrac{1}{2}\gamma_e \bar{a} &&\rightarrow -a' \\ \omega_2 &= \omega_{bb'} = \omega_e + \tfrac{1}{2}\gamma_e \bar{a} &&\rightarrow a' \\ \omega_3 &= \omega_{ba'} = \omega_e + \omega_n &&\rightarrow \omega_n \\ \omega_4 &= \omega_{ab'} = \omega_e - \omega_n &&\rightarrow -\omega_n \end{aligned} \tag{34}$$

The resonant frequencies become $\pm a' = \pm\gamma_e \bar{a}/2$ and $\pm\omega_n$ when ω_e is taken as the origin of the spectrum for convenience.

For simplicity we again assume axial symmetry for the g-tensor and the dipolar tensor. Then:

$$\mathcal{H}_1(\Omega) = \mathcal{D}^2_{0,0} S_z [\mathcal{F} + D' I_z] + (\mathcal{D}^2_{0,1} I_+ S_z - \mathcal{D}^2_{0,-1} I_- S_z) D \tag{35}$$

where

$$D = -2\pi \xi_i D_i^0 \tag{36a}$$

where $\xi_i = \frac{1}{2\pi}|\gamma_e|\gamma_i \hbar$ and D_i^0 is given in XVIII.85. Also

$$D' = -(8/3)^{\frac{1}{2}} D \tag{36b}$$

Note that one has

$$[\mathcal{H}_1^{\times}]_{\alpha\alpha',\beta\beta'} = [\mathcal{H}_{1\alpha\beta}\delta_{\alpha'\beta'} - \mathcal{H}_{1\beta'\alpha'}\delta_{\alpha\beta}] \tag{37}$$

where α, α', β and β' are eigenstates of $\mathcal{H}_0$. $\mathcal{H}_1^{\times}$ may be represented as a simple Hermitian matrix in the space of transitions 1-4:

$$\mathscr{H}_1^{\times}(\Omega) = \begin{array}{c} \\ \\ \\ \\ \end{array}\!\!\begin{pmatrix} \overset{1}{\mathscr{D}^2_{0,0}(\mathscr{F}+\frac{1}{2}D')} & \overset{2}{0} & \overset{3}{s} & \overset{4}{s^*} \\ 0 & \mathscr{D}^2_{0,0}(\mathscr{F}-\frac{1}{2}D') & s & s^* \\ s^* & s^* & \mathscr{D}^2_{0,0}\mathscr{F} & 0 \\ s & s & 0 & \mathscr{D}^2_{0,0}\mathscr{F} \end{pmatrix} \tag{38}$$

where $s = \frac{1}{2}\mathscr{D}^2_{0,1}D$. One must now develop eq. 19 for non-degenerate transitions $\lambda = 1$, 2, 3, and 4. We shall represent the appropriate coefficients as $C^L_{K,M}(\lambda)$. Then we obtain:

$$(2L+1)^{-1}[(\omega+a')-i\{T_2^{-1}+RL(L+1)\}]C^L_{0,0}(1)-(\mathscr{F}+\tfrac{1}{2}D')\sum_{L'}\begin{pmatrix} L & 2 & L' \\ 0 & 0 & 0 \end{pmatrix}^2 C^{L'}_{0,0}(1)$$
$$+\tfrac{1}{2}D\sum_{L'\neq 0}\begin{pmatrix} L & 2 & L' \\ 0 & 0 & 0 \end{pmatrix}\begin{pmatrix} L & 2 & L' \\ 0 & 1 & -1 \end{pmatrix}[C^{L'}_{0,1}(3)-C^{L'}_{0,-1}(4)] = q\omega_1 d\delta_{L,0} \tag{39a}$$

$$(2L+1)^{-1}[(\omega-a')-i\{T_2^{-1}+RL(L+1)\}]C^L_{0,0}(2)-(\mathscr{F}-\tfrac{1}{2}D')\sum_{L'}\begin{pmatrix} L' & 2 & L \\ 0 & 0 & 0 \end{pmatrix}^2 C^{L'}_{0,0}(2)$$
$$+\tfrac{1}{2}D\sum_{L'\neq 0}\begin{pmatrix} L & 2 & L' \\ 0 & 0 & 0 \end{pmatrix}\begin{pmatrix} L & 2 & L' \\ 0 & 1 & -1 \end{pmatrix}[C^{L'}_{0,1}(3)-C^{L'}_{0,-1}(4)] = q\omega_2 d\delta_{L,0} \tag{39b}$$

$$\frac{D}{2}\sum_{L'}\begin{pmatrix} L & 2 & L' \\ 0 & 0 & 0 \end{pmatrix}\begin{pmatrix} L & 2 & L' \\ 1 & -1 & 0 \end{pmatrix}[C^{L'}_{0,0}(1)+C^{L'}_{0,0}(2)] + (2L+1)^{-1}[(\omega-\omega_n)$$
$$-i\{T_2^{-1}+RL(L+1)\}]\,C^L_{0,1}(3)+\mathscr{F}\sum_{L'\neq 0}\begin{pmatrix} L & 2 & L' \\ 0 & 0 & 0 \end{pmatrix}\begin{pmatrix} L & 2 & L' \\ 1 & 0 & -1 \end{pmatrix}C^{L'}_{0,1}(3)=0 \tag{39c}$$

$$-\frac{D}{2}\sum_{L'}\begin{pmatrix} L & 2 & L' \\ 0 & 0 & 0 \end{pmatrix}\begin{pmatrix} L & 2 & L' \\ -1 & 1 & 0 \end{pmatrix}[C^L_{0,0}(1)+C^L_{0,0}(2)] + (2L+1)^{-1}[(\omega+\omega_n)$$
$$-i\{T_2^{-1}+RL(L+1)\}C^L_{0,-1}(4) + \mathscr{F}\sum_{L'\neq 0}\begin{pmatrix} L & 2 & L' \\ 0 & 0 & 0 \end{pmatrix}\begin{pmatrix} L & 2 & L' \\ -1 & 0 & 1 \end{pmatrix}C^{L'}_{0,-1}(4) = 0 \tag{39d}$$

While eqs. 39a and b are applicable for $L = 0$, eqs. 39c and d require $L, L' > 0$, and in all cases L and L' must be even and $L' = L \pm 2$ or L. Equations 39 represent four infinite sets of coupled equations (i.e. expansions in L) which are then coupled amongst each other due to the pseudo-secular contribution from the dipolar term.

The absorption is proportional to:

$$Z''_1 + Z''_2 = \frac{1}{\pi}\,\mathrm{Im}\,[C^0_{0,0}(1)+C^{(0)}_{0,0}(2)] \tag{40}$$

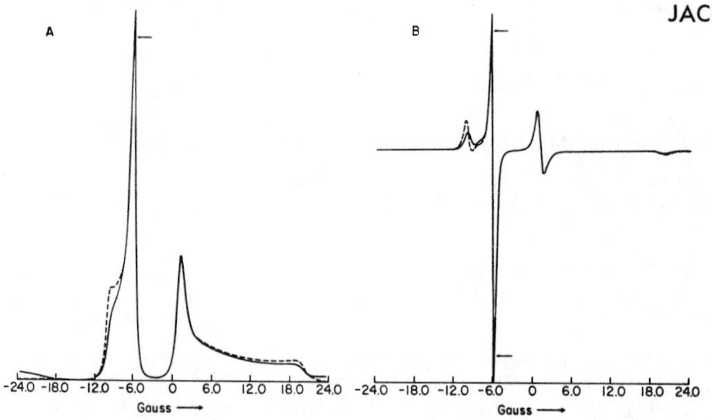

Fig. 3. Comparison for $S = \frac{1}{2}$, $I = \frac{1}{2}$ of lineshapes which include pseudo-secular contributions to lineshapes for which they are omitted $|\mathcal{F}|/R = 100$. $|\omega_n/\gamma_e| = 23.0$ G. All other parameters as in Fig. 4. A) Absorption; B) Derivative. _____ corresponds to inclusion of pseudo-secular terms, _ _ _ _ corresponds to their omission. [Spin parameters $(g_{\|}-g_{\perp})$, $|\omega_n/\gamma_e|$, $A_{\|}, A_{\perp}$ and the abscissas of graph scale to a typical ring proton case when divided by 4.6]. [By permission from Ref. 1.]

When the series of eqs. 39 are terminated for $L > n$, the coupled algebraic equations are of order $r=2(n+1)$.

We show in Fig. 3 a case where $|\mathcal{F}|/R = 100$ and $|D| << |a' \pm \omega_n|$. This result scales reasonably well to that for a typical aromatic ring proton (one need only divide the spin parameters by 4.6). The dotted lines are for neglect of pseudo-secular terms (i.e. let $C^L_{0,1}(3) = C^L_{0,-1}(4) = 0$ for all L). The minor pseudo-secular contributions can be accounted for by perturbation techniques (cf. Section 3)

ii) <u>One Nuclear Spin of $I = 1$; (^{14}N)</u>. The energy levels, and the 3 allowed and 6 forbidden transitions are shown in Fig. 2b for this case, and the details for axially symmetric tensors and for the true asymmetric nitroxide tensors are given elsewhere.[1,5] It is important to note that because $D \sim a$, pseudo-secular contributions must be explicitly included as they may not be handled by perturbation techniques. Typical simulations are shown in Figs. 4, 5 and 6 (see also Gordon, Ch. XIII). Fig. 4 gives the progress from liquid-like to solid-like spectra for axially symmetric tensors; Fig. 5 compares axially symmetric results for different models (a) Brownian dif-

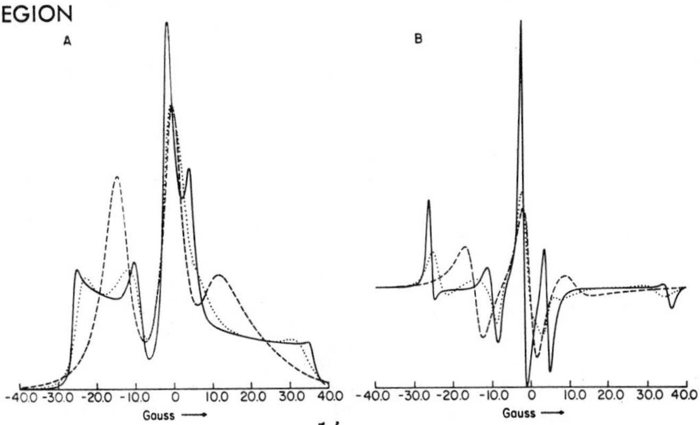

Fig. 4. Lineshapes for $S=\frac{1}{2}$, I=1 (^{14}N nucleus) with axially symmetric g-tensor and hyperfine tensor. A) Absorption; B) Derivative. All correspond to $g_{\parallel} = 2.00270$, $g_{\perp} = 2.00750$, $A_{\parallel} = 32$ G., $A_{\perp} = 6$ G., $|\omega_n/\gamma_e| = 0.36$ G. $(2/\sqrt{3})T_2^{-1}/|\gamma_e| = 0.3$ G. The $|\mathcal{F}|/R$ values are _ _ _ _2, ...15, _____100. [By permission from Ref. 1.]

fusion; (b) Free diffusion (i.e. approximate inclusion of reorientational effects, cf. Ch.VIII.5); (c) Jump diffusion. There are clearly significant differences. These models may be approximated by multiplying the rhs of eq. 21a by a "model parameter" B_L which has different L dependence for the different models.[5] Figure 6 gives a comparison between an experimental result and the best theoretical fit, which is obtained with model (b), for several values of $R_3/R_1 \equiv N$ (cf. eq. 21a). The complete anisotropic magnetic parameters are used in the simulations.

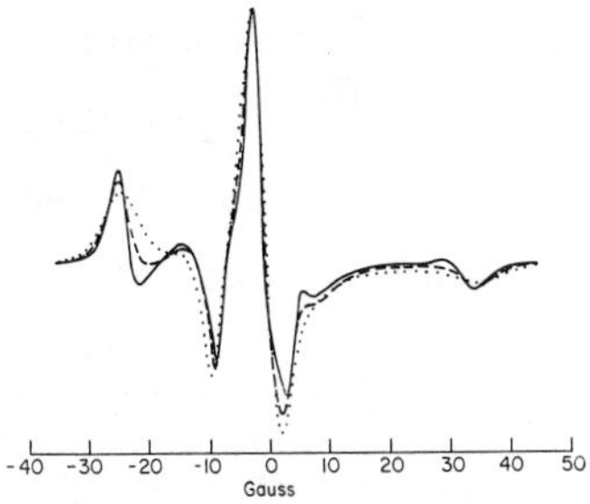

Fig. 5. Comparison of lineshapes for different reorientational models for $S = \frac{1}{2}$, I = 1 (nitroxide) with axially symmetric tensors given in Fig. 4. Solid line-Brownian Diffusion ($|\mathcal{F}(0)\tau_R| = 2.5$); Dashed line-free diffusion ($|\mathcal{F}(0)\tau_R| = 5/3$); Dotted line-jump diffusion ($|\mathcal{F}(0)\tau_R| = 7/6$). [By permission from Ref. 5.]

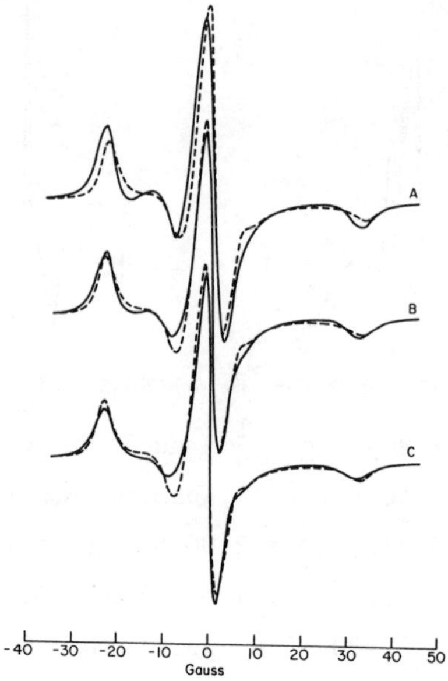

Fig. 6. Comparison of experimental (peroxylamine disulfonate in frozen D_2O at -60°C) lineshape— dashed line, with simulated lineshapes— solid line, for free diffusion model with $\mathcal{F}(0)\tau_R = 5/3$ and for $R_3/R_1 = N = 1$, 3 and 6 (from top to bottom). The complete anisotropic parameters are utilized in the simulations. [By permission from Ref. 5.]

XIV.3. SATURATION[1] (cf. Ch. XVIII).

In order to describe a saturated spectrum, one needs, according to eq. 21, expressions for the $[C_m^{(0)}]_{\lambda_j\pm}$. These may be obtained by taking the $\langle\lambda_j\pm|-|\lambda_j\pm\rangle$ matrix elements of eq. 2, and performing a derivation like that which leads to eq. 19. We obtain:

$$N_{m'}(in\omega + E_{m'})[C_{m'}^{(n)}]_{\lambda_j\pm} = \mp i d_{\lambda_j} N_{m'}([C_{m'}^{(n+1)}]_{\lambda_j^{\rightarrow}} - [C_{m'}^{(n-1)}]_{\lambda_j^{\leftarrow}})$$
$$-i\sum_m \int d\Omega\, G_{m'}^*(\Omega)\, [\mathcal{H}_1^{\times\,(\text{non-sec})} C_m^{(n)}]_{\lambda_j\pm,\,\lambda_j\pm} \qquad (41)$$

(Here the subscript $\lambda_j^{\rightarrow} \equiv \lambda_j$ refers to the $\langle\lambda_j -|-|\lambda_j +\rangle$ matrix element, while $\lambda_j^{\leftarrow}$ refers to the $\langle\lambda_j +|-|\lambda_j -\rangle$ matrix element). The superscript (non-sec) refers to the fact that only the non-secular part of $\mathcal{H}_1$ (including pseudo-secular terms) need be retained.

Equation 41 is also needed for problems involving no saturation, but a non-secular $\mathcal{H}_1$ which induces electron spin-flips.

It is often convenient to look at the difference:

$$[b_{m'}^{(n)}]_{\lambda_j} \equiv [C_{m'}^{(n)}]_{\lambda_j+} - [C_{m'}^{(n)}]_{\lambda_j-} \tag{42}$$

It follows that the Hermitian properties of σ and σ_0 as well as eqs. 4, 5, and 16 that:

$$N_{m'}[C_{m'}^{(n)}]_{ab} = \sum_m [C_m^{(-n)}]_{ba}^* \int G_m^*(\Omega)d\Omega . \tag{43}$$

A form of eq. 19 generalized to any n (not just n = 1) is sometimes needed. It is:

$$N_{m'}[(n\omega - \omega_\lambda)-iE_{m'}][C_{m'}^{(n)}]_{\lambda_j} + \sum_m \int d\Omega G_{m'}^*(\Omega)G_m(\Omega)[\mathcal{H}_1(\Omega)^\times C_m^{(n)}]_{\lambda_j} +$$

$$d_{\lambda_j} N_{m'}[b_m^{(n-1)}]_{\lambda_j} = q\omega_\lambda d_{\lambda_j}\delta_{m',0}\delta_{n,1} \tag{44}$$

In eqs. 41 and 44, one sees that it is only though the effects of the radiation field, where the strength of interaction with the spins is given by $d_{\lambda_j} \equiv \frac{1}{2}\omega_1$, that the nth harmonics $[C_{m'}^{(n)}]$ are coupled to harmonics $[C_m^{(n\pm1)}]$. An analysis of these equations leads to the result that the extent of coupling depends essentially on the ratio ω_1/ω_0, which is very small in the presence of large applied DC fields. Hence, it is sufficient for high-field saturation cases to retain only the n=1 terms (which include $[C_0^{(1)}]_{\lambda_j}$, the observed signal) and the n=0 terms (which include $[b_0^{(0)}]_{\lambda_j}$, the dc population differences). Higher harmonics become important in a variety of multiple resonance schemes, or experiments done at lower DC fields.

A. Rotationally Invariant T_1.

We consider a simple case of saturation. The unsaturated line shape is assumed to be due mainly to the secular part of an axially-symmetric g-tensor, while there is a rotationally invariant $T_1 = (2W_e)^{-1}$ where W_e is the lattice-induced electron spin-flip process (e.g. the spherically symmetric part of the spin-rotational interaction). This is introduced by replacing in the equation obtained from eq. 41 and 42, $E_{m'} \rightarrow E_{m'} + 2W_e$. We also include a rotationally invariant $T_2 \leq T_1$ by letting $iE_{m'} \rightarrow i\,(E_{m'} + T_2^{-1})$ in eq. 44.

For the secular perturbation of eq. 27, it is only necessary to consider $[C^{L}_{0,0}{}^{(1)}]_{\lambda_j}$ and its coupling to $[b^{L}_{0,0}{}^{(0)}]_{\lambda_j}$, and one obtains from eqs. 41-43:

$$[(\omega-\omega_0)-iRL(L+1)-iT_2^{-1}][C^{L}_{0,0}{}^{(1)}]_{\lambda_j} - (2L+1)\mathcal{F}\sum_{L'}\begin{pmatrix} L & 2 & L' \\ 0 & 0 & 0 \end{pmatrix}^2 \times$$

$$\times\, [C^{L'}_{0,0}{}^{(1)}]_{\lambda_j} + \frac{4|d_{\lambda_j}|^2}{RL(L+1)+2W_e}\,\mathrm{Im}[C^{L}_{0,0}{}^{(1)}]_{\lambda_j} = q\omega_\lambda d_{\lambda_j}\delta_{L,0} \quad (45)$$

Equation 45 with eq. 30 then determines the saturated spectrum. Note that the saturation term (viz. the last term on the lhs of eq. 45) for L = 0 is unaffected by the rotational motion, while for $L>0$ we find $W_e \rightarrow W_e + \frac{1}{2}RL(L+1)$, i.e. the rotational motion aids the spin relaxation by spreading the spins over all orientations, which, due to $\mathcal{H}_1(\Omega)$ have different "static" resonant frequencies.

In general, one sees that the effect of the saturation on the line shape is to broaden out the spectrum while acting to "wash out" the asymmetric appearance. This is similar to the effect of increasing T_2^{-1} for the unsaturated spectrum, (except for the reduction of intensity in the case of saturation). The results for $|\mathcal{F}|/R=100$ are shown in Fig. 7.

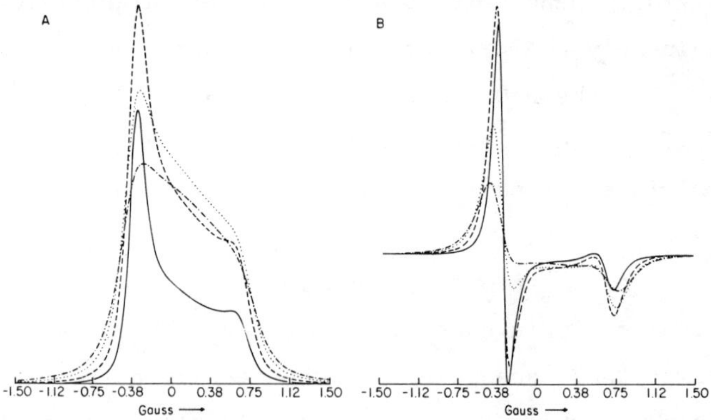

Fig. 7. Saturation of Single Line with rotationally invariant T_1, as a function of B_1 for $\mathcal{F}/R=100$. $g_{\parallel} = 2.00235$, $g_{\perp} = 2.00310$, $T_2 = T_1 = (2W_e)^{-1}$ and $(2/\sqrt{3})\,T_2^{-1}|\gamma_e| = 0.02$G. The different values of $(1/2)B_1$ are ______0.01 G, _ _ _0.025 G, ...0.050 G, _ . _ .0.075 G. A) Absorption, B) Derivative. [By permission from ref. 1.]

B. G-Tensor-(Axially Symmetric).

We now include the non-secular portion of the g-tensor to give:

$$\mathcal{H}_1 = \mathcal{F}[\mathcal{D}^2_{0,0}S_z - (\tfrac{3}{8})^{\frac{1}{2}}(\mathcal{D}^2_{0,1}S_+ - \mathcal{D}^2_{0,-1}S_-)] \qquad (46)$$

Equation 46 is substituted into eqs. 44, 41 and 42 respectively and leads to:

$$(-)^{K-M}[n\omega - \omega_0 - iT_2^{-1} - iRL(L+1)][C^{L(n)}_{-K,-M}]_{\lambda_j} - (2L+1)\mathcal{F}\sum_{L',K',M'}\begin{pmatrix}L & 2 & L'\\ K & 0 & K'\end{pmatrix}\begin{pmatrix}L & 2 & L'\\ M & 0 & M'\end{pmatrix}[C^{L'(n)}_{K',M'}]_{\lambda_j} + (-)^{K-M}d_{\lambda_j}[b^{L(n-1)}_{-K,-M}] + (2L+1)(\tfrac{3}{8})^{\frac{1}{2}}\mathcal{F}\sum_{L',K',M'}\begin{pmatrix}L & 2 & L'\\ K & 0 & K'\end{pmatrix}\begin{pmatrix}L & 2 & L'\\ M & -1 & M'\end{pmatrix}[b^{L'(n)}_{K',M'}]_{\lambda_j} - q\omega_0\delta_{n,0}\delta_{K,0}\delta_{M,1}(\tfrac{3}{8})^{\frac{1}{2}}\mathcal{F} = q\,\omega_\lambda d_{\lambda_j}\delta_{L',0}\delta_{n,1} \qquad (47)$$

and

$$(-)^{K-M}[\tfrac{1}{2}in\omega + \tfrac{1}{2}RL(L+1) + W_e][b^{L(n)}_{-K,-M}]_{\lambda_j} = -id_{\lambda_j}(-)^{K-M}([C^{L(n+1)}_{-K,-M}]_{\vec{\lambda}_j} - [C^{L(n-1)}_{-K,-M}]_{\overleftarrow{\lambda}_j}) + i(2L+1)(\tfrac{3}{8})^{\frac{1}{2}}\mathcal{F}\sum_{L',K',M'}\begin{pmatrix}L & 2 & L'\\ K & 0 & K'\end{pmatrix}\begin{pmatrix}L & 2 & L'\\ M & 1 & M'\end{pmatrix}[C^{L'(n)}_{K'M'}]_{\vec{\lambda}_j} + \begin{pmatrix}L & 2 & L'\\ M & -1 & M'\end{pmatrix}[C^{L'(n)}_{K'M'}]_{\overleftarrow{\lambda}_j} \qquad (48)$$

We now employ the high field, moderate saturation approximations, (i.e. $|\omega_1/\omega_0| << 1$, and $|\mathcal{F}/\omega_0| << 1$), as well as the ad hoc relaxation to thermal equilibrium assumption.[1] This leads to simplified coupled eqs. between the coefficients $C^{L(1)}_{0,0}$ and $b^{L(1)}_{0,0}$ and between $C^{L(0)}_{0,0}$ and $b^{L(0)}_{0,0}$ as a result of the non-secular part of eq. 46, while $C^{L(1)}_{0,0}$ and $b^{L(0)}_{0,0}$ are coupled via the saturating microwave field. However, the assumption that $|\mathcal{F}/\omega_0| << 1$, further allows one to employ second order perturbation theory and decouple $C^{L(1)}_{0,0}$ from $b^{L(1)}_{0,0}$ and $b^{L(0)}_{0,0}$ from $C^{L(0)}_{0,0}$. The resulting eqs. are:

$$\{(\omega - \omega_0) - i[T_2^{-1} + RL(L+1)]\}C^{L(1)}_{0,0} - (2L+1)\mathcal{F}\sum_{L}\begin{pmatrix}L & 2 & L'\\ 0 & 0 & 0\end{pmatrix}^2 C^{L'(1)}_{0,0} + d_{\lambda_j}b^{L(0)}_{0,0} + \left[\text{terms in } C^{L''(1)}_{0,0} \text{ of order } \frac{\mathcal{F}^2}{-\omega_0 + iRL(L+1)}\right] = q\omega_\lambda d_\lambda \delta_{L,0} \qquad (49)$$

$$\left[\frac{RL(L+1)}{2} + W_e\right]b^{L(0)}_{0,0} - 2d_{\lambda_j}\,\mathrm{Im}C^{L(1)}_{0,0} - (\tfrac{3}{8})(2L+1)\mathcal{F}^2\sum_{L',L''}\begin{pmatrix}L & 2 & L'\\ 0 & 0 & 0\end{pmatrix}\begin{pmatrix}L & 2 & L'\\ 0 & 1 & -1\end{pmatrix} \times \begin{pmatrix}L' & 2 & L''\\ 0 & 0 & 0\end{pmatrix}\begin{pmatrix}L' & 2 & L''\\ 1 & -1 & 0\end{pmatrix} \times \left[\frac{(2L+1)V(L,L')}{V(L,L')^2 + \left(\frac{2L+1}{2L'+1}\omega_0\right)^2} + \frac{(2L''+1)V(L'',L')}{V(L'',L')^2 + \left(\frac{2L''+1}{2L'+1}\omega_0\right)^2}\right] \times b^{L''(0)}_{0,0} = 0 \qquad (50)$$

where

$$V(L, L') = \tfrac{1}{2}RL(L+1) + W_e - \left(\frac{2L+1}{2L'+1}\right)[RL'(L'+1)+T_2^{-1}]. \quad (50a)$$

In eq. 49 we are neglecting terms of order $\dfrac{\mathcal{F}^2 C_{0,0}^{L''}(1)}{-\omega_0+iRL(L+1)}$ since when $RL(L+1)\ll|\omega_0|$, as is the case for slow tumbling, then these non-secular terms are of order $|\mathcal{F}/\omega_0|$ smaller than the secular terms in $\mathcal{F}$, i.e. we are neglecting the non-secular contributions to the unsaturated linewidths as compared to the secular contributions for slow tumbling. These non-secular terms must, however, be included in eq. 50 since this eq. predicts the T_1-type behavior and is not explicitly affected by the secular terms in $\mathcal{F}$. Note that for $RL(L+1)\ll|\omega_0|$, then $|V(L, L')|$, $|V(L'', L')|\ll \omega_0^2$ and the terms in V^2 in the denominators of the last terms on the lhs of eq. 50 may be omitted. Furthermore, in this limit a perturbation analysis of the coupling of $b_{0,0}^{L(0)}$ to $b_{0,0}^{L''(0)}$ in eq. 50 shows that it is sufficient to restrict the summation over L'' in this eq. to just $L'' = L$. This, then, just leaves the terms diagonal in $b_{0,0}^{L(0)}$ in eq. 50. However, these diagonal corrections are of order of magnitude $|\mathcal{F}/\omega_0|^2 R$ and are thus negligibly small compared to $(\tfrac{1}{2})RL(L+1)$. Thus, the contribution of the nonsecular terms in eq. 50 is negligible in our approximation except for the diagonal term for $L = 0$. This contribution to $L=0$ is readily calculated, and is

$$W_e^{(G)} = (9/10)(\mathcal{F}/\omega_0)^2 R \quad (51)$$

which is just the result obtained from relaxation theory for $|R/\omega_0|\ll 1$, even though we are now allowing for slow tumbling: $|\mathcal{F}/R|>1$. The net conclusion is $|\mathcal{F}|, R, \ll|\omega_0|$, the solution to the present case is just given by eq. 45 for the rotationally invariant T_1, but with $W_e\rightarrow W_e+W_e^{(G)}\delta_{L,0}$ in that eq., where W_e^{G} is given by eq. 51.

One again finds that the effect of saturation is to broaden out and to tend to reduce the asymmetry of the spectrum. The results for $|\mathcal{F}|/R=100$ are shown in Fig. 8.

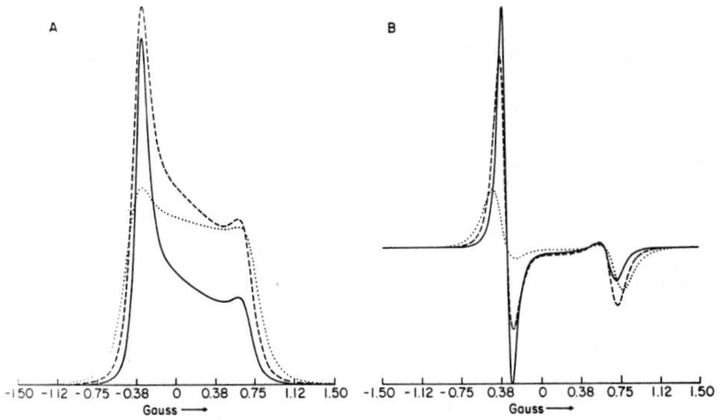

Fig. 8. Saturation of Single Line with g-tensor relaxation as a function of B_1 for $|\mathcal{F}|/R=100$. The different values of $(\frac{1}{2})B_1$, are _____ 3×10^{-6}G, _ _ _7.5×10^{-6}G, ... $.2 \times 10^{-5}$G. All other parameters as in Fig. 5, except $W_e = T_2^{-1} = 0$. [By permission from Ref. 1.]

XIV.4. TRIPLETS[6]

A. General Solutions

For triplets, we cannot necessarily utilize high-field approximations if the zero field splitting is large. It is thus better to employ linear response theory.

We start with the general expression for the imaginary part of the magnetic susceptibility resulting from a very weak linearly polarized rf field of frequency $\omega/2\pi$ being applied to the system:[7]

$$\chi''_{\alpha\alpha}(\omega) = \frac{\omega}{2NkT} \int_0^\infty (e^{i\omega t} + e^{-i\omega t}) \mathrm{Tr}\,[M_\alpha(t) M_\alpha] dt. \qquad (52)$$

The perturbation of the spins by the rf field is given by

$$\epsilon(t) = M_\alpha B_1 \cos\omega t, \qquad (53)$$

i.e. an oscillating field along the α = x, y, or z direction. We shall assume an essentially isotropic g-value and an ensemble of non-interacting triplets so:

$$\mathrm{Tr}\,[M_\alpha(t) M_\alpha] = \mathfrak{N} \gamma_e^2 \mathrm{Tr}\,[S_\alpha(t) S_\alpha] \qquad (52a)$$

We now separate $\mathcal{H}$ into

$$\mathcal{H} = \mathcal{H}_z + \mathcal{H}_1(\Omega)$$

where $\mathcal{H}_z$ is the Zeeman part of the Hamiltonian and $\mathcal{H}_1(\Omega)$ contains the orientation dependent zero-field splitting terms. Here $\mathcal{H}_1(\Omega)$ need not be small compared to $\mathcal{H}_z$. We again assume that the motional process modulating $\mathcal{H}_1(\Omega)$ can be described as a stationary Markoff Process. It then follows that the operator $S_\alpha(\Omega,t)$ obeys the stochastic Liouville equation of motion (cf. eq. 2):

$$\frac{\partial}{\partial t} S_\alpha(\Omega,t) = i\mathcal{H}^\times S_\alpha(\Omega,t) - \Gamma_\Omega S_\alpha(\Omega,t) \tag{54}$$

with initial condition:

$$S_\alpha(\Omega,0) = S_\alpha(0). \tag{55}$$

Let:

$$S_\alpha(s) \equiv \int_0^\infty e^{-st} S_\alpha(t)dt \tag{56}$$

be the Laplace Transform of $S_\alpha(t)$. Then equations 54 and 55 yield:

$$(s - i\mathcal{H}^\times + \Gamma_\Omega) S_\alpha(s,\Omega) = S_\alpha(0) \tag{57}$$

Now in equation 52a, the trace over orientational degrees of freedom is replaced by a classical average:

$$\overline{S_\alpha(t)S_\alpha} \equiv \int d\Omega\, S_\alpha(t,\Omega)\, P_0(\Omega)\, S_\alpha = \overline{S_\alpha(t)S_\alpha}. \tag{58}$$

It follows from eqs. 56 and 58 that the susceptibility, eq. 52 may be rewritten as:

$$\chi''(\omega) = \frac{\mathfrak{N}\gamma_e^2\omega}{2NkT} \mathrm{Tr}[(\overline{S_\alpha(i\omega)} + \overline{S_\alpha(-i\omega)})S_\alpha] \tag{59}$$

where $S_\alpha(\pm i\omega)$ are the Laplace transforms of the spin operator with $s = \pm i\omega$, and the trace is now only over spin degrees of freedom. The plus and minus signs are found to correspond to the two counter rotating components of the rf field. Note that it follows from eq. 53, the Hermiticity of $S_\alpha(t)$, and the fact that Γ_Ω is a real operator independent of spin that:

$$S_\alpha(i\omega)^\dagger = S_\alpha(-i\omega) \tag{59a}$$

where the dagger indicates Hermitian conjugate. Thus we require $\overline{S_\alpha(s)}$ given by:

$$\overline{S_\alpha(\Omega,s)} = \overline{[s - i\mathcal{H}^\times + \Gamma_\Omega]^{-1} S_\alpha(0)}$$

$$\equiv \int d\Omega\, [s - i\mathcal{H}^\times + \Gamma_\Omega]^{-1} P_0(\Omega)\, S_\alpha(0) \tag{60}$$

We expand $S_\alpha(\Omega,s)$ in a complete set of orthogonal eigenfunctions of Γ_Ω, i.e. $G_m(\Omega)$, with eigenvalues E_m (cf. eq. 16 and isotropic liquids):

$$S_\alpha(\Omega,\pm i\omega) = \sum_m a_{\alpha,m}(\pm\omega)\, G,(\Omega) \tag{61}$$

where $a_{m,\alpha}(\omega)$ are still operators in spin space. Clearly from eqs. 56, 58, and 61

$$\overline{S_\alpha(\Omega,\pm i\omega)} = a_{\alpha,0}(\pm\omega) \qquad (62)$$

when we take $P_0 \propto G_0 \equiv 1$.

When $B_1 \perp B_0$ we are interested in S_x. Now the non-zero matrix elements of S_x are $\langle\pm|S_x|0\rangle = \langle 0|S_x|\pm\rangle = \frac{\sqrt{2}}{2}$. It thus follows from eqs. 59 and 62 that

$$\chi''(\omega) = \frac{\sqrt{2}\,\omega}{2NkT}\mathfrak{N}\gamma_e^{\,2}R_e[\langle -|a_{x,0}(\pm\omega)|0\rangle + \langle 0|a_{x,0}(\pm\omega)|+\rangle] \qquad (63)$$

where the $a_{x,0}(\pm\omega)$ terms imply that the effects of both the rotating and counter-rotating components are to be added. Also, we have for the zero-field terms:

$$\mathcal{H}_1 = \sum_{m'}[\frac{1}{\sqrt{6}}D\mathcal{D}^2_{0,m'}(\Omega) + \frac{E}{2}(\mathcal{D}^2_{2,m'}(\Omega) + \mathcal{D}^2_{-2,m'}(\Omega))]A^{2,m'} \qquad (64)$$

where

$$A^{2,0} = \sqrt{6}\,(S_z^2 - \tfrac{1}{3}S^2)$$

$$A^{2,\pm 1} = \mp(S_\pm S_z + S_z S_\pm)$$

$$A^{2,\pm 2} = S_\pm^{\,2} \qquad (65)$$

As before, eqs. 57 and 61 yield sets of coupled differential equations for matrix elements of the coefficients: $a^L_{K,M}(\pm\omega)$. We note the following relation.*

---

*The modified coefficients $C^L_{KM}(\pm\omega)$ are introduced in this way to correspond with the coefficients as used in eq. 16 for the expansion of the density matrix rather than the magnetization. The change of sign in defining $C^L_{KM}(-\omega)$ as compared to $C^L_{KM}(\omega)$ does not correspond to the usage in eq. 16 unless eq. 14 and all those which follow have $q\omega_\lambda d_\lambda$ replaced by $q\omega d_\lambda$, i.e. the modified form required for a Bloch-Redfield type approach to yield the correct dependence of $\chi''(\omega)$ on ω as obtained naturally from linear response theory. A redefined $C'_m(-\omega) = -C_m(-\omega)$ would correspond to the exact usage in the previous sections.

$$C^{L}_{KM}(\pm\omega) = \pm i(-)^{K-M} a^{L*}_{-K-M}(\mp\omega) \tag{66}$$

and abbreviate the matrix elements of the $C^{L}_{K,M}(\pm\omega)$ as:

$$\begin{aligned}
&\langle -|C|0\rangle = C(1), \ \langle 0|C|-\rangle = C(-1), \ \langle 0|C|1\rangle = C(2) \\
&\langle 1|C|0\rangle = C(-2), \langle -1|C|1\rangle = C(3), \ \langle 1|C|-1\rangle = C(-3) \\
&\langle -1|C|-1\rangle = C(a), \langle 0|C|0\rangle = C(b), \ \langle 1|C|1\rangle = C(c).
\end{aligned} \tag{67}$$

This generates a set of coupled eqs. of order $(9/2)n + 2$ for the solution of the triplet problem ($E = 0$).[6] Typical simulations are shown in Figs 9 and 10. One can repeat this for $B_1 \| B_0$ where the

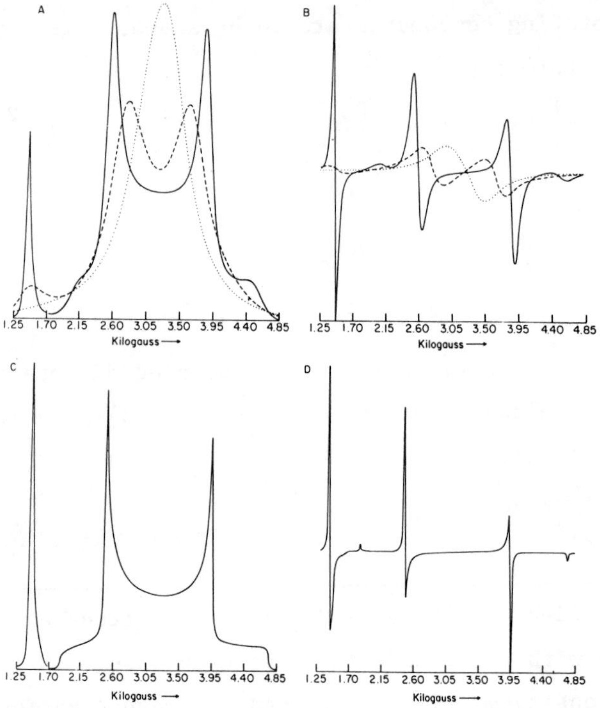

Fig. 9. Lineshapes as a function of D/R for a triplet with B_1 perpendicular to B_0. A) Absorption lineshapes. B) First derivative lineshapes. The different lines correspond to .....D/R=5, n = 6; _ _ _ _D/R = 20, n = 8; _____D/R = 200, n = 12; C and D) Rigid limit absorption and first derivative lineshapes. All plots are for D = 1,435 G., $\frac{\omega}{|\gamma_e|}$ = 3,300 G., a rotationally invariant $(\sqrt{3}\, T_2^{-1}/2)\,|\gamma_e|$ = 15 G and E = 0. The field B_0 is swept. [By permission from Ref. 6.]

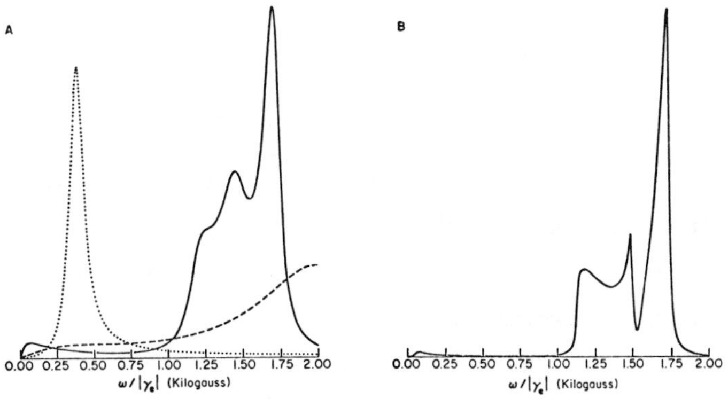

Fig. 10. Low field absorption lineshapes as a function of D/R for a triplet with B_1 perpendicular to B_0. A) The different lines correspond to .....D/R = 0.2, n = 2; _ _ _D/R = 20, n = 8; _____D/R = 200, n = 12. B) Rigid limit. All plots are for D = 1,435 G., B_0 = 300 G. and $T_2^{-1}/|\gamma_e| = \frac{2}{\sqrt{3}}$ 15 G. The frequency, $\omega = |\gamma_e|B$ is swept. [By permission from Ref. 6.]

spectrum is given by:

$$\chi''(\omega) = \frac{\mathfrak{N}\gamma_e^2\omega}{NkT}\,[\langle+|a_{z,o}(-\omega)|+\rangle-\langle-|a_{z,o}(-\omega)|-\rangle] \qquad (68)$$

with a somewhat different set of coupled equations.

B. Perturbation Theory

There are two limiting cases in which the coupled equations may be solved more simply.

i. High Field Case. Here D, $6R \ll \omega_0$. One can show that for this case the Δm = 2 transition is a simple Lorentzian of width $6R \equiv \tau_R^{-1}$ provided $D^2/\omega_0 R \ll 330$ both for $B_1 \perp B_0$ and $B_1 \| B_0$. Note that the observed derivative linewidth of the half-field line in a field swept experiment is given by:

$$\delta = (2\gamma_e)^{-1}[\frac{2}{\sqrt{3}}\{T_2^{-1} + 6R\}] \qquad (69)$$

where $2\gamma_e$ is the "effective" gyromagnetic factor and T_2^{-1} is the rotationally invariant width term.

ii. Fast Motional Case. Here $D \ll 6R$. This is the condition for conventional relaxation theory to apply. One obtains a single

Lorentzian line at $\omega = \omega_0$ with:

$$T_2^{-1} = \frac{\Delta^2}{20}[3J(0)+5J(\omega_0)+2J(2\omega_0)] \quad (70)$$

where $\Delta^2 = 2(D^2/3+E)$ and $J(\omega) = 12R/[(6R)^2+\omega^2]$. This result can also be obtained from a perturbation analysis of our coupled equations using only terms up through L = 2. Cases 1 and 2 are clearly not mutually exclusive.

C. Summary of Spectra

i. High-Frequency and Field Sweep ($\omega/|\gamma_e| \equiv B = 3,300$ G.) Figure 9 shows typical results when $B_1 \perp B_0$. As the rotational motion increases from the rigid limit, both the $\Delta m = 1$ and 2 lines initially broaden. However, fast motion leads to a narrowing of the $\Delta m = 1$ transition until it becomes Lorentzian with its absorption maximum at B = 3,300 G. and width given by eq. 70.

The $\Delta m = 2$ transition for $B_1 \| B_0$ gives results that are very similar to what is obtained for the $B_1 \perp B_0$ case.

ii. Low Field and Frequency Sweep ($B_0 = \omega_0/|\gamma_e| = 300$ G.). Typical results for this case are given in Fig. 10 which shows the $B_1 \perp B_0$ rf orientation. As the motion starts, the rigid limit spectra occurring in the region of $\omega \sim D = 1,435$ G. $|\gamma_e|$ are seen to broaden out. However, as the motion becomes rapid, one finds that for $B_1 \perp B_0$ a Zeeman line appears at $\omega \sim \omega_0 = 300$ G$|\gamma_e|$, while there is a negligible contribution in the region of $\omega \sim D$. In fact, in the motional narrowing region the width of the Zeeman line is found to be given by eq. 70, as it should. This, then, is a simple example of a more general phenomenon; that if, in the rigid limit the spins are quantized essentially in a molecular frame (the Zeeman field is only a perturbation), they will nevertheless appear to be quantized in the laboratory frame yielding the usual Zeeman line, when the tumbling rate of the molecule is fast compared to zero field splittings.

In the case of $B_1 \| B_0$, a Zeeman line cannot appear. Instead a line appears at $\omega = 0$ with width predicted by eq. 70. This line is apparent if $\chi''(\omega)/\omega$ is plotted instead of $\chi''(\omega)$. It is seen from eq. 52 that the latter must go to zero at $\omega = 0$. It is also found that for very slow motion $\chi''(\omega)/\omega$ shows a resonance at $\omega \sim 0$, which broadens out as does the regular line which occurs around $\omega \sim D$.

iii. Zero Field and Frequency Sweep. The results for this case are qualitatively similar to $B_1 \| B_0$ and low fields.

An analysis of $\chi''(\omega)/\omega$ shows two lines of equal intensity at $\omega = D$ and $\omega = 0$, for the rigid limit and for slow motion. These lines are found to be Lorentzian in shape with T_2^{-1} very well approximated by $T_2^{-1} = 4R$ for $|D/6R| \gg 1$. They correspond to the doubly degenerate $T_x \leftrightarrow T_z$ and $T_y \leftrightarrow T_z$ transitions occurring at $\omega = \pm D$ and to the $T_x \leftrightarrow T_y$ transition at $\omega = 0$. Here T_x, T_y, and T_z are the standard zero-field triplet wave functions with zero-field energies of D, D and 0 respectively. For fast motion a line narrows up at $\omega = 0$, with width as predicted by eq. 70. Of course, the $\omega = 0$ lines are suppressed since $\chi''(\omega)$ is studied in a real experiment rather than $\chi''(\omega)/\omega$.

Initial high field results, where $|D| \ll \omega_0$ (thus simplifying the analysis) have been discussed by Norris and Weissman[8] who found a Brownian diffusion-type model best fit their experiments.

We note in conclusion, that an important possibility for slow-tumbling experiments is that they may well prove useful in distinguishing between different models for molecular reorientation.[5]

XIV.5. ACKNOWLEDGEMENT

This work was supported in part by the Advanced Research Projects Agency and by the National Science Foundation.

References

1. J. H. Freed, G. V. Bruno, and C. Polnaszek, Materials Science Center, Cornell University, Report No. 1484 (Nov. 1970); Abstracts 2nd Symposium on Electron-Spin Resonance, Athens, Ga. Dec. 1970; J. Phys. Chem. 75, 3385 (1971).
2. R. Kubo, Advan. Chem. Phys. 16, 101 (1969).
3. R. Kubo, J. Phys. Soc. Japan 26 Supplement, 1 (1969).
4. A. R. Edmonds, "Angular Momentum in Quantum Mechanics" (Princeton University Press, Princeton, N. J., 1957).
5. S. Goldman, G. V. Bruno, C. Polnaszek, and J. H. Freed, J. Chem. Phys. 56, Jan. 15 (1972).
6. J. H. Freed, G. V. Bruno, and C. Polnaszek, J. Chem. Phys. 55, Dec. 1 (1971).
7. A. Abragam "The Principles of Nuclear Magnetism" (Oxford University Press, London, 1961) P. 101.
8. J. R. Norris and S. I. Weissman, J. Phys. Chem. 73, 3119 (1969).

ELECTRON SPIN RELAXATION IN LIQUID CRYSTALS

G.R. Luckhurst

Department of Chemistry, The University

Southampton SO5 9NH, England

XV.1. INTRODUCTION

The dominant spin relaxation process for many paramagnetic species in fluid solution results from the anisotropy in the magnetic interactions coupled to the molecular rotational diffusion.[1] Rotation of the molecule modulates the energies of the spin levels and also induces transitions between them. The line broadening caused by this mode of relaxation provides a useful technique for investigating the nature and rate of the reorientation process. Although the results of such studies have appeared they are invariably for systems in which the rotation is rapid.[2,3] As a consequence the line shape for non-degenerate transitions is Lorentzian and all of the dynamic information is contained in the linewidth. In general this width is determined by the Fourier transform of the correlation functions of certain rotation matrices, but in practice the dominant contribution comes from the transform at zero frequency. In other words only the correlation time for the molecular motion can be obtained from the linewidth.

The vast majority of investigations of the correlation time has been made in macroscopically isotropic systems where the linewidth can only provide information about the dynamics of the molecular environment. In contrast the environment of a paramagnetic solute in a liquid crystalline solvent is highly anisotropic, even though the rate of reorientation is fast. As a result of the macroscopic anisotropy in the system the linewidths depend on both the correlation time for the motion and on certain static properties. In this Chapter we shall investigate the nature of this dependence both theoretically and experimentally. However, before we can commence this investigation it is essential to understand some of the important properties of liquid crystals.[4,5]

XV.2. NEMATIC LIQUID CRYSTALS

The pair distribution function $n^{(2)}(\underset{\sim}{r}_1;\underset{\sim}{r}_2)$, plays a central role in the calculation of the thermodynamic properties of dense fluids. Of course $n^{(2)}(\underset{\sim}{r}_1;\underset{\sim}{r}_2)d\underset{\sim}{r}_1 d\underset{\sim}{r}_2$ is just proportional to the probability of finding molecules in the two volume elements $d\underset{\sim}{r}_1$ and $d\underset{\sim}{r}_2$ at positions $\underset{\sim}{r}_1$ and $\underset{\sim}{r}_2$. Provided the constituent molecules are spherically symmetric this distribution function depends only on the scalar of the difference $(\underset{\sim}{r}_1-\underset{\sim}{r}_2)$ i.e. on r. In most fluids $n^{(2)}(r)$ is extremely dependent on r in the region of small separations. However, for separations greater than several molecular diameters $n^{(2)}(r)$ tends rapidly to the limiting value of ρ^2 where ρ is just the number density. In other words, a dense fluid is generally characterised by short range order and long range disorder. When the molecular symmetry is less than spherical the definition of the distribution function must be extended to include angular as well as spatial correlations.[6,7] Thus $n^{(2)}(\underset{\sim}{r}_1,\Omega_1;\underset{\sim}{r}_2,\Omega_2)d\underset{\sim}{r}_1 d\underset{\sim}{r}_2 d\Omega_1 d\Omega_2$ is now proportional to the probability of finding molecules at $\underset{\sim}{r}_1$ and $\underset{\sim}{r}_2$ with orientations Ω_1 and Ω_2. Under certain conditions the pair distribution function for parallel molecules may be approximated by

$$n^{(2)}(\underset{\sim}{r}_1,\Omega;\underset{\sim}{r}_2,\Omega) \propto \frac{1}{r}\exp(-r/\xi). \qquad \text{(XV.1.)}$$

According to this expression the probability of finding two molecules with the same orientation decreases continuously with increasing separation. A pair of molecules will therefore remain parallel for separations much less than the coherence length[8] ξ but for separations greater than ξ the molecular orientations are uncorrelated. In isotropic liquids the coherence length is rarely greater than several molecular diameters. In other words, both the spatial and the orientational order in a fluid such as benzene, are essentially short range.

On the other hand, the crystalline state is characterised by the existence of long range order which is invariably destroyed at the melting points of most solids. However, this destruction does not always take place, and there exists a group of compounds for which the coherence length in the fluid is as large as 10^6 Å. This high degree of angular correlation confers on the liquid, optical properties which are more characteristic of a crystal even though the hydrodynamic behaviour is typical of a liquid.[4,5] This group of compounds was therefore given the descriptive although contradictory title of liquid crystals.[4] The liquid crystalline state, or mesophase, is not stable at all temperatures and in addition to the melting point there is another first order transition at which the orientational order vanishes and an isotropic fluid results.[4,5] There are a variety of liquid crystals which differ in their molecular organisation; however we shall only be concerned with the nematic mesophase. Nematic liquid crystals or nematogens are usually formed by elongated molecules such as 4,4'-dimethoxyazoxybenzene whose structure is shown in Figure 1. This particular nematogen melts at 118°C, when the spatial order is lost, to give a turbid fluid known as the nematic mesophase. At 135°C there is a transition to a clear isotropic fluid and this temperature is referred to as the nematic-isotropic transition point.

The long coherence length in the mesophase implies that the molecules tend to be arranged with their long axes parallel to one another. A highly idealised picture of this structure is shown in Figure 1. This picture is idealised for although ξ is extremely large the orientational order is not perfect because at any position $\mathbf{r}$ the molecular orientation fluctuates about a preferred orientation. The direction of alignment in a liquid crystal is usually described by a vector field $\mathbf{d}(\mathbf{r})$, known as the director, which is parallel to the preferred

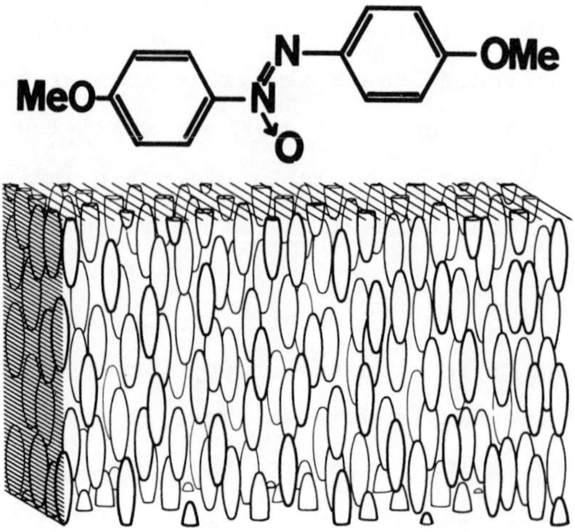

Figure XV.I. An idealised picture of the molecular organisation in a nematic mesophase such as that of 4,4'-dimethoxyazoxybenzene.

orientation at $\underset{\sim}{r}$. If, as is often supposed, the molecules possess cylindrical symmetry, presumably as a result of rapid rotation about their long axes, then their instantaneous orientations with respect to the director can be described by a single angle β. The time or ensemble average of any function of β can therefore be employed as a measure of the molecular orientational order. The even Legendre polynomials $P_L(\cos\beta)$ form a convenient set of functions for this description. The ensemble averages of the odd functions vanish because the energy of the system is invariant under the reversal of the orientation of a molecule. Experimental values of $\overline{P}_2(\cos\beta)$ can usually be obtained with the aid of a nuclear magnetic resonance experiment.[10] For a perfectly ordered system $\overline{P}_2$ is just 1 and vanishes for the other extreme of a disordered or isotropic fluid.

Within the bulk sample of a nematic mesophase, despite the high angular correlation, the director changes slowly but continuously from one region to the next. It is possible, however, to obtain a uniform alignment of the director in a number of ways. For example a magnetic field is capable of aligning most mesophases with the director parallel to the field.[11] The reason for this behaviour is quite straightforward. The diamagnetic susceptibility χ is anisotropic and as a consequence there is a magnetic contribution to the orientational free energy. This contribution takes the form[9]

$$F_{mag.} = -\frac{(\chi_{\parallel} - \chi_{\perp})}{3} B^2 P_2(\cos\beta), \qquad (XV.2.)$$

where β is the angle between the field direction and the molecular long axis. Now for a single molecule $F_{mag.}$ is negligible compared with the thermal energy even for quite intense magnetic fields. However, in a nematic mesophase the large coherence length enhances the anisotropy $(\chi_{\parallel} - \chi_{\perp})$ and as a result the director is completely aligned by magnetic fields as low as 2 kgauss.[11] In géneral the director is parallel to the field because most nematogens are aromatic and $\chi_{\parallel}$ is greater than $\chi_{\perp}$. The macroscopic alignment of the nematic mesophase with $\underset{\sim}{d}(\underset{\sim}{r})$ parallel to $\underset{\sim}{B}$ is therefore unavoidable in most magnetic resonance experiments. The director is also aligned by electric fields[12] as well as suitably pretreated surfaces.[13] Although the mechanism of the electric field alignment is beginning to be understood, we are largely ignorant of the origins of the surface alignment. The existence of the orientational order-disorder transition is a direct consequence of the anisotropy in the pair intermolecular potential.[14] Any theory of the nematic state must therefore start with a specific form for this pair potential. The intermolecular potential for two cylindrically symmetric molecules may be written, in general, as a sum of products of spherical harmonics:[15]

$$U_{ij} = 4\pi \sum_{L_1,L_2,n} u_{L_1,L_2:n}(r) Y_{L_1,n}(\Omega_i) Y^*_{L_2,n}(\Omega_j). \qquad (XV.3.)$$

The coefficients $u_{L_1,L_2:n}(r)$ depend only on the intermolecular separation r, and the molecular orientations Ω_i and Ω_j are defined with respect to a coordinate system containing the intermolecular vector.

Since the molecules as assumed to be both linear and symmetric the summation in equation (3) is restricted to even values of L. The calculation of the orientational properties of a dense fluid composed of molecules interacting with such a pair potential is readily handled within the conceptual framework of the molecular field theory.[16] In this approach the interaction of a particular molecule with the remainder is represented by an effective molecular field. The strength of the field for the liquid crystal problem is obtained by averaging over all orientations of the surrounding molecules, over all orientations of the intermolecular vector and finally over the intermolecular separations. If the approximations implicit in the Maier-Saupe theory[14] of liquid crystals are invoked when evaluating these three averages, then the anisotropic potential of a single molecule is[17]

$$U_i = \sum_L \bar{u}_L \bar{P}_L P_L(\cos\beta)_i. \qquad \text{(XV.4.)}$$

In this expression the averages $\bar{u}_L$ are defined by

$$\bar{u}_L = (1/\rho) \sum_n \int u_{LL:n}(r) n^{(2)}(r) d\underset{\sim}{r}, \qquad \text{(XV.5.)}$$

where β_i is the angle between the long molecular axis and the director[17]. The potential U_i for a single molecule is often referred to as the pseudo-potential.[18] If the orientational free energy is calculated as a function of temperature from the pseudo-potential given by equation (4) then a first order phase transition is found at which the orientational order is destroyed.[14] This transition is obtained even when the pseudo-potential consists of just the L = 2 term, a result which is thought to indicate that dispersion forces dominate the anisotropic intermolecular potential.[14] The theory may be used to calculate any orientational property of a single molecule by taking the appropriate Boltzmann average. For example, the molecular orientational order $\bar{P}_2$ is given by

$$\overline{P}_2 = \frac{\int P_2(\cos\beta_i)\exp\{-U_i(\cos\beta_i)/kT\}d\cos\beta_i}{\int \exp\{-U_i(\cos\beta_i)/kT\}d\cos\beta_i} . \qquad (XV.6.)$$

The theoretical temperature dependence of $\overline{P}_2$ is found to be in excellent agreement with experiment even when the pseudo-potential contains only the first two terms.[17]

Before proceeding to calculate the forms of the static and dynamic spin hamiltonian for a paramagnetic probe dissolved in a nematic mesophase we must consider the origin of the alignment of the probe by the mesophase. As long as the solute deviates from spherical symmetry it will be aligned as a result of its interaction with the molecular field generated by the surrounding molecules. Since, in electron resonance studies, the solute concentration is rather low, this field comes entirely from solvent molecules and as a consequence the molecular field is cylindrically symmetric about the director. In general the molecular orientation with respect to the director may be specified by the two spherical polar angles made by the director in the molecular coordinate system. The solute pseudo-potential may therefore be expanded as a sum of spherical harmonics:

$$U^{(2)} = \sum_{L,m} \left(\frac{4\pi}{2L+1}\right)^{\frac{1}{2}} C_{L,m} Y^{(2)}_{L,m} . \qquad (XV.7.)$$

Provided dipolar forces are unimportant in the solute-solvent interaction the coefficients $C_{L,m}$ will vanish unless L is even. The coefficients depend on both the solvent order $\overline{P_L(1)}$ and the solute-solvent pair potential. For example, when the symmetry of the solute is also cylindrical the $C_{L,m}$ vanish unless m is zero and then

$$C_{L,o} = \overline{u_L^{(1,2)}} \; \overline{P_L^{(1)}} , \qquad (XV.8.)$$

where $\overline{u_L^{(1,2)}}$ is given by an expression analogous to equation (5) for the solute-solvent interaction.[19]

XV.3. THE STATIC SPIN HAMILTONIAN

The spin hamiltonian for a given orientation of a paramagnetic species may be written, using irreducible spherical tensor notation,[20] as:

$$\mathcal{H} = \sum_{\mu;L,p} (-1)^p F_\mu^{(L,p)} T_\mu^{(L,-p)} \quad . \qquad \text{(XV.9.)}$$

In this expression the nature of the spin interaction is denoted by the subscript μ, the strength of the coupling is measured by $F^{(L,p)}$ and $T^{(L,-p)}$ is the appropriate combination of spin operators. Both F and T are expressed in the same coordinate system which is taken to be set in the molecule. In a fluid the molecular orientation is continually changing and so the spin operators fluctuate in time. This time dependence can be removed from T by transforming to a laboratory coordinate system with the aid of Wigner rotation matrices

$$\mathcal{H} = \sum_{\mu;L,p,q} (-1)^p F_\mu^{(L,p)} \mathcal{D}_{q,-p}^{(L)} T_\mu^{(L,q)} \quad . \qquad \text{(XV.10.)}$$

Provided the rate of molecular reorientation is sufficiently rapid the static spin hamiltonian is obtained from equation (10) by taking an ensemble average over all orientations:

$$\overline{\mathcal{H}} = \sum_{\mu;L,p,q} (-1)^p F_\mu^{(L,p)} \overline{\mathcal{D}_{q,-p}^{(L)}} T_\mu^{(L,q)} \quad . \qquad \text{(XV.11.)}$$

When the molecular motion is isotropic, in the sense that all orientations are equally probable, then the averaged rotation matrices vanish unless L is zero, which simply leaves the scalar spin hamiltonian $\mathcal{H}^{(o)}$. Of course, in the nematic mesophase the motion is highly anisotropic and the terms with non-zero L need not vanish. The static spin hamiltonian in equation (11) may be simplified if the laboratory frame contains the director as one of the axes. The average which must be evaluated is

$$\overline{\mathscr{D}^{(L)}_{q,-p}} = \int_0^{2\pi}\int_0^{\pi}\int_0^{2\pi} \mathscr{D}^{(L)}_{q,-p}(\alpha\beta\gamma)P(\alpha\beta\gamma)d\alpha\sin\beta d\beta d\gamma, \qquad \text{(XV.12.)}$$

where $P(\alpha\beta\gamma)$ is the angular distribution function. In the previous section we saw that the solute pseudo-potential, given by equation (7), is a function of just two angles which, in this notation, are β and γ. Since the angular distribution is related to the solute pseudo-potential by

$$P(\Omega) = \frac{\exp\{-U^{(2)}(\Omega)/kT\}}{\int\exp\{-U^{(2)}(\Omega)/kT\}d\Omega}, \qquad \text{(XV.13.)}$$

then $P(\Omega)$ is also independent of the Euler angle α. The integral over α can therefore be evaluated directly since the α-dependence of the rotation matrix is simply $e^{-iq\alpha}$. The average in equation (12) will therefore vanish unless q is zero[21] and so the static spin hamiltonian reduces to

$$\overline{\mathscr{H}} = \mathscr{H}^{(o)} + \sum_{\mu;L\neq o,p} (-1)^p F_\mu^{(L,p)} \overline{\mathscr{D}^{(L)}_{o,-p}} T_\mu^{(L,o)}. \qquad \text{(XV.14.)}$$

This spin hamiltonian is entirely equivalent to that for a static molecule in which all the magnetic interactions $\tilde{F}_\mu$ possess cylindrical symmetry with the only non-zero component $\tilde{F}_\mu^{(L,o)}$ given by

$$\tilde{F}_\mu^{(L,o)} = \sum_p (-1)^p F_\mu^{(L,p)} \overline{\mathscr{D}^{(L)}_{o,-p}}. \qquad \text{(XV.15.)}$$

The symmetry axes for the magnetic tensors of this fictitious static molecules are, of course, parallel to the director. When, as in most experiments, the magnetic field is parallel to the director the static spin hamiltonian is

$$\bar{\mathcal{H}} = \mathcal{H}^{(o)} + \sum_{\mu;L\neq o} \tilde{F}_{\mu}^{(L,o)} T_{\mu}^{(L,o)} . \qquad (XV.16.)$$

If the director can be orientated at some angle to the field then the angular dependence may be predicted by transforming the spin operators from a coordinate system containing the director to one containing the field. This gives the spin hamiltonian as

$$\bar{\mathcal{H}} = \mathcal{H}^{(o)} + \sum_{\mu;L\neq o,q} \tilde{F}_{\mu}^{(L,o)} \mathcal{D}_{q,o}^{(L)} T_{\mu}^{(L,q)} . \qquad (XV.17.)$$

Alignment of the director parallel to the magnetic field results in a reduction of the number,of terms in $\bar{\mathcal{H}}$ and possibly in the information content of the spectrum.

We shall now consider the form of the electron resonance spectrum for a species such as vanadyl acetylacetonate, which contains a single electron and a magnetic nucleus of spin I, when it is dissolved in a nematic mesophase. The dominant magnetic interactions in such a probe are the electron Zeeman coupling and the hyperfine interaction. Vanadyl acetylacetonate also possesses a small quadrupole coupling but provided the director is parallel to the magnetic field this small term has no effect on the electron resonance transition frequencies,[22] although it does modify the ENDOR frequencies. The specific form for the spin hamiltonian is then

$$\bar{\mathcal{H}} = \mathcal{H}^{(o)} + \tilde{g}^{(2,o)} \left(\frac{2}{3}\right)^{\frac{1}{2}} \beta B S_z +$$

$$\tilde{A}^{(2,o)} \left(\frac{2}{3}\right)^{\frac{1}{2}} \{I_z S_z - \tfrac{1}{4}(I_+ S_- + I_- S_+)\} , \qquad (XV.18.)$$

where the scalar spin hamiltonian is

$$\mathcal{H}^{(o)} = g\beta B S_z + a\, \underset{\sim}{I} \cdot \underset{\sim}{S} . \qquad (XV.19.)$$

The fictitious g and hyperfine tensors are related to the degree of solute alignment and the total tensors by equation (15). For most paramagnetic species the non-secular hyperfine terms may be safely handled using second order perturbation theory. This gives the resonant field for the transition with nuclear quantum number m as

$$B_m = B_o - \frac{\{a + (2/3)^{\frac{1}{2}}\tilde{A}^{(2,o)}\}}{\{g + (2/3)^{\frac{1}{2}}\tilde{g}^{(2,o)}\}}\frac{h}{\beta}m$$

$$- \frac{\{a - (1/6)^{\frac{1}{2}}\tilde{A}^{(2,o)}\}}{\{g + (2/3)^{\frac{1}{2}}\tilde{g}^{(2,o)}\}^2}\frac{h}{\omega_o\beta}\{I(I+1)-m^2\}, \quad (XV.20.)$$

where $B_o = h\omega_o/\{g+(2/3)^{\frac{1}{2}}\tilde{g}^{(2,o)}\}\beta$, (XV.21.)

ω_o is the klystron frequency and all hyperfine couplings are in frequency units. The partial alignment of the probe by the nematic mesophase produces two changes in the electron resonance spectrum. The g factor is altered from its isotropic value to

$$\bar{g} = g + \left(\frac{2}{3}\right)^{\frac{1}{2}} \tilde{g}^{(2,o)}, \quad (XV.22.)$$

and this shifts the centre of the spectrum. Secondly the spacing between the m and -m lines devided by 2m changes from the isotropic value a to

$$\bar{a} = a + \left(\frac{2}{3}\right)^{\frac{1}{2}} \tilde{A}^{(2,o)}. \quad (XV.23.)$$

These spectral changes are illustrated in Figure 2 which shows the spectrum of vanadyl acetylacetonate dissolved in the isotropic melt and nematic mesophase of 4,4' - diethoxyazoxybenzene. The observed increase in the g factor and the decrease in the hyperfine spacing are both consistent with the expected alignment of the complex with its plane parallel to the director. In order to determine the degree of solute alignment from the g factor shift it is necessary to know

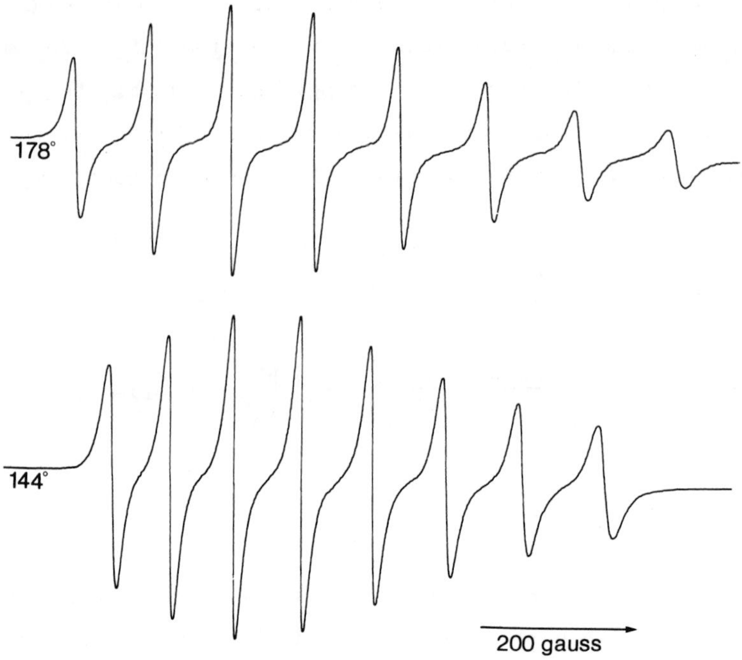

Figure XV.2. The electron resonance spectrum of vanadyl acetylacetonate dissolved in the isotropic melt (173°C) and nematic mesophase (144°C) of 4,4' - diethoxyazoxybenzene.

the isotropic g factor. This parameter can only be determined from measurements in the isotropic melt. In contrast the isotropic coupling constant can be obtained from the spectrum in the nematic mesophase. Inspection of the expression for the resonant fields shows that the difference $(B_m - B_{-m})$ is determined by $(a + (2/3)^{\frac{1}{2}}\tilde{A}^{(2,o)})$ whereas the sum $(B_m + B_{-m})$ yields $(a - (1/6)^{\frac{1}{2}}\tilde{A}^{(2,o)})$.[23] Such a determination avoids any assumption about the nature of environmental changes in the scalar parameters when the probe is dissolved in the mesophase. The total g and hyperfine tensors for vanadyl acetylacetonate are approximately cylindrical symmetric about the VO bond.[2] Determination of $\tilde{A}^{(2,o)}$ can therefore only yield the degree of alignment for this symmetry axis since from equation (15)

$$\tilde{A}^{(2,o)} = A^{(2,o)}\overline{\mathcal{D}^{(2)}_{o,o}} , \qquad \text{(XV.24.)}$$

where $\overline{\mathcal{D}^{(2)}_{o,o}}$ is just $\overline{P_2(\cos\beta)}$ and β is the angle between the VO bond and the director. Measurement of such ordering parameters for the solute is important since it permits a test of the pseudo-potential given in equations (7) and (8).[7]

The two spectra in Figure 2 also show a dramatic decrease in the linewidths on passing from the isotropic melt to the nematic meso-phase. In the next section we shall derive the dynamic spin hamiltonian needed to understand the changes in the electron spin relaxation processes.

XV.4. THE LINEWIDTH CALCULATION

The orientation of a solute molecule dissolved in a nematic mesophase is continually changing. When the solute is a paramagnetic probe, such as vanadyl acetylacetonate, these orientational fluctuations constitute the dominant spin relaxation process just as in normal isotropic fluids.[23] The molecular orientation in a nematic mesophase can change in two distinct ways. One of these is the re-orientation of the molecule with respect to the director and this is the most important motion for relaxation of the electron spin. The rate of this mode of molecular reorientation is fast in the sense that the linewidth may be obtained from Redfield's theory[24] of spin relaxation (c.f. section B of Chapter VIII). The first step in calculating the elements of the relaxation matrix is the determination of the dynamic spin hamiltonian $\mathcal{H}'(t)$. Provided the director is parallel to the magnetic field then $\mathcal{H}'(t)$ is just the difference between equations (9) and (14):

$$\mathcal{H}'(t) = \sum_{\substack{\mu;L\neq o,\\ p,q}} (-1)^p F_\mu^{(L,p)}\{\mathcal{D}^{(L)}_{q,-p}(t) - \overline{\mathcal{D}^{(L)}_{o,-p}}\,\delta_{oq}\}T_\mu^{(L,q)} . \qquad \text{(XV.25.)}$$

The time dependence of $\mathcal{H}'(t)$ is contained entirely in the Wigner rotation matrix and so the statistical part of the problem is the evaluation of the correlation function ψ (t). The dynamic process is taken to be stationary and so ψ can be written as:

$$\psi(t) = \overline{\{\mathcal{D}^{(L)}_{q,-p}(\Omega_o) - \overline{\mathcal{D}^{(L)}_{o,-p}}\delta_{oq}\}\{\mathcal{D}^{(L')*}_{q',-p'}(\Omega) - \overline{\mathcal{D}^{(L')*}_{o,-p'}}\delta_{oq'}\}},$$

$$= \{\overline{\mathcal{D}^{(L)}_{q,-p}(\Omega_o)\mathcal{D}^{(L')*}_{q',-p'}(\Omega)} - \overline{\mathcal{D}^{(L)}_{o,-p}}\;\overline{\mathcal{D}^{(L')*}_{o,-p'}}\delta_{oq}\;\delta_{oq'}\}$$

(XV.26.)

where Ω_o denotes the orientation at zero time and Ω that at time t. Strictly the correlation function depends on the values of L,L', p,p', q and q', but we shall not complicate the notation by labelling ψ with these six symbols at this stage. The first ensemble average in equation (26) for the correlation function may be written in terms of the known angular distribution $P(\Omega)$ and the conditional probability $P(\Omega_o|\Omega,t)$ as

$$\overline{\mathcal{D}^{(L)}_{q,-p}(\Omega_o)\mathcal{D}^{(L')*}_{q',-p'}(\Omega)} = \int \mathcal{D}^{(L)}_{q,-p}(\Omega_o)P(\Omega_o)d\Omega_o \int \mathcal{D}^{(L')*}_{q',-p'}(\Omega)P(\Omega_o|\Omega,t)d\Omega.$$

(XV.27.)

In order to evaluate the conditional probability and hence the correlation function we shall first adopt the strong collision model for molecular reorientation.[25]

The basic notion in the strong collision model is that the molecular orientations before and after a collision are quite uncorrelated. Of course the term collision does not necessarily mean direct molecular contact, it might, for example, imply a change in the intermolecular potential holding the molecule in a given orientation. We shall therefore use the word collision to denote a process which causes the molecular orientation to change. These collisions occur, on average, once every τ s. If the collisions can be described by a Poisson process then the probability of a collision occuring in the interval t since the last collision is just $(1 - e^{-t/\tau})$. If no collision has taken place in this time then the angle Ω is just Ω_o and so the probability of an orientation Ω is $\delta(\Omega_o - \Omega)$. Since the probability of no collisions before time t is $e^{-t/\tau}$ this gives a contribution to the conditional probability of $\delta(\Omega_o - \Omega)e^{-t/\tau}$. Similarly if a collision has occurred the probability of finding an orientation Ω is $P(\Omega)$ and the contribution to $P(\Omega_o|\Omega,t)$ is $P(\Omega)(1 - e^{-t/\tau})$. In total this gives the conditional probability as

$$P(\Omega_o|\Omega,t) = \delta(\Omega_o - \Omega)e^{-t/\tau} + P(\Omega)(1 - e^{-t/\tau}). \qquad \text{(XV.28.)}$$

The strong collision model is undoubtedly rather crude and cannot accurately describe reorientation in the mesophase. However, it does yield a conditional probability which has the correct limiting forms for both long and short times. In addition this expression for $P(\Omega_o|\Omega,t)$ is mathematically tractable and, as we shall see, leads to predictions which are in qualitative agreement with experiment.

Substitution of the conditional probability given by equation (28) into equation (27) leads to a correlation function ψ of the form.

$$\psi(t) = \{\overline{\mathcal{D}^{(L)}_{q,-p}\mathcal{D}^{(L')*}_{q',-p'}} - \overline{\mathcal{D}^{(L)}_{o,-p}}\;\overline{\mathcal{D}^{(L')*}_{o,-p'}}\,\delta_{oq}\,\delta_{oq'}\}e^{-t/\tau}. \qquad \text{(XV.29.)}$$

It is not obvious from this expression whether any restrictions can be placed on the values taken by either the subscripts or superscripts. However, the Clebsch-Gordan series:[20]

$$\mathcal{D}^{(L)}_{q,p}\mathcal{D}^{(L')}_{q',p'} = \sum_{L''} C(LL'L'';qq')C(LL'L'';pp')\mathcal{D}^{(L'')}_{q+q',\,p+p'}, \qquad \text{(XV.30.)}$$

can be used to simplify ψ. The coefficients in this series are the Clebsch-Gordan coefficients and are closely related to the 3j-symbols encountered in Chapter XI. The series is summed over values of L'' from $|L-L'|$ in steps of one to $|L+L'|$. The averaged product of two rotation matrices can be replaced by the average of a single rotation matrix with the aid of this series. The first average in equation (29) becomes

$$\overline{\mathscr{D}^{(L)}_{q,-p}\mathscr{D}^{(L')*}_{q',-p'}} = \sum_{L''} (-1)^{q'+p'} C(LL'L'';q-q')C(LL'L'';-pp')\overline{\mathscr{D}^{(L'')}_{q-q',p'-p}} , \qquad (XV.31.)$$

since

$$\mathscr{D}^{(L)*}_{qp} = (-1)^{q-p}\mathscr{D}^{(L)}_{-q,-p}. \qquad (XV.32.)$$

From our previous discussion of solute alignment in a nematic mesophase we know that the average $\overline{\mathscr{D}^{(L'')}_{q-q',p'-p}}$ will vanish unless L" is even and q-q' is zero. We now have

$$\overline{\mathscr{D}^{(L)}_{q,-p}\mathscr{D}^{(L)}_{q',-p'}}^{*} = \sum_{L''(\text{even})} (-1)^{q+p'} \delta_{qq'} C(LL'L'';q-q)C(LL'L'';-pp')\overline{\mathscr{D}^{(L'')}_{o,p'-p}}. \qquad (XV.33.)$$

One further simplification is possible, for the ensemble average of the rotation matrix when L" is zero is just $\delta_{pp'}$. The two Clebsch-Gordan coefficients for this first term can be evaluated directly since.

$$C(LL'0;q-q) = (-1)^{L-q}(2L+1)^{-\frac{1}{2}}\delta_{LL'} , \qquad (XV.34.)$$

and so

$$\psi(t) = \left\{(2L+1)^{-1}\delta_{LL'}\delta_{pp'}\delta_{qq'} + \sum_{\substack{L''\neq o \\ (\text{even})}}^{L+L'} (-1)^{q+p'}\delta_{qq'} C(LL'L'';q-q) \right.$$
$$\left. \times C(LL'L'';-pp')\overline{\mathscr{D}^{(L'')}_{o,p'-p}} - \overline{\mathscr{D}^{(L)}_{o,-p}}\,\overline{\mathscr{D}^{(L')*}_{o,-p'}}\delta_{oq}\delta_{oq'}\right\} e^{-t/\tau}. \qquad (XV.35.)$$

Of course, when the system is isotropic all of the averages vanish and we are left with the familiar correlation function[1]

$$\psi(t) = (2L+1)^{-1}\,\delta_{LL'}\delta_{pp'}\delta_{qq'}e^{-t/\tau}. \qquad \text{(XV.36.)}$$

The δ functions in this equation are of practical importance for they show that only cross terms between the same component of tensorial operators of the same rank make any contribution to the relaxation matrix. When the paramagnetic solute is subject to the anisotropic environment provided by the nematic mesophase some of these restrictions are removed. It is then possible to obtain cross terms between interactions of different rank and products of different components of the spatial tensors. These possibilities could, in principle, greatly increase the complexity of the relaxation matrix calculation. In practice the nature of the paramagnetic probe can be chosen to reduce these complications.

One of the most useful species for studying electron spin relaxation via molecular reorientation is vanadyl acetylacetonate.[1,2,3] The isotropic spectrum, shown in Figure 2, is particularly simple, and the determination of the linewidths is therefore quite straightforward. The same probe is also well suited for investigating both the static and dynamic properties of liquid crystals. The only interactions of any importance are the Zeeman and hyperfine terms, both of which have scalar and second rank components. In addition, solid state spectra show that the assumption of cylindrically symmetric g and hyperfine tensors is reasonable.[2] With these simplifications the dynamic perturbation, for vanadyl acetylacetonate dissolved in a nematic mesophase, is obtained from equation (25) as:

$$\mathcal{H}'(t) = \sum_{\mu;q} F_{\mu}^{(2,o)}\{\mathcal{D}_{q,o}^{(2)}(t) - \overline{\mathcal{D}_{o,o}^{(2)}}\delta_{oq}\}T_{\mu}^{(2,q)}, \qquad \text{(XV.37.)}$$

and the correlation function in equation (35) becomes

$$\psi_{qq'}(t) = \delta_{qq'}\{\tfrac{1}{5} + \sum_{L=2,4}(-1)^q C(22L;q-q)C(22L;oo)\overline{P_L} - \overline{P_2}^{\,2}\delta_{oq}\delta_{oq'}\}e^{-t/\tau}. \qquad \text{(XV.38.)}$$

In this equation the resulting $\overline{\mathcal{D}^{(L)}_{o,o}}$ have been replaced by Legendre polynomials of the same rank. The correlation function has now been labelled with the subscripts q and q' to emphasise its dependence on the components of the spin operators. Finally, because of the symmetry relation

$$C(LL'L'';qq') = (-1)^{L+L'-L''}C(LL'L'';-q-q') \qquad (XV.39.)$$

for the Clebsch-Gordan coefficients one can show that

$$\psi_{qq'}(t) = \psi_{-q-q'}(t) \quad . \qquad (XV.40.)$$

Even before we have calculated the linewidth it is possible to see from this form of the correlation function what information is potentially available from linewidth studies. Obviously one of the parameters which can be determined is τ, the time between collisions. Since our knowledge of the dynamic properties of a nematic mesophase is extremely sparse experimental values of this parameter are of great importance. The correlation function depends not only on $\overline{P}_2$, which can be measured from the positions of the spectral lines, but also on $\overline{P}_4$ which is not readily available. Experimental values for $\overline{P}_4$ are of some importance since they would enable us to determine how many terms must be retained in the expansion of the pseudo-potential as a sum of spherical harmonics.

The electron spin transitions for vanadyl acetylacetonate are non-degenerate and so only one element of the relaxation matrix is needed to calculate the width of the Lorentzian line. The evaluation of this matrix element is relatively straightforward, although the large vanadium hyperfine interaction does pose one problem. As a consequence of this large interaction the non-secular terms in the static spin hamiltonian cannot be ignored, and the simple product spin functions $|\pm\frac{1}{2},m\rangle$ are not eigenstates of $\overline{\mathcal{H}}$. It proves to be essential to allow for the mixing of the spin states by the non-secular term, $(a-(1/6)^{\frac{1}{2}}\tilde{A}^{(2,o)})/2$ is even greater than in the isotropic phase. It is possible, when calculating the linewidth, to allow for the non-secular terms in one of two equivalent ways. The first simply uses perturbation theory to obtain eigenstates of the static spin hamiltonian correct to second order in a/ω_o. These corrected spin functions are then employed with the same dynamic perturbation $\mathcal{H}'(t)$ to calculate the relaxation matrix element. The second, more elegant technique mentioned in Section 11 of Chapter X is to transform the dynamic spin hamiltonian by means of a unitary transformation:[2]

$$\tilde{\mathcal{H}}'(t) = \mathcal{H}'(t) + i\epsilon[F,\mathcal{H}'(t)] - \frac{\epsilon^2}{2}\left[[\mathcal{H}'(t),F],F\right] + \ldots\ldots, \qquad (XV.41.)$$

and calculate the relaxation matrix within the original basis $|\pm\frac{1}{2},m\rangle$. The form of the operator F and the magnitude of the coefficient ϵ are chosen so that the transformed static spin hamiltonian is diagonal within the basis $|\pm\frac{1}{2},m\rangle$ through order ϵ^3. For this particular problem

$$\epsilon F = i\left\{\frac{a-(1/6)^{\frac{1}{2}}\tilde{A}^{(2,o)}}{2\omega_o}\right\}\{S_-I_+ - S_-I_+\}, \qquad (XV.42.)$$

although it is convenient to retain only the linear term ϵ in $\tilde{\mathcal{H}}'(t)$. The calculation now becomes a matter of book-keeping, although some of the tedium may be relieved by neglecting the non-secular terms in the transformed perturbation. This approximation may not be too unrealistic[23] for the viscosity of the nematic mesophase of, for example, 4,4'-dimethoxyazoxybenzene is comparable to that of toluene at -60^oC and contributions to the linewidth in $J(\omega_o)$ are found to be unimportant for the latter case.[2]

The resultant linewidth[26] shows the expected cubic dependence on the nuclear quantum number:

$$T_2^{-1}(m) = A+Bm+Cm^2+Dm^3, \qquad (XV.43.)$$

where

$$A = \frac{2}{3}\left(\frac{\Delta g\omega_o}{\bar{g}}\right)^2 J_o + \frac{\Delta a^2}{4} I(I+1)J_1 + \frac{\Delta g\Delta a}{6\bar{g}} \bar{a}^* I(I+1)(3J_1-4J_o) , \tag{XV.44.}$$

$$B = \frac{4\Delta g\Delta a\omega_o}{3\bar{g}} J_o + \frac{\Delta a^2}{6}\frac{\bar{a}^*}{\omega_o} I(I+1)(3J_1-4J_o) - \frac{4}{3}\left(\frac{\Delta g\omega_o}{\bar{g}}\right)^2 \frac{\bar{a}}{\omega_o} J_o - \frac{\Delta a^2}{2}\frac{\overline{a^*}}{\omega_o} J_1 , \tag{XV.45.}$$

$$C = \frac{\Delta a^2}{12}(8J_o-3J_1) - \frac{4\Delta g\Delta a}{3\bar{g}} \bar{a} J_o + \frac{\Delta g\Delta a}{6\bar{g}} \bar{a}^* (4J_o-3J_1) , \tag{XV.46.}$$

and

$$D = \frac{\Delta a^2}{6}\frac{\bar{a}^*}{\omega_o}(4J_o-3J_1). \tag{XV.47.}$$

In these equations for the linewidth coefficients the spectral densities J_o and J_1 are the Fourier cosine transforms of the correlation functions $\psi_{oo}(t)$ and $\psi_{11}(t)$ at zero frequency. That is

$$J_o = (\frac{1}{5} + \frac{2}{7}\bar{P}_2 + \frac{18}{35}\bar{P}_4 - \bar{P}_2^2)\tau , \tag{XV.48.}$$

and

$$J_1 = (\frac{1}{5} + \frac{1}{7}\overline{P}_2 - \frac{12}{35}\overline{P}_4)\tau. \qquad \text{(XV.49.)}$$

In addition Δg and Δa are just the zeroth components, $F^{(2,o)}$, of the second rank g and hyperfine tensors i.e. $(3/2)^{\frac{1}{2}}g'_{zz}$ and $(3/2)^{\frac{1}{2}}A'_{zz}$.[27] The partially averaged quantities $\overline{g}$ and $\overline{a}$ are given in equations (22) to (24) and $\overline{a}*$ is defined by

$$\overline{a}* = \{a - (1/6)^{\frac{1}{2}}\tilde{A}^{(2,o)}\} . \qquad \text{(XV.50.)}$$

These equations were derived originally, without the correction terms, in $\overline{a}*/\omega_o$, by Glarum and Marshall.[23]

The linewidth coefficients depend in a fairly complicated fashion on the degree of solute alignment although there are certain simple limiting cases. For example, when the system is isotropic both $\overline{P}_2$ and $\overline{P}_4$ vanish and the spectral densities J_o and J_1 are equal to $\tau/5$. The expressions for the linewidth coefficients then reduce to those obtained by Wilson and Kivelson.[2] At the other extreme the solute may be completely aligned and this can occur in two quite distinct ways. Thus when a planar solute such as vanadyl acetylacetonate is completely aligned the angle between the director and the symmetry axis of the magnetic interactions is 90°. The spectral densities J_o and J_1 then vanish, together with the secular and pseudo-secular contributions to the linewidth. There is, however, a non-zero contribution from J_2 which, in this limit, is just $3\tau/8(1+\omega_o^2\tau^2)$ since J_2 is associated with the non-secular terms in the dynamic perturbation. Physically, even though the symmetry axis is constrained to be orthogonal to the director its orientation with respect to an axis at right angles to the director can change. This motion then modulates the coefficients of the non-secular spin operators $T^{(2,\pm 2)}$. The other case of complete alignment occurs when the symmetry axis is parallel to the director and then J_o, J_1 and J_2 all vanish. Such behaviour is to be expected since the only motion is rotation about the symmetry axis and the total spin hamiltonian is invariant under this rotation.

The effect of alignment on the linewidth coefficients between the two extremes of complete order and disorder is best seen by calculating the coefficients as a function of the alignment factor $\overline{P}_2$. In order to make the calculation we must estimate the other alignment factor $\overline{P}_4$. Since experimental values of $\overline{P}_4$ are not available they must be calculated from the solute pseudo-potential. If the series for the pseudo-potential given by equation (7) is restricted to those terms with L = 2 and it is further assumed that the molecule is cylindrically symmetric about the magnetic symmetry axis along the VO bond, then

$$U^{(2)} = C_{2,o}P_2^{(2)}. \qquad \text{(XV.51.)}$$

The required parameters $\overline{P}_2$ and $\overline{P}_4$ may now be calculated by evaluating the appropriate Boltzmann averages and in these calculations $C_{2,o}$ is taken to be positive in order for $\overline{P}_2$ to range from 0 to -0.5. The magnetic parameters required in the calculation were given the values:[2,27]

$$\begin{aligned} g &= 1.969, \\ a &= -1.853 \times 10^9 s^{-1}, \\ \Delta g &= -0.0318, \\ \Delta a &= -1.677 \times 10^9 s^{-1}, \\ \text{and} \quad \omega_o &= 59.69 \times 10^9 s^{-1}. \end{aligned} \qquad \text{(XV.52.)}$$

The calculated values of B and C, normalised to their isotropic values are plotted in Figure 3 as a function of the alignment $\overline{P}_2$. The most striking feature of the plots is the marked decrease in the linewidth parameters as the solute alignment increases. Since a similar decrease is also predicted for the A coefficient the theory accounts for the decrease in the linewidths, on partial alignment, observed for the spectra shown in Figure 2. This agreement with experiment shows that the correlation time does not increase dramatically on passing from the isotropic to the nematic mesophase,

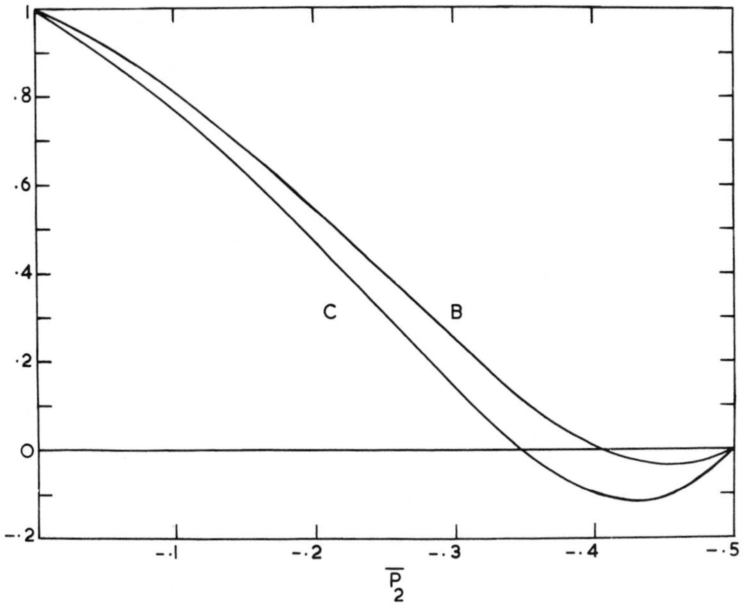

Figure XV.3. The theoretical dependence of the linewidth coefficients B and C for vanadyl acetylacetonate.

although it is not known if τ decreases on alignment.

For isotropic fluids the C coefficient is necessarily positive, but in the nematic mesophase the theory predicts that for a degree of alignment less than -0.35 C should be negative. This change in the sign of C occurs because the dominant contribution to this coefficient is proportional to $(8J_o - 3J_1)$ and the secular spectral density J_o decreases on alignment at a faster rate than the pseudo-secular term J_1 which is eventually dominent. The plots in Figure 3 show that B should also change sign but at a higher degree of solute alignment when $\bar{P}_2$ is -0.41. The reversal of the sign of B occurs because the zeroth-order secular contribution decreases at a faster rate than the negative correction terms in $\bar{a}^*/\omega_o$ which eventually outweigh it. Although the alignment of vanadyl acetylacetonate in the super-cooled mesophase of 4,4'-diethoxyazoxybenzene is appreciable it does not exceed -0.35 and so the expected sign reversals are not observed for this system.[28] However, the larger complex, vanadyl tetraphenylacetylacetonate is highly aligned in most mesophases[23] and the spectra of this probe do exhibit the theoretical dependence of B and C on the alignment factor $\overline{P}_2$. Typical spectra of the probe dissolved in the isotropic melt and nematic mesophase of 4,4'-dimethoxyazoxybenzene are shown in Figure 4. Just below the isotropic-nematic transition point, at 132°C, both B and C are small but positive. The two spectra at the lower temperatures, however, clearly show that C must be negative since the widths of the central lines are greater than the outermost spectral lines. Although B is negative at 116°C it is rather small, but has increased in magnitude in the spectrum obtained at 100°C. The degree of alignment at which C changes sign is observed to be less than that at which the inversion of B occurs. In addition to the qualitative success of the theory the experimental values of $\overline{P}_2$ at which C and then B vanish are in good agreement with the theoretical predictions.[23]

Although the assumption of effective cylindrical symmetry would appear eminently reasonable for vanadyl acetylacetonate, measurements of the spectra of the probe in a frozen nematic glass suggest that the ordering of the in-plane axes are markedly different.[29] At first sight this result is surprising, it is however supported by nuclear magnetic resonance studies of toluene[30] dissolved in the nematic mesophase of n-(4-ethoxybenzylidene)-4'-n-butylaniline which also revealed a profound difference in the ordering of the in-plane axes. If we accept this interpretation then the pseudo-potential for vanadyl acetylacetonate must be modified to allow for the departure from cylindrical symmetry. Provided only the second rank terms are retained in the expansion of $U^{(2)}$ then

$$U^{(2)} = \sum_m \left(\frac{4\pi}{5}\right)^{\frac{1}{2}} C_{2,m} Y_{2,m} \, . \qquad \text{(XV.53.)}$$

When the coefficient C_2 is evaluated in the principle molecular axis system the terms with $m = \pm 1$ vanish and the coefficients $C_{2,\pm 2}$ are identical. The pseudo-potential is then

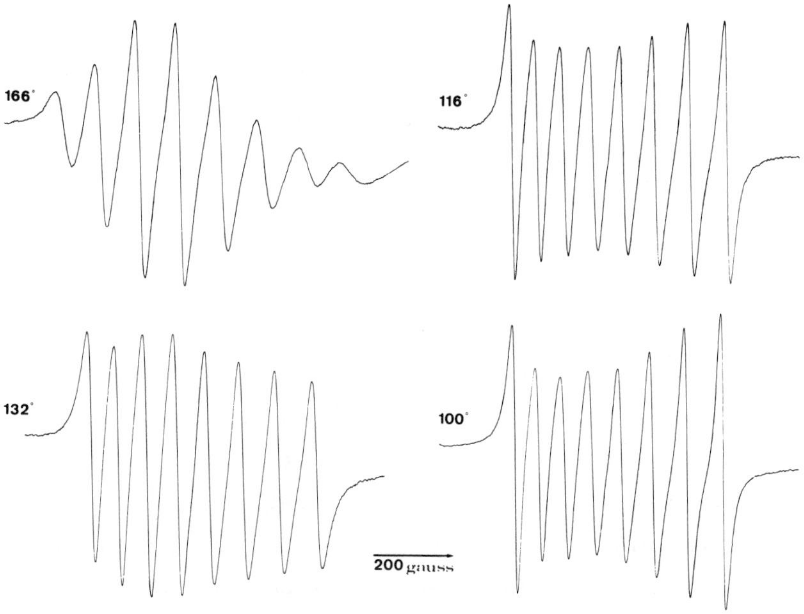

Figure XV.4. The electron resonance spectrum of vanadyl tetraphenylacetylacetonate dissolved in 4,4'-dimethoxyazoxybenzene.

$$U^{(2)} = C_{2,o}P_2(\cos\beta) + (3/2)^{\frac{1}{2}}C_{2,2}\sin^2\beta\cos2\alpha \,, \qquad \text{(XV.54.)}$$

and the angular distribution function is obtained from equation (13). Since we require the ensemble averages of P_2 and P_4, which depend only on the angle β, the dependence of $P(\Omega)$ on α may be integrated out to give the reduced angular distribution function:

$$P(\beta) = N\exp\{-C_{2,o}P_2(\cos\beta)/kT\}I_o\{-(3/2)^{\frac{1}{2}}C_{2,2}\sin^2\beta/kT\}, \qquad \text{(XV.55.)}$$

where I_o is a modified Bessel function of the first kind and N is the normalisation constant. Now in the previous calculations although the distribution function contains an arbitrary parameter, $C_{2,o}$, the dependence of the linewidth coefficients on $\overline{P}_2$ is independent of this parameter. This situation obtains because there is only one value of $\overline{P}_4$ associated with a given $\overline{P}_2$. The distribution function given in equation (55) contains two parameters;

as a consequence the value of $\overline{P}_4$ associated with a given $\overline{P}_2$ is not uniquely determined and will depend on the ratio $C_{2,o}/C_{2,2}$. Experimentally $C_{2,o}/C_{2,2}$ is found to be 1.05[29]; theoretically this ratio might be expected to be independent of both $C_{2,o}$ and the temperature. We have therefore repeated the calculation of the linewidth coefficients using this ratio and found that the conclusions reached with the aid of the single parameter distribution function are essentially unaffected.

Before concluding this section it is appropriate to comment on the validity of the application of Redfield's theory to the problem of spin relaxation in the mesophase. The lines in the isotropic spectrum shown in Figure 4 are extremely broad, and as the temperature is lowered towards the transition point there is a considerable increase in the widths. In addition the spacing between the lines decreases and the line shape is no longer Lorentzian. These observations suggest that the condition $|\mathcal{H}'|^2\tau^2 \ll 1$ is no longer satisfied and this might also be the case in the nematic mesophase. Since the bulk viscosity actually decreases below the transition point the rotational correlation time might not increase on passing from the isotropic melt to the nematic mesophase. Of course, on alignment the anisotropy in the spin hamiltonian is no longer averaged to zero and so there is a marked reduction in $|\mathcal{H}'|^2$. As a consequence provided the condition $|\mathcal{H}'|^2\tau^2 \ll 1$ is satisfied in the isotropic phase it should also be satisfied in the mesophase. In addition even if this condition is not met in the isotropic melt it might well be true in the mesophase.

XV.5. THE LINE SHAPE

The qualitative success of the theory of electron spin relaxation in nematic liquid crystals is encouraging and might have been expected to lead to more quantitative studies. Such experiments have not been reported and it is fairly easy to see the reason for this. Close examination of the spectra of vanadyl acetylacetonate dissolved in a nematic mesophase, shown in Figures 2 and 4, reveals an asymmetry in the individual line shapes. Although this asymmetry cannot be accounted for by Redfield's theory it is predicted[31] when the rate of molecular reorientation is slow in the sense that $|\mathcal{H}'|^2\tau\sim1$. However, as we saw in the last section Redfield's theory should be applicable to these systems, especially for vanadyl acetylacetonate dissolved in 4,4'-diethoxyazoxybenzene where the hyperfine lines are symmetric in the isotropic melt but not in the nematic mesophase. We must therefore look elsewhere for the source of the asymmetry which, as we shall see, is associated with the second mechanism responsible for fluctuations in the molecular orientation.

Although on average the director is aligned parallel to the magnetic field at any instant the components of $\underset{\sim}{d}$ orthogonal to the field do not vanish because of thermal fluctuations.[9] As the orientation of the director changes so does that of the molecule and in nuclear magnetic resonance these fluctuations are known to produce a powerful spin relaxation process.[32] In contrast, the molecular reorientation with respect to the director, which is responsible for electron spin relaxation, was thought to be too rapid to relax the nuclear spins.[33] However a more recent theory[34] of nuclear spin relaxation in the nematic mesophase has found it necessary to allow for relaxation caused by molecular reorientation with respect to the director. The fluctuations in the director are presumably slow on the electron resonance time scale and the electron spin simply experiences a static angular distribution of the director. The electron resonance spectrum of the probe is therefore a weighted sum of the spectra from each of these orientations.[35] The observation of spectra from a range of orientations produces, in general, a polycrystalline spectrum whose shape cannot be obtained analytically. However, for this particular problem the magnitude of the fluctuations are small and so the angular distribution function is rather steep. As a consequence it is possible to obtain a simple analytic expression for the line shape by using the following arguments.

The resonant fields $B_m(o)$ when the director is parallel to the magnetic field are given by equation (20), although when the director is at an angle θ to the magnetic field the resonant fields $B_m(\theta)$ are given by more formidable expressions.[2] However, for small values of θ the difference $\{B_m(\theta) - B_m(o)\}$ in the two fields is also small and so the line shape $L\{B_m(\theta),B,T_2^{-1}\}$ may be expanded in a Taylor series as

$$L\{B_m(\theta),B,T_2^{-1}\}=L\{B_m(o),B,T_2^{-1}\} + \{B_m(\theta)-B_m(o)\}\left(\frac{\partial L}{\partial B_m}\right)_{B_m=B_m(o)} + \ldots\ldots \qquad \text{(XV.56.)}$$

The retention of just two terms in the series in not a particularly good approximation when the director makes a single angle with the magnetic field, but is better if the total line shape h(B) is obtained by integrating $L\{B_m(\theta),B,T_2^{-1}\}$ over all orientations. Since there is no theoretical expression for the angular distribution function of the director, we shall adopt the following function

$$P(\theta) = \theta\exp(-\theta^2/\theta_o^2) , \qquad \text{(XV.57.)}$$

where θ_o is the root mean square fluctuation in θ. This function is qualitatively correct, and has the virtue of yielding integrals which may be evaluated analytically. The total line shape is therefore given by

$$h(B)=L\{B_m(o),B,T_2^{-1}\} + \left(\frac{\partial L}{\partial B_m}\right)_{B_m=B_m(o)} \frac{\int_o^{\pi}\{B_m(\theta)-B_m(o)\}\theta\exp(-\theta^2/\theta_o^2)d\theta}{\int_o^{\pi}\theta\exp(-\theta^2/\theta_o^2)d\theta} , \qquad \text{(XV.58.)}$$

provided the linewidth is orientation independent for then both L and $(\partial L/\partial B_m)_{B_m(o)}$ are also independent of the orientation. For small deviations of the director from alignment parallel to the magnetic field

$$B_m(\theta) = B_m(o) + \lambda_m \theta^2 , \qquad \text{(XV.59.)}$$

where

$$\lambda_m = \tfrac{1}{2}\Big[\frac{\tilde{g}_{\parallel}^2 - \tilde{g}_{\perp}^2}{\tilde{g}^2}\,B_o + \frac{(2\tilde{A}_{\parallel}^2\tilde{g}_{\perp}^2 - \tilde{A}_{\parallel}^2\tilde{g}_{\parallel}^2 - \tilde{A}_{\perp}^2\tilde{g}_{\perp}^2)}{\tilde{A}_{\parallel}\tilde{g}_{\parallel}^3}\,\frac{h}{\beta}\,m \tag{XV.59.}$$

$$+ \frac{h^2\tilde{A}_{\perp}^2\,(\tilde{A}_{\perp}^2\tilde{g}_{\perp}^2 + \tilde{A}_{\parallel}^2\tilde{g}_{\perp}^2 - 2\tilde{A}_{\parallel}^2\tilde{g}_{\parallel}^2)}{2\beta^2\tilde{g}_{\parallel}^4\tilde{A}_{\parallel}^2 B_o}\{I(I+1) - m^2\}$$

$$- \frac{(\tilde{A}_{\perp}^2 - \tilde{A}_{\parallel}^2)^2}{\tilde{A}_{\parallel}^2 B_o}\,\frac{\tilde{g}_{\perp}^2}{\tilde{g}_{\parallel}^4}\,\frac{h^2}{\beta^2}\,m^2\Big] . \tag{XV.60.}$$

In this rather cumbersome expression[36] for λ_m the Cartesian components of the magnetic tensors for the fictitious static species are denoted by the subscripts $\parallel$ and $\perp$. As a further consequence of the small magnitude of the fluctuations the two integrands in equation (58) are rapidly decreasing of θ and so the limit of the integrals may be safely extended to ∞. Evaluation of the two integrals together with the use of equation (59) gives the final line shape function as

$$h(B) = L\{B_m(o), B, T_2^{-1}\} - \lambda_m\theta_o^2\left(\frac{\partial L}{\partial B}\right)_{B_m = B_m(o)} . \tag{XV.61.}$$

In the case of vanadyl acetylacetonate the line shape is predicted to be the sum of a first derivative Lorentzian and a second derivative. The composite shape is therefore asymmetric as required, and the extent of the asymmetry depends on two factors, the anisotropy in the magnetic tensors, via λ_m and the magnitude of the fluctuations, via the root mean square angle θ_o. An electron resonance spectrum of partially aligned vanadyl acetylacetonate calculated from the theoretical line shape given in equation (61), with a root mean square fluctuation of 10° and for a degree of order $\overline{P}_2$ of -0.2, is shown in Figure 5. The linewidth was taken to be independent of the nuclear quantum number and so the apparent linewidth variation is caused entirely by

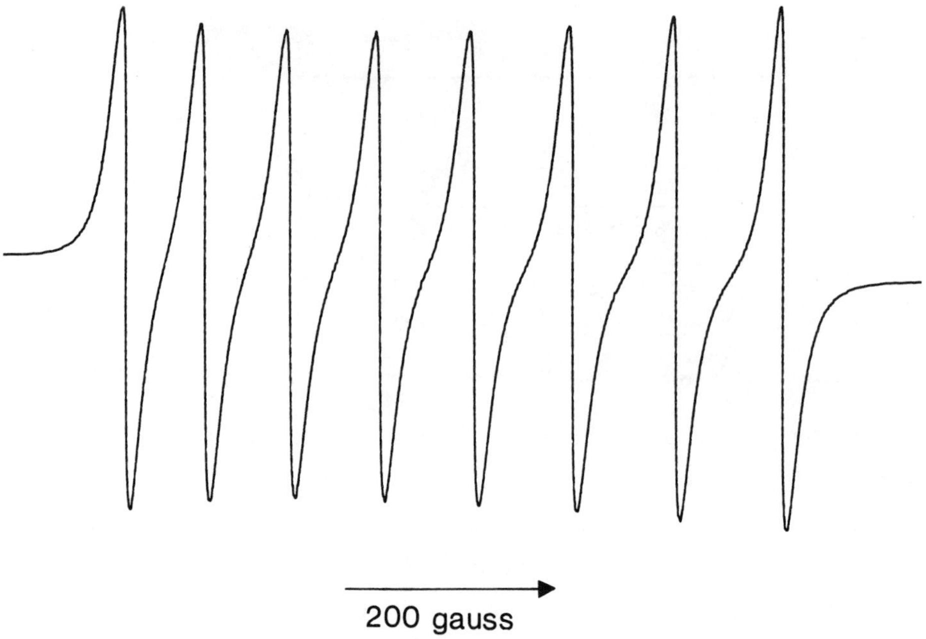

Figure XV.5. A theoretical electron resonance spectrum of a partially aligned vanadyl acetylacetonate molecule which reveals the asymmetry in the line shape caused by fluctuations in the director.

the dependence of λ_m on m. The line shape is clearly asymmetric and the extent of the asymmetry varies from one line to the next in agreement with experiment. Although this model does account for the asymmetry it also predicts a linewidth variation which is remarkably similar to that shown in Figure 4. Nevertheless the majority of the line broadening is caused by relaxation of the electron spin and in any quantitative experimental analysis of these effects due allowance must be made for the angular distribution of the director.

References

1. See, for example, (a) Chapter X as well as (b) A. Hudson and G.R. Luckhurst, Chem. Rev., 69, 191 (1969).

2. R. Wilson and D. Kivelson, J. Chem. Phys., 44, 154 (1966).

3. G.R. Luckhurst and J.N. Ockwell, Mol. Phys., 16, 165 (1969).

4. G.W. Gray, Molecular Structure and the Properties of Liquid Crystals(Academic Press, London, New York, 1962).

5. A. Saupe, Angew.Chemie (Internat. Edit.) 7, 97 (1968).

6. A.D. Buckingham, Discuss. Faraday Soc., 43, 205 (1967).

7. D.H. Chen, P.G. James and G.R. Luckhurst, Mol. Cryst. and Liq. Cryst., 8, 71 (1969).

8. P.G. de Gennes, Phys. Lett., 30a, 454 (1969).

9. P.G. de Gennes, Mol. Cryst. and Liq. Cryst., 7, 325 (1969).

10. J.C. Rowell, W.D. Phillips, L.R. Melby and M. Panar, J. Chem. Phys., 43, 3442 (1965).

11. W. Maier, Z. Naturf., 2a, 458 (1947).

12. E.F. Carr, Adv. Chem. Ser., 63, 76 (1967); G.R. Luckhurst, Chem. Phys. Lett., 9, 289 (1971).

13. P. Chatelain, Acta Cryst., 1, 315 (1948).

14. W. Maier and A. Saupe, Z. Naturf., 13a, 564 (1958); 14a, 882 (1959); 15a, 287 (1960).

15. J.A. Pople, Proc. Roy. Soc., 221A, 498 (1954).

16. See, for example, T.J. Krieger and H.M. James, J. Chem. Phys., 22, 796 (1954); R. Brought, Phase Transitions (W.A. Benjamin, Inc., New York, 1965).

17. R.L. Humphries, P.G. James and G.R. Luckhurst, Trans. Faraday Soc., to be submitted.

18. P.G. James and G.R. Luckhurst, Mol. Phys., 19, 489 (1970); 20, 761 (1971).

19. R.L. Humphries, P.G. James and G.R. Luckhurst, Faraday Soc. Symposium on Liquid Crystals, 1971.

20. M.E. Rose, Elementary Theory of Angular Momentum (John Wiley and Sons, New York, 1957).

21. S.H. Glarum and J.H. Marshall, J. Chem. Phys., 44, 2884 (1966).

22. H. R. Falle and G.R. Luckhurst, J. Mag. Res., 3, 161 (1970).

23. S.H. Glarum and J.H. Marshall, J. Chem. Phys., 46, 55 (1967).

24. A.G. Redfield, Adv. Mag. Res., 1, 1 (1965).

25. J. Roberts and R.M. Lynden-Bell, Mol. Phys., 21, 689 (1971).

26. S.A. Brooks, G.R. Luckhurst, G.F. Pedulli and J. Roberts, to be published.

27. Our definition of the anisotropic components Δg and Δa, differs from that used elsewhere. In Kivelson and Wilson's notation $\Delta g_{K.W.} = -(3/2)^{\frac{1}{2}}\Delta g$ and $b_{K.W} = -(3/2)^{\frac{1}{2}}\Delta a$ whereas in Glarum and Marshall's notation $\Delta g_{G.M} = -(1/6)^{\frac{1}{2}}\Delta g$ and $b_{G.M} = -(1/6)^{\frac{1}{2}}\Delta a$.

28. G.R. Luckhurst and J.N. Ockwell, unpublished results.

29. G. Heppke and F. Schneider, Ber. Bunsenges. phys. Chem., 75, 61 (1971).

30. P. Diehl, H.P. Kellerhals and W. Niederberger, J. Mag. Res., 4, 352 (1971).

31. H. Sillescu and D. Kivelson, J. Chem. Phys., 48, 3493 (1968); H. Sillescu, ibid, 54, 2110 (1971).

32. P. Pincus, Solid State Comm., 7, 415 (1969).

33. J.W. Doane and D.L. Johnson, Chem. Phys. Lett., 6, 291 (1970).

34. C.C. Sung, Chem. Phys. Lett., 10, 35 (1971).

35. S.A. Brooks, G.R. Luckhurst and G.F. Pedulli, Chem.Phys.Lett., 11, 159 (1971).

36. R.M. Golding, Trans. Faraday Soc., 59, 1513 (1963). The expression given in this paper differs from that given here because amongst other things the last term in equation (60) is ignored.

TWO PROBLEMS INVOLVING ESR IN LIQUID CRYSTALS

J. I. Kaplan

Battelle, Columbus Laboratories

Columbus, Ohio 43201

Two problems will be considered:

I. The ESR lineshape of free radicals in non-viscous liquid crystals which are in the nematic phase[1] and constrained to form a thin film. The non-viscous liquid crystal implies the free radicals are tumbling so rapidly that one need deal only with an angularly averaged hamiltonian.

II. The theory for ESR lineshape of free radicals in bulk samples of liquid crystals in the nematic phase where the viscosity is other than in the non-viscous limit. In this problem one must deal with an angular dependent hamiltonian.

XVI. 1. THIN FILM ESR (RAPID TUMBLING)

The free radical probe that will form the model for our discussion is VAAC (vanadylacetylacetonate). We assume (and separate experiments support this) that VAAC aligns with the local nematic direction.[2] As described in a paper by G. E. Fryburg and E. Gelerinter[3] the hamiltonian for VAAC in the laboratory coordinate system with the magnetic field making an angle θ with the molecular axis is

$$\mathcal{H}(\theta) = \omega_o S^z\left[1 + 1/3 \frac{\Delta g}{g}\left(3 \cos^2\theta - 1\right)\right] + a\, I\cdot S$$
$$+ \frac{b}{3} I^z S^z (3 \cos^2\theta - 1) + \mathcal{H}_{\text{non-secular}} \qquad (1)$$

where

$$b = A_{\|} - A_{\perp} \qquad a = 1/3\ (A_{\|} + 2A_{\perp})$$
$$g = 1/2\ (g_{\|} + 2g_{\perp}) \qquad \Delta g = g_{\|} - g_{\perp} \qquad (2)$$

and $A_{\|}$, $A_{\perp}$ ($g_{\|}$, $g_{\perp}$) are the components of the hyperfine (g tensor) interaction in the coordinate system of the molecule.

Using the spherical harmonic addition theorem[4] one can re-write Eq. 1 in a coordinate system where the local nematic axis makes an angle θ with the magnetic field direction, the result is

$$\overline{\mathcal{H}}(\theta) = \omega_o S^z\left[1 + 1/3 \frac{\Delta g}{g} \sigma_z (3 \cos^2\theta - 1)\right] + a\, I\cdot S$$
$$+ \frac{b}{3} I^z S^z (3 \cos^2\theta - 1)\, \sigma_z \qquad (3)$$

where $\overline{\mathcal{H}}$ indicates an additional weighted angular average about the local nematic axis. This weighting is parametrized on the RHS of Eq. 3 as

$$\sigma_z = 1/2\ \langle 3 \cos^2\theta - 1\rangle \qquad (4)$$

$$\langle A\rangle = \frac{\int e^{-E(\theta)/kT} A\, d\Omega}{\int e^{-E(\theta)/kT}\, d\Omega} \qquad (5)$$

where E(θ) is the energy of a molecule as measured from the local nematic axis. We refer to E(θ) as the MS energy as Meir and Saupe[5] were the first to develop a theory for this energy. Note that for a normal liquid $\sigma_z = 0$.

If we ignore (for simplicity) the off diagonal terms in Eq. 3, the resonant frequencies (one for each hyperfine line) are given as

$$\omega'_{o,m_I}(\theta) = +\, m_I\left[a + \frac{b}{3}\sigma_z(3\cos^2\theta - 1)\right] + \omega_o\left[1 + 1/3\,\frac{\Delta g}{g}\,\sigma_z(3\cos^2\theta - 1)\right] \tag{6}$$

For a thin film (or near the boundary of a containing surface) it will be found that the angle θ between the directions of the local nematic axis and the magnetic field is position dependent (in the bulk $\theta = 0$). To obtain the position dependence of θ we follow closely a similar calculation as reported in a review paper of P. Pincus[6] which itself is based on some earlier work of F. C. Frank.[7] As discussed in Ref. 6, the position dependent free energy density is given as

$$\mathfrak{F}_{NonMag} = 1/2\left[K_{11}(\nabla\cdot\underset{\sim}{n})^2 + K_{22}(\underset{\sim}{n}\cdot\nabla\times\underset{\sim}{n})^2 + K_{33}(\underset{\sim}{n}\times\nabla\times\underset{\sim}{n})^2\right]$$

$$\mathfrak{F}_{Mag} = -\,1/2(\underset{\sim}{n}\cdot\underset{\sim}{H})^2\ \chi_a \tag{7}$$

where the K_{11}'s are elastic constants, χ_a is the diamagnetic susceptibility, and $\underset{\sim}{n}$ is a unit vector in the direction of the local nematic axis.

Suppose we consider first the geometry shown in Fig. 1 (i.e., the liquid crystal contained by a single surface. The arrow indi-

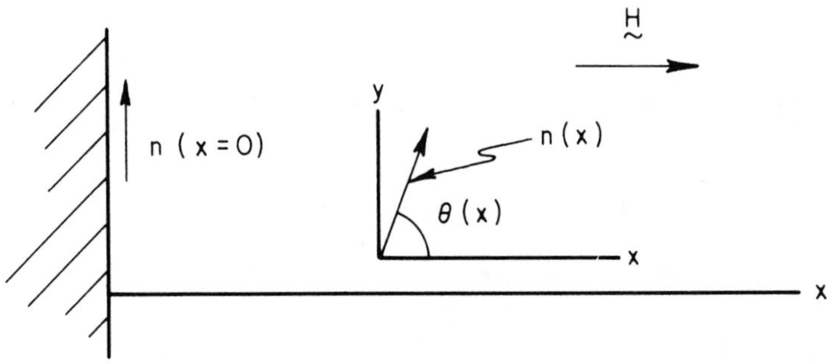

FIGURE 1
One Sided Container

cates the local direction of the nematic axis and in addition, we define (this is the most general experimental case) that at x = 0 $\theta(x = 0) = \pi/2$. Writing

$$\begin{aligned} n_x &= \cos\theta \\ n_y &= \sin\theta \end{aligned} \tag{8}$$

the total free energy is given as

$$F = 1/2 \int_0^\infty dx\left[\left(K_{11}\sin^2\theta + K_{33}\cos^2\theta\right)\frac{\partial^2\theta}{\partial x^2} - \chi_a H^2\cos^2\theta\right] \tag{9}$$

Assuming $K_{11} = K_{33} = K$ one obtains on minimizing F

$$-2\xi\frac{\partial^2\theta}{\partial x^2} + \sin\theta \cos\theta = 0 \tag{10}$$

where $\xi^2 = (K/\chi_a)H^{-2}$.

This equation can be simply integrated and one obtains

$$\tan\theta/2 = e^{-x/\xi} \tag{11}$$

which can be reexpressed as

$$\cos^2\theta = \tanh^2 x/\xi \tag{12}$$

Therefore

$$\begin{aligned} \omega'_{o,m_I}(x) &= \omega_o\left[1 + 1/2\,\frac{\Delta g}{g}\,\sigma_z(3\tanh^2 x/\xi - 1)\right] \\ &+ \frac{b}{3}\,m_I\,\sigma_z(3\tanh^2 x/\xi - 1) \end{aligned} \tag{13}$$

The ESR spectrum is thus given as

$$A(\omega) = \sum_{m_I=-I}^{m_I=I}\int_0^\infty g_{m_I}\left(\omega - \omega'_{o,m_I}(x)\right)dx \tag{14}$$

where

$$g_{m_I}(x) = \frac{\tau_{m_I}}{1 + x^2 \tau_{m_I}^2} \tag{15}$$

and τ_{m_I} are experimental values obtained from bulk experiments.[8]

Next consider the liquid crystal contained by two surfaces (i.e., a film). Here as shown in Fig. 2 we again have the differential equation

$$- 2\xi^2 \frac{d^2\theta}{dz^2} + \sin\theta \cos\theta = 0 \tag{16}$$

with boundary conditions

$$\begin{aligned} \theta &= \pi/2 \qquad & z &= \pm d/2 \\ \frac{d\theta}{dz} &= 0 \qquad & z &= 0 \end{aligned} \tag{17}$$

Integrating once we obtain

$$\xi^2\left(\frac{d\theta}{dz}\right)^2 = 1/2\left[\sin^2\theta - \sin^2\theta_{min}\right] \tag{18}$$

where θ_{min} is the angle of z = 0 and which by symmetry must be

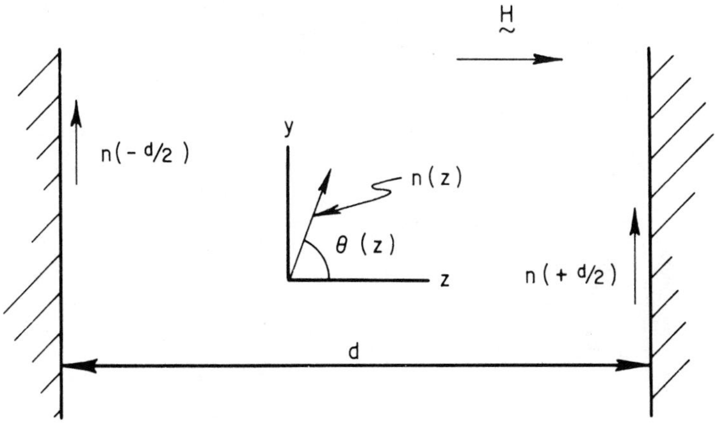

FIGURE 2

Film of Width d

the minimum angle. Letting $k^{-1} = \sin\theta_{min}$ and integrating again one obtains

$$\frac{d}{2} - z = \frac{2k^2\xi^2)^{1/2}}{(k^2-1)^{1/2}} \int_0^{\pi/2-\theta} \frac{d\theta'}{\left[1-\frac{k^2}{k^2-1}\sin^2\theta'\right]^{1/2}} \qquad (19)$$

$$= \frac{(2k^2\xi^2)^{1/2}}{(k^2-1)^{1/2}} F\left(\pi/2-\theta,\left(\frac{k^2}{k^2-1}\right)^{1/2}\right) \qquad (20)$$

where F is the elliptic integral of the first kind.[9]

The question now arises as to whether the nematic axis varies spatially for all values of the magnetic field (the situation for the one-sided container) or whether it remains rigidly parallel to the sides of the containing surfaces up to a specific value of H (called H_{crit}) and only above this value does it begin to vary spatially. To investigate this point set z = 0 in Eq. 19 (recall that the maximum variation of the nematic axis will be at z = 0) and then look for solutions of

$$\frac{d}{2} = \frac{(2k^2\xi^2)^{1/2}}{(k^2-1)^{1/2}} \int_0^{\pi/2-\theta_{min}} \frac{d\theta'}{\left[1-\frac{k^2}{1-k^2}\sin^2\theta'\right]^{1/2}} \qquad (21)$$

where $\Delta = \pi/2 - \theta_{min}$ is very small. Expanding the denominator in Eq. 21 one obtains

$$\frac{d}{2} = \frac{(2k^2\xi^2)^{1/2}}{(k^2-1)^{1/2}} \int_0^{\Delta} d\theta\left[1 + 1/2\,\frac{k^2}{k^2-1}\sin^2\theta + \ldots.\right] \qquad (22)$$

Noting that in the limit $\Delta \to 0$

$$\frac{k^2}{k^2-1} \simeq \frac{1}{\Delta^2} \qquad (23)$$

$$\sin^2\theta \simeq \theta^2$$

Eq. 22 can be evaluated as

$$\frac{d}{2} = (2\xi^2)^{1/2}\left[\sum_n \frac{(2n-1)!!}{n!2^n(2n+1)}\right]$$

$$\frac{d}{2} = (2\xi^2)^{1/2}\ \pi/2 \qquad (24)$$

or

$$H_{crit}\ d = \pi(K/\chi_a)^{1/2}\ \sqrt{2} \qquad (25)$$

Had we assumed $\theta(z = \pm\ d/2) = 0$ as the boundary condition we would have found

$$H_{crit}\ d = \pi(K/\chi_a)^{1/2} \qquad (26)$$

Substituting the solution of Eq. 20 (i.e., $\theta = f(z)$) into Eq. 6 and then substituting this result into Eq. 14 gives the ESR spectrum of a thin film. For magnetic fields up to H_{crit} the lineshape will not vary (we are excluding small values of H where $(\omega_o/a) >> 1$ may not hold) but above H_{crit} the lineshape will begin to broaden as the resonant frequencies become spatially dependent (inhomogeneous broadening). Thus, ESR should serve as a probe to study alignment in thin films.

Experiments made in our laboratory[10] indicate that VAAC cannot be used for such a probe because it adsorbs on the glass plates used to form the film. An additional experiment[11] to corroborate this unfortunate result is that without VAAC benzene does not form a continuous film when a steel ball is immersed first into benzene and then into mercury, but benzene with a small concentration of VAAC does form a continuous film. The inference is that the adsorbed VAAC stabilizes the film.

A search is now being made for free radical probes which do not adsorb on glass.

XVI. 2. SLOW TUMBLING ESR OF A FREE RADICAL IN A BULK LIQUID CRYSTAL

We will formulate a theory for the ESR spectra of free radicals in a bulk liquid crystal which is based on the theory of NMR as modified by hindered rotation.[12] Instead of the free radical molecular axis taking on a continuous angular distribution relative to the magnetic field direction (also the nematic axis direction) while tumbling, we assume that it can occupy only a finite number of angular positions (hindered rotation positions - see Fig. 3).

For each angle there will be the hamiltonian (see Eq. 1)

$$\mathcal{H}(\theta_n) = \omega_o\, S^z\left[1 + 1/3\,\frac{\Delta g}{g}\,(3\cos^2\theta_n - 1)\right] + a\, I\cdot S$$

$$+ \frac{b}{3}\, I^z S^z\,(3\cos^2\theta_n - 1) \tag{27}$$

$$+ \mathcal{H}_{\text{non-secular}} + \mathcal{H}_{\text{radio frequency}}$$

The non-secular terms will in general only make a slight correction to the lineshape but would vastly increase the numerical work, thus in what follows we will ignore their effect.

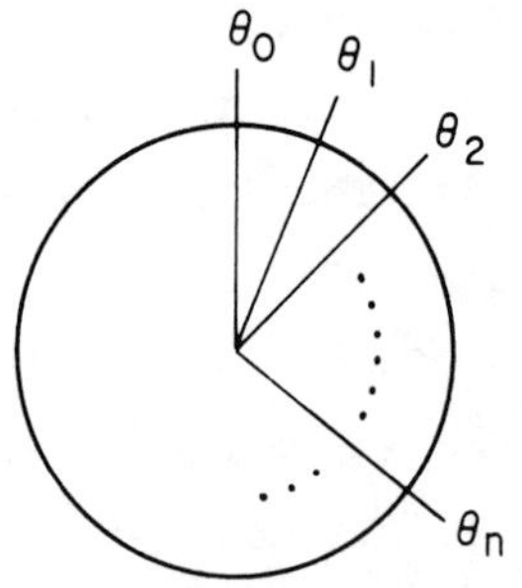

FIGURE 3

Hindered Rotation Positions of the Nematic Axis Relative to the Magnetic Field Direction

The density matrix ρ_n is associated with the hamiltonian $\mathcal{H}(\theta_n)$. Hopping between "hindered rotation positions" will then couple ρ_n to $\rho_{n\pm1}$. The coupling becomes simply expressed in the product representation

$$|M,m\rangle = |N\rangle \quad \text{where}$$
$$S^z|M\rangle = M|M\rangle \tag{28}$$
$$I^z|m\rangle = m|m\rangle$$

as

$$\langle N|\dot{\rho}_n|N'\rangle = -i\hbar^{-1}\langle N|[\mathcal{H}(\theta_n),\rho_n]|N'\rangle - \langle N|\rho_n|N'\rangle / T_2$$
$$-\left[\frac{\langle N|\rho_n|N'\rangle}{\tau_{n,\,n+1}} + \frac{\langle N|\rho_n|N'\rangle}{\tau_{n,\,n-1}} - \frac{\langle N|\rho_{n+1}|N'\rangle}{\tau_{n+1,\,n}} - \frac{\langle N|\rho_{n-1}|N'\rangle}{\tau_{n-1,\,n}}\right] \tag{29}$$

T_2 is to be thought of as a width parameter introduced so as to provide smoothing of the computer plots. The normalization of the ρ_n's is such that

$$\mathrm{Tr}\,\rho_n = 1 \tag{30}$$

which implies that (seen by taking the trace of Eq. 29)

$$\tau_{n,\,n+1} = \tau_{n+1,\,n}$$
$$\tau_{n,\,n-1} = \tau_{n-1,\,n} \quad \text{etc.} \tag{31}$$

For an isotropic liquid $\tau_{n,\,n+1} = \tau_o$ for all n, but this will not be so for a liquid crystal. The angular dependence of $\tau_{n,\,n+1}$ can be defined in two ways.

(I) As can be seen by a comparison of our approach with the formulation of ESR of liquids in the slow tumbling limit using the rotational diffusion approach (Chapter III Jensen, Ch. VIII Freed, Ch. XIII Gordon and Messenger) the diffusion constant D is seen to be proportional to $\tau^{-1}_{n,\,n\pm1}$. The Einstein relation $D = kT/6\pi\eta a$, where η is the viscosity and "a" the effective radius, allows one to write

$$\tau_{n,\, n\pm 1} \propto \eta/T \tag{32}$$

The generalization of Eq. 32 appropriate for a liquid crystal is then given as

$$\tau_{n,\, n+1} = \frac{A}{T}\left\{\eta_1 \cos^2\left(\frac{\theta_n + \theta_{n+1}}{2}\right) + \eta_2 \sin^2\left(\frac{\theta_n + \theta_{n+1}}{2}\right)\right\} \tag{33}$$

where η_1 and η_2 are experimentally known values (for a liquid crystal η_1 and η_2 will differ by about 20%).

(II) Using only the angular dependence of Eq. 33 one can write the general parametrized relation

$$\tau_{n,\, n+1} = \tau_1 \cos^2\left(\frac{\theta_n + \theta_{n+1}}{2}\right) + \tau_2 \sin^2\left(\frac{\theta_n + \theta_{n+1}}{2}\right) \tag{34}$$

Solving Eq. 29 one obtains the magnetization

$$\underset{\sim}{M}\,(\chi' \cos \omega t + \chi'' \sin \omega t)\, H_1 \quad ,$$

where χ' and χ'' are respectively the real and imaginary parts of the complex susceptibility, as

$$\underset{\sim}{M} = \sum_{n} \sum_{N,N'} C_n \langle N|\rho_n|N'\rangle\langle N'|m_{op}|N\rangle$$

where

$$C_n = e^{-E(\theta_n)}\, 2\pi \sin \theta_n \quad .$$

$E(\theta_n)$ is the MS energy and $2\pi \sin \theta_n$ is the solid angle.

In summary we will make a few comments on our procedure. It appears that our approach leads to essentially (ignoring the additional complications attendant to working with liquid crystals) the same matrix equations as derived by R. Gordon and T. Messenger (Chapter XIII, Eq. 3.4) using the finite difference approach to the solution of the spin diffusion equation. Our matrices (because we have ignored non-secular terms) for S = 1/2, I = 7/2 can be factored into a number of uncoupled matrices of order $(\frac{N}{4} + 1)$ and $2(N/4 + 1)$ where N is the number of divisions into which we divide 2π. Finally, as also commented on by R. Gordon and T. Messenger, we can generalize Eq. 29 by allowing for big hops, i.e., allowing for ρ_n to couple directly to $\rho_{n\pm\Delta N}$ for $\Delta N > 1$.

XVI. 3. ACKNOWLEDGEMENT

This work was supported by the Office of Naval Research under Contract No. N00014-69-C-0218.

REFERENCES

1. Read the Introduction in Chapter XII by G. R. Luckhurst
2. S. H. Glarum and J. H. Marshall, J. Chem. Phys. 46, 55 (1967).
3. G. C. Fryburg and E. Gelerinter, J. Chem. Phys. 52, 3378 (1970).
4. M. E. Rose, "Elementary Theory of Angular Momentum", John Wiley and Sons p. 60 Eq. 4.28 (1957).
5. W. Maier and A. Saupe, Z. Naturforsch 14a, 882 (1959).
6. P. Pincus, J. Appl. Phys. 41, 974 (1970).
7. F. C. Frank, Dis. Faraday Soc. 25, 1 (1958).
8. A more exact theory would have to include the dependence of τ_{m_I} on θ.
9. M. Abramowitz and I. Stegan "Handbook of Mathematical Functions" Dover Publications p. 587.
10. E. Drauglis, C. M. Allen, W. H. Jones, Jr., N. F. Hartman, J. M. Genco and J. F. Howes, Thin Film Rheology of Boundary Lubricating Surface Films, Third Quarterly Report to Naval Air Systems Command.
11. Private communication N. F. Hartman
12. J. I. Kaplan, J. Chem. Phys. 28, 278 (1958).

THE ESR LINE SHAPE OF TRIPLET EXCITONS IN DISORDERED SYSTEMS : THE ANDERSON THEORY APPROACH

J.P. Lemaistre and Ph. Kottis

Centre de Mécanique Ondulatoire Appliquée du

C.N.R.S., 23, rue du Maroc, Paris XIX, France

XVII. 1. INTRODUCTION

The ESR triplet line-shape of randomly oriented molecular aggregates in the solid phase reflects static and dynamic properties of the excited triplet state. By static properties we refer to the spin hamiltonian parameters describing the fine and hyperfine structure, while by dynamic property we refer to the lifetime for the triplet exciton transfer caused by resonance exchange forces between the units of the aggregate.

In aromatic aggregates, the line-shape of which will be the subject of this lecture, the anisotropy of the ESR absorption determines "widths" of the order of 3,000 G. Such a broad absorption would make line-shapes of disordered systems entirely inadequate for the study of the parameters of the spin hamiltonian and, a fortiori, for the investigation of exciton diffusion in the aggregate.

The technique of triplet line-shapes in glassy solutions, introduced by van der Waals and de Groot[1], offers, however, two irreplaceable advantages : a) the solid phase quenches radiationless transitions and allows one to obtain excited electronic triplet (phosphorescent) states of long life times ($0.01<\tau_{ph}<20$ sec.).

b) glassy solutions remove the need of host crystal matrices of adequate symmetry.

Furthermore, an analysis of symmetry properties of the spin hamiltonian associated with excited triplet states, allows us to calculate the resonance field density function $g_{\omega}(B)$. This function provides evidence for the existence of a magnetoselection phenomenon, characterized by the presence of sharp pseudo-lines in the derivative ESR spectrum of randomly oriented systems. As a consequence of the magnetoselection, the derivative line-shape of poly-oriented systems has, for certain values of the magnetic field (the so-called resonance stationary fields), the same properties as that of an assembly of systems oriented in a host crystal, which is the technique introduced by Hutchison and Mangum[2)]. Therefore, the theory of the magnetoselection allows one to simulate the ESR line-shape, by considering only three oriented families of the molecules in the sample, and thereby to remove the inherent difficulties of disordered systems. Magnetoselection will be used here to calculate the line-shape of the so-called coherent exciton (collective electronic excitation) and that of incoherent (hopping) excitons in molecular aggregates embedded in an amorphous solid matrix[3)].

The aim of this chapter is first to summarize the correlation between the line-shape and the parameters of the spin hamiltonian and then to show how the loss of phase of a collective excitation (the so-called coherent exciton) coupled to a thermal bath, may lead to temporary localization of the exciton and to its incoherent motion through many channels of adiabatic transfer[4)]. Localization and incoherent motion are two prerequisites for the description of the exciton as a short memory stochastic process and for the use of Anderson's theory of stochastic resonance absorption in the Markov motion model.

This lecture is divided in three parts. In the first one, the static problem is treated : the line-shape is correlated to the parameters of the spin hamiltonian and to the symmetry of the absorbing systems. Theoretical and experimental evidence of the magnetoselection phenomenon are given.

In the second part, in order to respect the spirit of this volume, localization and randomness of the motion of the exciton(s) are taken for granted : the line-shape is calculated in a phenomenological way as resulting from one, or many hopping excitons in thermal equilibrium. Then, Anderson's theory is applied to excitons moving through one or many discrete transfer channels[3-5]. In this theory we take complete account of the non-secular effets, but we introduce them in the diagonal elements of the relaxation matrix A, as "frequency shifts" and additional terms in the transition matrix Π, thus reducing drastically the dimension of the matrix A.

This approximation, instead of using the density matrix technique, is first justified on pragmatic grounds : simulation of a line-shape in triplet polyoriented systems needs no less than 100x100 points of integration for each value of B, then 2,000 G must be swept for each trial value of the transfer rate. Consequently, the computation time becomes prohibitive. Secondly, this approximation is justified by a theoretical reason : the transfer mechanism in the solid phase is not well understood ; a single parameter τ_c, used to express the random motion in the density matrix technique, has been found insufficient to simulate triplet exciton line-shapes.

In the third part of this chapter, we start from the total hamiltonian of an aggregate and its collective excitation states coupled explicitly to a dissipative medium. According to this coupling, the resulting states may be non-stationary and described as localized states diffusing adiabatically, with different rates through the manifold of low frequency vibronic levels of the phosphorescent state. It is shown how the intra-aggregate couplings and the temperature are controlling the exciton diffusion whose rates are calculated. Therefore, a new picture arises for the temperature dependence of the triplet exciton line-shape : a line-shape due to many excitons in thermal equilibrium, with the temperature modifying the statistical weights of the different excitons.

XVII. 2. THE STATIC LINE-SHAPE

A. The Magneto-Selection Theory

In this section, we discuss briefly the line-shape of a stationary triplet state. The spin observables of a molecular triplet state are described by the spin hamiltonian[6].

$$H_S(\theta,\phi,B) = \beta \mathbf{S}.\mathbf{g}.\mathbf{B}. + \mathbf{S}.\mathbf{T}.\mathbf{S} + g\beta\,\beta_p \sum_N g_N\, \mathbf{I}_N.\mathbf{A}^N.\,\mathbf{S} \qquad (2.1)$$

g and T are the fine coupling tensors for the Zeeman and dipolar interactions. The A^N are the hyperfine coupling tensors. For ESR experiments at the X band frequency ($\omega \sim 10^{10}$Hz), the dipolar interaction introduces the main anisotropy ($\Delta B_T \sim 2{,}000$ G). The **g** tensor is considered isotropic ($\Delta B_g \sim 10$ G). The hyperfine couplings, whose total magnitude is of the order of 10 G, are introduced as an adiabatic perturbation. In expression (1), the angles θ and ϕ determine the orientation of the magnetic field B in the coordinate system which diagonalizes the dipolar tensor **T** (the symmetry axes of the molecule, if any).

In experiments with a constant frequency ω and with varying B, we define a resonance field density function :

$$g_\omega(B) = \int d\phi \int d\theta \cdot \sin\theta \cdot \delta\{B - B^r(\theta,\phi)\} \qquad (2.2)$$

$B^r(\theta,\phi)$ is one of the resonance fields derived from(1), δ is the Dirac function. Instead of obtaining $B^r(\theta,\phi)$ from (1), we solve only the fine part and consider the hyperfine perturbation as a gaussian energy spread on the electronic spin levels i (i = 1,2,3).

Then, the density function (2) is given a more tractable form[7,8]:

$$\gamma_\omega(B) = \sum_i \int d\phi \int d\theta \cdot \sin\theta \cdot \exp\{-2\frac{\{B-B_i^r(\theta,\phi)\}^2}{\lambda_i^2}\} \qquad (2.3)$$

The summation is carried over the electron spin resonances i (two $\Delta m = 1$ and one "$\Delta m = 2$"). We present the variation of γ_ω (B) in Fig. 1A, and that of its derivative in Fig. 1C. The anisotropy of the resonance field B^r (θ,ϕ) is presented in Fig. 1B, with the transitions of constant frequency ω indicated below. Fig. 1 summarizes the magneto-selection phenomenon : the pseudo-lines observed in the derivative spectrum are due to the singularities of $\gamma_\omega(B)$. The physical origin of these singularities is the stationary behaviour of the energy levels and, hence, of B^r (θ,ϕ), when one of the canonical axes of T becomes quasi-parallel to B. When the molecules investigated have sufficient symmetry, the canonical axes of T coincide with the molecular axes and the lines in Fig. 1C are due uniquely to three families of oriented molecules as indicated in Fig. 1D. The selection of these three families in the ESR spectrum of disordered systems is the magneto-selection phenomenon. The latter proved to be very helpful in investigating static and dynamic properties of excited electronic states in polyoriented solutions[1-2,7-9].

B. Line-Shape and Symmetry of the Triplet State

As an example of static properties, let us indicate how the stationary resonance fields, of Fig. 1C, allow a straightforward derivation of the fine structure parameters of expression (1) :

$$T_{XX} = (g_{XX}\beta)^2 \ (6\omega)^{-1} \ \{(H_X^{III})^2 - (H_X^{II})^2\} \qquad (2.4)$$

$$g_{XX} = (2\beta H_X^I)^{-1} \ \{\omega^2 - (T_{YY} - T_{ZZ})^2\}^{1/2} \qquad (2.5)$$

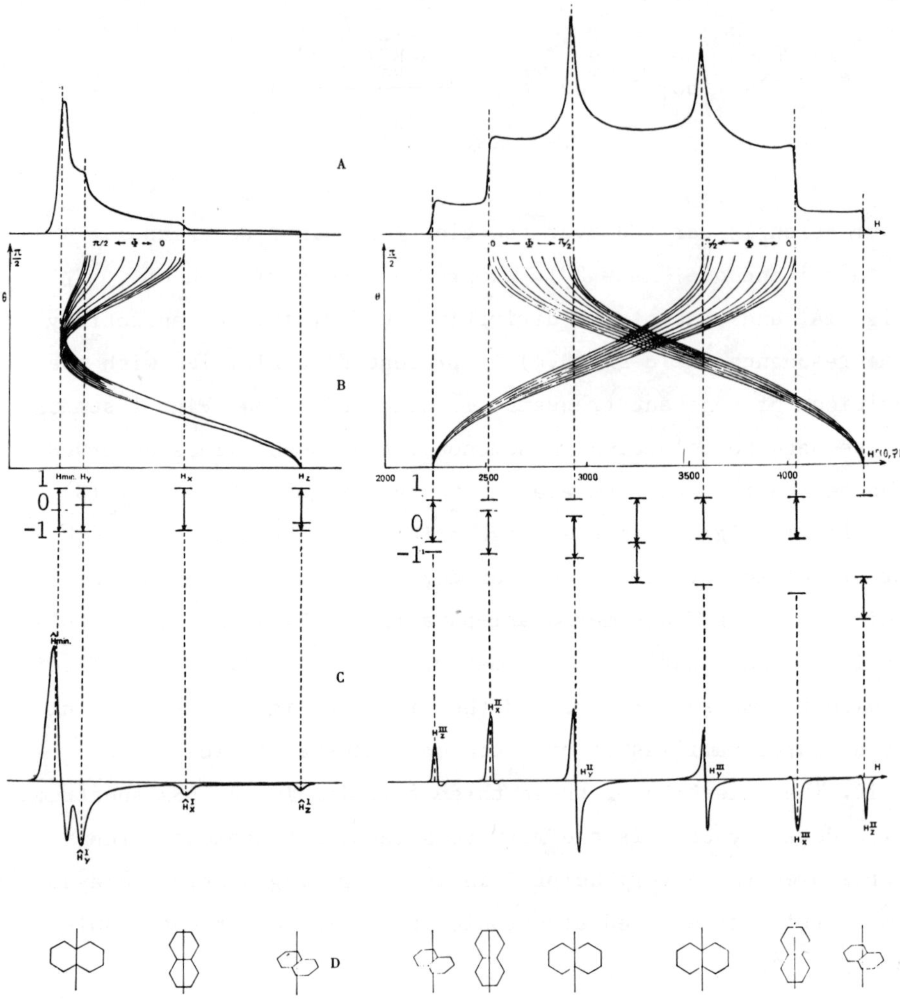

Fig. 1 - Magnetoselection.

A) Variation of the resonance field density function $\gamma_\omega(H)$ for naphthalene with $\omega = 0.31\ cm^{-1}$. B) The anisotropy of the resonance field with, below, the transitions corresponding to the singularities of $\gamma_\omega(H)$. C) The derivative of $\gamma_\omega(H)$ showing the pseudo-lines due to magnetoselected molecules. D) The three families of molecules in the polyoriented assembly causing the structure of the ESR spectrum.

The other components of T and **g** are obtained by cyclic permutation of X, Y and Z. In order to avoid confusion with notations in original publications, the resonance stationary fields (rsf) in (4) and (5), and on the figures, will be indicated by H, not by B. The principal values of the traceless tensor **T** are related to the two definitions of the zero-field splitting (zfs) parameters[1,2] :

$$T_{ZZ} = -Z = \frac{2}{3} D \; ; \; T_{YY} = -Y = -\left(\frac{D}{3} + E\right) \; ;$$

$$T_{XX} = -X = -\left(\frac{D}{3} - E\right) \qquad (2.6)$$

The line-shape of Fig. 1 is typical of a system of D_{2h} symmetry (naphthalene) with $T_{XX} \neq T_{YY} \neq T_{ZZ}$. For systems of D_{3h} symmetry, as triptycene-like trimers considered in the next section (with $T_{XX} = T_{YY} \neq T_{ZZ}$), the peaks of the srf H_X and H_Y of each transition fuse to a single peak to give the following pattern : two symmetrical peaks in the $\Delta m = 2$ region, two large and two small peaks in the $\Delta m = 1$ region[9,12].

From these elementary considerations, it is clear that in the trimer of D_{3h} symmetry, cf. Fig 4A, we have two definite line-shapes when exciton motion occurs : the line-shape reflects the site's C_{2v} symmetry when the residence time k^{-1} of the exciton is larger than the local coherence time T_2 ; while the line-shape reflects D_{3h} symmetry when : $k \gg g\beta\{B^r(\theta_i,\phi_i) - B^r(\theta_j,\phi_j)\}$; where $B^r(\theta_i,\phi_i)$ is the resonance field of the site i (with i,j = 1,2,3) cf. Fig. 17 of ref. 9, also Fig. 5 in this work.

C. The Magneto-Photo-Selection

In the last subsection, the importance of the three families of molecules, oriented in the sample with respect to B, has been pointed out. A second test of this property has been to bring to the phosphorescent state (through a radiative electronic transition followed by a non-radiative one) selectively one of the three families, using UV light polarized parallel to B. Assume that, for symmetry reasons, the moment of the electronic radiative transition is polarized along one of the molecular axes, say the axis OY, cf. Fig. 1D. Then, the excitation of the sample with UV light polarized along B, will excite selectively a single family. The ESR lineshape of such a sample will be very similar to that of an oriented assembly. This experiment has proved to be very sensitive and gave rise to the Magneto-photo-selection technique. In Fig. 2, we give the first theoretical results and experimental data of a magneto-photo-selection experiment [10,11]. One of the immediate results is the knowledge of the polarization of the electronic radiative transition.

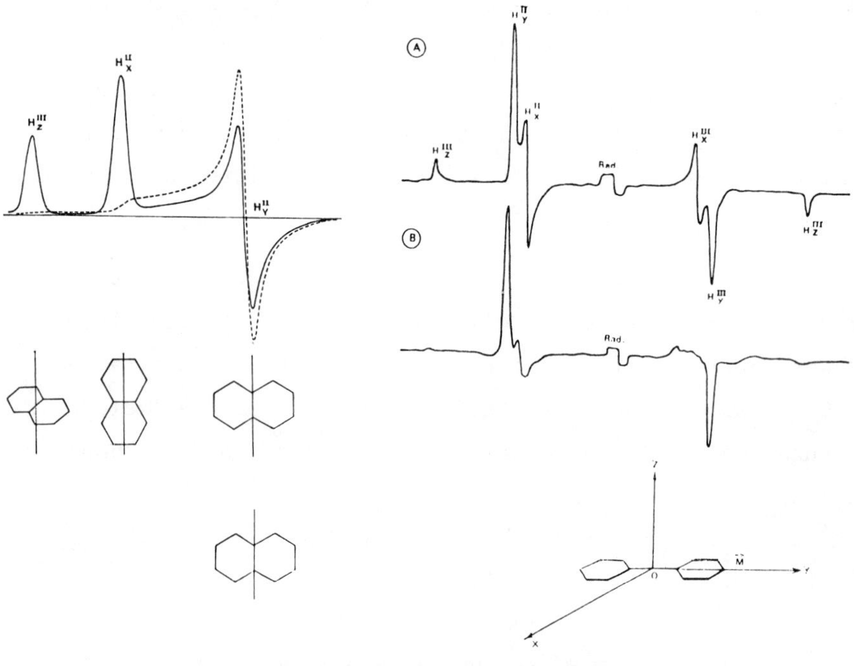

Fig. 2 - Magneto-photo-selection.

Left side : calculated $\Delta m = 1$ spectra of a triplet assembly excited with isotropic (solid line) and with parallel polarized UV light (dotted line). Below are indicated the molecules excited in each experiment. Right side : $\Delta m = 1$ spectra obtained in magneto-photo-selection. A) with isotropic UV light. B) with parallel polarized UV light, showing that the radiative transition of the system investigated lies along its OY axis.

XVII. 3. THE INCOHERENT EXCITON LINE-SHAPE

1. THE EXPERIMENTAL PROBLEM

In recent years, many investigations have been reported on ESR line-shapes of excited triplet molecular aggregates in a solid phase. Such aggregates consist of electronically equivalent parts not conjugated with each other. For clarity of the exposition, the aggregate is called the system and its parts are called subsystems. Typical examples of the systems investigated are Triptycene-like trimers[12)], spirane dimers[13)], chromophoric polymers[14)], physical polymers or dimers[15)].

A common feature of the reported ESR spectra is that they have a very unusual and strongly temperature dependent shape. What is highly suggestive is that at low temperature , 20°K, the line-shape reflects the local symmetry of the subsystem, with the pseudo-lines in Fig. 1 broadened. At higher temperature, 77°K, on the other hand, the line-shape reflects the symmetry of the system, cf Fig. 3.

2. THE SINGLE-CHANNEL TRANSFER MODEL

In order to describe these features of a non-stationary electronic triplet state, the hopping exciton was first used in the framework of Anderson's theory and in that of the density matrix[16,17)]. For instance, in the trimer of Fig. 4A, each subsystem constitutes a site in which the exciton relaxes (it is localized) for a residence time $(2k)^{-1}$, with k noting the hopping probability from one site to another. Because the triplet spin is strongly coupled, through the dipolar interaction, to the electronic wave-function of the subsystem, the three differently oriented naphthalenes are magnetically non-equivalent sites, i.e. they have different spin functions and resonance frequencies in the presence of an external magnetic field B. Therefore, the exciton hopping causes a stochastic

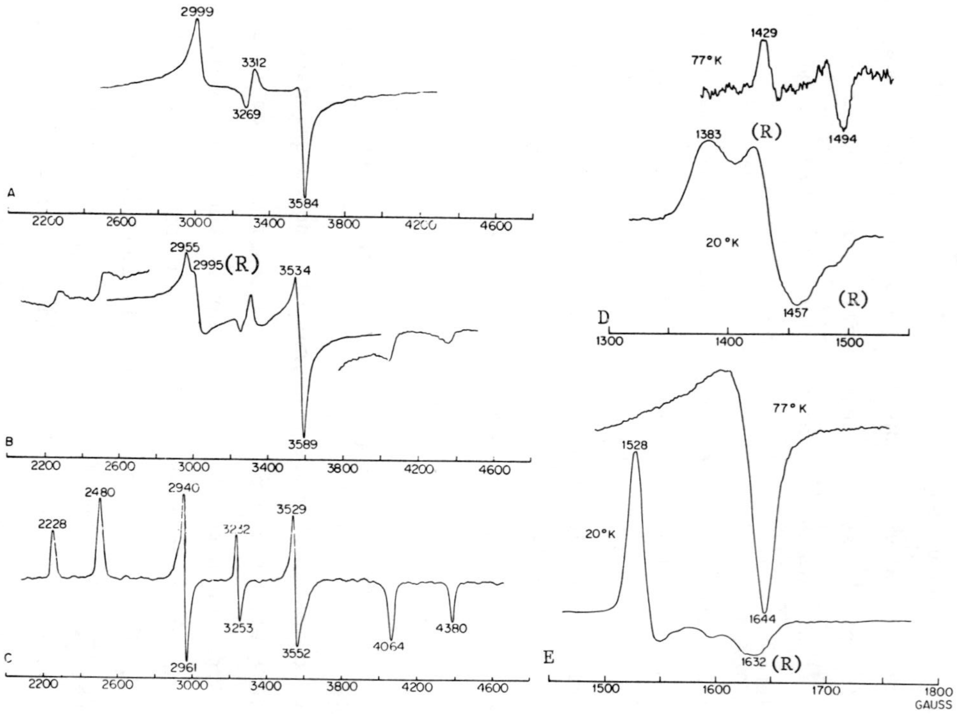

Fig. 3 - Triplet line-shapes in a glassy solution for high and low magnetic field[12).]

A) TBT at 77°K. B) TBT at 20°K. C) Naphthalene at 77°K. D) Spectra of Triptycene with U||H. E) Spectra of TBT with U||H.

modulation of the spin hamiltonian (2.1) among its three values $H_s(\theta_i, \phi_i)$, with i = 1,2,3. This modulation changes the static line-shape of polyoriented aggregates.
As mentionned in XVII.2, in the slow limit case ($k^{-1} >> T_2$), each subsystem absorbs almost independently, as in the static case, and the line-shape reflects the symmetry C_{2v} of the subsystem , with an additional half-width k. In the rapid limit case, the line-shape reflects the symmetry D_{3h} of the system and equation (2.4) allows the derivation of its spin hamiltonian parameters.

In order to apply Anderson's theory[18)], the following assumptions define our model of the absorbing system : 1°) The symmetry is conserved when the system is embedded in an amorphous solid matrix. 2°) The sites i of the system are electronically equivalent and their spin hamiltonians are described with the same local tensor T(i). 3°) the hopping probability k is isotropic and, this defines the single channel transfer model, for a given temperature k(T) is the same for all the molecules of the excited triplet population. 4°) The temperature dependent perturbation is considered to modulate the spin dipolar interaction stochastically and the modulation of the Zeeman term is ignored. 5°) The duration Λ^{-1} of a hop from site to site is considered very small compared to k^{-1} ($\Lambda >> k$), so that the spin hamiltonian evolution may be considered as a Markov process. 6°) The line-shape function $I_B(\omega)$ of the system, under conditions of "slow passage" and in the absence of saturation, may be calculated from formula (3.1),

$$I_B(\omega) = \mathrm{Tr}\left|\int \mu(t)e^{-i\omega t}dt\right|^2 \qquad (3.1)$$

$\mu(t)$ is the radiative dipole moment matrix in the Heisenberg representation :

$$i\hbar\,\dot{\mu} = \left[H_s(t),\, \mu\right] \qquad (3.2)$$

$H_s(t)$ is the spin hamiltonian whose time dependence is postulated in the above model, assumptions 2 to 5. This model, because of as-

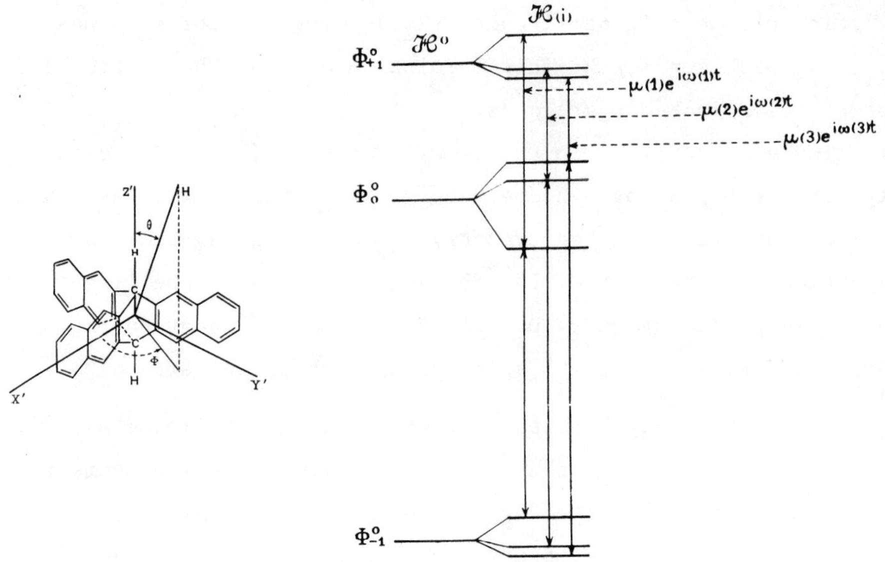

Fig. 4 - Tribenzotriptycene as a system of three sites.

Right side : a three-site adiabatic model for the spectral line $\phi^o_+ \leftrightarrow \phi^o_o$ (the other two lines are far out of resonance).

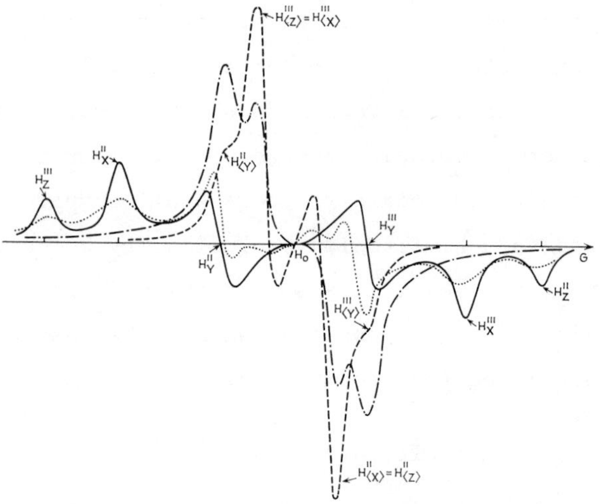

Fig. 5 - Exciton line symmetry vs transfer rate in a SCT Model

Calculated line for polyoriented systems composed of two sites having a common OY axis and two axes permuted. At slow transfer the line (——) reflects the site C_{2v} symmetry. At rapid transfer the line (----) reflects the cylindrical symmetry of the system ($\langle x\rangle = \langle z\rangle$). The other two lines correspond to intermediate transfer cases.

sumption 3 is referred to as the Single Channel Transfer Model. It allows one to derive a line-shape function, $I_B(\theta,\phi,\psi,\omega,k)$, which depends on a single parameter k, for all systems having the same orientation relative to the laboratory coordinate system. The line-shape function for the assembly of poly-oriented systems is then obtained by summing over all possible orientations (θ,ϕ,ψ).

A. The Exciton Line-Shape of a Trimer

We calculate here, in detail, the exciton line-shape of the Tribenzotriptycene (TBT) which consist of three naphthalene molecules, not conjugated, bound together to form a molecule of D_{3h} symmetry, cf. Fig. 4A. The triplet electron spin levels of the subsystems present the following characteristics when investigated at the X band or at a higher frequency ($\omega \geq 10^{10}$ Hz) : a) Second order perturbation theory gives the electron spin levels and the transition probabilities with an accuracy better than 5%. This property allows us to approximate the denominators of the second order terms by field independent quantities : 0.5ω and ω for $\Delta m = 2$ transitions, ω and 2ω for $\Delta m = 1$ transitions. Thus, the resonance frequency depends linearly on the field B. b) For each value of B, no more than one of the three possible transitions can be in resonance[19]. Besides the separation of the relaxation function $G(\tau)$ into three independent relaxation functions $g_{ij}(\tau)$ (one for each spectral line), the property b will be utilized in order to include the non-secular term in Anderson's theory.

Under these conditions, the three spectral lines (one $\Delta m = 2$ and two $\Delta m = 1$) can be calculated independently. The intensity distribution (1) is, therefore, reduced to the calculation of three autocorrelation functions (or relaxation functions as referred to in Chap. II) of the transition moments $\mu_{ij}(t)$. The moment $\mu_{ij}(t)$, which is in resonance at ω, is perturbed in a very complex manner by the fluctuation [20] of the spin dipolar energy $H^{ss}(t)$, cf. Fig. 4B. The other two non-resonating transitions contribute far in the wings of the spectral line ω_{ij} and will be introduced as perturbations when the non-secular term will be considered.

The auto-correlation function of $\mu_{ij}(t)$ has been calculated first by Anderson for the case when the fluctuation of $H^{ss}(t)$ is described as a Markov process and when its non-secular part can be neglected[18]. The latter is never the case for triplet excitons in which $H^{ss}(t)$ is equally important and does not commute with the Zeeman term. The contrasting aspect of the triplet exciton motion, compared to equivalent problems in NMR or in ESR of tumbling free radicals, cf. Chap. III and XIII, is the importance of the time dependent part (to give an idea of the order of magnitude we recall that the site frequencies may be separated by $\sim$1,000 G and that the non-secular part of $H^{ss}(t)$ causes a frequency shift of the order of 100 G). Because of the characteristics a and b, indicated above, that present the aromatic systems, expression (1) reduces to :

$$I_{\omega}^{i\to j}(B) \;\alpha\; \left|\int \mu_{ij}(t) e^{-i\omega t} dt\right|^2 \qquad (3.3)$$

$i\leftrightarrow j$ is the transition in quasi-resonance at constant frequency ω. $\mu_{ij}(t)$ is the instantaneous moment : i.e. the matrix element of the operator $H_1 = -\mu.U$ which couples the fluctuating states, cf. Fig. 4B. μ is the operator $-g\beta S$ associated with a state of spin $|S| = 1$, U is the unit vector of the polarized r.f.f. (the most informative data are obtained for $U \parallel B$ for $\Delta m = 2$ transitions and for $U \perp B$ for $\Delta m = 1$ transitions .

It is shown in Chap. II that in the high temperature approximation and for a sample in a stationary state, expression (3) may be written, dropping the i,j, :

$$I_{\omega}(B)\,\alpha\; \mathrm{Re}\int_0^{\infty} g(t) e^{i\omega t} dt \qquad (3.4)$$

$$\text{with } g(t) = [\![\overset{\star}{\mu}(o)\mu(t)]\!]_{Av.} = \sum_{m=1}^{3} \left[\mu^{(m)}(o).\mu^{(m)}(t)\right]_{Av.}$$

The average of the product $\overset{\star}{\mu}(o)\mu(t)$, over all the paths $[o,t]$, is taken on the three ensembles m, associated with the three sites (the ensemble associated with site 1 is composed, in our model, of trimers which have the same orientation, carry at t=o the exciton on site 1 and are in quasi-resonance at ω). Since the three sites are electronically equivalent and the high temperature approximation is valid, the statistical weights W_i of the three ensembles are equal :

$$W_1 = W_2 = W_3 = 1/9 \tag{3.5}$$

The time dependence of μ(t) is obtained through (2) with :

$$H_s(t) = H^z + H^{ss}(t) \tag{3.6}$$

After an exciton transfer from site n to site m, the reorganization of the electron cloud (electronic relaxation) is a very fast process. Therefore, the transfer appears in ESR as a sudden change of the parameters related to the electronic coordinates, the tensor T, from its components in the site n to its components in the site m :

$$T(n) \xleftrightarrow{k} T(m) \tag{3.7}$$

The three values of 7 are given in Ap. 1, as a function of the orientation of the trimer in a laboratory coordinate system {x,y,z} that has its z axis along the direction of the magnetic field.

In order to solve Eq. 2, we use the approach suggested by Anderson : the spin hamiltonian 6 is split in one unperturbed part H_s^o and one perturbing part $H_s^p(t)$:

$$H_s(t) = H_s^o + H_s^p(t) \tag{3.8}$$

with $$H_s^o(\theta,\phi,B) = H^Z + \langle H^{ss}(t)\rangle \tag{3.9}$$

and $$H_s^p(t) = H^{ss}(t) - \langle H^{ss}(t)\rangle = H^{\star}(t) + H^{\star\star}(t) \tag{3.10}$$

H_s^o is time independent, it includes the Zeeman interaction and the site average of the dipolar interaction. The latter is not zero, as in the liquids, and reflects the symmetry of the aggregate. In Anderson's terminology $H^{\star}$ and $H^{\star\star}$ are respectively the secular and the non-secular parts of $H_s^p(t)$ with zero site averages :

$$\langle H^{\star}\rangle = o \qquad \langle H^{\star\star}\rangle = o \tag{3.11}$$

They give, respectively, only diagonal and only off-diagonal matrix elements in the basis of the eigenkets of H_s^o denoted $\{\phi_+^o, \phi_o^o, \phi_-^o\}$.

B. The Secular Approximation

In order to integrate Eq. 2, we take only the secular part of $H_s^p(t)$. Thus, the truncated spin hamiltonians of the sites commute,

$$H_s(i) = H_s^o + H^{\star}(i) \tag{3.12}$$

Then, the integration of 2, for instance for the high field transition $\phi_+^o \to \phi_o^o$, gives :

$$\mu(t) = \mu^o \exp\{i(\omega^o + t^{-1}\int_o^t \omega^{\star}(t')dt')t\} \tag{3.13}$$

with $$\mu^o = \langle\phi_+^o|S.U|\phi_o^o\rangle$$

$$\omega^o = \{\langle\phi_+^o|H_s^o|\phi_+^o\rangle - \langle\phi_o^o|H_s^o|\phi_o^o\rangle\} \tag{3.14}$$

$$\omega^{\star}(t) = \{\langle\phi_+^o|H^{\star}(t)|\phi_+^o\rangle - \langle\phi_o^o|H^{\star}(t)|\phi_o^o\rangle\}$$

ω^o and $\omega^{\star}(i)$ are the contributions of H_s^o and $H^{\star}(i)$, respectively, μ^o being the transition moment in the absence of the perturbation due to the transfer.

Then, in a Markov process approximation, the absorption intensity may be given in the form :

$$I_\omega(B)\alpha \ \mathrm{Re}\int M_s^\star(o).M_s(t)dt \tag{3.15}$$

$$\text{with } M_s^\star = \{W_1\mu^o, W_2\mu^o, W_3\mu^o\}^\star ; \ M_s(t) = \exp(A_s t)M_s^\star(o) \tag{3.15'}$$

M_s is a column vector with equal components μ^o. The three components of $M_s(t)$ give the evolution of the transition moments in the three ensembles, the index s stands for secular. Following Sack and Abragam[21)] (see also chap. II) expression 15 integrates to :

$$I_\omega(B)\alpha \ - \mathrm{Re}\{M_s^\star . A_s^{-1} . M_s\} \tag{3.16}$$

with the following definitions :

a) $A_s = i\Delta\omega + \Pi$ is the relaxation matrix calculated by Anderson in the secular approximation. In the case of the trimer of Fig. 4A, A_s is a 3x3 matrix and has the explicit form :

$$A_s = \begin{pmatrix} i\{\omega^s(B,\theta,\phi_1)-\omega\}+\Pi_{11} & \Pi_{12} & \Pi_{13} \\ \Pi_{21} & i\{\omega^s(B,\theta,\phi_2)-\omega\}+\Pi_{22} & \Pi_{23} \\ \Pi_{31} & \Pi_{32} & i\{\omega^s(B,\theta,\phi_3)-\omega\}+\Pi_{33} \end{pmatrix} \tag{3.17}$$

Π is the "transition" matrix. Owing to the symmetry of the system, its elements obey the relations :

$$\Pi_{12} = \Pi_{13} = \Pi_{23} = k \qquad \Pi_{11} = -(\Pi_{12} + \Pi_{13} + P_o) = -(2k + P_o) \tag{3.18}$$

Π_{11} is the probability per second for the ensemble 1 to be out of resonance at the frequency ω. Then, P_o accounts for the mean probability of the ensemble being out of resonance by other means than electronic transfer : hyperfine interaction, spin-lattice relaxation or spin-spin relaxation through the neglected dipolar interaction $H^{\star\star}(t)$. Therefore, P_o will be noted as depending also on k.

b) $\omega^s(B,\theta,\phi_i)$ is the secular part of the static resonance frequency of the site i, derived from 12.

In the absence of transfer, k=0, formula 16 reproduces three Lorentzian lines of half-width P_o, centred at $\omega^s(i)$, all of them proportional to $W_i|\mu^o|^2$, with i=1,2,3. In slow or intermediate transfers, the lines broaden because of random modulation of the resonance frequency[22]. For a rapid transfer, the lines coalesce to a single line whose intensity is proportional to $3W|\mu^o|^2$ and centred at the average of the frequencies $\omega^s(i)$, which is also the resonance frequency of the unperturbed hamiltonian H_s^o. Indeed taking account of 11, the average of $\omega(i)$ is :

$$<\omega^s(i)> = \omega^o + <\omega^{\star}(i)> = \omega^o \qquad (3.19)$$

The half-width of the coalesced line is given by the well-known expression :

$$\lambda^s(k) = (Q/2k) + P_o(k) \qquad (3.20)$$

with Q demoting the mean squared value of $H^{\star}$, i.e.

$$Q = \frac{1}{3}\sum_{i=1} \{\omega^{\star}(i)-\omega\}^2$$

The second order contributions to the resonance frequencies and to the transition moments due to $H^{\star\star}(i)$, are denoted by $\omega^{\star\star}(i)$ and $\mu^{\star\star}(i)$, respectively. Then, Anderson's secular approximation is expected to describe the hopping exciton line-shape with a success which improves when the following conditions are satisfied :

$$\omega^{\star\star}(i) << \omega^{\star}(i) \qquad \mu^{\star\star}(i) << \mu^o \qquad (3.21)$$

For the two $\Delta m=1$ transitions, which give the high field line-shape (with the spin almost quantized along B and $\omega^{\star\star}$, and $\mu^{\star\star}$, going to zero with increasing B) conditions 21 are satisfied for all the orientations, except for H_y^{II} and H_y^{III}, for which $\omega^{\star}=0$ and $\omega^{\star\star}$ ∿50 G.
For $\Delta m=2$ transitions, which give the low field line-shape (whose presence is proof that the spin is not quantized along B) conditions 21 are never realized, except at the limit of very fast transfer $(k>>\omega)$

for which $\omega^{**}(t)$ and $\mu^{**}(t)$ are averaged out. Then, the $\Delta m=2$ line-shape is due to the residual dipolar interaction of the unperturbed hamiltonian H_s^o. If, for symmetry reasons, H_s^o reduces to the Zeeman term, then the $\Delta m=2$ transitions are strictly forbidden in fast transfer as in fast tumbling, cf. Fig. 6A.

In conclusion, the secular approximation simulates fairly well the high field line-shape of the triplet exciton in the single channel transfer, but it cannot represent the low field part whose dominant term H^{**} it excludes.

C. The Adiabatic Approximation

In order to include $H^{**}(t)$ in the integration of the radiative dipole 2, we introduce the adiabatic model which we define as follows : during the residence time of the exciton in the site i, the average evolution of $\mu(t)$ is described by the static hamiltonian $H_s(i)$, instead of a slowly varying one (this approximation rather than a phase discontinuity during the hop, introduces site average discontinuities). Then, the intensity (16) again takes the form :

$$I_\omega(B)\alpha - \mathrm{Re}\{\overset{\star}{M}.A^{-1}.M\} \qquad (3.22)$$

with the following changes with respect to formulae (15') and (17) :

a) the relaxation matrix (17) contains the exact static frequencies : $\omega(i) = \omega^o+\omega^{\star}(i)+\omega^{**}(i)$, instead of only their secular part $\omega^o+\omega^{\star}(i)$.

b) Similarly, $\overset{\star}{M}$ and M have as components the static transition moments : $\mu(i) = \mu^o+\mu^{**}(i)$, instead of only the secular part μ^o.

In the absence of transfer, formula (22) reproduces three Lorentzian lines of half-width P_o, centred at the correct resonance frequencies $\omega(i)$, with intensities correctly proportional to $W_i|\mu(i)|^2$, i = 1,2,3.

For low and intermediate values of k, the lines broaden for two reasons : a) resonance frequency modulation which is included in the secular approximation. b) random change in the spin orientation. This

broadening is not included in the secular treatment, it may be crucial especially when the sites are degenerate, cf. Fig. 6.
For a fast transfer, the coalesced line has a correct intensity proportional to $|\langle\mu(i)\rangle|^2 = |\mu^o|^2$, but an incorrect position :

$$\langle\omega(i)\rangle_i = \omega^o + \langle\omega^{\star}(i)\rangle_i + \langle\omega^{\star\star}(i)\rangle_i = \omega^o + \langle\omega^{\star\star}(i)\rangle_i \qquad (3.23)$$

The limit (23) is not one of the eigenfrequencies of H_s^o. Because of the calculation of the $\omega(i)$'s by second order perturbation theory, the site average of the second order terms cannot cancel out, cf. expression (24) below.

It is well known that the adiabatic model is a fairly good approximation as long as the transfer rate k remains small compared to the resonance frequencies of the system. When k becomes comparable to ω, expression (22) reproduces neither the frequency shift nor the non-adiabatic broadening[21)].

In conclusion expression (22) simulates the low field line-shape of the exciton for small values of k. For large values of k, because the frequency shift is very important, and anisotropic, the line-shape of polyoriented systems is not merely shifted but distorted (cf. Fig. 2 of ref. 16a).

D. A Line-Shape Formula for the General Case

We have purposely stressed on the nature and the importance of the limitations of Anderson's theory when used in conjunction with the adiabatic model. In this sub-section we propose a general line-shape formula which includes averaged non-secular contributions in the diagonal elements of the relaxation matrix. This formula allows one to include the frequency shift and the non-adiabatic broadening and to extend Anderson's theory to all values of k.

The restriction to small values of k, allowed us to approximate the non-secular contribution of $H^{\star\star}(t)$ by its static values $H^{\star\star}(i)$. For instance, for the transition $\phi_+^o \leftrightarrow \phi_o^o$ at a frequency ω, the contribution of $H^{\star\star}(t)$ to the site frequencies was approximated by (3.24),

$$\omega^{\star\star}(i) = |\langle\phi_+^o|H^{\star\star}(i)|\phi_o^o\rangle|^2 \omega^{-1} + |\langle\phi_+^o|H^{\star\star}(i)|\phi_-^o\rangle|^2 (2\omega)^{-1} \qquad (3.24)$$

Following Anderson[23)] and Abragam (p.444) it is possible, for larger values of k, to avoid the incorrect coalescence limit (23) and the lack of non-adiabatic broadening in (22), by introducing averaged contributions of $H^{\star\star}$ in the relaxation matrix used in (22), i.e. :
a) in the form of an effective non-secular contribution in the site frequency $\omega^{\star\star}(k,i)$, the frequency shift, and b) in the form of a supplementary width P(k) accounting for the so-called non-adiabatic broadening[18,21)]. These two contributions are taken, respectively, as the real and the imaginary part of the averaged expression :

$$\sum_m{}' \int \overline{\overline{e^{i\omega_{mn}t}\langle n|H^{\star\star}(o)|m\rangle\langle m|H^{\star\star}(t)|n\rangle dt}} \; + \qquad (3.25)$$

$$+ \quad \sum_m{}' \int \overline{\overline{e^{i\omega_{1m}t}\langle 1|H^{\star\star}(o)|m\rangle\langle m|H^{\star\star}(t)|1\rangle dt}}$$

The double bar implies the ensemble average : the average on the paths $[o,t]$ and on the initial values of $H^{\star\star}(o)$. For the transition $\phi_+^o \leftrightarrow \phi_o^o$ we consider, $n = \phi_+^o$, $1 = \phi_o^o$, $m = \{\phi_+^o, \phi_o^o, \phi_-^o\}$ and the real and imaginary parts of (25) reduce, respectively, to zero and to (24) for $k \ll \omega$. We choose a simpler example, a two-site case, in order to discuss the validity of our model for any value of k, especially in the limiting cases.
Contributions from (25) are included in the relaxation matrix in the following way :

a) The diagonal elements of the transition matrix Π are given the form (3.26),

$$\Pi_{11} = -\Pi_{12} - P(k) - P_o \qquad (3.26)$$

P(k) is an additional probability for the ensemble i to be out of resonance, owing to spin transitions from the resonating levels and induced by $H^{**}(t)$. From (25), and for the spectral line $\phi_+^o \leftrightarrow \phi_o^o$, we derive (3.27),

$$P(k) = 0.5\{P(k)_-^+ + P(k)_o^+\} + 0.5\{P(k)_+^o + P(k)_-^o\} \qquad (3.27)$$

$$\text{with } P(k)_o^+ \simeq \{|\langle\phi_+^o|H^{**}(i)|\phi_o^o\rangle|^2\}_{Av.}\ \{2k/(\omega^2+4k^2)\}$$

ω denotes the resonance frequency of $\phi_+^o \leftrightarrow \phi_o^o$. Similar expressions are obtained for $P(k)_-^o$ and $P(k)_-^+$.

b) To the site frequencies are added effective non-secular contributions (3.28),

$$\omega^{**}(i,k) \simeq |\langle\phi_+^o|H^{**}(i)|\phi_o^o\rangle|^2\{\omega(\omega^2+4k^2)^{-1}\}$$
$$+|\langle\phi_+^o|H^{**}(i)|\phi_o^o\rangle|^2\{2\omega(4\omega^2+4k^2)^{-1}\} \qquad (3.28)$$

Following (26) and (28), the relaxation matrix is written :

$$A_{ns} = \begin{pmatrix} i\{\omega(1,k)-\omega\}-k-P(k)-P_o & k \\ k & i\{\omega(2,k)-\omega\}-k-P(k)-P_o \end{pmatrix} \qquad (3.29)$$

with $\omega(i,k) = \omega^o + \omega^*(i) + \omega^{**}(i,k)$, $i = 1,2$.

The matrix (29) leads to an expression similar to (22):

$$I_\omega(B) \alpha - \mathrm{Re}\{\overset{*}{M}.A_{ns}^{-1}.M\} \qquad (3.30)$$

This line-shape formula has been used to produce simulations in the single channel transfer model. It gives the exact intensities and positions of the lines.

The intensity distribution 30 reproduces the main features of the perturbation caused by a random transfer between sites with non-commuting spin hamiltonians.

If the following abbreviations are adopted[24) :

$$x_i = \omega(i,k)-\omega \; ; \; \mu^o = \langle\mu(i)\rangle_i \; ; \; \mu_+^2 = \langle|\mu(i)|^2\rangle_i$$

$$\mu_-^2 = 0.5\{|\mu(1)|^2-|\mu(2)|^2\} \; ; \; W_\pm = 0.5\{W(1)\pm W(2)\}$$

$$\lambda = P(k)+P_o \; ; \; \mu_i^R = \mathrm{Re}\{\mu(i)\} \; ; \; \mu_i^I = \mathrm{Im}\{\mu(i)\}$$

Equation (30) takes the explicit form (3.32) :

$$I_\omega(B) \; \alpha \; \frac{(k+\lambda)(x_1+x_2)\{(x_2W(1)|\mu(1)|^2+x_1W(2)|\mu(2)|^2)-2kW_-(\mu_1^R\mu_2^I-\mu_2^R\mu_1^I)\}}{(k+\lambda)^2(x_1+x_2)^2 + \{-x_1x_2+\lambda(2k+\lambda)\}^2} +$$

$$+ \frac{-\{x_1x_2+\lambda(2k+\lambda)\}\{\lambda(W(1)|\mu(1)|^2+W(2)|\mu(2)|^2)+2k(W_-\mu_-^2+W_+2|\mu^o|^2)\}}{(k+\lambda)^2(x_1+x_2)^2 + \{x_1x_2+\lambda(2k+\lambda)\}^2}$$

The limiting cases of (32) are : 1°) The two spin hamiltonians commute, then $\mu(1) = \mu(2)$. The expression (32) reduces to that derived by Sack and Abragam, with P_o accounting for the finite half-width at each limit. It is very important to note that in the investigated systems, the width at the two limits ($k \ll T_2^{-1}$ and $k \gg \omega$) are not determined by relaxation, but by the unresolved hyperfine interaction, and the spread[25)] in the values X,Y,Z, and P_o is given a value that reproduces these widths[26)]. 2°) The two spin hamiltonians do not commute, then $\mu(1) \neq \mu(2)$ and we consider the following limiting cases : slow transfer, intermediate transfer[27)] and rapid transfer.

a) Slow transfer : $k \ll |x_1-x_2|$ and $k \ll \omega$. Expression (32) reproduces two Lorentzian lines, with half-widths $k+\lambda$, centred at the static resonance fields : $B^r(\theta_1,\phi_1)$ and $B^r(\theta_2,\phi_2)$:

$$I_\omega(B) \alpha \; W(i)|\mu(i)|^2\{(k+\lambda)/(x_i^2+(k+\lambda)^2)\} \qquad (3.33)$$

$$i = 1,2.$$

b) Intermediate transfer : $k \gg |x_1-x_2|$ but $k \ll \omega$. Neglecting λ, as compared to k, the line-shape reduces to two Lorentzians (3.34),

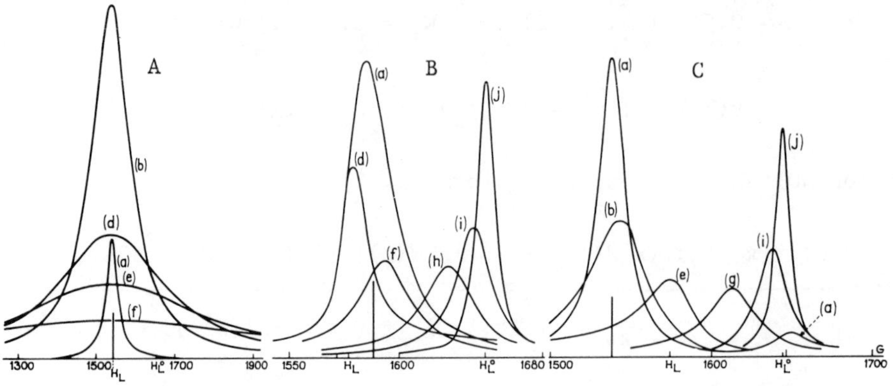

Fig. 6 - SCT Model calculations for oriented systems[9) $\Delta m=2$ lines for site canonical orientations. A) H is along the symmetry axis of TBT. B) H is along a site OX axis. C) H is along a site OY axis. Lines are calculated for increasing values of k (in 10^6 sec^{-1}): from 25(a) to 50,000 (j). The vertical lines represent the static absorption intensities, reduced by a factor of 10 in B and C and of 100 in A in which a is also reduced by a factor of 10.

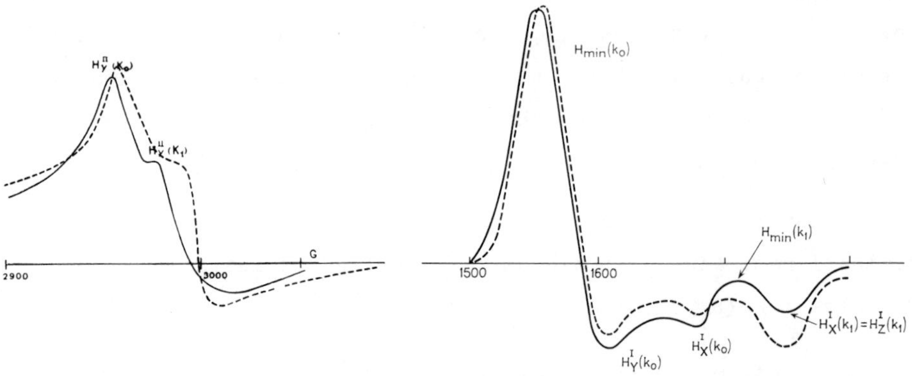

Fig. 7 - SCT Model Simulations for polyoriented systems. Low field line-shapes with U||H : A) The static case. B) k = 50. C) (a) 750, (b) 1650, (c) 2500, (d) 3000, (e) 50.000. The experimental spectrum is indicated in dotted line.

$$I_\omega(B)\ \alpha\ \frac{\{W_+2|\mu^o|^2+W_-\mu_-^2\}\{(-x_1x_2/2k)+\lambda\}}{\{-(x_1x_2/2k)+\lambda\}^2+\{(x_1+x_2)/2\}^2}+ \tag{3.34}$$

$$+\frac{\lambda}{k}\frac{\{W(1)|\mu(1)|^2+W(2)|\mu(2)|^2\}\{-(x_1x_2/2k)+\lambda\}}{\{-(x_1x_2/2k)+\lambda\}^2+\{(x_1+x_2)/2\}^2}$$

because of $k<<\omega$, the two lines are centred at the resonance field B_L which corresponds at the adiabatic coalescence (a.c.) frequency $\omega_{ac}= 0.5|\omega(1)+\omega(2)|$, with a half-width :

$$\lambda^{ns}(k) = \frac{x_1^2+x_2^2}{2}\ \frac{1}{2k}+\lambda \tag{3.35}$$

The index ns stands for non-secular treatment. The first right-hand term is the adiabatic broadening, the second term is the so-called non-adiabatic broadening due to the non-commutation of the spin hamiltonians (fluctuation of the spin functions, cf. Abragam p. 445).

c) Rapid transfer : $k>>|x_1-x_2|$ and $k \gtrsim \omega$. The center of gravity of the line shifts to (3.36),

$$<\omega(i,k)>_i = \omega^o+<\omega^{**}(i,k)>_i \tag{3.36}$$

$<\omega^{**}(i,k)>_i$ is the frequency shift, the imaginary part of (25). The center of gravity reaches the limiting value ω^o (one of the eigenfrequencies of H_s^o) for $k>>\omega$. The line-width has a maximum non-adiabatic contribution for $2k\sim\omega$ or for $k\sim\omega$.

The three cases above are illustrated in Fig. 6 .

E. Simulation of the Experimental Data

Using Eq.(30) for three sites, line-shapes for single orientations and simulations of spectra of polyoriented systems are presented in Figs. 6 and 7 . If we ignore the extra peaks noted R, simulations of the experimental data of Tribenzotriptycene glassy solutions, recorded

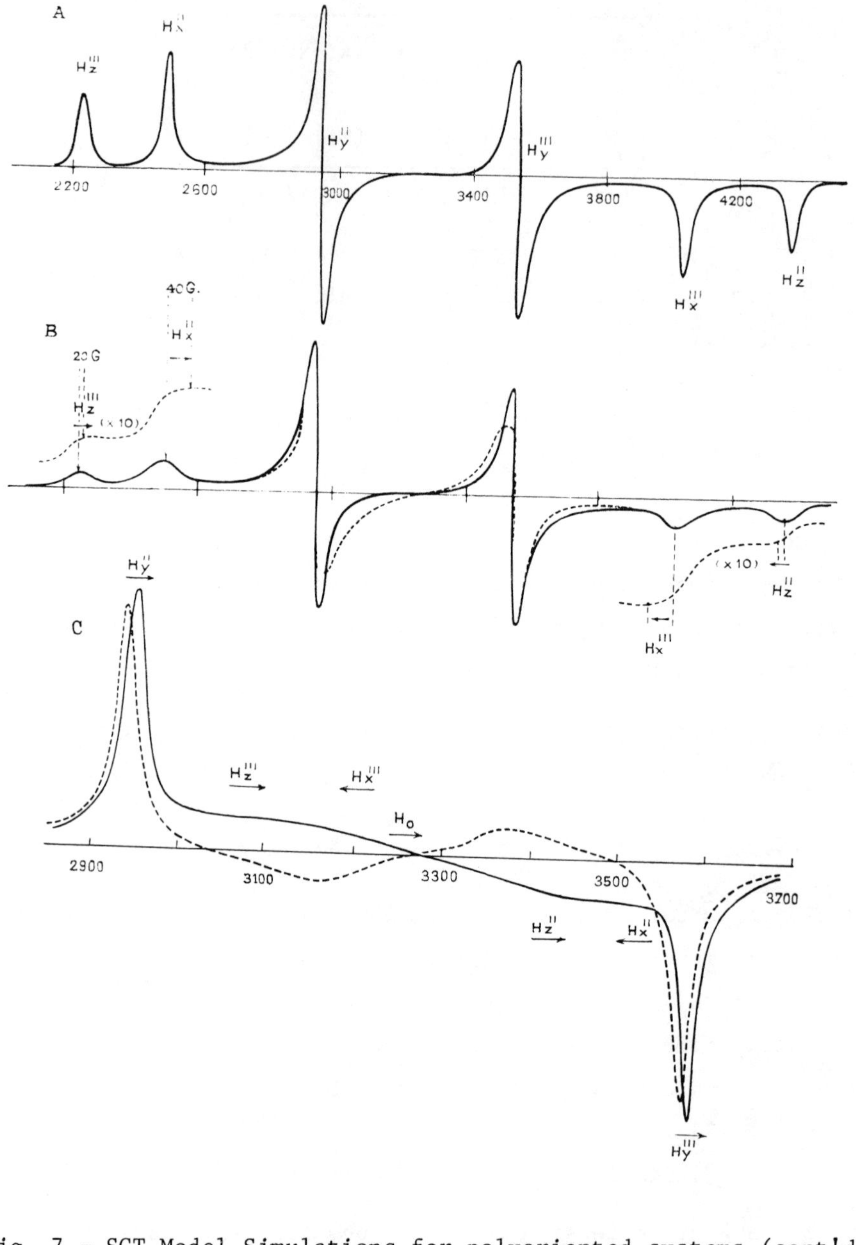

Fig. 7 - SCT Model Simulations for polyoriented systems (cont'd)

High field line-shapes with U1H. A) The static case. B) k = 30 (solid line), k = 150 (dotted line). C) k = 750 (dotted line), k = 2500 (solid line).

at $T_1 = 20°K$ and at $T_2 = 77°K$, are very satisfactory and confirm the existence of the exciton. These results show that the exciton transfer varies by two orders of magnitude from one temperature to the other :

$$5\times10^{7} < k(T_1) < 10^{8}\ \text{sec}^{-1} \quad ; \quad 10^{9} < k(T_2) < 5\times10^{9}\ \text{sec}^{-1} \qquad (3.37)$$

The agreement of the results derived from simulations with $\Delta m=2$ and with $\Delta m=1$ spectra is evidence that the extension of Anderson's theory to include the non-secular term is a realistic approximation (note that the non-secular part is controlling the $\Delta m=2$ absorption, while, for $\Delta m=1$ transitions, it contributes only to second order corrections).

The inability of the single channel transfer model to account for the presence of the extra peaks, is not a proof that Anderson's theory cannot simulate exciton line-shapes, but rather that of the need of a better understanding of the mechanism by which the temperature perturbs the ESR absorption in the solid phase. This problem is tackled in the last section (the Kaplan-Alexander density matrix method, extended to strongly non-commuting hamiltonians, gave no different results and no clue for the presence of the extra peaks in triplet exciton line-shapes[17]).

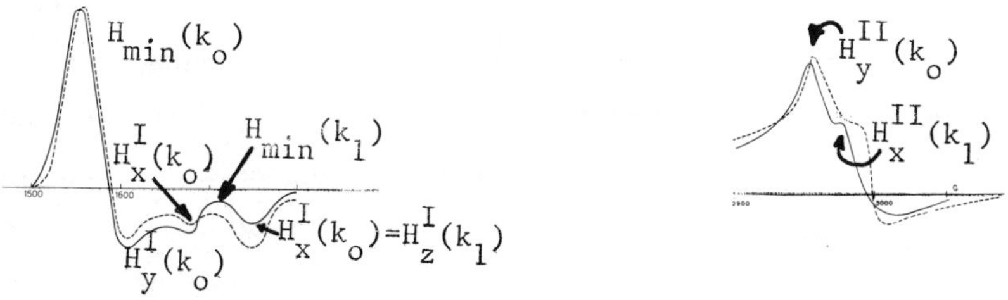

Fig. 8 - TCT Model simulations for polyoriented systems. Solid line: simulations with $k_o=5\times10^7\ \text{sec}^{-1}$, $k_1=2.5\ 10^9\ \text{sec}^{-1}$ and $W_o=W_1$. Dotted lines: experimental data of low and high field spectra at $T=20^oK$, showing extra peaks R. (in $\Delta m=2$ simulations, account was taken of the gaussian distribution of the hyperfine levels[26]).

XVII. 4. EXCITONS IN THERMAL EQUILIBRIUM

In section XVII.3, we presented simulations of temperature dependent triplet ESR spectra, using a single-channel transfer (SCT) model for the exciton : for a given temperature T, the rate k(T) was assumed to be the same in all the systems of the excited triplet population. The SCT model gave satisfactory results[8,16] for spectra recorded at the temperature $T_1 = 20°K$ and $T_2 = 77°K$.

However, although reproducing correctly the changes of the static spectrum of Fig. 1, this model of exciton failed to account for the extra peaks R,cf. Figs 3. These peaks are reproducible in all systems investigated and cannot be attributed to impurities. Furthermore, the peaks R present the following peculiar properties : a) They are present at spectra recorded at low temperature. b) They lie in the region of a rapid exciton. c) At intermediate temperatures, $20°K<T<77°K$, the intensity of the extra peaks increases without significant narrowing, while the intensity of the slow exciton diminishes without further broadening or shifting, cf. Fig. 2 of ref. 5. The property c seems now to be quite general in the lineshapes of aggregates. It shows that the extra peaks R are involved in the triplet exciton ESR absorption and that these peaks, which we identify as belonging to rapid excitons lines, do not originate from the coalescence of the slow exciton line, as it occurs in the usual pattern of motional narrowing. Therefore, it appears clearly that in molecular aggregates the temperature dependence of the transfer rate is not of the form $k(T) = k_o \exp(-\frac{\Delta E}{KT})$, with ΔE being an activation energy. As shown in section XVII.3, the latter form leads to typical pattern of variation of the line with increasing temperature : broadening → shifting → coalescence → narrowing. It will be shown in this section that this pattern is to be excluded for molecular aggregates, possessing low frequency modes, in which excitons arise from electronic adiabatic resonance interactions.

1. THE MULTI-CHANNEL TRANSFER MODEL

In order to elucidate the mechanism whereby the ESR is perturbed, we use, as in section XVII.3, a phenomenological approach : we consider an exciton transfer through two channels : i) in the first channel, the exciton transfers adiabatically through the vibrationless triplet state of the subsystems, with a probability k_o. ii) In the second channel, the exciton transfers adiabatically through a thermally activated triplet state of the subsystems, with a probability $k_1 > k_o$ because of the easier tunneling, or larger overlap, in the activated state[4a).]

The two channels are considered in thermal equilibrium and their statistical weights are temperature dependent : $W_{oi}(T)$, $W_{1i}(T)$ (the first subscript stands for channel, the second for site). In order to calculate the line-shape in a two-channel transfer (TCT) model, we use again Anderson's theory to calculate the auto-correlation function (3.5) for a dimer-aggregate. Now, however, in a two site case, we consider the time evolution in four ensembles with four transition probabilities : U_o, U_1, k_o, k_1 (instead of two ensembles and one transition probability k we used in a SCT model). The constant weights W_i used in Anderson's theory are now temperature dependent : they are the Boltzmann populations established through the thermal intra-subsystems transitions $U_{01}(T)$ and $U_{10}(T)$, cf. Fig. 9B. Therefore, the line-shape function in a TCT model is expected to be analogous to (3.30) except that : a) It depends on two temperature independent transfer rates k_o and k_1. b) It contains explicitly the temperature through the weights of the two channels. c) The row vector $\overset{\star}{\mathbf{M}}$, the matrix **A** and the column vector **M**, have doubled dimension : 4, 4x4 and 4, respectively. The TCT line-shape function is then written :

$$I_\omega (k_o, k_1, T, B) \alpha - \text{Re} \left[\overset{\star}{\mathbf{M}} . \mathbf{A}^{-1} . \mathbf{M} \right] \qquad (4.1)$$

If we assume that the thermal transitions U_{01} and U_{10} conserve the spin state[4,5)], then the spin evolution (3.15') is decoupled into two single-channel transfers with rates k_o and k_1. With this assumption, $\overset{\star}{\mathbf{M}}$, A and **M** take the simplified form :

$$\overset{\star}{\mathbf{M}} = \{W_{o1}(T)\overset{\star}{\mu}(1), \quad W_{o2}(T)\overset{\star}{\mu}(2), \quad W_{11}(T)\overset{\star}{\mu}(1), \quad W_{12}(T)\overset{\star}{\mu}(2)\} \tag{4.2}$$

$$\mathbf{A} = \begin{pmatrix} A\ (k_o) & & o & o \\ & & o & o \\ o & o & & \\ o & o & A\ (k_1) & \end{pmatrix} \qquad \mathbf{M} = \begin{pmatrix} \mu(1) \\ \mu(2) \\ \mu(1) \\ \mu(2) \end{pmatrix} \tag{4.3}$$

A (k_i), with i = 0, 1, is the 2x2 matrix used in the SCT model. The weights are a normalized Boltzmann population and obey relations :

$$\frac{W_{o1}}{W_{11}} = \frac{U_1}{U_o} = \exp\left(+ \frac{\Delta E_w}{KT}\right) \quad ; \quad W_{o1} + W_{11} = 1/6 \tag{4.4}$$

ΔE_w is the vibronic energy difference between the slow and fast excitons (of the order of 20 cm^{-1}, much larger than the ESR energies $\omega \sim 0.3\ cm^{-1}$). Using the decoupled channel approximation (2,3), the line-shape function, in a multi-channel transfer model, is written :

$$I_\omega\ (k_o,\ k_1, \ldots\ k_i,\ T,\ B) = \sum_i W_i\ (T)\ I_\omega\ (k_i,\ B) \tag{4.5}$$

In (5), $I_\omega(k_i,B)$ is the line-shape of the ith exciton, calculated in (3.30), and W_i (T) is the population of the channel i through which the exciton diffuses. Let us discuss some implications of the MCT model (5). At low temperatures ($KT < \Delta E_w$), the lines corresponding to the weakly populated fast excitons k_i will appear if their peaks are not blurred by the strongly populated slow exciton line. In Fig. 9C, we present the superposition of a slow and a rapid exciton line : the total absorption of the weakly populated rapid exciton k_1 is small, but this absorption occurs in a small region (narrowing) of

B and its intensity may be large. Let us compare an ideal case of two coalesced exciton lines with predominant adiabatic linewidths, the ratio p of their maxima is :

$$p \simeq \frac{k_1}{k_o} \exp\left(-\frac{\Delta E_w}{KT}\right)$$

So that the rapid exciton line will appear unless the Boltzmann factor is much smaller than $\frac{k_o}{k_1}$.

According to MCT Model (5) when the temperature varies, the lines corresponding to excitons k_o and k_i loose or gain intensity but neither of them broadens, narrows or shifts. This is in agreement with the observed behaviour of the lines at the intermediate temperatures : the peaks R in the spectrum of TBT, identified as $H_{min}(k_1)$ and $H_Y^{II}(k_1)$, take intensity away from the corresponding peaks of the slow exciton $H_{min}(k_o)$ and $H_Y^{II}(k_o)$. The Multi-Channel Model (5) explains the two main features observed in the ESR absorption of aromatic aggregates : a) The existence of the extra peaks as being the stationary peaks of the fast exciton line-shape, and b) The unusual behaviour of the line-shape as being a consequence of a thermal equilibrium between different excitons.

Simulations of the experimental data are presented in Fig. 8, see also ref. 5. Their very satisfactory agreement shows that the temperature perturbation implied in Eq. (5) is much more realistic than that assumed in a SCT model. Theoretical evidence for a MCT mechanism will be given in the next sub-section.
The reason of the success of the SCT model in simulating the data at the temperatures T_1 and T_2, is obvious when considering Eq. (5). For instance, in the $\Delta m = 2$ line-shape at $T_1 = 20°K$, the slow exciton is the dominant term and portrays the line-shape (the rapid exciton peak R appears because at this region the slow exciton absorption is fortunately almost zero). At 77°K the dominant terms originate from rapid excitons, the slow exciton appears only as a

noise broadening cf. Figs 3B and 7E. Therefore, the average values (3.37) hide two discrete transfer rates of the exciton : one in the slow channel, with $k_o \simeq 5\times10^7\ sec^{-1}$ and another one in the rapid channel with $k_1 \simeq 2.5\times10^9\ sec^{-1}$, the temperature dependence being now included in the Boltzmann population of the channels.

2. THE VIBRONIC COUPLING APPROACH OF THE EXCITON

The excited vibronic triplet states of molecular aggregates form narrow bands ($\Delta E \sim 10\ cm^{-1}$), owing to the small overlap of the wavefunctions of the subsystems [28,31]. These are exciton states and arise from adiabatic vibronic resonance couplings of the localized states (only localized states of the same quantum number are coupled). In the so-called weak coupling case [29], the triplet exciton states n of the aggregate have the form (4.6),

$$\Psi_w^{(n)} = \{\sum_i C_{w,i}^{(n)}(Q,R) \mid \phi_{T_m}^i \prod_{i\neq j} \phi_{S_o}^j >\} \exp\left(-i \frac{E_w^{(n)} t}{\hbar}\right) \qquad (4.6)$$

where $|\phi_{T_m}^i \prod_{i\neq j} \phi_{S_o}^j>$ is the ith electronic state which localizes the triplet exciton on the site i with a spin component m, w is a vibronic quantum number to be specified later. Q and R are nuclear coordinates which describe the intra-site and the inter-site nuclear motion respectively, n = 1 denotes the phosphorescent state manifold.

The states (6) are often referred to as coherent excitons (i.e. the states **k** in a molecular crystal at low temperature). The latter have the property of stationary delocalized states and do not imply any physically observed localization (vibronic relaxation). Because of the weak coupling, which gives rise to narrow bands, when the system is coupled to a dissipative medium, its local modes induce strong thermal transitions among states (6). Therefore, the aggregate ceases to be a conservative system and its description by stationary states is no longer valid : the hamiltonian of the system has to be written explicitly time dependent, then one of the possible description of the excited states is the hopping exciton.

In this subsection, the ESR line-shape will be derived from such a hamiltonian and the Multi-Channel Transfer Model will appear as a direct consequence of the adiabatic vibronic coupling.

A. The Coherent Exciton States

Let us consider a dimer-aggregate, for instance the dihydro-pentacene molecule, obtained from a TBT molecule in which we replace one naphthalene by two hydrogen atoms on the saturated carbons. Our system A-B consists now of two electronically equivalent subsystems, naphthalenes A and B, differently oriented. In the absence of dissipative medium, the vibronic hamiltonian for the π states of the system is written ;

$$H_v = H^A + H^B + T_N^A + T_N^B + H_m^A + H_m^B + H^{AB} \qquad (4.7)$$

- $H^{A(B)}$ is the "electronic hamiltonian"[30] of the subsystem A(B). It depends on the electronic coordinates $q_{A(B)}$, on the intra-molecular vibrational coordinates $Q_{A(B)}$ and on the coordinates of the center of gravity (librations) $R_{A(B)}$ of the subsystems. The motion defined by the degenerate librations, R_A and R_B, play a crucial role in the mechanism of the triplet exciton localization (by exciton-libration coupling) and of its random motion, as was predicted by Sternlicht and McConnel[31].

- $T_N^{A(B)} = T_N(Q_{A(B)}) + T_N(R_{A(B)})$, is the kinetic energy operator for the degrees of freedom $Q_{A(B)}$ and $R_{A(B)}$.

- $H_m^{A(B)} = H^z_{A(B)} + H^{ss}_{A(B)}$, are the Zeeman and dipolar spin interaction operators in the subsystem A(B). The spin interaction between A and B is neglected.

- H^{AB} $(q_A, q_B, Q_A, Q_B, R_A, R_B)$ is the coulombic, time independent, interaction between A and B. In the adiabatic approximation which is valid for the first librational levels of the triplet exciton [5], the off-diagonal elements of H^{AB} depend only on q_A, q_B and on the distance L_{AB} of the centers of gravity of A and B.

The molecular functions $\phi^{i}_{T_m}$ and $\phi^{j}_{S_o}$, used to build the localized excitons in (6), are Born-Oppenheimer (B.O.) states and depend parametrically on $Q_{A(B)}$ and on $R_{A(B)}$. For instance, for the subsystem A, the B.O. states will be noted A_{S_o} for the ground singlet state, and A_{T_m} for the excited triplet state. They are solutions of the electronic hamiltonian H^A,

$$H^A|A_{S_o}(q_A, Q_A, R_A)\rangle = E^A_{S_o}(Q_A, R^o_A)|A_{S_o}(q_A, Q_A, R_A)\rangle$$

$$H^A|A_{T_m}(q_A, Q_A, R_A)\rangle = E^A_{T_m}(Q_A, R^o_A)|A_{S_o}(q_A, Q_A, R_A)\rangle \tag{4.8}$$

Q^o and R^o note the equilibrium nuclear configuration in the ground state. For the system A-B, the triplet electronic localized states will be noted as fallows :

$$|A_{T_m} B_{S_o}\rangle \quad \text{and} \quad |A_{S_o} B_{T_m}\rangle \tag{4.9}$$

where m and n indicate the triplet spin components in each site. Then, the exciton vibronic triplet states,for the dimer-aggregate, are written, dropping the phase factor $\exp(-\frac{iE_w t}{\hbar})$:

$$\Psi_w = \mathcal{A}\sum_{m,n}\{C^m_A(Q,R)|A_{T_m} B_{S_o}\rangle + C^n_B(Q,R)|A_{S_o} B_{T_n}\rangle\} \tag{4.10}$$

where $\mathcal{A}$ is the antisymmetrizer which generates, from one combination of localized states, the proper antisymmetrized function with regard to the full permutation group.

The vibronic coefficient of (10) must satisfy the secular equation

$$(H_V - W)\Psi = 0 \tag{4.11}$$

Introducing symmetric (q^+, r^+) and antisymmetric (q^-, r^-) reduced nuclear coordinates, the vibronic coefficients, C(Q,R), have been obtained according the Witkowski and Moffitt and Fulton and Gouterman method, in the high field approximation. The decoupling of the linear term of the exciton-nuclear motion has been obtained using the contact transformation technique[5].

Omitting $\mathcal{H}$, the exciton states of the first two bands (the band index is the quantum number w of the libration states, cf. Fig. 7A), have the explicit form[37] :

$$\Psi^{\pm}_{o,w,w',i} = \{\chi_o(q^+-q^o)\chi_o(q^--q^o)\chi_w(r^--r^o)\chi_{w'}(r^+-r^o)|A_{T_i} B_{S_o}> \pm\chi_o(q^+-q^o)\chi_o(q^-+q^o)\chi_w(r^-+r^o)\chi_{w'}(r^+-r^o)|A_{S_o} B_{T_i}>\} \quad (4.12)$$

where {i = +1,0,-1} denote the spin states. The χ's are displaced harmonic ascillators, associated with the high frequency modes $q^\pm$ and with the low frequency modes $r^\pm$ of the distorted configurations. We are interested only in the low vibronic states (susceptible to be thermally populated at temperatures below 77°K). Therefore, we consider states with zero quantum number for $q^\pm$ and small quantum numbers for $r^\pm$. The q^o and r^o are displacements, in the space $\{q^+, q^-\}$ $\{r^+, r^-\}$, of the oscillators in the excited sites. The antisymmetric coordinates q^- and r^- define the "active modes" (for this definition see refs 44).

In the weak coupling case, the potential energy curve along the active mode r^- presents a double minimum cf. Fig. 7A. This double minimum is often given a dynamic interpretation : as two energy wells where the exciton is trapped (it relaxes) owing to its coupling with the nuclear motion. In the absence of a dissipative medium, which could randomize the degrees of freedom $r^\pm$, the dynamic interpretation of weak coupling states in terms of localization and excitation transfer (from well to well) is erroneous.

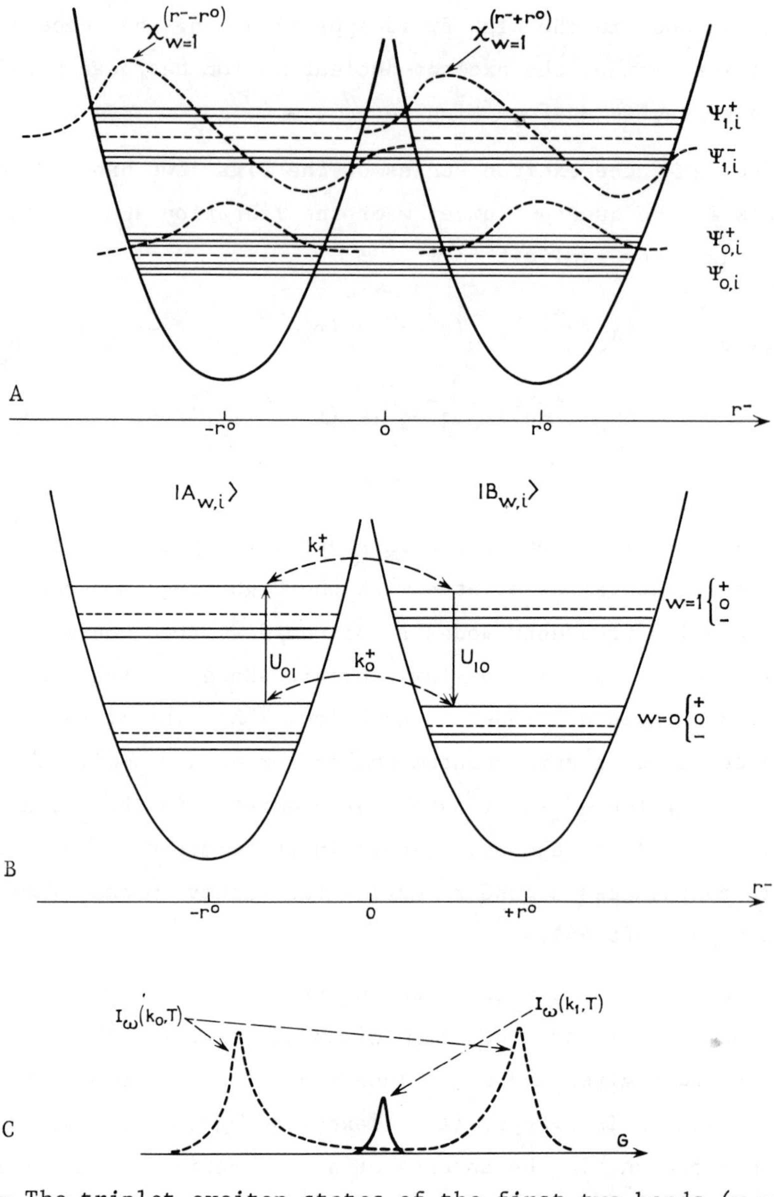

Fig. 9 - The triplet exciton states of the first two bands (w = 0, w = 1) and the potential energy curve along the libration active mode.

A) The two displaced oscillators are in phase and couple to form collective excitation states $\Psi^{\pm}$. B) The two displaced oscillators are out of phase Λ times par sec., the exciton is described by localized states $|A_{w,i}\rangle$ diffusing with rates k_w^i. C) The ESR line in a TCT Model as a superposition of two exciton lines.

Indeed, expression 12 shows that the coherent excitons are delocalized : they are symmetric and antisymmetric combinations of two degenerate (displaced) configurations and they imply neither localization nor excitation transfer followed by a displacement of nuclei. The ESR line of a coherent exciton will be observed almost at the center of gravity of the lines of the subsystems[32], but the position of this line and its width cannot be interpreted in terms of fast motion coalescence and narrowing.

We recall that in using Anderson's theory, in the Single or Multi-Channel Transfer Model, we assumed two fundamental characteristics for the exciton : a) the triplet exciton is localized in subsystem states (or in loges A or B, see chap. VI of ref. 33), which are not eigenstates of the system. b) the consequent motion of the exciton is not periodic but incoherent : i.e. there is no collective excitation of the two subsystems as in (12), or back-and-forth oscillations with a well defined period[34].

The experimental data for the systems investigated are consistent with assumptions a) and b), since :

a') the ESR reflects the symmetry of the subsystem, which means localization. b') the broadening and variation of the lines with temperature, are evidence of incoherent motion of the triplet spin. This implies a random time dependence of the microscopie hamiltonian[34]. In order to introduce this dependence, we consider the local couplings between the bath and the subsystem A(B), when the temperature increases. The local couplings are thermally modulated and introduce a finite electronic correlation time Λ^{-1} between the sites phase in 12. This finite correlation time may decouple the coherent exciton in hopping exciton, Fig. 7. Ignoring now any a priori assumptions on localization and transfer, the exciton states are investigated from the vibronic hamiltonian. This will be done by considering the two parameters : the coherent coupling V^{AB} between the two subsystems which establishes their collective excitation[33] and the electronic correlation time Λ^{-1} introduced by the coupling of the system to the bath.

B. The Incoherent Exciton States

We write down the spinless part of the π-vibronic hamiltonian (7), in order to calculate the non-stationary triplet vibronic states of the system A-B coupled to a dissipative medium (i.e. the thermal bath S that constitute the local modes in the amorphous solid matrix). Then, a spin hamiltonian (4.13) is derived for each band w, by first order perturbation theory :

$$H_s^w(t) = \langle\Psi_w(t)|H_A^z + H_A^{ss} + H_B^z + H_B^{ss}|\Psi_w(t)\rangle \tag{4.13}$$

In (13) the subscript A(B) means that the spin interactions depend also on space variables over which the integration is carried out. For the system A-B coupled to a bath S, the spinless vibronic hamiltonian is written as follows :

$$H_v(t) = H_v^A + H_v^B + H^{AB} + H^{AS}(t) + H^{BS}(t) \tag{4.14}$$

where $H_v^{A(B)}$ is a π-hamiltonian for the subsystem A(B). $H^{AS}(t)$ is an adiabatic local vibronic interaction (it does not change the quantum numbers of H_v^A) between A and the bath S. $H^{AS}(t)$ determines a shift and a random modulation of the localized triplet state, with an uncertainty broadening width Λ (in Sewell's theory Λ^{-1} is the coherence time of $H^{AS}(t)$ or a "period" of the lattice motion).
The hamiltonian (14) is again split into an unperturbed and a perturbing part $H_v^p(t)$. The latter induces transitions, if any, among the states of H_v^o.

$$H_v(t) = H_v^o + H_v^p(t) \tag{4.15}$$

where $$H_v^o = \langle H_v(t)\rangle_T = H_v^A + \overline{H}^{AS} + H_v^B + \overline{H}^{BS} + \overline{H}^{AB} \tag{4.16}$$

and $$H_v^p(t) = H_v(t) - H_v^o \tag{4.17}$$

with $$\langle H_v^p(t)\rangle_T = 0$$

The bar implies an average over one "coherent period" T of the exciton ; with $T = (V_w^{AB})^{-1}$, V_w^{AB} being the Davydov splitting in the band w.

The matrix element of $\overline{H}^{AB}$, which appears in(16),plays a crucial role in the determination of the nature of the states of H_v^o (i.e. collective or localized excitation states). Indeed, H^{AB} does not commute with $H^{AS}(t) + H^{BS}(t)$,it will be, therefore, affected by their random changes. Then, the states of (15) will be quite different according to the averaging out or not of $\overline{H}^{AB}$, hence, according to the relative value of the couplings Λ and V^{AB} of a subsystem with the bath and with the other subsystem, respectively. The resulting motion for H^{AB} is obtained in the Heisenberg representation :

$$i\hbar \dot{H}^{AB} = [H_v^A + H_v^B + H^{AS}(t) + H^{BS}(t), H^{AB}] \qquad (4.18)$$

Let us denote the distorted localized excitons, i.e. the components of (12) without the spin states, symbolically as follows :

$$|A_w> \equiv \chi_o(q^+ - q^o)\chi_o(q^- - q^o)\chi_w'(r^+ - r^o)\chi_w(r^- - r^o)|A_T B_{S_o}> \qquad (4.19)$$

The left hand symbol means that the triplet exciton is on site A and its active mode r^- is on the wth state (the mode r^+ gives no contribution in the coupling V^{AB} and its quantum number is ignored). Then, the matrix element of H^{AB} between the states (19), has the form :

$$V_w^{AB}(t) = V_w^{AB} \exp(-\frac{i\Delta Et}{\hbar}) \exp(-\frac{i}{\hbar}\int_o^t \delta E(t')dt) \qquad (4.20)$$

where $\Delta E = (E_T^A + \overline{E}_{AS}) - (E_T^B + \overline{E}_{BS})$

$\delta E = E_{AS}(t) - E_{BS}(t)$

The energy of the ground state is taken as the origin $(E_{S_o}^A = E_{S_o}^B = 0)$. $E_T^{A(B)}$ is the energy of the triplet isolated subsystem A(B). $\overline{E}_{AS}$ is the static contribution (shift) of $H^{AS}(t)$. $E_{AS}(t)$ is the random modulation of the localized exciton energy, with $<E_{AS}(t)>_T = 0$. Then,

$\Delta E + \delta E(t)$ is the instantaneous energy difference between the localized excitons states.

In the absence of a thermal bath, these states are degenerate, $E_T^A = E_T^B$, and V^{AB} is time independent. Then the exciton states of the system are the symmetric and antisymmetric combinations $\Psi^{\pm}$ in 12.

In the presence of a solvent, and at very low temperature at which the random modulation can be neglected $\delta E(t) \sim o$, we distinguish three situations :

1°) $\overline{E}_{AS} = \overline{E}_{BS}$: The subsystems have homogeneous environments, then the exciton states are $\Psi^{\pm}$ as in (12), except for a shift $\overline{E}_{AS}$ in the levels of the system.

2°) $\overline{E}_{AS} \neq \overline{E}_{BS}$, with $|\overline{E}_{AS} - \overline{E}_{BS}| = \Delta E < V^{AB}$: The subsystems have slightly inhomogeneous environments. The exciton states are delocalized and, using notation (19), have the form :

$$\Psi_W^1 = a(\Delta E)|A_W> + b(\Delta E)|B_W>$$
$$\Psi_W^2 = b(\Delta E)|A_W> - a(\Delta E)|B_W> \qquad (4.21)$$
$$\text{with} \qquad |a|^2 \neq |b|^2$$

3°) $\overline{E}_{AS} \neq \overline{E}_{BS}$, with $\Delta E \gg V_W^{AB}$: The subsystems have strongly inhomogeneous environments. The exciton states of the system are localized (quasi-stationary) states equivalent to those for localized electrons in disordered materials, predicted in the Anderson and Mott theories, or to physically trapped electrons in molecular amorphous solids[36]. This mechanism of decoupling a collective excitation must be important in disordered systems at low temperature. In this case the exciton transfer must be temperature activated and present a large distribution by its dependence on ΔE (cf. Eq.29), which we excluded in a pure resonance case, cf. XVII.4.1.

In the presence of solvent, and at high temperature, the degeneracy of the localized states is lifted at each instant. We will discuss the exciton states in the case of subsystems with homogeneous environments. ($\Delta E = 0$). $\delta E(t)$ is a random function with a correlation time Λ^{-1}, consequently so is the resonance coupling $V_w^{AB}(t)$ through equation (20). To simplify, let us consider that Λ does not depend on the band index w and investigate the exciton states of the system in the two limiting cases[38] which are : a) The weak coupling case ($V^{AB}\Lambda^{-1} >> 1$, or the coherence time of the bath is long compared to the coherent exciton period T) and b) The strong coupling case ($V_w^{AB}\Lambda^{-1} << 1$, or the coherence time of the bath is very short compared to T).

1°) The weak coupling : the average of expression (20), over one period T is not affected :

$$\langle V_w^{AB}(t)\rangle_T \simeq V_w^{AB} \tag{4.22}$$

Then the hamiltonians (16-17) reduce, for the description of the excited system in band w spanned by the localized excitons (19), to the operators :

$$H_v^o = H^A + H^B + H^{AB} \tag{4.23}$$

$$H_v^p(t) = H^{AS}(t) + H^{BS}(t)$$

The states of H_v^o are coherent exciton states (12). $H_v^p(t)$ does not induce transitions among these states which are delocalized states (with a width $\Lambda << V_w^{AB}$). Consequently, the hopping exciton picture is inconsistent with conditions (22-23).

2°) The strong coupling : the average of $V_w^{AB}(t)$ over one period T vanishes :

$$\langle V_w^{AB}(t)\rangle_T = 0 \tag{4.24}$$

Then, the hamiltonians (16-17) reduce, for the excitation of the band w, to two operators

$$H_v^o = H_v^A + H_v^B \tag{4.25}$$

$$H_v^p(t) = H^{AB}(t) \tag{4.26}$$

The states of (25) are localized exciton states, they are coupled by (26) which causes their random transfer with a rate k_w.
Thus, the description of the band w by the incoherent (hopping) exciton picture :

$$|A_w> \overset{k_w}{\longleftrightarrow} |B_w> \tag{4.27}$$

is permitted only when conditions (24-25) are satisfied, i.e. when the system is coupled with a dissipative medium which randomizes the resonance interaction. The same criteria of weak and strong coupling are used in Holstein's theory for the description of small polaron states[35]. Note that (22,24) is a state property and not a system property. Therefore some bands may be described by hopping excitons (24), some others by coherent excitons (22).
Thus, the origin of the two characteristics, localization and random motion we assumed in section XVII.3 for the exciton are shown to be properties of the hamiltonian (14). The third characteristic, which allowed us the use a Markov process formalism, was that the duration of a hop, i.e. Λ^{-1}, is much smaller than the residence time k_w^{-1} of the exciton. k_w will be calculated in the next subsection. It will be shown that strong coupling implies $k_w \ll \Lambda$.

C. The Factors controlling the Exciton Diffusion

From expression (27) we see that the spin hamiltonian time dependence (13), derived from the total hamiltonian, is that assumed in our model in section XVII.3, expression (7).
The transfer rate k_w^i, between two triplet excitons in the band w and with a spin component i, is derived in a straightforward manner. It is the mean transition probability[39] between two quasi-degenerate states, $|A_{w,i}>$ and $|B_{w,i}>$, induced by a random coupling $H^{AB}(t)$:

$$k_w^i = \mathrm{Re}\int_0^\infty \overline{<A_{w,i}|H^{AB}(o)|B_{w,i}><B_{w,i}|H^{AB}(t)|A_{w,i}>} \exp(-\tfrac{i}{\hbar}\,\Delta E_i^{AB} t) \tag{4.28}$$

using for $H^{AB}(t)$ an exponential correlation funtion, $\exp(-\frac{t}{\tau_c})$ with $\tau_c = \hbar\Lambda^{-1}$ (the width Λ is now divided by 2π), we obtain :

$$k_w^i = \frac{2\pi}{h} \frac{|V_{w,i}^{AB}|^2 \Lambda(T)}{(\Delta E_i^{AB})^2 + \Lambda^2(T)} \tag{4.29}$$

k_w^i is the transfer rate through the channel w and the spin state i ($\phi_i^A \leftrightarrow \phi_i^B$, cf. Fig. 9B). ΔE_i^{AB} is the lack of resonance introduced by different magnetic energies and environmental interactions in the two sites. By definition of the strong coupling, we see from expression (29) that $k_w^i \Lambda^{-1} \ll 1$, which is the condition for using the Markov process formalism.

$V_{w,i}^{AB}$ is the electron exchange integral K_{AB}, diminished by the overlap integrals accounting for the relaxation possibility of the different degrees of freedom of the system[3] :

$$V_{w,i}^{AB} = K_{AB} \cdot S_o^2 \cdot S_w \cdot F_{ii} \tag{4.30}$$

S_o^2 is an overlap integral, $\langle\chi_o(\bar{q}-q^o)|\chi_o(\bar{q}+q^o)\rangle$ cf. (12), associated with the intramolecular degrees of freedom. At temperature between 4° and 77°K where ESR experiments are performed, only the zero-point state of these degrees of freedom is populated. Therefore, S_o^2 remains constant at these temperatures. S_w corresponds to low frequency oscillator overlap, $S_w = \langle\chi_w(\bar{r}-r^o)|\chi_w(\bar{r}+r^o)\rangle$. As Marechal and Witkowski[40] pointed it out, S_w and, consequently, the transfer rate k_w^i, increases dramatically with w. On the other hand, because of their low frequency, the states χ_w are thermally populated. This explains the temperature effect on the exciton diffusion : i.e. the existence of many diffusion channels with different rates k_w^i and the justification for a multi-channel transfer model.

F_{ii} is an overlap integral of the two spin functions i, $F_{ii} = \langle\phi_i^A|\phi_i^B\rangle$. This quantity may be strongly orientation dependent because of the different spin polarization in the two sites. The anisotropy of F_{ii}

is negligible in the high field approximation, in which the spin is quantized along B and $<\phi_i^A|\phi_i^B> \simeq 1$, or in the rapid transfer in which the spin has no time to relax in the two sites. On the contrary, in zero-field experiments, F_{ii} might vary from zero to one, as a function of the mutual orientation of the subsystems (see discussion in ref. 3).

In conclusion, equations (18,20,24,25) show that the existence of incoherent excitons in molecular aggregates in the solid phase, implies the averaging out of the resonance vibronic coupling and, therefore, the coupling of the system to a dissipative medium. The existence of low frequency nuclear motion and that of a very weak resonance coupling between triplet subsystems[28], explain the success of a multi-channel mechanism for the description of the exciton diffusion.

Expression (29) shows a temperature dependence of the transfer rate through the coupling $\Lambda(T)$. Simulations including this variation will be published soon.

APPENDIX

Expression of the site dipolar interactions in the laboratory coordinate system, with the third Euler angle $\Psi = 0$.

$$H^{ss}(i) = -\Big[T_{xx}^{(i)}S_x^2 + T_{yy}^{(i)}S_y^2 + T_{zz}^{(i)}S_z^2 + T_{xy}^{(i)}(S_xS_y + S_yS_x) + T_{xz}^{(i)}(S_xS_z + S_zS_x) + T_{yz}^{(i)}(S_yS_z + S_zS_y)\Big]$$

$T_{xx}^{(i)} = (Z\cos^2\phi_i + X\sin^2\phi_i)\cos^2\theta + Y\sin^2\theta$, $T_{yy}^{(i)} = Z\sin^2\phi_i + X\cos^2\phi_i$,

$T_{zz}^{(i)} = \sin^2\theta\,(Z\cos^2\phi_i + X\sin^2\phi_i) + Y\cos^2\theta$, $T_{xy}^{(i)} = \cos\phi_i\sin\phi_i\cos\theta(X-Z)$,

$T_{xz}^{(i)} = \cos\theta\sin\theta(Z\cos^2\phi_i + X\sin^2\phi_i - Y)$, $T_{yz}^{(i)} = \cos\phi_i\sin\phi_i\sin\theta(X - Z)$

θ is the angle between H and the common axis OZ', ϕ is the angle between the molecular axis OX'_i and the projection of H in the plane $X'_iOY'_i$, Fig. 4. For TBT we assume $\phi_{2,3}=\phi_1 \pm 120°$; $Z = -0.067\ cm^{-1}$; $X = 0.047\ cm^{-1}$; $Y = 0.020\ cm^{-1}$.

REFERENCES

1) J.H. van der Waals and M.S. de Groot, Mol. Phys. (1959) 2, 333 ; (1959) 3, 190.

2) C.A. Hutchison and B.W. Mangum, J. Chem. Phys. (1961), 34, 908.

3) Ph. Kottis, Chem. Phys. Letters, (1970) 6, 133.

4) Ph. Kottis, J. Chim. Phys. (1970) Spécial Edition on Radiationless Transitions, p. 119.
J.P. Le Falher, J.P. Lemaistre and Ph. Kottis, Chem. Phys. Letters (1970) 4, 491.

5) J.P. Lemaistre and Ph. Kottis, Ann. Phys. (to be published).

6) A. Abragam and M.H.L. Pryce, Proc. Roy. Soc. (London), (1951), A250, 135.

7) Ph. Kottis, Thèse de Spécialité, Sorbonne June 1962.

W.A. Yager, E. Wasserman et R.M.R. Cramer, J. Chem. Phys. (1962) 37, 1148.

8) Ph. Kottis and R. Lefebvre, J. Chem. Phys. (1963) 39, 393 ; (1964) 41, 379.

E. Wasserman, L.C. Snyder et W.A. Yager, J. Chem. Phys. (1964) 41, 1763.

For the derivation of hyperfine structure parameters in magnetoselection, see the Fluorine Labelling Method, J. Mispelter, J.Ph. Grivet and J.M. Lhoste, Mol. Phys. (1971).

9) Ph. Kottis, Ann. Phys. (1969) 4, 459. In this work, a resonance field density function $\gamma_\omega(B)$ is derived for two tensors, g and T, whose principal axes are not the same.

10) Ph. Kottis and R. Lefebvre, J. Chem. Phys. (1964) 41, 3660.

11) J.M. Lhoste, A. Haug and M. Ptak, J. Chem. Phys. (1966) 44, 654.
M.A. El-Sayed and S. Siegel, J. Chem. Phys. (1966) 44, 1416 ; See also ref. 9 p. 489.

12) M.S. de Groot and J.H. van der Waals, Mol. Phys. (1963) 6, 545 ; ibid. J. Chim. Phys. (1964) 61, 1643.

13) R.D. Cowell, G. Urry and S.I. Weissman, J. Chem. Phys. (1963) 38, 2028.
A.L. Shain, J.P. Ackerman and M.W. Teague, Chem. Phys. Letters (1969) 3, 550.

14) R.F. Cozzens and R.B. Fox, J. Chem. Phys. (1969) 50, 1952.

15) M. Schwoerer and H.C. Wolf, XIVth Colloque Ampere Ljubljana (1966) ; ibid. Mol. Cryst. (1967) 3, 177 ; ibid. The Triplet State, Beirut (1967), U. Press Cambridge.

16) Ph. Kottis, Phys Letters (1965) ; ibid. XIVth Colloque Ampere, Ljubljana (1966) ; ibid. J. Chem. Phys. (1967) 47, 509.

17) A. Hudson and A.D. Mclachlan, J. Chem Phys. (1965) 43, 1518.

18) P.W. Anderson, J. Phys. Soc. Japan, (1954) 9, 316.
P.W. Anderson and P.R. Weiss, Rev. Mod. Phys. (1953) 25, 269.

19) This is not true in the two-quantum transition field H_o where the two transitions $\Delta m = 1$ are in resonance (see M.S. de Groot and J.H. van der Waals, Physica (1963) 29, 1128). In H_o, our treatment of independent spectral lines breaks down, see $\Delta m = 1$ simulations.

20) This fluctuation is due to energy exchange of the subsystems with the thermal bath considered explicity in section 4.

21) A. Abragam, The Principles of Nuclear Magnetism, Oxford (1961), Chap. X.
R.A. Sack, Mol. Phys. (1958) 1, 163.

22) N. Bloembergen, E.M. Purcell and R.V. Pound, Phys. Rev. (1948) 73, 679.

23) P.W. Anderson, private communication.

24) By putting $W_1 \neq W_2$, we allow the sites to have different weights at equilibrium. Hence, the W's might depend on temperature.

25) E. Wasserman, L.C. Snyder and W.A. Yager, J. Chem. Phys. (1964) 41 , 1763.

26) Formula (3.32) is derived for sites with one hyperfine level. In order to take into account the gaussian distribution of the hyperfine levels, a gaussian convolution of width $\lambda_G = 10$ G, must be applied to (3.32). This convolution will give a gaussian line for the static case and a quasi-gaussian line in the slow motion case, cf. Figs 7A and 7B.

27) The intermediate case gives a flat line with very little information. Magnetoselection "piles up" these lines in such a way as to give an informative line-shape.

28) J. Jortner, S.A. Rice and J.L. Katz, J. Chem. Phys. (1965) 42, 309.

29) A. Witkowski and M. Moffitt, J. Chem. Phys. (1960) 33, 872.

30) H.C. Longuet-Higgins, Advan. Spectry (1961) 2, 429.

31) H. Sterlicht and H.M. McConnell, J. Chem. Phys. (1961) 35, 1793.

32) von D. Haarer and H.C. Wolf, Mol. Crystals, (1970) 10, 359.
H. Haken and P. Reineker in Excitons, Phonons and Magnons ed. A. Zahlan (Cambridge U.P., London 1968).

33) C. Aslangul, R. Constanciel, R. Daudel and Ph. Kottis, Adv. Quant. Chem. (1971) vol. 6.

34) The excited state may also be time-dependent, oscillating with a well defined period $T \sim (V^{AB})^{-1}$, when an interference between two closely spaced states is induced by coherent excitation. However in this case, the hamiltonian of the system is time independent. There cannot be random motion of the spin and consequent broadening of the ESR line.

35) T. Holstein, Ann. Phys. (N.Y.) 1959 8, 343.
G.L. Sewell, Phys. Rev. (1963) 129, 597.

36) P.W. Anderson, Phys. Rev. (1958) 109, 1492
N.F. Mott, Contemp. Phys. (1969) 10, 125.
J.P. Jardin and Ph. Kottis, C.R. Ac. Sc. Paris (1971) 273, 135.

37) At this stage of calculation, electrostatic site-cage local interactions have been considered. The latter decouple the relative motion mode of the sites and give the double potential energy well along the coordinate r^-. These two wells correspond to the same libration mode r^- in the two distorted configurations which localize the exciton on one site, cf. ref.5, also refs 44.

38) No confusion must be made between intrasystem weak coupling defined in 29, which does not change the delocalized character of the exciton, and the incoherent coupling Λ, which localizes and causes random motion of the exciton, see also ref. 41-43.

39) V. Ya. Gamurar, Yu. E. Perlin and B.S. Tsukorblat, Soviet Phys., Solid state (1969) 11, 970.

40) Y. Marechal and A. Witkowski, Theoret. Chim. Acta (1964) 2, 453.

41) S. Leach, II Intern. Symp. on Organic Solid State Chemistry (1970), Rehovot Israel.

42) A.H. Francis and C.B. Harris, Chem. Phys. Letters (1971) 9, 181.

43) P. Avakian, V. Ern, R.E. Merrifield and A. Suna, Phys. Rev. (1968) 165, 974.

44) A. Witkowski, Roczniki Chemii, 1961, 35, 1399.

R.L. Fulton and M. Gouterman, J. Chem. Phys., 1961, 35, 1059.

Th. Förster, In Modern Quantum Chemistry III, p. 105, Academie Press N.Y. 1965.

R. Lefebvre and M. Garcia Sucre, Int. J. Quant. Chem., 1967, IS, 339.

ESR SATURATION AND DOUBLE RESONANCE IN LIQUIDS

Jack H. Freed

Department of Chemistry, Cornell University

Ithaca, New York 14850

XVIII.1. INTRODUCTION TO SATURATION: A SIMPLE LINE[1,2]

The well-known result from the steady-state (s.s.) solution of the Bloch Equations is that the absorption is given by the y-component of magnetization $\breve{M}_y$ in the rotating frame:

$$\breve{M}_y = \frac{\gamma H_1 T_2}{1 + (T_2 \Delta\omega)^2 + \gamma^2 H_1^2 T_1 T_2} M_o \qquad (1)$$

with M_o the equilibrium magnetization. When we switch to a quantum mechanical description, we can calculate:

$$M_{\pm} = M_x \pm iM_y = (\breve{M}_x \pm i\breve{M}_y)e^{\pm i\omega t} \qquad (2)$$

statistically from its associated quantum mechanical operator

$$\mathcal{M}_{\pm} = \mathfrak{N}\hbar\gamma_e S_{\pm} \qquad (3)$$

where $\mathfrak{N}$ is the concentration of electron spins, by taking a trace of the spin density matrix $\sigma(t)$ with the spin operator $S_{\pm}$:

$$M_{\pm}(t) = \mathfrak{N}\hbar\gamma_e \mathrm{Tr}[\sigma(t)S_{\pm}] \qquad (4)$$

The trace is invariant to a choice of zero-order basis states. The equation of motion for $\sigma(t)$ is taken to be the relaxation matrix form given by Eq. VIII-20, and we shall neglect effects of higher order than $R^{(2)}$.

We have from Eq. VIII-20a, that when $\mathcal{H}_1(t) = 0$, so $R = 0$,

$$\sigma_{\alpha\alpha'}(t) = e^{-i\omega_{\alpha\alpha'}t}\sigma_{\alpha\alpha'}(0) \quad . \tag{5}$$

Thus if $\sigma_{\alpha\alpha'}(0) \neq 0$, then $\sigma_{\alpha\alpha'}(t)$ will be oscillatory. Now suppose we have only a simple line with $\omega_0 = \omega_{ab}$ where a and b are the $M_s = +\frac{1}{2}$ and $-\frac{1}{2}$ levels, and there are no other spin levels. Then

$$\langle b|S_-|a\rangle = \langle a|S_+|b\rangle = 1 \tag{6}$$

and

$$\mathrm{Tr}[\sigma(t)S_+] = \sigma(t)_{ba}S_{+ab} = \sigma(t)_{ba} \tag{7}$$

with

$$\sigma_{ba}(t) = \exp[(-i\omega_{ba} + R_{ba,ba})t]\sigma_{ba}(0). \tag{8}$$

Since ReR is negative, $\sigma_{ba} \to 0$ for $t \gg |\mathrm{ReR}|^{-1}$. Thus, there will be no steady state absorption unless we include effects of the rf field. So we add to the Hamiltonian:

$$\hbar\varepsilon(t) = \tfrac{1}{2}\hbar\gamma_e B_1 [S_+e^{-i\omega t} + S_-e^{+i\omega t}] \tag{9}$$

which is the interaction of the spin with a rotating field $\vec{B}_1 = B_1(\cos\omega t\hat{i} + \sin\omega t\hat{j})$. Then for our simple line the $\langle b|-|a\rangle$ matrix element of eq. VIII-20 is :

$$\dot{\sigma}_{ba} = (i\omega_o + R_{ba,ba})\sigma_{ba} - id(\sigma_{bb}-\sigma_{aa})e^{i\omega t} \tag{10}$$

where

$$d = \tfrac{1}{2}\gamma_e B_1 \tag{11}$$

Now the power absorbed from the rotating field is just:

$$P = \omega H_1\breve{M}_y = \frac{-\omega H_1 i}{2}[M_+e^{-i\omega t} - M_-e^{i\omega t}] \tag{12}$$

where from eq. 4 $M_\pm \propto \mathrm{Tr}_\sigma[\sigma(t)S_\pm]$ and S_{+ab} requires $\sigma(t)_{ba}$ in the trace. Thus only the component of $\sigma(t)_{ba}$ oscillating as $e^{i\omega t}$ will give a net time-averaged power absorption. So, let

$$\sigma_{ba} = Ze^{i\omega t} \tag{13}$$

and assume Z is time independent to achieve the steady state solution, which is:

$$(\Delta\omega + iR_{ba,ba})Z = d(\sigma_{bb} - \sigma_{aa}) \tag{14}$$

where $(\sigma_{bb} - \sigma_{aa})$ is the population difference in the two states. Now note that σ is Hermitian, so $\sigma_{ab} = \sigma^*_{ba}$ and

$$\sigma^*_{ab} = Z^* e^{-i\omega t} . \tag{15}$$

Thus

$$P \propto \text{Im} Z \equiv Z'' . \tag{16}$$

We may begin to suspect that Z plays the role of $\breve{M}_+$ (while Z^* is $\breve{M}_-$). Also

$$R_{ba,ba} = -(1/T_2)_{ba} = -(1/T_2)_{ab} .$$

We are now writing the equation of motion (eq. VIII-20) as

$$\dot{\sigma} = -i[\mathcal{H}_o + \epsilon(t), \sigma] + R\sigma . \tag{17}$$

We now need the diagonal spin-density-matrix elements σ_{bb} and σ_{aa}, which in steady state are not oscillating in time. We get from eq. 17:

$$R_{aa,aa}\sigma_{aa} + R_{aa,bb}\sigma_{bb} = di(Z - Z^*) = -2d\,\text{Im}Z \tag{18a}$$

$$R_{bb,aa}\sigma_{aa} + R_{bb,bb}\sigma_{bb} = 2d\,\text{Im}Z . \tag{18b}$$

Note that

$$R_{aa,bb} = R_{bb,aa} = 2J_{ab,ab} = W_{ab} = W_{ba} \tag{19a}$$

while

$$R_{aa,aa} = -\sum_{\gamma \neq a} W_{a\gamma} . \tag{19b}$$

where W_{ab} is the transition-probability from state b to state a, which leads to spin relaxation (cf. Eq. VIII.48).

For simplicity let $\gamma = b$ only, (i.e., our simple line). Then we have

$$W_{ab}(\chi_a - \chi_b) = 2dZ'' \tag{20}$$

where we have made the ad hoc replacements:

$$\sigma_{aa} \rightarrow \chi_a = \sigma_{aa} - \sigma_{oaa} \tag{21a}$$

$$\sigma_{bb} \rightarrow \chi_b = \sigma_{bb} - \sigma_{obb} \tag{21b}$$

so that the effects of the W_{ab}, etc. is to lead to thermal equilibrium in the absence of $\epsilon(t)$. σ_o is the equilibrium value of σ. Now eq. 14 is rewritten as:

$$(\Delta\omega - iT_2^{-1})Z + d(\chi_a - \chi_b) = q\,\omega_o d \tag{22}$$

where the high temperature approximation:[1, 2]

$$\sigma_{oaa} - \sigma_{obb} \cong \frac{e^{-E_a/kT} - e^{-E_b/kT}}{\sum_{\alpha} e^{-E_\alpha/kT}} \cong \frac{-\hbar\,\omega_{ab}}{kTA} \equiv -q\,\omega_o \tag{23}$$

has been used. Here A is the number of spin states (2 in our example). We now need to solve the coupled equations:

$$\begin{pmatrix} \Delta\omega & T_2^{-1} & d \\ -T_2^{-1} & \Delta\omega & 0 \\ 0 & -2d & W_{ab} \end{pmatrix} \begin{pmatrix} Z' \\ Z'' \\ (\chi_a - \chi_b) \end{pmatrix} = \begin{pmatrix} q\,\omega_o d \\ 0 \\ 0 \end{pmatrix} \tag{24}$$

This gives:

$$Z' = \Delta\omega T_2 Z'' \tag{24a}$$

$$Z'' = \frac{qd\,\omega_o T_2}{1 + \Delta\omega^2 T_2^{\,2} + 4d^2 T_2 T_1} \tag{24b}$$

where $T_1 \equiv (2W_e)^{-1}$ and

$$(\chi_a - \chi_b) = q\,\omega_o 4d^2 \frac{T_2 T_1}{1 + \Delta\omega^2 T_2^{\,2} + 4d^2 T_2 T_1} \;. \tag{24c}$$

These results are very similar to steady state solutions of the Bloch Eqs. and we can get correspondence if:

$$2M_o = q\,\omega_o = \frac{\hbar\,\omega_o}{AkT} = \sigma_{oaa} - \sigma_{obb} \tag{25a}$$

$$T_2 = (T_2)_{ab}, \quad T_1 = (T_1)_{ab} \tag{25b}$$

$$\gamma\vec{H} = -\omega_o\vec{k} + 2|d|\{\hat{i}\cos\omega t + \hat{j}\sin\omega t\} \tag{25c}$$

$$Z' = \breve{M}_x, \quad Z'' = \breve{M}_y, \quad Z = \breve{M}_+ \tag{25d}$$

$$\chi_a - \chi_b = 2(M_0 - M_Z) \tag{25e}$$

The above treatment, is based on the high field approximation $|B_o| >> |B_1|$, as well as the fast motional condition $|\mathcal{H}_1|\tau_c << 1$.

It also requires that $|\gamma B_1|\tau_c << 1$ in order that the R matrix is not significantly affected by the presence of the rf field. This can be seen as follows. When we have a time-dependent Hamiltonian,

$$E(t) = \mathcal{H}_o + \epsilon(t)$$

we must define a new interaction representation:

$$\sigma^\ddagger = U(t)\sigma(t)U^{-1}(t) \tag{26a}$$

$$\mathcal{H}_1^\ddagger(t) = U(t)\mathcal{H}_1(t)U^{-1}(t) \tag{26b}$$

where the unitary operator $U(t) \equiv U(t,0)$ is a solution of the differential equation:

$$\frac{d}{dt}U(t,0) = iU(t,0)E(t) \tag{27}$$

with $U(t_o,t_o) = 1$ and $U(t-\tau,0) = U(t,0)U(t-\tau,t)$. Its solution is the time-ordered exponential:

$$U(t,0) = \exp_o\left(\int_o^t iE(t')dt'\right) . \tag{28}$$

Since U is unitary, the differential equation for U^{-1} is obtained by taking Hermitian conjugates of Eq. 27. Thus:

$$\dot{U}^{-1} = -iE(t)U^{-1} \tag{29}$$

Then we have in the interaction representation:

$$\dot{\sigma}^\ddagger = -i\mathcal{H}_1^{\ddagger\times}\sigma^\ddagger \tag{30}$$

as before (cf. Eq. VIII-5). Thus, to second order (assuming $K_1 = 0$, cf. eq. VIII-13a) the cumulant average is, in the long time limit:

$$\begin{aligned}\dot{\sigma}^\ddagger &\cong \int_o^\infty d\tau \langle \mathcal{H}_1^\ddagger(t)^\times \mathcal{H}_1^\ddagger(t-\tau)^\times\rangle_c \sigma^\ddagger(t)\\ &= -U(t)\int_o^\infty \{\langle \mathcal{H}_1(t)^\times [U(t-\tau,t)\mathcal{H}_1(t-\tau)U^{-1}(t-\tau,t)^\times\rangle_c \sigma\}\\ &\qquad \times d\tau U^{-1}(t)\end{aligned} \tag{31}$$

Then

$$\dot{\sigma} = -i[E(t),\sigma] - \int_o^\infty d\tau \langle \mathcal{H}_1(t)^\times [U(t-\tau,t)\mathcal{H}_1(t-\tau)U^{-1}(t-\tau,t)]^\times\rangle_c \sigma(t) \tag{32}$$

Now the integral in eq. 32 is non-negligible only for $\tau \leqslant \tau_c$, but since $|\varepsilon|\tau_c << 1$ we can in this interval write

$$U(t-\tau) = e^{-i\mathcal{H}_0\tau} + O|\varepsilon(t)|\tau_c \cong e^{-i\mathcal{H}_0\tau} \tag{33}$$

which when substituted into eq. 32 gives the desired result. A similar analysis applies for the higher order cumulants in the long time limit yielding R to all orders.

The next most complicated case, is a simple line, coupled by relaxation to other spin eigenstates:

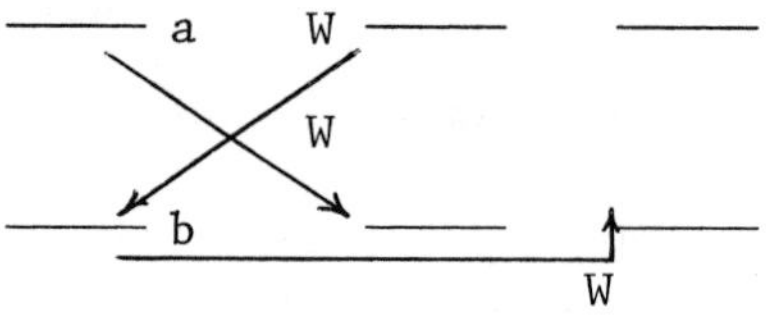

Then we have:

$$\sum_{\alpha\neq a} W_{a\alpha}(\chi_a - \chi_\alpha) = 2d\,\mathrm{Im}Z \tag{34a}$$

$$\sum_{\alpha\neq b} W_{b\alpha}(\chi_b - \chi_\alpha) = -2d\,\mathrm{Im}Z \quad . \tag{34b}$$

And, for $\alpha \neq a, b$ we get A-2 equations:

$$\sum_{\alpha}{}' W_{\alpha\beta}(\chi_\alpha - \chi_\beta) = 0 \quad \beta \neq \alpha \quad . \tag{34c}$$

In eq. 34c we have assumed all transitions other than $a \rightarrow b$ are too far off-resonance to have any appreciable off-diagonal density matrix elements; i.e. they are not excited by the rf field. The conservation of probability is:

$$\mathrm{Tr}\sigma = \mathrm{Tr}\sigma_0 = 1, \text{ or } \mathrm{Tr}\chi = 0 \quad . \tag{35}$$

This is needed, because the above set of A equations are not all linearly independent. We can write these A equations in matrix notation as:

$$\overline{W}\,\overline{\chi} = \overline{U} \quad . \tag{36}$$

When the rank of $\overline{W}$ is A-1, then replacement of any one equation by eq. 35 yields the matrix $\overline{W}^i$, which is now non-singular, and we have:

$$\overline{\chi} = (\overline{W}^i)^{-1}\overline{U}^i \quad . \tag{36a}$$

Proper solutions of this $\overline{W}$ inversion are crucial in all saturation and double resonance analyses. It is possible to obtain solutions in the form:[1,5]

$$(\chi_a-\chi_b) = \Omega_{ba,ba}V_{ba}[q\omega_o - (\chi_a-\chi_b)] \tag{37}$$

and from eq. 34

$$dZ'' = V_{ba}[q\omega_o - (\chi_a-\chi_b)] \tag{38}$$

where

$$V_{ba} = 2d^2T_2/(1+T_2{}^2\Delta\omega^2) \tag{39}$$

and

$$\Omega_{ba,ba} = 2C_{ba,ba}/C \tag{40}$$

where C is any cofactor of $\overline{W}$, (they are all equal, as may be shown from the properties of $\overline{W}^1$); and $C_{ba,ba}$ is the double cofactor of $\overline{W}$ obtained as the (signed) determinant resulting when the ath and b^{th} rows and columns are deleted from $\overline{W}$. More generally we write

$$\Omega_{\alpha\beta,\gamma\delta} = 2C_{\alpha\beta,\gamma\delta}/C \tag{40a}$$

where $C_{\alpha\beta,\gamma\delta}$ is the double cofactor of $\overline{W}$ obtained by deleting the αth and β^{th} rows and the γth and δ^{th} columns of $\overline{W}$ and giving it the correct sign.

The net result is to obtain our earlier results of eq. 24 but now

$$T_1 \rightarrow \tfrac{1}{4}\Omega_{ba,ba} \equiv \tfrac{1}{4}\Omega_{ba} \tag{41}$$

where Ω_{ba} is the saturation parameter for the b $\leftrightarrow$ a transition. It is not a simple T_1, nor decay time. In fact there are as many as (A-1) different, non-zero decay constants in the transient solution (which come from diagonalizing the $\overline{W}$ matrix).

This Ω_{ba} may be regarded as a steady-state self-impedance representing the response of the b $\leftrightarrow$ a transition to the application of an rf field. I.e., we rewrite eqs. 37 and 38 as:

$$(\chi_a - \chi_b) = \Omega_{ba} \cdot (dZ'') \tag{42}$$

and make the electrical circuit analogy by letting $(\chi_a - \chi_b) = E$, $\Omega_{ba} = R$, and $dZ'' = I$. Thus we see that inducing a resonant transition is formally equivalent to inducing a current flow, which causes a voltage drop $(\chi_a - \chi_b)$ proportional to the resistance Ω_{ba}.

XVIII.2. ELDOR[3]

Now we introduce a second ESR microwave field. Assume there are only two transitions of interest.

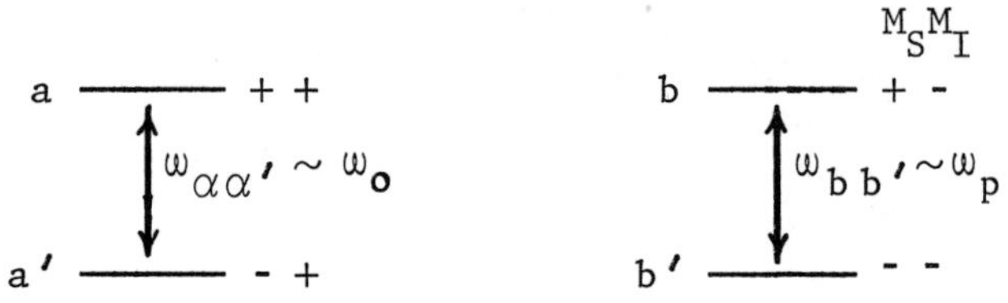

Now we have

$$\varepsilon(t) = \tfrac{1}{2}\gamma_e B_o [S_+ \exp(-i\omega_o t) + S_- \exp(+i\omega_o t)]$$

$$+ \tfrac{1}{2}\gamma_e B_p [S_+ \exp(-i\omega_p t) + S_- \exp(+i\omega_p t)] \quad . \tag{43}$$

We are looking to the applied fields to generate s.s. off-diagonal density matrix elements as a result of the resonance phenomena. We assume

$$|\gamma_e B_o|, \ |\gamma_e B_p|, \ |R| \ll |\omega_{aa'} - \omega_{bb'}| \sim |a| \quad , \tag{44}$$

so the hyperfine lines always remain well separated. Then we may have $\omega_{aa'} - \omega_o = \Delta\omega_o \sim 0$, while $|\omega_{aa'} - \omega_p| \sim |a|$ and $\omega_{bb'} - \omega_p = \Delta\omega_p \sim 0$, while $|\omega_{bb'} - \omega_o| \sim |a|$. Thus, the important elements are:

$$\sigma_{a'a} = \chi_{a'a} = Z_{a'a} \exp(i\omega_o t) \equiv Z_o \exp(i\omega_o t) \tag{45a}$$

$$\sigma_{b'b} = \chi_{b'b} = Z_{b'b} \exp(i\omega_p t) \equiv Z_p \exp(i\omega_p t) \tag{45b}$$

We obtain from eq. 17:

$$[\Delta\omega_o - i/T_{2,o}]Z_o + d_o(\chi_a - \chi_{a'}) = q\omega_{aa'} d_o \cong q\omega_e d_o \tag{46a}$$

$$[\Delta\omega_p - i/T_{2,p}]Z_p + d_p(\chi_b - \chi_{b'}) = q\omega_{bb'}d_p \cong q\,\omega_e d_p \tag{46b}$$

Also, the analogues of eqs. 34 are now:

$$\sum_{\alpha\neq a} W_{a\alpha}(\chi_a - \chi_\alpha) = 2d_o Z_o'' \tag{47a}$$

$$\sum_{\alpha\neq a} W_{a'\alpha}(\chi_{a'} - \chi_\alpha) - -2d_o Z_o'' \tag{47b}$$

$$\sum_{\alpha\neq b} W_{b\alpha}(\chi_b - \chi_a) = 2d_p Z_o'' \tag{47c}$$

$$\sum_{\alpha\neq b'} W_{b'\alpha}(\chi_{b'} - \chi_\alpha) = -2d_p Z_o'' \;. \tag{47d}$$

These equations may be rewritten in matrix form as:

$$(\overline{K} + i\overline{R})\overline{Z} = \overline{D\chi} + \overline{Q} \tag{48a}$$

$$(\overline{W}^j)(\overline{\chi}) = -2\overline{D}^{trj}\overline{Z}'' \tag{48b}$$

where

$$\overline{Q} \cong q\omega_e \begin{pmatrix} d_o \\ d_p \end{pmatrix} \tag{49a}$$

$$K = \begin{pmatrix} \Delta\omega_o & 0 \\ 0 & \Delta\omega_p \end{pmatrix} \tag{49b}$$

$$-R = \begin{pmatrix} T_{2,o}^{-1} & 0 \\ 0 & T_{2,p}^{-1} \end{pmatrix} \tag{49c}$$

$$-\overline{D} = \begin{pmatrix} d_o & -d_o & 0 & 0 \\ 0 & 0 & d_p & -d_p \end{pmatrix} . \tag{49d}$$

$\overline{W}^j$ is a 4 x 4 transition probability matrix in the space of the 4 spin eigenstates with the jth row replaced by ones, and $\overline{D}^{trj}$ is the transpose of $\overline{D}$ with the jth row replaced by zero. $\overline{Z}$ is a vector in the 2 dimensional space of induced transitions. The formal solution is given by:

$$\overline{Z}'' = \overline{M}^{-1}(-\overline{R}^{-1})\overline{Q} \tag{50a}$$

$$\overline{Z}' = (-\overline{R}^{-1})\overline{KZ}'' \tag{50b}$$

$$\overline{D\chi} = -\overline{S}\,\overline{Z}'' \tag{50c}$$

where

$$\bar{M} = 1 + (\bar{R}^{-1}\bar{K})^2 + (-\bar{R}^{-1})\bar{S} \tag{51a}$$

and

$$\bar{S} = 2[\bar{D}(\bar{W}^j)^{-1}\bar{D}^{trj}] \quad . \tag{51b}$$

Suppose $d_p = 0$. One recovers the single-line, simple saturation result and by comparison, we find

$$S_{o,o} = d_o^2\Omega_{aa',aa'} \equiv d_o^2\Omega_{o,o} \tag{52}$$

One finds more generally:[1]

$$S_{i,j} = d_i d_j \Omega_{i,j} \tag{53}$$

where $\Omega_{i,j}$ is a cross-impedance (cross-saturation parameter) which is determined solely by the spin relaxation processes and represents the impedance at transition i from an external disturbance (e.g., a resonant rf field) on the transition j. [It is obtained by eq. 40a with $\alpha\rightarrow\beta$ being the ith transition and $\gamma\rightarrow\sigma$ the jth transition.]

Thus eq. 50c is a generalization of eq. 42 for the single resonance case. In fact it gives

$$(\chi_a-\chi_{a'}) = d_o\Omega_{o,o}Z_o'' + d_p\Omega_{o,p}Z_p'' \tag{54a}$$

$$(\chi_b-\chi_{b'}) = d_p\Omega_{p,p}Z_p'' + d_o\Omega_{p,o}Z_o'' \quad . \tag{54b}$$

with an electrical circuit analogy similar to that of eq. 42. It follows from eqs. 49-53 that:

$$M = \begin{pmatrix} 1 + \Delta\omega_o^2 T_{2,o}^2 + d_o^2 T_o \Omega_o & d_o d_p \Omega_{o,p} T_{2,o} \\ d_p d_o \Omega_{p,o} T_{2,p} & 1 + \Delta\omega_p^2 T_{2,p}^2 + d_p^2 T_{2,p}\Omega_p \end{pmatrix}, \tag{55}$$

where we have let $\Omega_{o,o} = \Omega_o$ and $\Omega_{p,p} = \Omega_p$. Then from eq. 50a:

$$Z_o'' = q\omega_e T_{2,o} d_o \frac{1 - \xi_o/\Omega_{p,o}}{1 + \Delta\omega_o^2 T_{2,o}^2 + d_o^2 T_{2,o}(\Omega_o - \xi_o)} \tag{56}$$

with

$$\xi_o = d_p^2 T_{2,p}\Omega_{o,p}\Omega_{p,o}/(1 + \Delta\omega_p^2 T_{2,p}^2 + d_p^2 T_{2,p}\Omega_p) \quad . \tag{56a}$$

Now consider some special cases. Let us have $\Delta\omega_p = 0$ (represented by a superscript r) and very strong saturation of the pump mode:

$$d_p^2 T_p \Omega_p \gg 1 \quad . \tag{57}$$

Then:

$$\xi_o^r (d_p^2 \to \infty) = \Omega_{o,p}\Omega_{p,o}/\Omega_p \tag{58}$$

which is just relaxation determined. We now let $T_{2,p} = T_p$, etc. Then

$$Z_o'' = q\omega_e T_o d_o \frac{(\Omega_p - \Omega_{o,p})/\Omega_p}{1 + T_o^2 \Delta\omega_o{}^2 + d_o^2 T_o(\Omega_o\Omega_p - \Omega_{o,p}\Omega_{p,o})/\Omega_p} \quad . \tag{59}$$

If we also introduce the generalized no-saturation condition for the observing mode:

$$d_o^2 T_o[(\Omega_o\Omega_p - \Omega_{o,p}\Omega_{p,o})/\Omega_p] \ll 1 \tag{60}$$

one has the simple result that:

$$Z_o'' = \frac{T_o q\omega_e d_o}{1 + \Delta\omega_o^2 T_o^2}[1 - \frac{\Omega_{o,p}}{\Omega_p}] \quad . \tag{61}$$

Since Ω_p is always positive,[1] it follows from eq. 61 that for $\Omega_{o,p} > 0$ the signal is reduced by the presence of the resonant pump field, while for $\Omega_{o,p} < 0$ the signal is amplified. The limiting (but not realistic) case for eq. 61 occurs when W_n is very strong and W_e is negligible. (Here W_e and W_n are respectively the lattice-induced electron-spin flip and nuclear-spin flip rates.) Then the case for the energy levels shown:

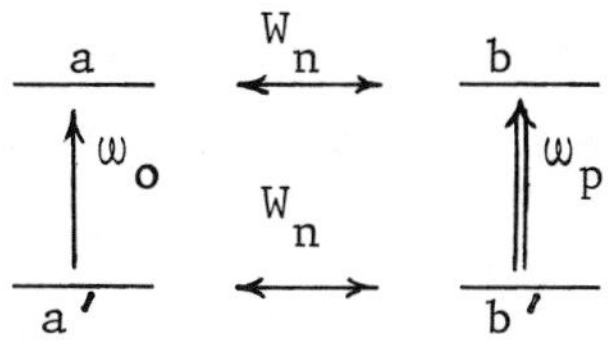

is easily understood. Let P_i be the population of the ith state. Then saturation by ω_p causes $P_b = P_{b'}$; a strong W_n causes $P_a = P_b$ and $P_{a'} = P_{b'}$ leading to a reduction in intensity of the observed signal. This extreme will be seen to be equivalent to

$\Omega_{o,p} = \Omega_o = \Omega_p$.

There are actually 2 effects that can be seen in ELDOR.

Effect 1. The no-saturation effect discussed above is a polarization effect (not unlike an Ovehauser effect in NMR) but the two transitions involved have no level in common, and this places special requirements on the relaxation processes in order to obtain significant effects.

Effect 2. It is important only when Z_o'' is being saturated. It reflects the fact that the induced absorption mode Z_p'' acts as an induced transition which, in conjunction with lattice-induced transitions, can facilitate the rate of energy transferred from the observing radiation field to the lattice via the spin systems.

Effect 1 is the main effect in ELDOR, while the analogue to effect 2 is the dominant one in ENDOR.

XVIII.3. ENDOR[1,4,5]

We again consider our 4 level system, but now:

$$\begin{aligned}\varepsilon(t) = {} & \tfrac{1}{2}\gamma_e B_e [S_+ \exp(-i\omega_e t) + S_- \exp(+i\omega_e t)] \\ & + \tfrac{1}{2}\gamma_n B_n [I_+ \exp(-i\omega_n t) + I_- \exp(+i\omega_n t)] \\ & + \tfrac{1}{2}\gamma_e B_n [S_+ \exp(-i\omega_n t) + S_- \exp(i\omega_n t)] \\ & + \tfrac{1}{2}\gamma_n B_e [I_+ \exp(-i\omega_e t) + I_- \exp(i\omega_e t)] \quad . \end{aligned} \tag{62}$$

In eq. 62, the microwave field at frequency ω_e is to induce electron-spin flips, while the rf field at frequency ω_n is to induce nuclear-spin flips. Thus the last term in eq. 62 can be neglected as being too far off resonance to affect the nuclear spins. We neglect the 3rd term in eq. 62 for simplicity, even though it does have a non-trivial effect on the effective transition moment of the nuclear spins.[1] Let us assume the following four-level system:

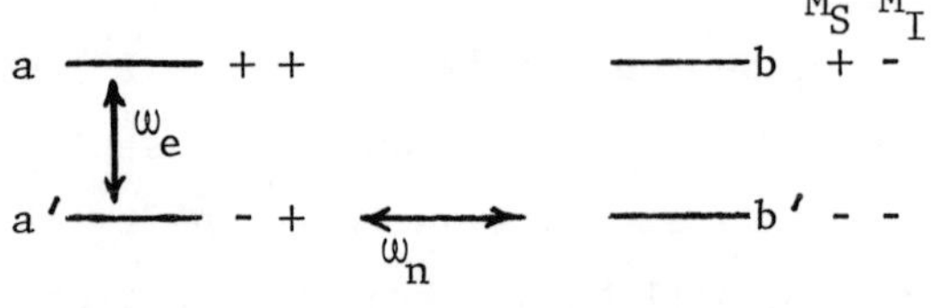

with

$$\Delta_e \equiv \omega_e - \omega_{aa'} \approx 0 \tag{63a}$$

$$\Delta_n \equiv \omega_n - \omega_{a'b'} \approx 0 \quad . \tag{63b}$$

Then, for assumptions similar to the ELDOR case, we expect important s.s. off-diagonal density-matrix elements:

$$\chi_{a'a} = Z_{a'a} e^{i\omega_e t} \equiv Z_e e^{i\omega_e t} \tag{64a}$$

$$\chi_{b'a'} = Z_{b'a'} e^{i\omega_n t} \equiv Z_n e^{i\omega_n t} \quad . \tag{64b}$$

We obtain the series of equations:

$$[\Delta_e - i/T_e]Z_e + d_e(\chi_a - \chi_{a'}) + d_n Z_{b'a} = q\omega_e d_e \tag{65a}$$

$$[\Delta_n - i/T_n]Z_n + d_n(\chi_{a'} - \chi_{b'}) - d_e Z_{b'a} = q\omega_n d_n \tag{65b}$$

$$[\Delta_e + \Delta_n - i/T_{b'a}]Z_{b'a} - d_e Z_n + d_n Z_e = 0 \quad . \tag{65c}$$

Note the appearance of

$$\chi_{b'a} = Z_{b'a} e^{i(\omega_e+\omega_n)t} \equiv Z_x e^{i(\omega_e+\omega_n)t} \quad . \tag{66}$$

This is an overtone term: a 2 quantum effect. Also:

$$\sum_{\alpha\neq a} W_{a\alpha}(\chi_a - \chi_\alpha) = 2d_e Z''_{a'a} \tag{67a}$$

$$\sum_{\alpha\neq a'} W_{a'\alpha}(\chi_{a'} - \chi_\alpha) = -2d_e Z''_{a'a} + 2d_n Z''_{b'a'} \tag{67b}$$

$$\sum_{\alpha\neq b} W_{b\alpha}(\chi_b - \chi_\alpha) = 0 \tag{67c}$$

$$\sum_{\alpha\neq b'} W_{b'\alpha}(\chi_{b'} - \chi_\alpha) = -2d_n Z''_{b'a'} \quad . \tag{67d}$$

Again we may write these equations in the matrix form given by eqs. 48 with the formal solution given by eqs. 50 and 51. Note that the K or coherence matrix is:

$$K = \begin{pmatrix} \Delta_e & 0 & d_n \\ 0 & \Delta_n & -d_e \\ d_n & -d_e & \Delta_e+\Delta_n \end{pmatrix} \quad . \tag{68}$$

and is no longer diagonal. Also intensities are proportional to

$$Q = q\begin{pmatrix} \omega_e d_e \\ \omega_n d_n \\ 0 \end{pmatrix} , \tag{69}$$

but because $\frac{\omega_e}{\omega_n} \sim \frac{1}{660}$ for protons, we may usually set

$$Q \cong q\begin{pmatrix} \omega_e d_e \\ 0 \\ 0 \end{pmatrix} , \tag{69a}$$

which amounts to neglecting the analogue of effect 1 in the ENDOR case.

Neglect of Coherence Effects. The coherence effects arise from the off-diagonal elements in the K matrix or in other words the contribution from Z_x. Consider the case of exact resonance, when $\Delta_e = \Delta_n = 0$, since this is the condition under which double-resonance effects will be maximized. Equations 65-67 and 50-51 then yield:

$$Z'^r_e = Z'^r_n = Z''^r_x = 0 \tag{70a}$$

$$Z''^r_e = \frac{q\omega_e d_e T_e}{1+d_e^2(\Omega_e - \xi_e^r)T_e + d_n^2 T_x T_e} \tag{70b}$$

where

$$\xi_e^r = \frac{T_n d_n^2 (T_x + |\Omega_{e,n}|)^2}{1+d_n^2 \Omega_n T_n + d_e^2 T_x T_n} \tag{70c}$$

Thus from eq. 70b when

$$1 + d_e^2 \Omega_e T_e \gg d_n^2 T_x T_e \tag{71}$$

$$(\text{and } \xi_e^r \not\approx \Omega_e)$$

the coherence effect on Z''^r_e may be neglected. ξ_e^r leads to an enhancement of a saturated ESR signal, since it effectively reduces the saturation parameter Ω_e. Now when

$$1 + d_n^2 \Omega_n T_n \gg d_e^2 T_x T_e \tag{72}$$

it follows from eq. 70c that the ratio ξ_e/Ω_e will not be affected by d_e, and further if

$$|\Omega_{e,n}| \gg T_x \tag{73}$$

we may completely neglect the coherence effects.

If there is appreciable saturation and

$$d_e^2 \sim d_n^2 \tag{74}$$

then we can replace eqs. 71-73 with the simpler set of conditions:

$$\Omega_e, \Omega_n, |\Omega_{e,n}| \gg T_x \tag{75}$$

for the neglect of coherence effects. The inequalities of eq. 75 are fulfilled if the T_1's or saturation parameters are much larger than the T_2's or inverse linewidths.

Now our solutions for Δ_e, $\Delta_n \approx 0$ are:

$$Z_e'' = \frac{q\omega_e d_e T_e}{1+(\Delta_e T_e)^2 + (\Omega_e - \xi_e) T_e d_e^2} \tag{76a}$$

$$\xi_e = \frac{d_n^2 (\Omega_{e,n})^2 T_n}{1+(\Delta_n T_n)^2 + d_n^2 T_n \Omega_n} \quad . \tag{76b}$$

If the ENDOR spectrum is monitored after subtraction of the ESR signal; then for $\Delta_e = 0$ and $\Omega_e T_e d_e^2 \gg 1$ we have

$$Z''^{r}_{ENDOR} - Z''^{r}_{ESR} = q\omega_e d_e \left(\frac{\Omega_{e,n}^2}{\Omega_e^2}\right) \frac{d_n^2 T_n}{1+(\Delta_n T_n)^2 + \left[1 - \frac{\Omega_{e,n}^2}{\Omega_e \Omega_n}\right] T_n \Omega_n d_n^2} \quad . \tag{77}$$

Thus the signal strength is proportional to $(\Omega_{e,n}/\Omega_e)^2$ and the shape is a Lorentzian of width T_n^{-1} and (modified) saturation parameter:

$$\Omega_n \left(1 - \frac{\Omega_{e,n}^2}{\Omega_n \Omega_e}\right) \quad . \tag{78}$$

The percent enhancement of an ESR line due to ENDOR is then, from eq. 77,

$$\%\ \text{enh} = \frac{Z''_{ENDOR} - Z''_{ESR}}{Z''_{ENDOR}} = \frac{\xi_e}{\Omega_e - \xi_e} \xrightarrow[\Delta_n \to 0]{d_n^2 \to \infty} \left[\frac{\Omega_n \Omega_e}{\Omega_{e,n}^2} - 1\right]^{-1} \quad . \tag{79}$$

XVIII.4. GENERAL APPROACH

One finds that, in general, multiple-resonance ESR experiments in liquids may be expressed in the matrix form eq. 48 with formal solution given by eqs. 51-53.[1] In this formal solution, $\bar{Z}$ is a vector in the space of all induced transitions and $\bar{\chi}$ is a vector in the space of all spin eigenstates. The only requirement is that a raising convention apply. This is the requirement that all induced transitions in the space of $\bar{Z}$ are those in which there is (are) increase(s) in spin quantum number but no decrease(s) in spin quantum number.[5] This requirement is often met for ENDOR and ELDOR experiments, but it sometimes requires neglect of some multiple quantum transitions. If it is not met, then a somewhat more complex form of eqs. 48 and 51-53 could become necessary. Also, in summary, the validity of the general relaxation eq. 17 for well separated hyperfine lines requires, that,

$$|\gamma_e B_o|,\ |\gamma_i B_o|,\ |\gamma_e \bar{a}_i|,\ \tau_c^{-1},\ \gg \varepsilon(t),\ |R| \tag{80}$$

where τ_c refers to the relevant correlation time(s).

XVIII.5. TRANSITION PROBABILITIES

Consider now the general 4-level system with all types of spin-lattice relaxation transitions:

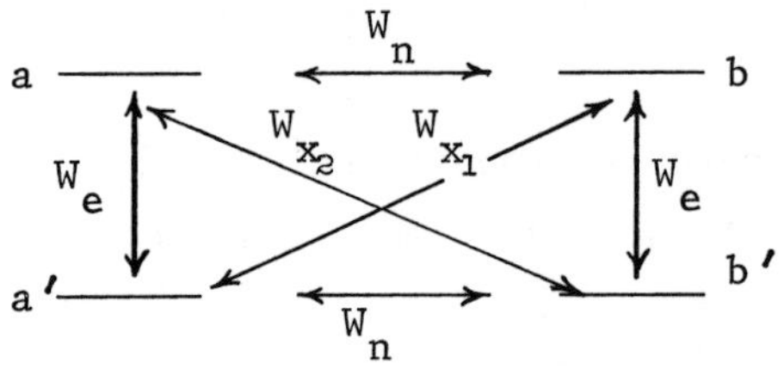

We can solve for C_{ii} and $C_{ij,kl}$ the cofactors and double cofactors of the $\bar{W}$ matrix to obtain all the $\Omega_{ij,kl}$.

A. ELDOR— Generalized No Saturation of Observing Mode.

We have from eq. 61 that the signal reduction is given by:

$$R \equiv \frac{\%\ \text{reduction}}{100} = \frac{\Omega_{o,p}}{\Omega_p} = \frac{W_n^2 - W_{x_1}W_{x_2}}{W_e(2W_n+W_{x_1}+W_{x_2})+(W_n+W_{x_1})(W_n+W_{x_2})}\ . \tag{81}$$

Clearly if $W_n^2 > W_{x_1} W_{x_2}$ one has a reduction in signal, while if $W_n^2 < W_{x_1} W_{x_2}$ there will be an enhancement.

a. Let $W_{x_1} = W_{x_2} = 0$ (i.e., only pseudosecular dipolar terms important). Then

$$R = \frac{W_n}{2W_e + W_n} = \frac{b}{2+b} \text{ where } b \equiv \frac{W_n}{W_e}$$

or a reduction.

b. Let $W_{x_1} = W_n = 0$ (i.e., isotropic hyperfine modulation). Then $R = 0$ i.e., no effect.

c. Let $W_{x_2} = 4W_n$, $W_{x_1} = \frac{2}{3}W_n$ (dipolar, extreme narrowing). Then

$$R = -\frac{b}{4+5b}$$

or an enhancement.

In the case of solids, one can also examine ELDOR enhancements for forbidden ESR transitions.[6]

B. ENDOR—Limiting Enhancements.

We have from eq. 79 that the % enhancement E is given by:

$$1 + E^{-1} = \frac{\Omega_n \Omega_e}{\Omega_{e,n}^2} =$$

$$[W_n(2W_e+W_{x_1}+W_{x_2})+(W_e+W_{x_1})(W_e+W_{x_2})] \times$$
$$[W_e(2W_n+W_{x_1}+W_{x_2})+(W_n+W_{x_1})(W_n+W_{x_2})] \times$$
$$[W_{x_2}(W_e+W_n+W_{x_1})+W_eW_n]^{-2} \quad . \tag{82}$$

a. Let $W_{x_1} = W_{x_2} = 0$ (i.e., only pseudo-secular dipolar terms important). Then

$$E = \frac{1}{2[2 + b + b^{-1}]} \quad \begin{array}{l} \xrightarrow{b\gg1} \frac{1}{2}b^{-1} \\ \xrightarrow{b\ll1} \frac{1}{2}b \\ \xrightarrow{b=1} \frac{1}{8} \end{array}$$

b. Let $W_{x_1} = W_n = 0$ (i.e., isotropic hyperfine modulation). Then

$$E = \frac{W_{x_2}}{W_e} \quad .$$

This would theoretically be a most effective ENDOR mechanism if W_{x_2} were large.

c. Case b but now the a↔b ENDOR transition is saturated. Then E = 0.

d. Let $W_{x_2} = 4W_n$, $W_{x_1} = \frac{2}{3}W_n$ (dipolar, extreme narrowing).

$$E = \frac{b[22.5 + 60b + 40b^2]}{6 + b[25 + 34b + 15b^2]} \qquad \xrightarrow{b>>1} (8/3)b \qquad \xrightarrow{b<<1} 3.75b \qquad \xrightarrow{b=1} 1.53$$

This is also a very effective ENDOR mechanism if b>1.

e. Case d but now the a↔b ENDOR transition is saturated

$$E = \frac{b[2.5 + 10b + 10b^2]}{6 + b[45 + 84b + 45b^2]} \qquad \xrightarrow{b>>1} 2/9 \qquad \xrightarrow{b<<1} (5/12)b \qquad \xrightarrow{b=1} 1/8$$

C. Expressions for Transition Probabilities.[1]

We assume a single set of completely equivalent nuclei with total nuclear spin quantum number J and total z component M. (We do not explitly indicate the distinction between degenerate states of the same values of J and M. Note that there will often be degenerate states for a given set of values of J and M. However, it is possible for dipolar terms (but not quadrupolar terms) to order the degenerate states according to a parameter κ or $J^{(\kappa)}$ such that the values of J <u>and</u> κ are preserved.[1] We do not explicitly indicate κ in the equations below.)

i. Nuclear-Spin Transitions

a. Dipolar:

$$W_{(M_S,M) \rightarrow (M_S,M\pm1)} = \tfrac{1}{2} j^D(0)\,[J(J+1)-M(M\pm1)] \tag{83}$$

where the electron-nuclear dipolar spectral density $j^D(0)$ is:

$$j^D(0) = \frac{1}{5}\gamma_e^2\gamma_n^2\hbar^2\sum_m |D^m|^2\tau_R \tag{84}$$

with τ_R the rotational correlation time, and it is assumed $|\omega_n\tau_R| \ll 1$. The dipolar coefficients are:[7]

$$D^{(m)} = \left(\frac{6\pi}{5}\right)^{\frac{1}{2}} \langle\psi_e|r'^{-3}Y_{2,m}(\theta',\varphi')|\psi_e\rangle \tag{85}$$

where θ', and φ' and r' are spherical polar co-ordinates which define the position of the unpaired electron with respect to a nucleus in the molecular co-ordinate frame.

b. Quadrupolar. (For this case only we consider a single nucleus of Spin I):

$$W_{(M_S, M) \to (M_S, M\pm 1)} = 2j^Q(0)\,[I(I+1) - M(M\pm 1)][2M\pm 1]^2 \tag{86a}$$

$$W_{(M_S, M) \to (M_S, M\pm 2)} = 2j^Q(0)\,[I(I+1) - M(M\pm 1)][I(I+1) - (M+1)(M\pm 2)] \tag{86b}$$

where

$$j^Q(0) = \frac{\tau_R}{80}\,\frac{e^2Q^2}{\hbar^2 I^2(2I-1)^2}\sum_m |\nabla\epsilon^{(m)}|^2 \tag{87}$$

with electric-field-gradient irreducible-tensor components:[7]

$$\nabla\epsilon^{(0)} = -\left(\frac{3}{2}\right)^{\frac{1}{2}}\langle\psi_e|V'_{ZZ}|\psi_e\rangle \tag{88a}$$

$$\nabla\epsilon^{(\pm 1)} = \pm\langle\psi_e|V'_{XZ}\pm iV'_{YZ}|\psi_e\rangle \tag{88b}$$

$$\nabla\epsilon^{(\pm 2)} = -\tfrac{1}{2}\langle\psi_e|V'_{XX}-V'_{YY}\pm 2iV'_{XY}|\psi_e\rangle\ . \tag{88c}$$

ii. Electron-Spin Transitions

$$W_{(\mp, M)\to(\pm, M)} = 2j^D(0)M^2 + 4j^{(DG_2)}(\omega_o)B_oM + 2j^{(G_2)}(\omega_o)B_o^2 + W_e^{SR} + X\ . \tag{89}$$

Here

$$j^D(\omega_o) = j^D(0)\,[1+\omega_o^2\tau_R^2]^{-1}\ . \tag{90}$$

The g-tensor spectral density is:

$$j^{(G_2)}(\omega_o) = \frac{1}{20}B_o^2\hbar^{-2} \times \left\{\sum_{k=1}^{3}(g_k)^2 - 3g_s^2\right\}\frac{\tau_R}{1+\omega_o^2\tau_R^2}\ . \tag{91}$$

The g-tensor-dipolar cross-term spectral density is:

$$j^{(DG_2)}(\omega_o) = -\frac{1}{10}\gamma_e B_o \gamma_n \sum_m D^{(m)}g^{(m)}\,\frac{\tau_R}{1+\omega_o^2\tau_R^2} \tag{92}$$

with $g^{(0)} = 6^{-\frac{1}{2}}[2g'_z - (g'_x + g'_y)]$ and $g^{(\pm 2)} = \frac{1}{2}(g'_x - g'_y)$. The spin-rotational contribution to W_e is in a semi-classical treatment:

$$W_e^{SR} = \frac{IkTC^2}{\hbar^2}\left(\frac{\tau_J}{1+\omega_o^2\tau_J^2}\right) \tag{93}$$

where I is the moment of inertia, C is the spin-rotational constant of the radical (and we have assumed both to be isotropic), and τ_J is the correlation time for the angular momentum, (cf. Atkins, Ch. XI). In liquids, usually $\tau_J \ll \tau_R, \omega_o^{-1}$. One has for a Stokes-Einstein model

$$\tau_R = 4\pi\eta a^3/3kT \tag{94a}$$

$$\tau_J = [6IkTr_R]^{-1} \tag{94b}$$

More generally (Atkins Ch. XI) $\overline{C} \cong -2\overline{A}\Delta\overline{g}$ where A is the inverse moment of inertia tensor and $\Delta\overline{g} = \overline{g} - 2.00231$. Then we have (for axially symmetric $\overline{A}$):

$$W_e^G \cong \sum_i (g_i - g_s)^2/40\tau_R \quad (\text{for } \omega_o^2\tau_R^2 \gg 1) \tag{95a}$$

$$W_e^{SR} \cong \sum_i (g_i - g_e)^2/18\tau_R \quad . \tag{95b}$$

If these are the dominant terms in W_e, then:

$$W_e \propto \tau_R^{-1} \tag{96a}$$

or

$$W_e \equiv A\ T/\eta \tag{96b}$$

Usually $\tau_R > \omega_o^{-1}$ for free radicals in liquids below room temperature since at X-band $\omega_o^{-1} \cong 1.7 \times 10^{-11}$ sec. Then pseudo-secular dipolar terms dominate in ELDOR or ENDOR. So, from eqs. 83 and 84:

$$W_n \propto \tau_R$$

or

$$W_n \equiv B\ \eta/T \quad .$$

Then

$$b = \left(\frac{B}{A}\right)\left(\frac{\eta}{T}\right)^2 \quad . \tag{97}$$

If we let

$$\eta \propto Te^{W/kT} \qquad W > 0$$

we get b increasing significantly with decreasing T. This usually leads to better ELDOR and ENDOR signals at reduced temperatures.

iii. Combined Electron-Spin-Nuclear-Spin Transitions. (Cross-Relaxation)

$$W_{M_s, M \to M_s \pm 1,\ M \mp 1} = \left[\tfrac{1}{3}j^{(D)}(\omega_o) + \tfrac{1}{2}j^{I}(\omega_o)\right] \times [J(J+1) - M(M\mp 1)]. \tag{98}$$

$$W_{M_s, M \rightarrow M_s \pm 1, M \pm 1} = 2 j^{(D)}(\omega_o)[J(J+1) - M(M\pm 1)] \quad . \tag{99}$$

The isotropic dipolar spectral density is:

$$j^I(\omega_o) = \gamma_e^2 [\langle a(t)a(t+\tau)\rangle - \bar{a}^2] \frac{\tau_c}{1+\omega_o^2 \tau_c^2} \tag{100}$$

Note that for $j^D(\omega_o) \ll j^D(0)$ (and small $j^I(\omega_o)$) $W_x \ll W_n$ and pseudo-secular terms dominate as noted above.

XVIII.6. HEISENBERG SPIN EXCHANGE AND CHEMICAL EXCHANGE

Heisenberg spin exchange is a very important radical-concentration dependent relaxation mechanism in normal liquids. It is probably the dominant one for $S = \frac{1}{2}$. It may be analyzed by a simple model which also serves as a simple example of the stochastic Liouville approach.[8] We assume radicals exist either as well separated "monomers" or as interacting pairs or "dimers" each with mean lifetimes τ_2 and τ_1 respectively, and with density matrices ρ and σ respectively. The equations of motion are then:

$$i\dot{\rho} = \mathcal{H}_T^{(1)\times}\rho + i\frac{2}{\tau_2}\mathrm{Tr}_s\sigma - i\frac{2}{\tau_2}\rho \tag{101}$$

$$i\dot{\sigma} = (\mathcal{H}_T^{(1)\times} + \mathcal{H}_T^{(2)\times} + \mathcal{H}_J^{\times})\sigma - i\tau_1^{-1}(\sigma - \rho \times \rho) \tag{102}$$

where $\mathcal{H}_T^{(1)}$ is the spin Hamiltonian and $\mathrm{Tr}_s\sigma = \frac{1}{2}(\mathrm{Tr}_1\sigma + \mathrm{Tr}_2\sigma)$ is a symmetrized trace over each of the two components of the interacting dimer. Also

$$H_J = J\, S^{(1)} \cdot S^{(2)} \tag{103}$$

where J is twice the exchange integral. One obtains a steady-state solution for σ in the rotating frame. It is then possible to show that when:

$$|J|,\ \tau_1^{-1} \gg |a_i|,\ \omega_1 \tag{104}$$

Eq. 101 is well approximated by:

$$i\dot{\rho} = \mathcal{H}_T^{(1)\times}\rho + i\omega_{HE}[\mathrm{Tr}_s(P\rho \times \rho P) - \rho] \tag{105}$$

where

$$\omega_{HE} = \frac{1}{\tau_2}\left[\frac{J^2\tau_1^2}{1 + J^2\tau_1^2}\right] \tag{106}$$

is the Heisenberg exchange frequency. In eq. 105 we have neglected a frequency shift term which is readily shown to be zero in the high temperature approximation, (i.e., $A\rho \cong 1 + \rho'$ with $|\rho'| \ll 1$, cf. eq. 23). Here P is the operator which permutes electron spins. The derivation of eq. 105 is based on the fact that for spins $S = \frac{1}{2}$:

$$H_J^{\times} = \tfrac{1}{2}JP^{\times} \quad . \tag{107}$$

For simple Brownian diffusion of the radicals in solution we have:

$$T_2^{-1} = 4\pi Df\mathfrak{N} \tag{108a}$$

$$T_1^{-1} = (6D/d^2)fe^u \tag{108b}$$

where $\mathfrak{N}$ is the density of radicals, the diffusion coefficient is $D = kT/6\pi a\eta$, and d is the interaction distance for exchange. The factors f and fe^u are introduced for charged radicals to take account of Coulombic and ionic atmosphere effects.[8,9]

The result, eq. 105 means that Heisenberg exchange appears as a simple exchange process analogous to chemical exchange processes for which the well-known Kaplan-Alexander[10] method applies. We let

$$\Phi_H(\chi) \equiv \omega_{HE}[\mathrm{Tr}_s(P\sigma \times \sigma P) - \sigma] \tag{109}$$

and add this relaxation term to eq. 17. One then finds that for well separated hyperfine lines the T_2 contributions are:

$$T_{2,HE}^{-1}(ESR,\lambda) = \left(\frac{A-2D_\lambda}{A}\right)\omega_{HE} \tag{110}$$

$$T_{2,HE}^{-1}(NMR) = \tfrac{1}{2}\omega_{HE} \quad . \tag{111}$$

Here D_λ is the degeneracy of the λth transition, and the T_2^{-1}(NMR) is the width contribution to a well-resolved ENDOR line. The diagonal elements of eq. 109 yield:

$$\tfrac{1}{2}\omega_{HE}[(\chi_{\alpha\mp} - \chi_{\alpha\pm}) + (\chi_+ - \chi_-)] \tag{112}$$

where

$$\chi_\pm = \frac{2}{A}\sum_\gamma \chi_{\gamma\pm} \tag{113a}$$

and

$$\chi_+ + \chi_- = 0 \tag{113b}$$

The notation $\alpha\pm$ in eqs. 112 and 113 refers to the αth nuclear-spin configuration and $M_s = \pm$.

The steady-state solution of eq. 112 in eq. 17 is:

$$\chi_{\alpha+} - \chi_{\alpha-} = 2\chi_+ \tag{114}$$

i.e., differences in population between all pairs of levels differing only in M_s are equal. The unlinearized rate equations yield the s.s. result that all the ratios $\frac{\sigma_{\alpha-}}{\sigma_{\alpha+}}$ are equal.

If in chemical exchange (CE) (i.e., electron transfer), the predominant NMR relaxation of the diamagnetic radical precursors is the CE process, then CE appears to be just like HE in magnetic resonance experiments on the radicals.[4]

We now consider the $\overline{W}$ matrix including eq. 112. Note first that

$$[\Phi_{HE}(\chi)]_{\alpha^+\alpha^+} + [\Phi_{HE}(\chi)]_{\alpha^-\alpha^-} = \omega_{HE}\Big\{[\tfrac{1}{2}(\chi_{\alpha^-} - \chi_{\alpha^+}) + \chi_+] + [\tfrac{1}{2}(\chi_{\alpha^+} - \chi_{\alpha^-}) + \chi_-]\Big\} = 0 \tag{115}$$

so that each pair of rows of $\overline{W}$ labeled α^+ and α^- are linearly dependent. Thus, while $\overline{W}^{HE}$ is an A x A matrix, it is of rank A/2, i.e., HE does not act to change $(\chi_{\alpha^+} + \chi_{\alpha^-})$ but rather to equate all $(\chi_{\alpha^+} - \chi_{\alpha^-})$. One must add W_n or W_x terms to reduce this high order singularity.

One may alternatively employ another method. Sufficient conditions for this method are:

1. All spin-flip relaxation transitions are of W_e, W_n or ω_{HE}-type (i.e., no W_x).

2. a. $W_{(+,M)\to(-,M)} = W_{(-,M)\to(+,M)}$

   b. $W_{(+,M)\to(+,M\pm1)} = W_{(-,M)\to(-,M\pm1)}$

Then we may define a $\frac{1}{2}$A dimensional square matrix W (which is usually non-singular or readily separated into non-singular components) according to

$$[\hat{W}\hat{\chi}]_\lambda \equiv [W\chi]_{\lambda+} - [W\chi]_{\lambda-} \quad . \tag{116}$$

This reduced eigenstate space is found to include only the $\hat{\chi}_\lambda \equiv$

$\chi_{\lambda+} - \chi_{\lambda-}$, which are closely related to pure ESR transitions.

Then if only ESR transitions are induced, we find

$$S_{\lambda,\eta} = 4d_\lambda d_\eta(\hat{W}^{-1})_{\lambda,\eta} = d_\lambda d_\eta \Omega_{\lambda,\eta} \quad . \tag{117}$$

(This method can also be generalized for ENDOR). Using this method one can then prove:[8]

$$\Omega_\lambda = \frac{2}{W_e D_\lambda} \frac{1 + D_\lambda b''}{1 + \frac{1}{2}Ab''} \tag{118}$$

with

$$b'' = \frac{\omega_{HE}}{AW_e} \tag{118a}$$

$$T_{1,\lambda} = \tfrac{1}{4}(D_\lambda \Omega_\lambda) \tag{119}$$

and

$$\Omega_{i,j} = \frac{2}{W_e} \frac{b''}{1 + \frac{A}{2} b''} \quad i \neq j \quad . \tag{120}$$

Equation 118 illustrates the "shorting-out" effect spin exchange has in coupling the different hyperfine lines (cf. eq. 114) without directly leading to electron-spin flips. It follows from eqs. 56 and 118-120 that

$$R \equiv \frac{Z''_{ESR} - Z''_{ELDOR}}{Z''_{ESR}} \tag{121}$$

is (for $\Delta\omega_o = 0$ and the no-saturation condition of eq. 60):

$$R^{-1} = \Omega_p/\Omega_{o,p} + [(1 + \Delta\omega_p^2 T_p^2)/T_p \Omega_{o,p}]dp^{-2} \tag{122}$$

and

$$R_\infty^{-1} = \Omega_p/\Omega_{o,p}$$
$$= \frac{1 + D_p b''}{D_p b''} = (D_p b'')^{-1} + 1 \quad . \tag{123}$$

Here R_∞ is defined in the same manner as the asymptotic R of Eq. 81. Eauation 123 shows how Heisenberg spin exchange is effective in enabling significant ELDOR reduction factors.

Now for ENDOR, and a single nucleus of spin $I = \frac{1}{2}$, with $W_x = 0$ one has

$$\Omega_e = \frac{1}{W_e} \frac{[2W_e + (W_n + \omega_{HE}/2)]}{W_e + (W_n + \frac{\omega_{HE}}{2})} \tag{124a}$$

$$\Omega_n = \frac{1}{W_n} \frac{[2W_n + (W_e + \omega_{HE}/2)]}{[W_e + \frac{\omega_{HE}}{2} + W_n]} \tag{124b}$$

$$\Omega_{e,n} = (W_e + W_n + \frac{\omega_{HE}}{2})^{-1} \tag{124c}$$

and

$$\xi_e^r(d_n \to \infty) = \frac{W_n}{[2W_n + W_e + \frac{\omega_{HE}}{2}][W_e + W_n + \frac{\omega_{HE}}{2}]} \tag{125}$$

In general if $W_n = 0$ and $W_x = 0$, $\xi_e^r(d_n \to \infty) = 0$ even for more than one magnetic nucleus. Thus Heisenberg exchange is not an effective ENDOR mechanism, i.e. it is ENDOR "inactive", although it is ELDOR "active".

In conclusion, we note that the characteristic behavior of the various relaxation mechanisms as differently manifested in line width, saturation, ELDOR, and ENDOR is a potentially useful approach to separate out the many possible components of relaxation in a particular paramagnetic system.

We present in the Table a summary of these characteristics. This summary should, however, be used with caution.

XVIII.7. ACKNOWLEDGMENT

This work was supported in part by the Advanced Research Projects Agency and by the National Science Foundation.

Table I
ESR Linewidth and Relaxation Mechanisms

	Linewidths			Saturation-(Ω)		ELDOR	ENDOR		
	Nuclear spin Dependence	Field-Frequency Dependence	Temperature-Viscosity	Activity	Nuclear Spin Dependence	Activity	Activity	Linewidth Contribution	%Enhancement (maximum)
1) G-Isotropic Secular Only	None	Quadratic	As τ_c	None	None	None	None	None	None
2) Dipole-Isotropic									
a) Secular	M^2	None	As τ_c	None	——	None	None	Yes	——
b) Non-secular	M^2	$[1+\omega_o^2\tau_c^2]^{-1}$	As $\tau_c[1+\omega_o^2\tau_c^2]^{-1}$	W_{x_1}	Yes	None unless $W_{x_2} \neq 0$	W_{x_1}	Yes	Goes as W_{x_1}/W_e
3) Isotropic G-Dipole X-Term Secular Only	M	Linear	As τ_c	None	——	None	None	None	——
4) G-Anisotropic									
a) Secular	None	Quadratic	$\tau_R \propto \eta/T$	None	——	None	None	None	——
b) Non-secular	None	$\omega_o^2[1+\omega_o^2\tau_R^2]^{-1}$	$\tau_R[1+\omega_o^2\tau_R^2]^{-1}$	W_e	None	W_e	W_e	Yes	W_e
5) Dipole-Anisotropic									
a) Secular ($S_Z I_Z$)	M^2	None	τ_R	None	——	None	None	Yes	——
b) Pseudo-secular ($S_Z I_\pm$)	M^2	None	τ_R	W_n	Yes	W_n-Reduction	W_n	Yes	$W_n/W_e \sim 1$
c) Non-secular									
i) $S_\pm I_Z$	M^2	$[1+\omega_o^2\tau_R^2]^{-1}$	$\tau_R[1+\omega_o^2\tau_R^2]^{-1}$	W_e	M^2	W_e	W_e	Yes	W_e
ii) $S_\pm I_\pm, S_\pm I_\mp$	M^2	$[1+\omega_o^2\tau_R^2]^{-1}$	$\tau_R[1+\omega_o^2\tau_R^2]^{-1}$	W_{x_1}, W_{x_2}	Yes	W_x-Enhancement	W_{x_1}, W_{x_2}	Yes	Goes as W_x/W_e
6) Anisotropic G-Dipole X-Term									
a) Secular	M	Linear	τ_R	None	——	None	None	None	None
b) Non-secular	M	$\omega_o[1+\omega_o^2\tau_R^2]^{-1}$	$\tau_R[1+\omega_o^2\tau_R^2]^{-1}$	W_e	M	W_e	W_e	Yes	W_e

7) Quadrupolar a) Secular	No width contribution	——	τ_R	None	——	None	None	Yes	——
b) Pseudo-secular	M^2 and M^4	None	τ_R	W_x	Yes	W_n-Reduction	W_n	Yes	$W_n/W_e \sim 1$
8) Spin-Rotation a) Secular	None	None	$\tau_J \propto T/\eta$	None	——	None	None	None	——
b) Non-secular	None	None	$\tau_J \propto T/\eta$	W_e	None	W_e	W_e	Yes	W_e
9) Intra-molecular spin-orbit processes a) Secular	None	None	Independent	None	——	None	None	None	——
b) Non-secular	None	None	Independent	W_e	None	W_e	W_e	W_e	W_e
10) Zero-Field Splitting $S>\frac{1}{2}$ a) Secular	None	None	τ_R	None	——	None	None	None	——
b) Non-secular	None	$[1+b^2\omega_o^2\tau_t^2]^{-1}$	$\tau_R[1+b^2\omega_o^2\tau_t^2]^{-1}$	W_e-type	None	W_e-type	W_e-type	Yes	W_e-type
INTERMOLECULAR									
11) Heisenberg Spin Exchange (+Electron Transfer)	Symmetric Dependence on D_M	None	$T/\eta[1+(J\tau_1)^2]^{-1}$ $\tau_1 \propto \eta/T$ for HE	ω_{EX}	Some dependence on D_M	ω_{EX}-Reduction	No	Yes	Decreases Enhancements
12) Dipole-Dipole a) Secular $S_{1Z}S_{2Z}+S_{1\pm}S_{2\mp}$ like	Symmetric Dependence on D_M	None	τ_t	None	——	None	None	None	——
b) Pseudosecular $S_{1\pm}S_{2\mp}$-unlike	Symmetric Dependence on D_M	None	τ_t	Analagous to ω_{EX}	Some dependence on D_M	Reduction Analagous to ω_{EX}	No	Yes	Decreases Enhancements
c) Nonsecular $S_{1\pm}S_{2Z}+S_{1\pm}S_{2\pm}$	Symmetric Dependence on D_M	$[1+b^2\omega_o^2\tau_t^2]^{-1}$ b = 1 or 2	$\tau_t[1+b^2\omega_o^2\tau_t^2]^{-1}$	W_e	Some dependence on D_M	W_e	W_e	Yes	W_e

References

1. J. H. Freed, J. Chem. Phys. 43, 2312 (1965).
2. A. Abragam, "The Principles of Nuclear Magnetism" (Oxford University Press, London, 1961).
3. J. S. Hyde, J. C. W. Chien, and J. H. Freed, J. Chem. Phys. 48, 4211 (1968).
4. J. H. Freed, J. Phys. Chem. 71, 38 (1967).
5. J. H. Freed, D. S. Leniart, and J. S. Hyde, J. Chem. Phys. 47, 2762 (1968).
6. G. Rist and J. H. Freed (to be published).
7. J. H. Freed and G. K. Fraenkel, J. Chem. Phys. 39, 326 (1963).
8. M. P. Eastman, R. G. Kooser, M. R. Das, and J. H. Freed, J. Chem. Phys. 51, 2690 (1969).
9. M. P. Eastman, G. V. Bruno, and J. H. Freed, J. Chem. Phys. 52, 2511 (1970).
10. J. I. Kaplan, J. Chem. Phys. 28, 278, 462 (1958); S. Alexander, ibid. 37, 966, 974 (1962).

SUBJECT INDEX